Geo-Information for Watershed Processes

Geo-Information for Watershed Processes

Editors

Walter Chen
Fuan Tsai

MDPI

Basel • Beijing • Wuhan • Barcelona • Belgrade • Novi Sad • Cluj • Manchester

Editors

Walter Chen
Department of Civil Engineering
National Taipei University of Technology
Taipei, Taiwan

Fuan Tsai
Center for Space and Remote Sensing Research
National Central University
Taoyuan, Taiwan

Editorial Office
MDPI
St. Alban-Anlage 66
4052 Basel, Switzerland

This is a reprint of articles from the Special Issue published online in the open access journal *ISPRS International Journal of Geo-Information* (ISSN 2220-9964) (available at: https://www.mdpi.com/journal/ijgi/special_issues/watershed).

For citation purposes, cite each article independently as indicated on the article page online and as indicated below:

Lastname, A.A.; Lastname, B.B. Article Title. *Journal Name* **Year**, *Volume Number*, Page Range.

ISBN 978-3-0365-8588-8 (Hbk)
ISBN 978-3-0365-8589-5 (PDF)
doi.org/10.3390/books978-3-0365-8589-5

Contents

Preface

We are delighted to present this Reprint, which brings together a collection of insightful scientific works in various fields and centers around the fascinating concept of watersheds and their profound significance in shaping the natural and human landscapes across the globe. This compilation aims to provide a comprehensive overview of cutting-edge research and advancements in the subjects covered.

As Guest Editors, our motivation for curating this Reprint stems from the desire to foster knowledge dissemination and facilitate collaboration among experts and enthusiasts in these areas. We firmly believe this collection will serve as a valuable resource for our audience, encompassing researchers, practitioners, educators, and students.

This compilation delves into the essential role of watersheds as the fundamental unit for evaluating landscape conditions and comprehending the environmental impacts of both natural phenomena and human activities. The articles featured in this Reprint explore diverse facets of watershed processes, and the interdisciplinary nature of watershed analysis is evident in the involvement of experts and researchers from a broad spectrum of fields. This collaborative effort underscores the urgency and relevance of incorporating geo-information in watershed studies to address pressing sustainability issues.

Through this Special Issue, we aim to foster knowledge exchange and encourage innovative research from diverse perspectives. We extend our heartfelt gratitude to all the contributors who have enriched this Reprint with their original work and valuable insights. Additionally, we would like to express our appreciation for the support and assistance we received from our colleagues and the editorial team throughout this endeavor.

We hope this compilation serves as a source of inspiration and prompts further exploration in the realms of watershed analysis, making a significant impact on scientific advancement and environmental stewardship.

Walter Chen and Fuan Tsai
Editors

International Journal of *Geo-Information*

MDPI

Article

Application of Hydro-Based Morphological Models for Environmental Assessment of Watersheds

Silvio Luís Rafaeli Neto [1], Vanessa Jutel dos Santos [1], Emili Louise Diconcili Schutz [1], Leticia Margarete de Moliner [1], Cristiane Gracieli Kloth [1], Daiane Teixeira Schier [1], Arsalan Ahmed Othman [2,3], Veraldo Liesenberg [1] and Polyanna da Conceição Bispo [4,*]

1 Graduate Program in Environmental Sciences, Santa Catarina State University (UDESC), Av. Luiz de Camões 2090, Lages 88520-000, SC, Brazil
2 Iraq Geological Survey, Al-Andalus Square, Baghdad 10068, Iraq
3 Iraq Geological Survey, Department of Petroleum Engineering, Komar University of Science and Technology, Sulaimaniyah 460013, Iraq
4 Department of Geography, School of Environment, Education and Development, University of Manchester, Oxford Road, Manchester M13 9PL, UK
* Correspondence: polyanna.bispo@manchester.ac.uk

Abstract: Hydro-based morphological models are representations of the terrain related to the flow or storage of water in the landscape. However, their application in the context of an integrated environmental assessment has been scarcely explored in the literature, despite the well-known importance of water for ecosystems and land use planning. Here, we derive the HAND and TWI models, which present solid conceptual bases based on water–landscape relationships from digital terrain models. We aim to present these models as useful representations in the environmental assessment of watersheds as they are relatively easy to generate and interpret. To this end, we applied these models in a Brazilian watershed and evaluated their spatial and reciprocal occurrence in the hydrological landscape through geographic entities and their spatial relationships with other landscape elements such as land use. We argue that HAND and TWI are simple hydrological-based models with robust premises that can reveal intrinsic relationships between relief parameters and water, providing new perspectives for the environmental assessment of small watersheds. Their outcomes have tremendous implications for land management initiatives. Our results show that geometric signatures of the TWI appeared through all the structural units of the hydrological landscape. The plateau areas were most prone to water accumulation/soil saturation, followed by floodplains, hillslopes, and ecotones. Thus, there is a tendency for the greatest geometric signatures of water accumulation/soil saturation entities to be located near the higher-order channels as well as the greatest geometric signatures of the floodplains. Agriculture and planted forests increased with distance, while the areas occupied by forest remnants tended to decrease within a range of up to 50 m from channels. However, they were also found within 50 m around the springs, whereas open fields, urban areas, and water bodies remained stable. We argue that HAND and TWI are simple hydrological-based models with robust premises that can reveal intrinsic relationships between the relief parameters and water, providing new perspectives for the environmental assessment of small watersheds whose outcomes have tremendous implications for land management initiatives.

Keywords: HAND; topographic wetness index; land use; hydrological landscape; topographic footprint; geometric signature

Citation: Neto, S.L.R.; dos Santos, V.J.; Schutz, E.L.D.; de Moliner, L.M.; Kloth, C.G.; Schier, D.T.; Othman, A.A.; Liesenberg, V.; Bispo, P.d.C. Application of Hydro-Based Morphological Models for Environmental Assessment of Watersheds. *ISPRS Int. J. Geo-Inf.* **2023**, *12*, 314. https://doi.org/10.3390/ijgi12080314

Academic Editors: Wolfgang Kainz, Walter Chen and Fuan Tsai

Received: 17 May 2023
Revised: 26 July 2023
Accepted: 28 July 2023
Published: 31 July 2023

1. Introduction

The relief records information about the topographic evolution dynamics, and its geomorphometric study can be applied for different purposes such as hydrological, geomorphological, and landscape analysis [1,2]. Interestingly, geomorphometric records are also helpful for ecosystem management applications, land use planning as well as the assessment and perception of risks [3], besides landforms and soil mapping, modeling the occurrence of landslides, and hazard mapping in steep terrains, erosion, and deposition, mass balance modeling on glaciers and hydrological applications involving channels and floodplains [4].

This study's motivation comes from natural hazard mitigation in several watersheds in Southern Brazil caused by the incorrect use of landscape such as removing vegetation in permanent preservation areas and constructing residences in areas prone to flooding events. We argue that the link between relief and water can be explored through spatial representations that provide parameters on water action in the environment simply and rapidly without using complex cascade models. This aspect is fundamental in any environmental analyses since anthropogenic interferences are highly conditioned by the water in the landscape, possibly resulting in problems such as supply, natural disasters, pollution, and diseases. Thus, the hydrological landscape [5] plays an important role in the environmental assessment of watersheds as it relates the physical space's structural units with the water's dynamics in that environment.

This paper aims to explore the so called topographic wetness index (TWI) [6] and the height above the nearest drainage (HAND) [7–9] models as another layer of information about the geomorphological agents that condition the anthropogenic activities in the structuring units on the landscape. Such knowledge may support environmental analysis in land management initiatives and improve the structuring of the problems and the consequent search for solutions. This work contributes by bringing these hydro-based morphological models from their single dimensions [7,10–21] to the context of the environmental analysis of watersheds. Thus, a hypothesis was formulated that the combined use of HAND and TWI models allows for even better characterization of the landscape and benefits land management initiatives. Such models can help identify the different geomorphological strata that form the landscape such as water accumulation zones, hillslopes, and plateaus, with reflections in land use planning to preserve sensitive ecological zones and prevent the occurrence of disasters caused by floods, extreme runoffs, debris flows, and mass movements.

In this paper, we briefly describe the conceptual approach in Section 2. Section 3 is divided into two subsections, where the first refers to the study area description. Then, the steps related to land use map generation are provided. Next, the DEM datasets and their processing steps are described. The three forthcoming subsections detail the determination of both the HAND and TWI models and the adopted data strategy analysis. Finally, we present and discuss the results of both the HAND and TWI models by thoroughly considering the landscape characterization.

2. The Conceptual Approach

The environmental assessment of watersheds seeks to identify, formulate, and structure the problems through data, information, and knowledge concerning the problem domain. It comprises the intelligence stage of a decision-making process [22] regarding planning anthropogenic activities in the watershed. The more elaborated this stage is, the better the chances of generating choices of viable solutions and of choosing the best possible alternative.

Chorley and Kennedy [23] mentioned the systems approach concept's applicability in analyzing complex geographic systems in which watersheds are the most viable research unit. In this approach concept, the watershed comprises two subsystems: the cascade and morphological systems. The cascade system represents the dynamic portion of the watershed, in which the state of the entities that compose it changes frequently. It is part of the system where the flow of matter or energy occurs such as the hydrological cycle and the movement of people and animals in its most diverse manifestations [23].

On the other hand, the morphological system represents the static part of the system, represented by entities in which the attributes of position and conformation do not frequently change such as the drainage network, the sub-basins (or small watersheds), and the relief itself. These entities are sometimes called features [23]. Both cascade and morphological systems are modeled by their strategies. In the case of cascade systems, mathematical models are used, called scientific models [24]. In the case of morphological systems, their modeling occurs through database modeling techniques, the most frequent of which is object orientation [25]. Spatial decision support systems (SDSS) [24,26,27] would be the technologies that deal with this conceptual universe; geographic information systems (GIS) [25], for example, a type of SDSS that works on the morphological portion of the geographic system.

The socio-economic and environmental consequences of population growth are a tacit finding for many watersheds subject to anthropogenic occupations, especially in developing countries where police and monitoring initiatives are inefficient and ineffective. In addition, houses and buildings near watercourses and springs can worsen water pollution and soil erosion, as they often lack permission from the local authorities. These structures are also associated with an increased risk of natural disasters.

These events are commonly caused by an evident lack of planning and non-continuity of occupational policies by public managers in developing countries [28,29]. All these issues may bring complexities to the system involved, together with the particular idiosyncrasies of each site, making a complete understanding of the problems and the search for their solutions challenging [30–33].

A watershed represents an environmental (geographical, hydrological, and ecological) system where topological relationships emerge, subject to being mapped through a collection of continuous geometries [34], sometimes perceived with distinct patterns, abstracted through discrete parameters, at an appropriate degree of completeness for interpreting and describing the processes and functions of these systems [35]. Watersheds are arrangements of spatial entities [36] or terrain objects [37] such as depressions, peaks, ridge lines, course lines, and break lines, with intrinsic topologies such as contiguities, adjacencies, proximities, and contingencies, among others.

The definition of these entities depends on the most appropriate conceptual data model for solving a problem and how they are represented on maps and in geographical databases. For example, geometric signatures of spatial entities [34] are usually coded in GIS through vector data structures for features such as point, line, and polygon, or even matrix/raster data structures for continuous geographical phenomena such as digital terrain models (DTM) [25]. Here, the difference between DTM and DEM merely lies in the meaning of the z attribute. In the former, the z attribute refers to the continuous variable terrain height in relation to a local or geodesic topographic reference, and the latter to the other variables of continuous spatial phenomena such as temperature, soil moisture, and parametric indices, among others.

Features identified on the terrain, revealed through DEM or based on a land use classification using remote sensing data, are relatively common in environmental assessments that are geographical in scope. However, the metrics usually applied have little or no relationship with the natural occurrence of water in the landscape.

The gravitational potential of water is a key element in the evolution of the landscape as it is the main element responsible for water flow in the soil and subsoil and can be applied in many models [38]. The relief, the soil, and its properties act as intervening agents in water and energy distribution, redistribution, and accumulation [1,5,37]. This behavior differs between the floodplain, hillslope, and plateau structural units of the hydrological landscape [5,39], and hydrological connectivity between them may eventually exist [39]. The gravity gradients between points on the terrain are the main physical agents that cause the flow or stationarity of water in the landscape, so the relief is considered the main conditioner of that behavior. These gradients can be absolute or relative, depending on the altimetric reference taken as a measure [37,40]. Thus, the landscape's evolution in watersheds results from complex interactions of multiscale (topographic, climatic, tectonic, and anthropogenic) systems that determine the surface and subsurface properties [37].

The multi-parametric nature of a DTM can be explored in multiple layers [1]. Hydro-based morphological models add relatively new layers of metrics based on the relation between relief and water. The TWI model, for example, explores the tendency for soil saturation in the terrain's low-sloping or concave surfaces through the topographic index (Equation (1)).

$$\lambda = ln\ (a/tan\beta) \quad (1)$$

where the λ parameter is the topographic index of a point or cell on landscape. The a parameter is the contributing area per unit contour length, that is $a = A/L$ [41], where A is the upslope contributing area, and L is the perimeter of a. The a parameter reflects the amount of water that may be captured during surface or subsurface runoff or groundwater flow events. The bigger a is, the more volume of water may flow to the point or cell on the landscape. Bigger a values are expected close to watercourses, whereas smaller values are expected near the watershed perimeter. The β parameter is the terrain gradient or local slope and captures the potential gravitational gradient of a water particle to move in/on a landscape. The higher β, the larger the vertical differences between two points in relation to its horizontal distance. Therefore, local sites with a higher β attribute tend to produce higher waterflow surface velocities.

Therefore, parameter a defines the amount of water input to a point or cell and β defines its potential water output. The parameter λ indicates the presence of potential water accumulation on the surface, soil saturation, or wetland areas within the watershed's environmental assessment scope. High λ values indicate places with large areas of upslope drainage and low terrain gradient. Thus, high TWI values (λ) are expected in flat areas or areas with low slopes and high upslope contributing areas, which should occur in the low regions of the watershed close to the higher-order channels. On the other hand, low TWI values are expected close to the headwaters of the watershed or in the top regions of the hydrological landscapes that compose its relief structure. The water flow at these locations tends to be exclusively vertical, since water particles present little or no lateral gradient. In saturated soil zones, the vertical flow depends on the subsoil's physical properties, and there may or may not be percolation. The contour lines of these zones characterize their geometric signatures and can vary according to the volume of water present and the level of water in the ground reservoir [42]. For this reason, these zones are also known as variable flow areas (VFAs), where the anaerobic conditions of the wetlands offer conditions for the production of biogenic methane, an important element for both the local and global carbon cycle [43].

This concept was initially presented within the hydrological model TOPMODEL (topography model) [6] to estimate a hydrograph of the runoff produced by a watershed in a river control section based on a rainfall and evapotranspiration dataset. The distributed parameter of the TOPMODEL is the topographic index (TI), calculated based on the natural logarithm of the ratio between the specific contributing area (runoff watershed area/length of the edge of the entry cell—obtained in the DTM pre-processing by the routine flow accumulation and by the resolution of the cell, respectively) and the local slope of the cell [41].

The HAND model uses the drainage network to reference the heights of points on the ground. Figure 1 illustrates the methodological concept first presented by Rodda [44] for mapping the flood extent and later explored by Rennó et al. [7]. Any point on the surface has a vertical distance to a watercourse. The choice of what watercourse and what cell in the watercourse should be used for distance measurement is made based on a horizontal distance. The horizontal distance depends on the algorithm for defining the flow direction such as D-8 [41] and D-Infinity [44].

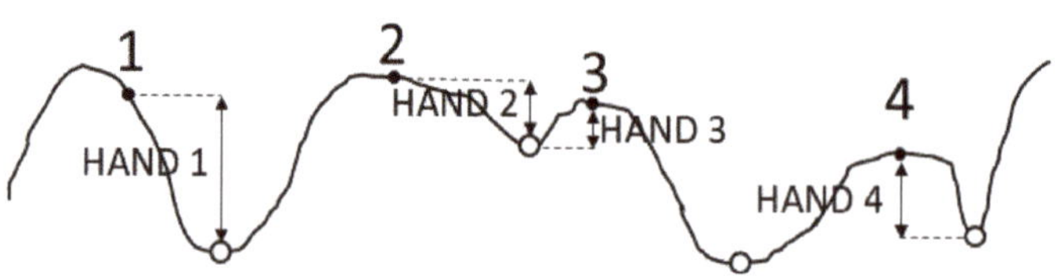

Figure 1. Schematic representation of the HAND model concept over a hypothetical elevation profile.

The idea has been evaluated in determining the geometric signatures of floods [19,45–47] and classifying the hydrological landscape [7,8,18]. It is based on the local gravitational potential of a water particle, corresponding to HAND. The HAND DEM, therefore, is a discrete 2D representation of the problem's domain, usually a watershed or stretches of floodplains.

3. Materials and Methods

3.1. Study Area Description

The case study was conducted in the Canoinhas River watershed with a size of 1440 km^2, axial length of 83.22 km, average width of 17.30 km, and drainage density of 1.13 km/km^2. The compacity index 2.71 indicates that the Canoinhas River watershed is elongated and has low susceptibility to maximum rainfall along all of its lengths, which limits the occurrence of floods on its plains to frontal rain events.

The Canoinhas River is a fifth-order channel with a length (the area and perimeter attributes were obtained from the attribute table in GIS. These values are automatically calculated when the layer is generated) of 189.08 km and a sinuosity of 2.27 km/km. It flows into the Rio Negro River, a tributary of the Iguaçu River (26°30′ S; 50°20′ W). The watershed is part of the continental hillslope in Hydrographic Region 5 (RH5), called "Planalto de Canoinhas", and belongs to the macro-region of the northern plateau of the Santa Catarina State, Brazil. The above sea level altitude ranged from 757 to 1344 m.

The vegetation comprises fields, tropical, and subtropical perennial and floodplain forests, with the North Plateau region being the most expressive forest center in Latin America [48]. The relief comprises strongly to gently undulating, mountainous, and flat strata. Soil depths vary from less than 60 cm to 150 cm, with textures ranging from clayey to very clayey. According to EMBRAPA [49], the most common soil types are described in Table 1 and further illustrated in Figure 2.

Table 1. Soil types and their attributes in the Canoinhas watershed.

Soil Type	Area		Texture	Relief	Depth (cm)	Drainage
	(km^2)	(%)				
Cambisols	370	25.7%	Clayey to very clayey	Strongly undulating to gently undulating	60–150	Moderately Drained
Humic and slightly humic gley soil	187	13.0%	Clayey to medium	Flat	<60	Poorly to Very Poorly Drained
Bruno/Red latosols	498	34.6%	Very clayey	Gently undulating	>150	Well drained
Litholic soils	319	22.2%	Very clayey to clayey	Hilly to gently undulating	<60	Moderately Drainage
Nitosols	64	4.4%	Very clayey	Gently undulating	>150	Well drained

The soils of the study area show a texture varying from clayey to very clayey (Table 1; Figure 2). Latosols and nitosols are the most profound soil types, occupying ~39% of the watershed and presenting the best drainage conditions. Allied with its gentle topography, these attributes favor agricultural activities in small rural properties, typically smaller than 30 hectares in the study area.

Humic and slightly humic gley soil occupies ~13% and is located in flat areas near the Canoinhas River. The flat relief, the shallow depth, and the clayey texture assure little drainage capacity. Litholic soils are characterized by little depth, sometimes presenting a stony ground surface and moderate drainage capacity, despite being clayey or very clayey. These soils occur in hilly or gently undulating regions (Figure 2).

Cambisols occupy different landscape strata (strongly undulating to gently undulating relief) and have moderate drainage capacity at medium depths (60 to 150 cm). This class and latosols and nitosols occupy about 65% of the study area. In summary, various soil types located on different layers of the landscape, with a varied drainage capacity and a relief amplitude of 847 m, assure exploitation in different ways.

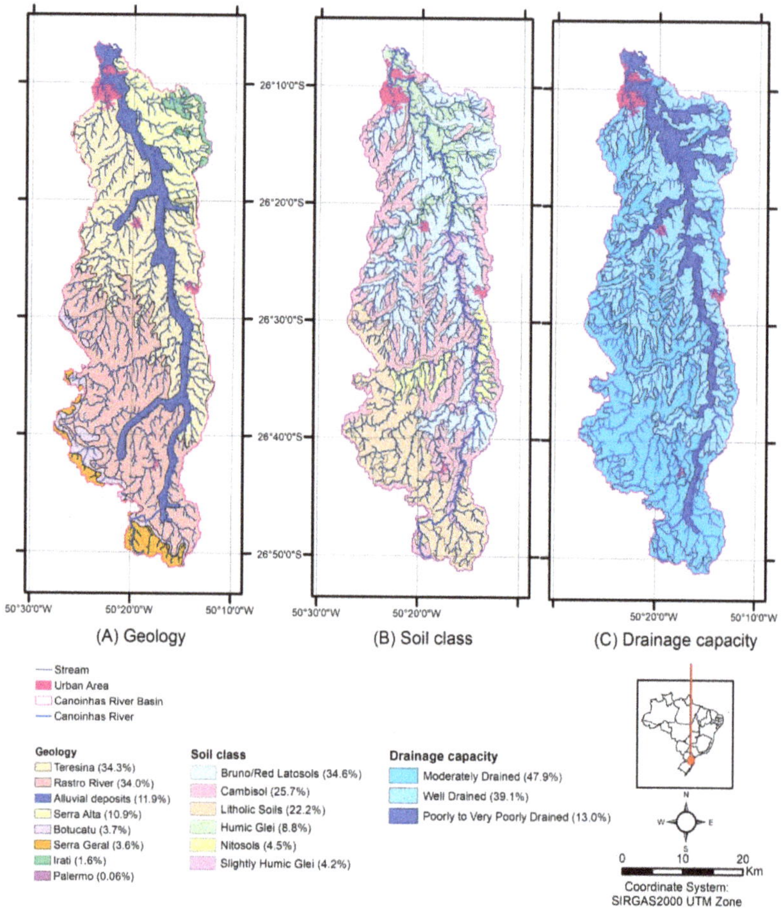

Figure 2. Geology (**A**), soil class (**B**), and drainage capacity (**C**) of the Canoinhas River watershed.

The Canoinhas River watershed is considered to be medium-sized [48] and partially covers five municipalities such as Monte Castelo (source), Canoinhas (mouth), Major Vieira, Papanduva, and Três Barras (Table 2). The urban centers of all the municipalities are found within the watershed. According to AMPLANORTE [50], these municipalities also show high municipal human development index (MHDI) scores, with the highest percentage of residents established in the urban area. The percentage of residents in the rural area is also relatively expressive, keeping a mix of rural and urban landscapes (Table 2). The essential economic activity is related to the pulp and paper industry, wood, and energy production. The region is also facing a gradual increase in the establishment of new industries and, consequently, urban growth is generating conflicts and environmental issues, which is why it was selected as a case study.

3.2. Land Use Maps

A Sentinel-2 scene with an acquisition date of 23 February 2021 was used to generate the land use map (Figure 3). The scene was chosen from the Earth Explorer, linked to the Copernicus program of the European Space Agency (ESA). First, all spectral bands were resampled at a spatial resolution of 10 m, roughly relative to the horizontal 1:50,000 map scale, which is compatible with regional studies [51].

Table 2. Information about the municipalities that compose the Canoinhas River watershed.

Parameters	Municipality				
	Monte Castelo	**Major Vieira**	**Papanduva**	**Três Barras**	**Canoinhas**
Population (inhab)	8346	7479	17928	18129	52765
Urban population (%)	58%	40%	51%	85%	74%
Rural population (%)	42%	60%	49%	15%	26%
Land area (km^2)	561	521	765	437	1148
Area covered by the watershed (%)	93%	93%	12%	40%	14%
Population density (inhab.km^{-2})	14.6	14.2	24.0	41.4	46.3
MHDI * (2010)	0.675	0.690	0.704	0.706	0.757

* MHDI: municipal human development index.

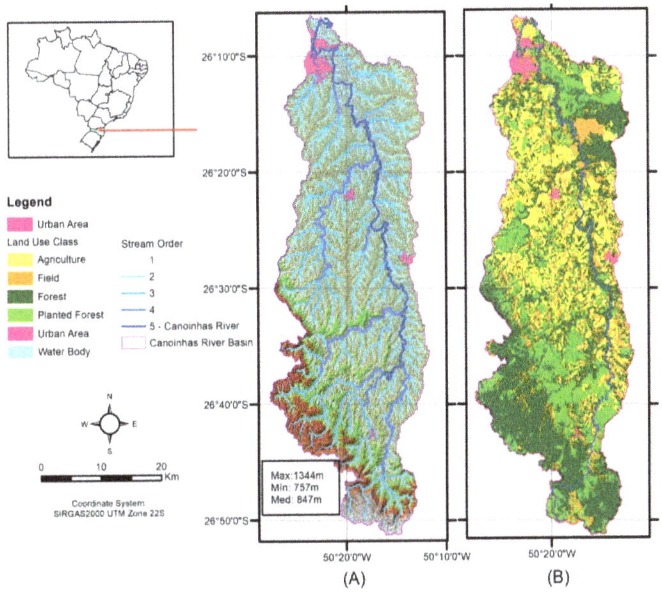

Figure 3. Representation of the study area by the digital model of the shaded terrain with the ordering of channels by the Strahler method (**A**) and land use classes (**B**). The study area is in the Northern Plateau of Santa Catarina State (Southern Brazil).

Afterward, samples were collected through the study area, and the supervised maximum likelihood was applied. Finally, a majority filter was applied to remove the isolated pixels, smoothing the classes' irregular boundaries. Further corrections were performed by comparing the classified map with the MapBiomas collection [52] of Brazilian land cover and land use, referring to 2020, in which the classification is originally from Landsat mosaics using the dynamic and procedural methodology in Google Earth Engine.

Six land use classes were defined based on the identifiable entities in the scenes and of interest to the environmental assessment of the watershed (Table 3). These classes also have a very important spatial distribution in the area. The samples of the land use class were selected through photo interpretation of the form, size, tone, color, texture, and pattern (Table 3) [53]. We used fieldwork measurements and high spatial images from different data sources such as ArcGIS and Google Earth Pro. Similar procedures were previously adopted in another regional research [54,55].

Table 3. Subsets of the six mainland cover classes in the selected study area with RGB composition bands 4, 3, 2.

Land Cover Classes	Subset (1200 m × 800 m)
Agriculture: associated with annual agricultural activities containing maize and other crops such as beans, yerba mate, and tobacco.	
Forest: natural forest remnants belonged to the mixed ombrophilous forest under different successional forests, in addition to natural fields.	
Urban Area: urban occupation in the watershed encompassing the urban centers of Canoinhas, Três Barras, Major Vieira, Monte Negro, and Papanduva.	
Field: represents pastures formed of native or planted annual or perennial grasses, besides small-sized vegetation.	
Planted Forest: forest plantations occupied by either *Pinus* spp. and *Eucalyptus* spp. stands.	
Water: liquid surfaces such as lakes, ponds, streams, and rivers.	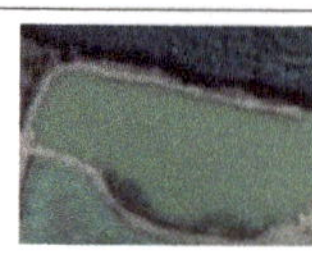

3.3. *Digital Terrain Model (DTM)*

A total of 94 digital terrain models (DTMs) with a spatial resolution of one meter were mosaiced [56]. Afterward, the DTM was used to generate the relief information needed to build the HAND and TWI morphological models. The DTM was pre-processed in ArcGIS®10.5, with the help of the Spatial Analyst, 3D Analyst, ArcHydro Tools, and Geo-HMS Tools extensions. The Geo-HMS tool was used to pre-process the DTM, which consisted of filling depressions, defining flow directions (D8 algorithm [41]), flow accumulation, defining the synthetic drainage network by testing different thresholds, and outlining the watershed by setting the Canoinhas River as an outlet. The thresholds were evaluated by comparing water streams with the support of visual interpretation over high spatial resolution images (i.e., ~0.40 m) freely available from the State Government [56].

The spatial resolution chosen to apply the TWI model in the Canoinhas River watershed was 20 m, and to apply the HAND model, it was 5 m. The 20 m resolution is relative to the roughly horizontal 1:100,000 scale, and the 5 m resolution to 1:25,000. Both were from resampling the original 1 m DTM (relative to a 1:5000 horizontal scale). The 20 m resolution was an arbitrary choice to identify horizontal TWI entities in sufficient detail and considerably improve the computational performance. Interestingly, the original DTM of the watershed with a 1 m spatial resolution accounted for 46.08 GB in the Canoinhas River watershed, while the DEM of the TWI with 20 m resolution accounted for 1.44 GB with 3.6×10^6 cells.

To classify the landscape based on HAND, we empirically considered the possible effects of discretization and smoothing of the DTM [57], the probable effects of the vertical resolution, and the fractal dimension of the hydrographic network of the Canoinhas River watershed. According to Gharari et al. [18], a spatial resolution of 20 m would be enough to classify landscapes using the HAND model. In this way, a 20 m spatial resolution for identifying horizontal TWI entities and a 5 m spatial resolution for landscape classification would be compatible resolutions with environmental analyses for planning purposes at a regional scale.

The different spatial resolutions from datasets were disregarded since the tabulation of the pair of information was considered using the zonal statistics tool available in a GIS. Specific aspects are discussed in forthcoming sections.

3.4. Topographic Wetness Index (TWI)

The TWI model was processed according to Quinn et al. [41] and subsequently classified according to the propensity of a cell for water accumulation on the surface or soil saturation (Table 4).

Table 4. Classes of the TWI values and propensity for water accumulation on the surface or soil saturation.

Class	Characteristics	Water Accumulation Propensity/Soil Saturation
−0.56 to 6.40	Small contributing area and steep slopes	Low
6.40 to 8.00	Average contributing area and average slopes	Average
8.00 to 25.69	Large contributing area and low local slopes (flatter areas)	High

A TWI parameter is assigned to each MDT cell. Cells with the same TWI show the same tendency for water accumulation or soil water saturation. The scale and resolution of the data affect the distribution of the TWI statistics in the landscape [13,58]. A horizontal resolution corresponds to the elementary size of a surface measured by a remote sensor [57] and should be sufficient to capture the minor features identifiable in the object space to portray them unmistakably [34,53]. In practical terms, geographic features that identify areas of water accumulation or soil saturation result from grouping neighboring cells with the same value.

At high MDT resolutions, 1 m, for example, the slight variations in the local slope parameter β (Equation (1)) between neighboring cells can assign quite a different TWI to each, with the cells having a similar a parameter. This can generate an inconsistency in the information about the areas prone to water accumulation, since this effect depends on the water table, of which no significant variations are expected at short distances. The floodplain areas (lowlands), which tend to present the highest TWI values, may present a large variation in the TWI, without significant variation in soil moisture [11,59]. This effect can exist in finer scales (or large scales), where small variations in relief can be perceived without necessarily being proportional to small variations in soil moisture. This effect also tends to produce an overabundance of TWI in the landscape without necessarily aiding in identifying zonal features of interest to decision-making processes at the regional planning level. Therefore, data and information saturation can impede the extraction of relevant information. Thus, the dimensions of the spatial domain of the problem and the most appropriate horizontal resolution have to be observed to show the geographical features of interest. Even in course scales (or small scales), it is possible to extract relevant information from the terrain for practical purposes [60]. For example, Gharari et al. [18] used a 10 m resolution for the TWI model to compare it with the HAND. The HAND model presented the best performance.

3.5. Height above Nearest Drainage (HAND)

The HAND model can be applied in any terrain and translate hydrological meanings [8]. The model has received great attention from researchers due to its simple concept, relatively simple implementation, availability in open-source software like TerraHidro [61], and large application possibilities.

The HAND model [7] was presented to outline the structural units of the hydrological landscape to map terra-firme environments in the Cuieiras Biological Reservation (Amazon), based on the shuttle radar topographic model (SRTM) DEM. These classes were subsequently validated by Nobre et al. [8] over an area in the lower Rio Negro watershed (Amazonia) using a set of field observation points. The most evident signs of the structural units of the hydrological landscape are lowland, valley side, and upland [5]. However, other classes may be needed to describe these units better such as waterlogged, ecotone, slope, and plateau [8]. Interestingly, the class intervals to outline the structural elements of the hydrological landscape have still been scarcely studied in the literature. Furthermore, the transitions between the classes are usually not abrupt, and the attribution of a point of the landscape to a certain class would imply the use of diffuse techniques [18]. In this paper, we used the thresholds suggested by Rennó et al. [7] and Nobre et al. [8], whose results are shown in Table 5.

Table 5. The HAND classes for classifying the hydrological landscape in the Canoinhas River watershed.

Class	Hydrological Landscape Unit
HAND < 5.3 m	Floodplain
5.3 ≤ HAND ≤ 15 m	Ecotone
HAND > 15 m and ≥7.6% slope	Slope/Hillslope
HAND > 15 m and <7.6% slope	Plateau

Source: adapted from Rennó et al. [7].

Floodplains are located at the lowest heights of the terrain, close to the natural drainage channels, presenting low slopes and a groundwater table close to the surface, which can saturate the land [5,8]. Ecotones represent smooth concave surfaces of transition between the floodplains and hillslopes, which mark the vadose regions [62] (i.e., the first landscape entities in which the soil is unsaturated, far from the channels). Slopes or hillslopes mark the surfaces with a height gradient in the landscape, which are well-drained and strongly interact with surface runoff. Plateaus are distributed in the high regions of the landscape, generally surrounded by hillslopes and with low slopes.

We chose these classes for the Canoinhas River watershed due to the greater number of classes and the general topographic characteristics of the watershed, especially the fact that 63% of the area of the watershed is composed of flat (slopes up to 3%) and slightly undulating terrain (slopes ranging from 3 to 8%), which would indicate the possibility of there being clear vadose zones (Figure 1).

Considering that the model's basic premise is the vertical distance of a point on the landscape from the nearest drainage channel, one has to assume that the model is sensitive to the number of channels generated by the channel definition operation on the DTM. This is a very common operation, which has the contributing area of the first grid cell as a user-defined parameter to which the channel is assigned. The area threshold (AT) parameter affects the degree of fractal discretization of the synthetic hydrographic network. Therefore, the AT of 0.5 km^2 was adopted, corresponding to 0.035% of the total area of the Canoinhas River watershed. This value is the approximate mean of the watershed area of a sample set of sources of first-order natural water courses recognized in high-resolution orthoimages available for the entire Santa Catarina State and are freely available for the general community [56].

The features attributed to the natural hydrographic network were identified by analyzing the form, size, tone, color, texture, and pattern [53]. The aim was to keep the geometric structure of the synthetic hydrographic network as close as possible to the natural geometric structure [18]. Therefore, the samples for calculating were based on AT, and the synthetic hydrographic network generated produced a set of linear features originating in the first pixel attributed to the flow accumulation algorithm.

Regarding the spatial resolution of the MDT, it is assumed that the better the resolution, the more accurate the vertical distances corresponding to the HAND. Furthermore, if the model is used to determine the extent of flood zones, it is valid to consider the best resolution available, especially in urban areas. However, the model here in this paper was only used to classify the hydrological landscape at the planning level, which could perhaps be sufficient for the 10 m resolution as used by Gharari et al. [18]. Therefore, we opted for the 5 m resolution resampled from the 1 m MDT.

3.6. Spatial Relationships

Spatial relationships are properties of spaces that do not undergo variations with their deformation. For example, a sub-watershed always belongs to its main watershed, independently of the distortions of scale. Similar affirmations can be made for relationships of continencies, intersections, contours, proximities, and adjacencies. These properties are studied in topology and used for validating data, modeling the integrated behavior of different features, data editing productivity, and query optimization [25].

The classes of features identified in the TWI and HAND morphological models, land use, and slope maps were assessed regarding spatial relationships. The relationships explored in this study were intersection, buffer clip, symmetrical differences, and proximity. These operations are used in GIS by means of predicates or mathematical operations such as the Euclidian distance between geometries of different classes, for example. Predicates are functions for comparing a given condition between pairs of features belonging to different classes coded as points, lines, or polygons. These functions can return a TRUE (T) condition when the condition under analysis is met or FALSE (F) otherwise.

The GIS approach used in this study (ArcGIS 10.5.1) adopts the dimensionally extended nine intersections model (DE-9IM [63]). The DE-9IM matrix presents nine possibilities of spatial relationships involving the features' interior, exterior, or edge. Point-type features have an interior and exterior but no edges. Similar to polygon-type features, line-type features have an interior, exterior, and edge. The final configuration of the matrix will be used to return the features of the classes involved in the operation or parts that satisfy the conditions expressed in the matrix, which can then be analyzed separately from the rest. The intersection operation returns the intercept features, which helps identify features that somehow share the same space, whether fully or partially. Their original geometries are maintained, which differs from the clip operation. This builds a group of features using the boundaries of other features and only returns the geometry relative to the portion in common. In both the intersection and clip operations, the attributes of both features can be kept in the resulting features. The buffer operation is normally used to analyze the area around features, especially regarding the proximity, continence, or intersection. The symmetrical difference operation chooses non-overlapping parts of the geometries, which operate inversely from the intersection. It is useful, for example, when excess information needs to be eliminated from the analysis, leaving only that which is actually of interest.

The operations above-mentioned were applied in the environmental assessment of the watershed. The operationalized classes of polygon features were land use, slope, TWI, HAND, the order of the channels, and the areas around the points of origin of the first-order channels. As the Brazilian Forest Act prescribes, these points are supposed to be the sources of the main watercourse [64]. In addition, areas of 0–15 m, 0–30 m, 0–50 m, and 0–100 m around the synthetic drainage features of the watershed were also defined. Finally, the surrounding features thus defined were operated with the features from the land use classes.

4. Results and Discussion

4.1. Topographic Wetness Index (TWI)

The TWI values vary considerably over the landscape according to the geomorphological characteristics of the terrain [65,66]. This is due to TWI parameters a and $tan\ \beta$ (Equation (1)) where a depends on the position of a point on surface, being near or far from the watercourse and $tan\ \beta$ depends on the slope. Values of a are expected to be higher the closer points are to the watercourse, since a has a superficial dimension. Parameter a also depends on the relief conformation because the shape of the watershed area A (Equation (1)) varies with relief configuration. The parameter β is essentially a parameter linked to the relief configuration. Therefore, the slope distribution in the watershed should affect the occurrence and distribution of $tan\ \beta$ in the landscape. In this regard, the geometric signatures show that the watershed presents 40.8 hectares of highly mountainous relief, 12.2% of highly undulating and mountainous relief, and the average slope is 0.09 m.m^{-1}. About 63% of the watershed area is on flat and gently undulating relief. This reveals that the relief of the watershed is quite diversified and well-distributed in the range from 0 to 20% slope (Table 6).

Table 6. Slope classes in the Canoinhas River watershed [67].

Slope (%)	Class	km^2	%
0–3	Flat relief	378.6	26.3
3–8	Slightly undulating relief	532.7	37.0
8–20	Undulating relief	345.4	24.0
20–45	Highly undulating relief	176.9	12.2
45–75	Mountainous relief	6.6	0.5
>75	Highly mountainous relief	0.0	0.0
	Total	1440.2	100

Some studies have shown that the TWI values that indicate prone to developing water accumulation or soil saturation in the landscape are above 8.0 [17,18,65,66]. For example, Figure 4 shows TWI variations ranging from −0.56 to 25.69, with a mean of 6.00 and a standard deviation of 3.00.

A TWI ranging from −0.56 to 6.40 represents 49.1% of the watershed, whereas 6.40 to 8.00 with 27.5%, and 8.00 to 25.69 with 23.2%, respectively. Similarly, DuPage city's values varied from −7 to 29.5 and from −5 to 288 in Will County [17]. Values below 6.00 were considered by Meles et al. [59] as very dry regions, those from 6.00 to 8.60 as dry, and those above 16.60 as moderately wet to very wet. Values above 8.00 were validated in the field by Schier [66] as being prone to flooding in the city of Lages, SC, Brazil.

Mapping the dry and wet zones of the landscape provides relevant information from a regional and local planning point of view. Drylands reveal regions available for land use for agricultural, forestry, and urban activities. However, these regions may be prone to mass movements during extreme precipitation events or be the first regions of the watershed to suffer water deficit during periods of drought. For example, we witnessed during the drought that hit the state in 2020 in the Canoas River watershed Management Committee that the cities located in the headwaters, close to the watershed, were the ones that suffered most from the lack of water, requiring effective help and assistance actions from the authorities. Wetlands, on the other hand, indicate areas that tend to be better able to withstand periods of drought. Indeed, Biffi and Neto [67] demonstrated that during a drought in 2005, the lower-lying areas of the relief in the apple production region of Santa Catarina State exhibited higher production levels than the higher-lying areas.

In general, terrains in the State of Santa Catarina tend to present a predominance of areas with little tendency for water accumulation/soil saturation in watersheds due to the predominance of steeper slopes. For example, in the case of the Canoinhas River watershed, 38% of the area with a flat terrain (slopes from 0 to 3%) presented TWI geometric signatures higher than 8.00, representing only 10% of the total area of the watershed. The fact that not all areas with slopes from 0 to 3% presented a TWI > 8 does not imply that these areas are not prone to water accumulation/soil saturation, especially if they are located close to drainage channels, where the water table is expected to be closer to the surface. Hence, adding the topographic variable HAND helps categorize the landscape based on the vertical distance between a point on the surface and the nearest drainage, making it an important consideration.

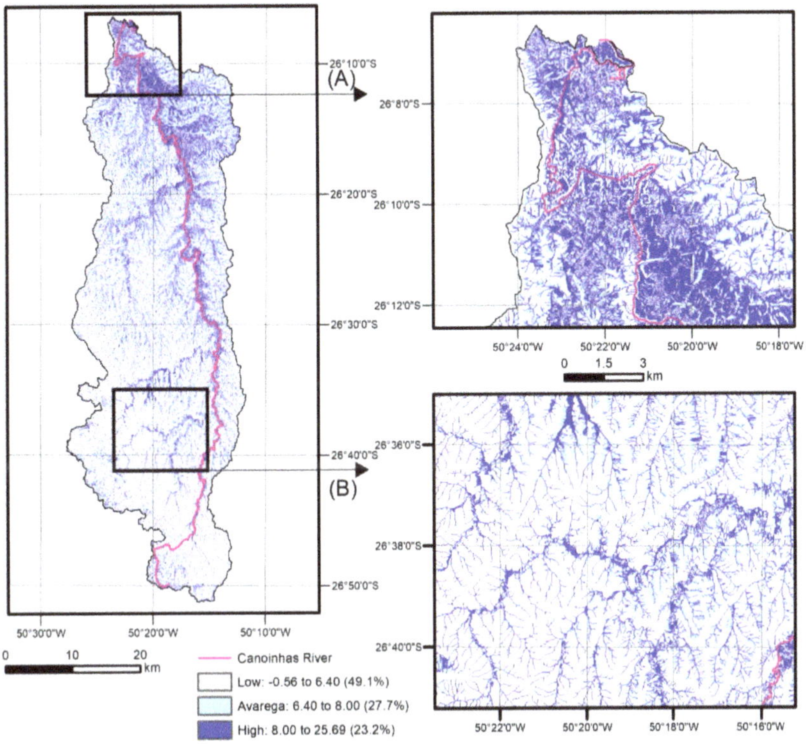

Figure 4. The TWI spatial distribution in the Canoinhas River watershed. The larger-size TWI spatial entities are located in the lower part of the watershed (**A**), while the smaller entities are located in the higher regions (**B**).

4.2. Height above Nearest Drainage (HAND)

The relative gravitational potential of a water particle is determined in the HAND model by the vertical distance to the nearest drainage. Various results were found by Nobre et al. [8] for the Rio Negro River watershed in Amazonia. The predominantly sedimentary constitution of that region found 20.9% of the area occupied by floodplains, 26.9% by ecotones, 9.1% by hillslopes, and 24.8% by plateaus. In the Canoinhas River watershed, we found 4.74% occupied by floodplains, 2.67% by ecotones, 38.04% by hillslopes, and 54.55% by plateaus (Table 7; Figure 5).

Table 7. Hydrological landscape classes in the Canoinhas River watershed.

Class	Hydrological Landscape Unit	Area (km^2)	%
HAND < 5.3 m	Floodplain	68.33	4.74
5.3 ≤ HAND ≤ 15 m	Ecotone	38.42	2.67
HAND > 15 m and ≥7.6% slope	Hillslope	547.93	38.04
HAND > 15 m and <7.6% slope	Plateau	785.64	54.55

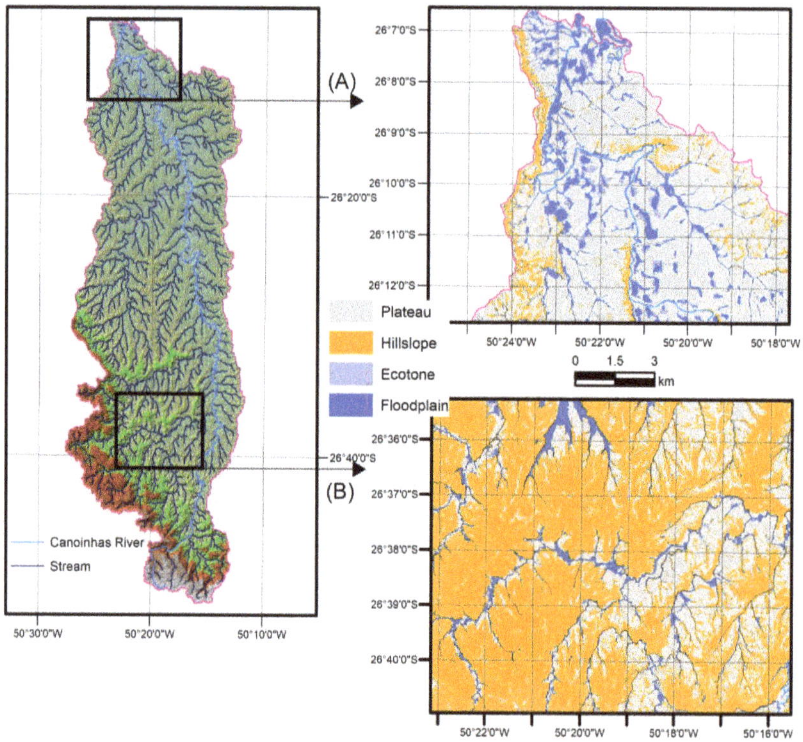

Figure 5. The HAND model classes in the Canoinhas River watershed. The lower region of the watershed has a wide area of plateaus and floodplains (**A**), while the higher regions have a predominance of hillslopes (**B**).

These results may be influenced by the DTM resampling from higher to lower resolutions due to the joint effect of discretization and smoothing [57]. However, the immediate consequence is the loss of detailed information on the relief and, consequently, on identifiable spatial entities, especially through the discretization effect.

The quality of the vertical variable is also a point to be considered. Smoothing the DTM tends to reduce the altimetric amplitude and with that, there is a loss of vertical accuracy. Smoothing should also affect the presumed gravitational gradients and runoff paths to the nearest channel. Thus, better altimetric determinations of the gravity gradients and of the extent of the surface runoff are expected in DTMs with higher spatial and vertical resolutions.

The fractal dimension of the hydrographic network of a watershed [60] makes the HAND model dependent on the area threshold (AT, presented in Section 3.5) to generate a channel [18]. The AT affects the size of the network similarly to the effect commonly associated with changes of scale in maps [60]. In a hydrographic network with a Hortonian structure [68], the channels' orders parametrize the channels' average lengths and watershed areas. Thus, morphometric parameters such as bifurcation ratio, length ratio, and the ratio of areas between the channels of contiguous orders are affected by the degree of fractal discretization of the synthetic hydrographic network extracted from a DTM. The algorithm considers the AT as an area value arbitrarily attributed as an injunction factor regarding the flow accumulation model. The z attributes of the flow accumulation DEM express the number of cells that flow to each cell of a DTM.

In a typical hydrological landscape, the subsurface water sheet tends to lie closer to the topographic surface on the floodplains and further away from the plateaus [5,69]. The continuity or not of that sheet throughout the landscape depends on the composition of the soil profiles. In general, water accumulation or soil saturation in the landscape is expected on the floodplains and plateaus due to the low slopes of the terrain, leading to expected susceptibility to flooding and sediment depositions on the floodplains.

The geometric signatures of floodplains appeared in the Canoinhas River watershed throughout the entire hydrographic network (Figure 6). The most expressive floodplain signatures tend to be closer to the higher-order channels, which would be expected due to the lowest slope class characteristics of the watershed. These signatures can guide the planning of land occupation, whether for agricultural activities or urban occupation, for example.

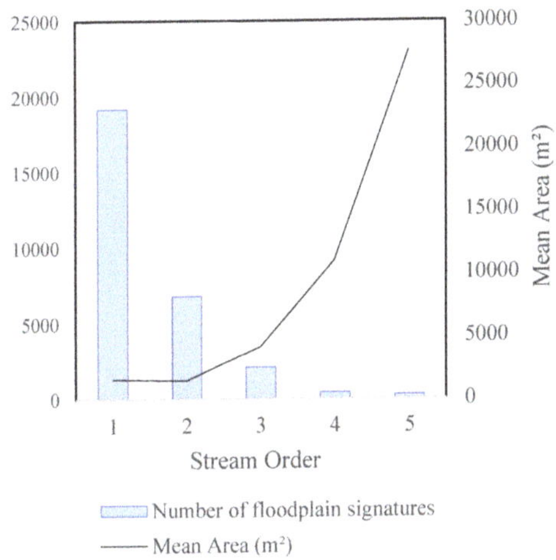

Figure 6. Geometric signatures of the floodplain class throughout a 100 m buffer from the Canoinhas River watershed hydrographic network.

The geometric signatures of the hillslopes were shown to be quite diversified in form and size. It is common to find fragments of hillslopes surrounded by plateaus on the same slope. This occurs due to changes in the local slope used to classify the model. These fragments can even be considered irrelevant from the viewpoint of the environmental analysis of the watershed, since few or no specific decision-making processes would be expected in adhering to this level of analysis.

4.3. Zones with a Propensity for Water Accumulation/Soil Saturation and Hydrological Landscapes Classes

The occurrence of zones with a high propensity for water accumulation or soil saturation (TWI > 8) concerning the HAND classes has been studied in the literature. Unlike the HAND model, whose manifestation of the vertical distances in the model depends on the degree of fractal discretization of the synthetic hydrographic network, the TWI model is manifested throughout the whole watershed precisely because the parameter depends solely on variables of the relief (a and β). The degree of fractal discretization of the synthetic hydrographic network may be responsible for some regions of the watershed that could be classified as floodplains (more discrete) being classified as hillslopes or plateaus (less discrete), which can add bias in the TWI occurrence analysis in the classes of the hydrological landscape. With this caveat, it was observed that in the Canoinhas River watershed, there is a tendency for the most prominent water accumulation or soil saturation entities to lie close to the higher-order channels (Figure 7).

There is a tendency for an exponential increase in the floodplain areas and the areas with high water accumulation or soil saturation as the order of the channels increases. This may be explained by the fractal dimension of the hydrographic network, the cumulative order of the channels using the Strahler method in measuring the ramifications, and the erosion, transportation, and fluvial sedimentation processes. The sedimentary areas tend to occupy the lowest regions of the relief and contribute to the formation of the floodplains as structural units of the hydrological landscape. These zones should receive special attention in the watershed's environmental planning due to their riparian importance or their tendency to flood.

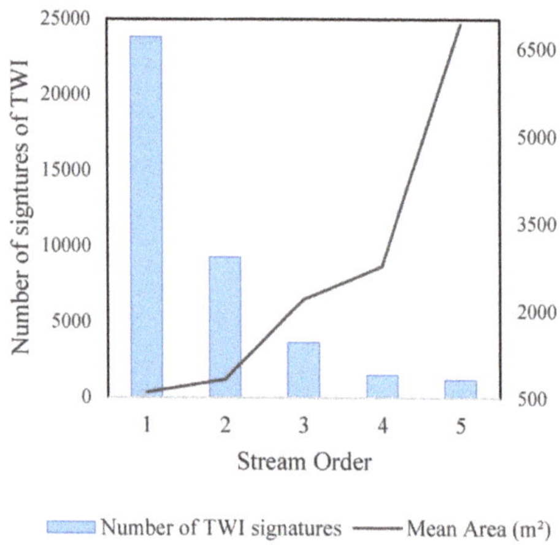

Figure 7. Geometric signatures of TWI throughout the 100 m buffer from the Canoinhas River watershed hydrographic network show the zones that are highly prone to developing water accumulation or soil saturation on the floodplains.

The areas prone to developing water accumulation or soil saturation were distributed throughout all units of the hydrological landscape (Figure 8). The plateau areas were the most indicated areas prone to developing water accumulation or soil saturation, followed by the floodplains, hillslopes, and ecotones. The means of the areas indicate that the number of geometric signatures followed that same trend. This shows the importance of plateaus as structural elements of the landscape for surface water storage. The hillslopes were also shown to be relevant as landscape units with a tendency to accumulate water. These regions can contain topographic footprints that indicate wetlands or waterlogged zones, which can play a relevant ecological role in the watershed. From an urban occupation perspective, these zones are potential areas of flooding caused by surface runoff from intense rainfall, hence the need to pay attention to local micro-drainage systems [70,71].

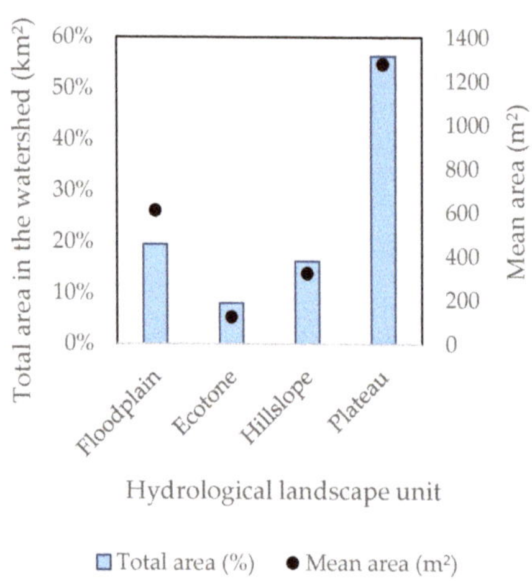

Figure 8. Geometric areas associated with a high propensity for developing water accumulation or soil saturation in the units of the hydrological landscape in the Canoinhas River watershed.

4.4. Land Use

Land use factors have been employed as parameters for assessing the levels of environmental degradation or preservation [31–33,72,73]. Some of these factors are recognized as modifying agents of the hydrological responses of the watershed and can act over water infiltration and percolation in the ground, surface, and subsurface runoff, and evapotranspiration, among others. As a result, the hydric balance of the watershed can be affected, the storage capacity of the aquifers can be reduced, and the occurrence of natural disasters from landslides, mudslides, and flooding can be intensified.

The image classification of the Canoinhas River watershed produced geometries associated with agriculture, water bodies, urban areas, fields, forests, and planted forests. The geometric signatures of the respective classes appeared throughout the whole spatial domain of the watershed and the structural units of the hydrological landscape (Figure 9). The Forest class is the one that covers the most significant part of this domain, occupying 33% of the total area of plateaus and 61% of the hillslope area. However, there is a clear decreasing trend in the geometric signatures of this specific class further away from the hydrographic network, up to around 50 m (Figure 9).

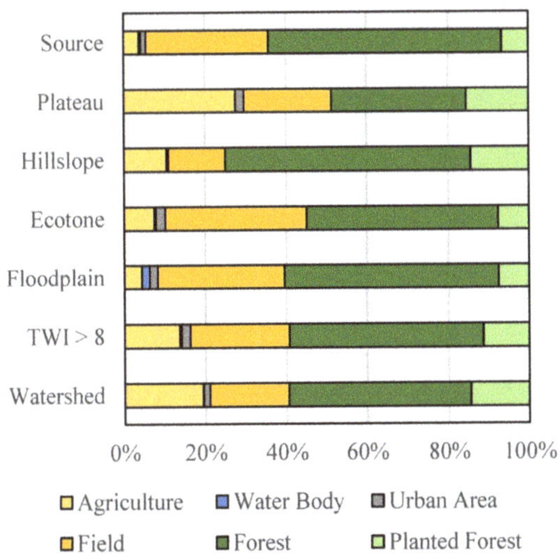

Figure 9. Land use in the Canoinhas River watershed in the classes of the hydrological landscape, and in the areas prone to developing water accumulation or soil saturation.

The Field class is the second most frequent and represents, together with the Forest and Water classes, the most sensitive geographical entities from an environmental quality and preservation viewpoint. This shows that the Canoinhas River watershed still has wide vegetated areas that warrant attention from a preservation viewpoint and wide field areas subject to agroforestry exploitation. Most field and native forest areas are distributed on floodplains, ecotones, and hillslopes. A substantial portion is also situated in areas prone to water accumulation or soil saturation.

Agriculture, Urban Areas, and planted forests represent the geographical entities associated with anthropogenic activities in the watershed. The geometric signatures of the agriculture class occupy 19% of the total area of the watershed, and the planted forest class occupies 15%. Agricultural activities are also found in 4% of the surrounding areas within 50 m of the source areas of the watershed. These areas are still occupied by entities of the planted forest (7%), grasslands (30%), and forests (58%).

The areas occupied by agriculture entities and planted forest classes around the hydrographic network tend to increase with distance, while the areas occupied by the forest class tend to decrease (Figure 10). A drainage network is a component of the morphological subsystem of the watershed that scarcely alters its position, geometry, and attribute characteristics, thus being considered fixed bases in the landscape and potentially applicable as both horizontal and vertical references of alterations in the state of the GIS of the chosen watershed. Thus, the hydrographic network is a potential indicator of the behavior of dynamic geographic systems such as the expansion or contraction of anthropogenic activities that are spatial in scope in one urban area [74].

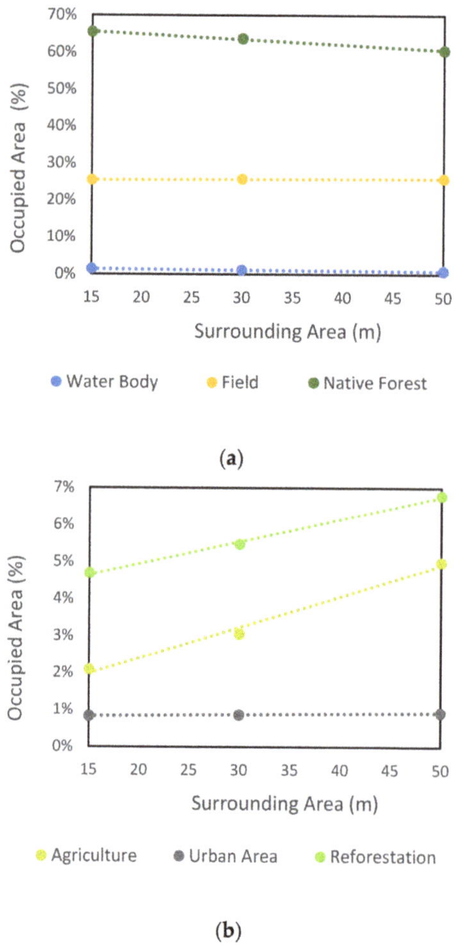

Figure 10. Areas occupied by the geometric signatures of land use in the 0–15 m, 0–30 m, and 0–50 m surrounding areas in the Canoinhas River watershed. Land use consists of classes of natural use (**a**) and classes with human activities (**b**).

The Urban Area class in the Canoinhas River watershed occupies a small total area. However, it is found on floodplains, ecotones, and plateaus and in areas prone to developing water accumulation or soil saturation. The plateaus can be considered as safe environments for urban equipment in terms of the risks of flooding disasters. On the other hand, the floodplains and areas prone to developing water accumulation/soil saturation warrant attention concerning the macro- and micro-drainage systems, especially in the vectors of urban expansion. Low-impact development (LID) measures should be considered to mitigate the effects of localized flooding caused by intense rainfall in the consolidated areas where the occupation has reached an irreversible state and address water quality in urban ecosystems [69–78].

4.5. Further Research Perspectives

Land use factors have been employed to assess environmental degradation or preservation [31–33]. Some of these factors are recognized as modifying agents of the hydrological responses of the watershed and can act over water infiltration and percolation in the ground, surface, subsurface runoff, and evapotranspiration, among others. As a result, the hydric balance of the watershed can be affected, the storage capacity of the aquifers can be reduced, and the occurrence of natural disasters from landslides, mudslides, and flooding can be intensified.

The relationship between water and landscape is a natural relationship for shaping the land surface, sustaining lives, and conditioning human activities. Hydro-based hydrological models such as TWI and HAND bring in their conceptual framework assumptions of this relationship, which can be explored practically for regional or local planning purposes to pursue environmentally sustainable development. They can be applied from a DEM since their conceptual basis rests on the physical functions of water on the land, its geographical position in the terrain, and its energy potential in the landscape. These models are conceptual abstractions built using the hydrological landscapes' structural units that form the watershed's geomorphology.

Interestingly, these models can be used to recognize landforms that condition the behavior of cascading systems such as the hydrological cycle, the dynamics of land use by cities, agriculture, cattle ranching, and planted forests, among others, for example, to preserve sensitive ecological zones and prevent the occurrence of disasters caused by floods from extreme runoffs, debris flows, and mass movements. Therefore, the water–landscape relationship can be explored with these models without the need for scientific modeling of the cascade hydrological systems, whose complexity sometimes makes its application unfeasible in countries like Brazil due to the lack of specialists or reliable data for effective modeling of the systems.

Regional-scale maps have been found to be incompatible with local urban analysis as it demands detailed DEMs for more accurate TWI and HAND models. Therefore, further assessments should consider, for example, the efficacy of the HAND model to represent the flood extent in flat areas under high spatial resolution DEMs such as those extracted from high point density cloud points acquired by airborne LIDAR.

Furthermore, a sensitivity effect of algorithms to define flow direction and flow accumulation from DEM, since the HAND model is spatially dependent on the channel's flow paths, is also needed. On the other hand, the TWI model tends to show a superabundance of water accumulation/soil saturation in high-resolution DEM. Therefore, efforts may drive toward a better choice of DEM resolution or DEM cell size to delineate features on the landscape that better represent the phenomena, and therefore helps land managers and environmental agencies support decisions for better use of the environment.

5. Conclusions

This research evaluated HAND and TWI morphological models as an important source of information on the geomorphological agents that condition the anthropogenic activities in the structuring units of the landscape. Our results showed that geometric signatures of the TWI emerged through all of the structural units of the hydrological landscape, with values between −0.56 and 25.69, where values above 8.0 represent areas prone to developing water accumulation or soil saturation. The plateau areas were the ones that most indicated that condition, followed by the floodplains, hillslopes, and ecotones. In such areas, plateau areas are suggested as structural elements of the landscape for surface water storage. However, this distribution depends on local relief characteristics, and more detailed studies are strongly encouraged.

In the Canoinhas River watershed, there is a tendency for the largest geometric signatures of water accumulation or soil saturation entities to be located close to the higher-order channels, along with the largest geometric signatures of the floodplains. The area around the drainage network within 50 m of these channels showed that the areas occupied by entities of the Agriculture and Planted Forest classes tended to increase with distance while the areas occupied by the Forest class tended to decrease. On the other hand, the Grasslands, Urban Areas, and Water-related classes remained stable. Some agricultural and forestry activities were also found within 50 m of the source areas, which shall be considered in the future by environmental agencies.

HAND and TWI are hydrological-based models that are relatively simple to formulate but have robust assumptions, which can be applied based on available DEMs. Their conceptual basis rests on the physical functions of water on the land, its geographical position in the terrain, and its energy potential in the landscape. These models are ultimately conceptual abstractions built using the hydrological landscapes' structural units that form the watershed's geomorphology.

Studying how land and water are related in morphological models like HAND and TWI can help us better understand and evaluate watersheds using freely available remote-sensing data sources. Factors like terrain, soil, and water quality all play a role in how people use the land and could drive decision-makers to use the landscape better. Hydrological-based models can be an easy way to analyze how all of these different factors interact and where environmental agencies must pay some attention.

Author Contributions: Conceptualization, Silvio Luís Rafaeli Neto and Vanessa Jutel dos Santos; Methodology, Silvio Luís Rafaeli Neto and Vanessa Jutel dos Santos; Software, Silvio Luís Rafaeli Neto, Vanessa Jutel dos Santos, Emili Louise Diconcili Schutz, and Leticia Margarete de Moliner; Validation, Silvio Luís Rafaeli Neto; Formal analysis, Silvio Luís Rafaeli Neto, Vanessa Jutel dos Santos, Emili Louise Diconcili Schutz, Leticia Margarete de Moliner, Cristiane Gracieli Kloth, Daiane Teixeira Schier, Veraldo Liesenberg, and Polyanna da Conceição Bispo; Investigation, Silvio Luís Rafaeli Neto; Resources, Silvio Luís Rafaeli Neto, Vanessa Jutel dos Santos, Emili Louise Diconcili Schutz, Leticia Margarete de Moliner, Cristiane Gracieli Kloth, Daiane Teixeira Schier, Veraldo Liesenberg, and Polyanna da Conceição Bispo; Data curation, Silvio Luís Rafaeli Neto; Writing—original draft preparation, Silvio Luís Rafaeli Neto, Vanessa Jutel dos Santos, and Emili Louise Diconcili Schutz; Writing—review and editing, Silvio Luís Rafaeli Neto, Vanessa Jutel dos Santos, Emili Louise Diconcili Schutz, Leticia Margarete de Moliner, Cristiane Gracieli Kloth, Daiane Teixeira Schier, Arsalan Ahmed Othman, Veraldo Liesenberg, and Polyanna da Conceição Bispo; Supervision, Silvio Luís Rafaeli Neto, Veraldo Liesenberg, and Polyanna da Conceição Bispo; Project administration, Silvio Luís Rafaeli Neto and Polyanna da Conceição Bispo; Funding acquisition, Silvio Luís Rafaeli Neto, Veraldo Liesenberg, and Polyanna da Conceição Bispo; Visualization, Silvio Luís Rafaeli Neto, Vanessa Jutel dos Santos, Emili Louise Diconcili Schutz, Leticia Margarete de Moliner, Cristiane Gracieli Kloth, Daiane Teixeira Schier, Arsalan Ahmed Othman, Veraldo Liesenberg, and Polyanna da Conceição Bispo. All authors have read and agreed to the published version of the manuscript.

Funding: This manuscript received financial support from Coordenação de Aperfeiçoamento de Pessoal de Nível Superior (CAPES PROAP/AUXPE) and FAPESC (2017TR1762, 2019TR828). Veraldo Liesenberg is supported by CNPq (313887/2018-7, 317538/2021-7).

Data Availability Statement: The dataset can be made available upon request.

Acknowledgments: This research received support from Fundação de Amparo à Pesquisa e Inovação do Estado de Santa Catarina (FAPESC PAP/UDESC).

Conflicts of Interest: The authors declare no conflict of interest.

References

1. MacMillan, R.A.; Jones, R.K.; McNabb, D.H. Defining a Hierarchy of Spatial Entities for Environmental Analysis and Modeling Using Digital Elevation Models (DEMs). *Comput. Environ. Urban Syst.* **2004**, *28*, 175–200. [CrossRef]
2. Rahmati, O.; Kornejady, A.; Samadi, M.; Nobre, A.D.; Melesse, A.M. Development of an Automated GIS Tool for Reproducing the HAND Terrain Model. *Environ. Model. Softw.* **2018**, *102*, 1–12. [CrossRef]
3. Keller, E.; Adamaitis, C.; Alessio, P.; Anderson, S.; Goto, E.; Gray, S.; Gurrola, L.; Morell, K. Applications in Geomorphology. *Geomorphology* **2020**, *366*, 106729. [CrossRef]
4. Gruber, S.; Peckham, S. Land-Surface Parameters and Objects in Hydrology. *Dev. Soil Sci.* **2009**, *33*, 171–194. [CrossRef]
5. Winter, T.C. The Concept of Hydrologic Landscapes. *J. Am. Water Resour. Assoc.* **2001**, *37*, 335–349. [CrossRef]

6. Beven, K.J.; Kirkby, M.J. A Physically Based, Variable Contributing Area Model of Basin Hydrology. *Hydrol. Sci. Bull.* **1979**, *24*, 43–69. [CrossRef]
7. Rennó, C.D.; Nobre, A.D.; Cuartas, L.A.; Soares, J.V.; Hodnett, M.G.; Tomasella, J.; Waterloo, M.J. HAND, a New Terrain Descriptor Using SRTM-DEM: Mapping Terra-Firme Rainforest Environments in Amazonia. *Remote Sens. Environ.* **2008**, *112*, 3469–3481. [CrossRef]
8. Nobre, A.D.; Cuartas, L.A.; Hodnett, M.; Rennó, C.D.; Rodrigues, G.; Silveira, A.; Waterloo, M.; Saleska, S. Height above the Nearest Drainage—A Hydrologically Relevant New Terrain Model. *J. Hydrol.* **2011**, *404*, 13–29. [CrossRef]
9. Cuartas, L.A.; Tomasella, J.; Nobre, A.D.; Nobre, C.A.; Hodnett, M.G.; Waterloo, M.J.; de Oliveira, S.M.; von Randow, R.d.C.; Trancoso, R.; Ferreira, M. Distributed Hydrological Modeling of a Micro-Scale Rainforest Watershed in Amazonia: Model Evaluation and Advances in Calibration Using the New HAND Terrain Model. *J. Hydrol.* **2012**, *462–463*, 15–27. [CrossRef]
10. Mattila, U.; Tokola, T. Terrain Mobility Estimation Using TWI and Airborne Gamma-Ray Data. *J. Environ. Manag.* **2019**, *232*, 531–536. [CrossRef]
11. Guan, M.; Liang, Q. A two-dimensional hydro-morphological model for river hydraulics and morphology with vegetation. *Environ. Model. Softw.* **2017**, *88*, 10–21. [CrossRef]
12. Mohamedou, C.; Tokola, T.; Eerikäinen, K. LiDAR-Based TWI and Terrain Attributes in Improving Parametric Predictor for Tree Growth in Southeast Finland. *Int. J. Appl. Earth Obs. Geoinf.* **2017**, *62*, 183–191. [CrossRef]
13. Drover, D.R.; Jackson, C.R.; Bitew, M.; Du, E. Effects of DEM Scale on TWI Spatial Distribution Effects of DEM Scale on the Spatial Distribution of the TOPMODEL Topographic Wetness Index and Its Correlations to Watershed Characteristics. *Hydrol. Earth Syst. Sci. Discuss* **2015**, *12*, 11817–11846. [CrossRef]
14. Rózycka, M.; Migoń, P.; Michniewicz, A. Topographic Wetness Index and Terrain Ruggedness Index in Geomorphic Characterisation of Landslide Terrains, on Examples from the Sudetes, SW Poland. *Z. Fur Geomorphol.* **2017**, *61*, 61–80. [CrossRef]
15. Riihimäki, H.; Kemppinen, J.; Kopecký, M.; Luoto, M. Topographic Wetness Index as a Proxy for Soil Moisture: The Importance of Flow-Routing Algorithm and Grid Resolution. *Water Resour. Res.* **2021**, *57*, e2021WR029871. [CrossRef]
16. Riittersfl, K.H.; O'neill, R.V.; Jones, K.B. Assessing habitat suitability at multiple scales: A landscape-level approach. *Biol. Conserv.* **1997**, *81*, 191–202. [CrossRef]
17. Ballerine, C. *Topographic Wetness Index Urban Flooding Awareness Act Action Support Will and DuPage Counties*; Illinois State Water Survey—Prairie Research Institute—University of Illinois: Champaign, IL, USA, 2017; Volume 22.
18. Gharari, S.; Hrachowitz, M.; Fenicia, F.; Savenije, H.H.G. Hydrological Landscape Classification: Investigating the Performance of HAND Based Landscape Classifications in a Central European Meso-Scale Catchment. *Hydrol. Earth Syst. Sci.* **2011**, *15*, 3275–3291. [CrossRef]
19. Dantas, A.A.R.; Paz, A.R. Use of HAND Terrain Descriptor for Estimating Flood-Prone Areas in River Basins. *Rev. Bras. Ciências Ambient.* **2021**, *56*, 501–516. [CrossRef]
20. Bayat, M.; Ghorbanpour, M.; Zare, R.; Jaafari, A.; Thai Pham, B. Application of Artificial Neural Networks for Predicting Tree Survival and Mortality in the Hyrcanian Forest of Iran. *Comput. Electron. Agric.* **2019**, *164*, 104929. [CrossRef]
21. Gao, Y.; Yao, L.; bin Chang, N.; Wang, D. Diagnosis toward Predicting Mean Annual Runoff in Ungauged Basins. *Hydrol. Earth Syst. Sci.* **2021**, *25*, 945–956. [CrossRef]
22. Simon, H.A. *The New Science of Management Decision*, 1st ed.; Harper and Row: New York, NY, USA, 1960.
23. Chorley, R.; Kennedy, B. *Physical Geography: A System Approach*; Prentice-Hall: London, UK, 1971.
24. Neto, S.L.R.; Rodrigues, M. A Taxonomy of Strategies for Developing Spatial Decision Support Systems. In *Systems Development Methods for Databases, Enterprise, Modelling, and Workflow Management*; Wojtkowski, W., Wojtkowski, W., Wrycza, S., Zupancic, J., Eds.; Kluwer Academic/Plenum: New York, NY, USA, 1999; pp. 139–155.
25. Longley Paul, A.; Goodchild, M.F.; Maguire, D.J.; Rhind, D.W. *Sistemas e Ciência Da Informação Geográfica*, 3rd ed.; Bookman: Porto Alegre, Brasil, 2013.
26. Neto, S.L.R.; Sá, E.A.S.; Debastiani, A.B.; Padilha, V.L.; Antunes, T.A. Efficacy of Rainfall-Runoff Models in Loose Coupling Spacial Decision Support Systems Modelbase. *Water Resour. Manag.* **2019**, *33*, 889–904. [CrossRef]
27. Densham, P.J. Spatial Decision Support Systems. In *Geographical Information Systems. Vol. 1: Principles*; Wiley: Hoboken, NJ, USA, 1991; pp. 403–412. [CrossRef]
28. Kobiyama, M.; Mendonça, M.; Moreno, D.A.; Marcelino, I.P.V.d.O.; Marcelino, E.V.; Gonçalves, E.F.; Brazetti, L.L.P.; Goerl, R.F.; Molleri, G.S.F.; Rudorff, F.d.M. *Prevenção de Desastres Naturais Conceitos Básicos*; Organic Trading: Rosebery NSW, Australia, 2006; ISBN 858775503X.
29. Tram, V.N.Q.; Somura, H.; Moroizumi, T.; Maeda, M. Effects of Local Land-Use Policies and Anthropogenic Activities on Water Quality in the Upstream Sesan River Basin, Vietnam. *J. Hydrol. Reg. Stud.* **2022**, *44*, 101225. [CrossRef]
30. Pandey, S.; Kumar, P.; Zlatic, M.; Nautiyal, R.; Panwar, V.P. Recent Advances in Assessment of Soil Erosion Vulnerability in a Watershed. *Int. Soil Water Conserv. Res.* **2021**, *9*, 305–318. [CrossRef]
31. Sisay, G.; Gitima, G.; Mersha, M.; Alemu, W.G. Assessment of Land Use Land Cover Dynamics and Its Drivers in Bechet Watershed Upper Blue Nile Basin, Ethiopia. *Remote Sens. Appl.* **2021**, *24*, 100648. [CrossRef]
32. Wang, Y.; Liu, X.; Wang, T.; Zhang, X.; Feng, Y.; Yang, G.; Zhen, W. Relating Land-Use/Land-Cover Patterns to Water Quality in Watersheds Based on the Structural Equation Modeling. *Catena* **2021**, *206*, 105566. [CrossRef]

33. Ross, E.R.; Randhir, T.O. Effects of Climate and Land Use Changes on Water Quantity and Quality of Coastal Watersheds of Narragansett Bay. *Sci. Total Environ.* **2022**, *807*, 151082. [CrossRef] [PubMed]
34. Pike, R.J. The Geometric Signature: Quantifying Landslide-Terrain Types from Digital Elevation Models I. *Math Geol.* **1988**, *20*, 491–511. [CrossRef]
35. Schröder, B. Pattern, Process, and Function in Landscape Ecology and Catchment Hydrology—How Can Quantitative Landscape Ecology Support Predictions in Ungauged Basins? *Hydrol. Earth Syst. Sci.* **2006**, *10*, 967–979. [CrossRef]
36. Hengl, T.; Evans, I.S. Mathematical and Digital Models of the Land Surface. *Dev. Soil Sci.* **2009**, *33*, 31–63. [CrossRef]
37. Bishop, M.P.; Young, B.W.; Huo, D. Geomorphometry: Quantitative Land-Surface Analysis and Modeling. In *Reference Module in Earth Systems and Environmental Sciences*; Elsevier: Amsterdam, The Netherlands, 2018.
38. Mao, W.; Yang, J.; Zhu, Y.; Ye, M.; Liu, Z.; Wu, J. An Efficient Soil Water Balance Model Based on Hybrid Numerical and Statistical Methods. *J. Hydrol.* **2018**, *559*, 721–735. [CrossRef]
39. van Buuren, M.; Kerkstra, K. The Framework Concept and the Hydrological Landscape Structure: A New Perspective in the Design of Multifunctional Landscapes. Available online: https://doi.org/10.1007/978-94-011-2318-1_10 (accessed on 30 September 2022).
40. Neto, S.L.R.; Biffi, L.J. Aplicação de Um Modelo Linear Local Na Determinação de Alturas Ortométricas Referidas Ao Sistema Geodésico Brasileiro. *Bol. Goiano Geogr.* **2016**, *36*, 157–176. [CrossRef]
41. Quinn, P.K.B.; Chevallier, P.; Planchon, O. The Prediction of Hillslope Flow Paths for Distributed Hydrological Modelling Using Digital Terrain Models. *Hydrol. Process.* **1991**, *5*, 59–79. [CrossRef]
42. Ambroise, B.; Beven, K.; Freer, J. Toward a Generalization of the TOPMODEL Concepts: Topographic Indices of Hydrological Similarity. *Water Resour. Res.* **1996**, *32*, 2135–2145.
43. Dušek, J.; Dařenová, E.; Pavelka, M.; Marek, M.V. Methane and Carbon Dioxide Release from Wetland Ecosystems. In *Climate Change and Soil Interactions*; Elsevier: Amsterdam, The Netherlands, 2020; pp. 509–553. [CrossRef]
44. Tarboton, D.G. A New Method for the Determination of Flow Directions and Upslope Areas in Grid Digital Elevation Models. *Water Resour. Res.* **1997**, *33*, 309–319.
45. Nobre, A.D.; Cuartas, L.A.; Momo, M.R.; Severo, D.L.; Pinheiro, A.; Nobre, C.A. HAND Contour: A New Proxy Predictor of Inundation Extent. *Hydrol. Process.* **2016**, *30*, 320–333. [CrossRef]
46. Momo, M.R.; Pinheiro, A.; Severo, D.L.; Cuartas, L.A.; Nobre, A.D. Desempenho Do Modelo Hand No Mapeamento de Áreas Suscetíveis à Inundação Usando Dados de Alta Resolução Espacial. *Rev. Bras. Recur. Hídricos* **2016**, *21*, 200–208. [CrossRef]
47. Bhatt, C.M.; Srinivasa Rao, G. HAND (Height above Nearest Drainage) Tool and Satellite-Based Geospatial Analysis of Hyderabad (India) Urban Floods, September 2016. *Arab. J. Geosci.* **2018**, *11*, 600. [CrossRef]
48. Santa Catarina. Plano de Recursos Hídricos Da Bacia Hidrográfica Do Rio Canoinhas e Afluentes Catarinenses Do Rio Negro. Available online: https://www.infoteca.cnptia.embrapa.br/infoteca/bitstream/doc/964417/1/BPD-46-2004-Santa-Catarina-.pdf (accessed on 30 September 2022).
49. EMBRAPA SOLOS Distribuição Geográfica Dos Solos Do Estado de Santa Catarina. Available online: https://www.fapesc.sc.gov.br/wp-content/uploads/2021/02/pgrh-canoinhas_produto_3_etapa_c_final_rev21jul2020.pdf (accessed on 30 September 2022).
50. AMPLANORTE. Plano de Desenvolvimento Regional Do Planalto Norte Catarinense. 2017. Available online: https://www.amplanorte.org.br/cms/pagina/ver/codMapaItem/74869 (accessed on 30 September 2022).
51. Rao, P.; Wang, Y.; Liu, Y.; Wang, X.; Hou, Y.; Pan, S.; Wang, F.; Zhu, D. A Comparison of Multiple Methods for Mapping Groundwater Levels in the Mu Us Sandy Land, China. *J. Hydrol. Reg. Stud.* **2022**, *43*, 101189. [CrossRef]
52. Souza, C.M.; Shimbo, J.Z.; Rosa, M.R.; Parente, L.L.; Alencar, A.A.; Rudorff, B.F.T.; Hasenack, H.; Matsumoto, M.; Ferreira, L.G.; Souza-Filho, P.W.M.; et al. Reconstructing Three Decades of Land Use and Land Cover Changes in Brazilian Biomes with Landsat Archive and Earth Engine. *Remote Sens.* **2020**, *12*, 2735. [CrossRef]
53. Avery, T.E. *Interpretation of Aerial Photographs*, 3rd ed.; Burgess Publishing Company: Minneapolis, MN, USA, 1977.
54. Souza, C.F.; Liesenberg, V.; Schimalski, M.B.; Casemiro Soares, P.R. Evaluating the Monetary Environmental Compensation over a Hydroelectric Power Plant Based on Opportunity Cost Simulation, GIS, and Remote Sensing Images. *Remote Sens. Appl.* **2021**, *23*, 100573. [CrossRef]
55. Costa, J.d.S.; Liesenberg, V.; Schimalski, M.B.; de Sousa, R.V.; Biffi, L.J.; Gomes, A.R.; Neto, S.L.R.; Mitishita, E.; Bispo, P.d.C. Benefits of Combining Alos/Palsar-2 and Sentinel-2a Data in the Classification of Land Cover Classes in the Santa Catarina Southern Plateau. *Remote Sens.* **2021**, *13*, 229. [CrossRef]
56. SDS. *Levantamento Aerofotogramétrico Do Estado de Santa Catarina. Secretaria de Estado Do Desenvolvimento Econômico e Sustentável*; ENGEMAP: Assis, Brazil, 2013.
57. Wolock, D.M.; McCabe, G.J. Differences in Topographic Characteristics Computed from 100- and 1000-m Resolution Digital Elevation Model Data. *Hydrol. Process.* **2000**, *14*, 987–1002. [CrossRef]
58. Wolock, D.M.; Price, C.V. Effects of Digital Elevation Model Map Scale and Data Resolution on a Topography-Based Watershed Model. *Water Resour. Res.* **1994**, *30*, 3041–3052. [CrossRef]
59. Meles, M.B.; Younger, S.E.; Jackson, C.R.; Du, E.; Drover, D. Wetness Index Based on Landscape Position and Topography (WILT): Modifying TWI to Reflect Landscape Position. *J. Environ. Manag.* **2020**, *255*, 109863. [CrossRef] [PubMed]
60. Gyasi-Agyei, Y.; Willgoose, G.; de Troch, F.P. Effects of Vertical Resolution and Map Scale of Digital Elevation Models on Geomorphological Parameters Used in Hydrology. *Hydrol. Process.* **1995**, *9*, 363–382. [CrossRef]

61. Camara, G.; Vinhas, L.; Ferreira, K.R.; de Queiroz, G.R.; de Souza, R.C.M.; Monteiro, A.M.V.; de Carvalho, M.T.; Casanova, M.A.; de Freitas, U.M. *TerraLib: An Open Source GIS Library for Large-Scale Environmental and Socio-Economic Applications*; Springer: Berlin/Heidelberg, Germany, 2008; pp. 247–270. [CrossRef]
62. Nobre, A.D. A new landscape classification: The HAND Model. Available online: https://hess.copernicus.org/preprints/8/C2446/2011/hessd-8-C2446-2011.pdf (accessed on 30 September 2022).
63. ESRI. *ArcGis Release 10.5.1*; ESRI: Redlands, CA, USA, 2017.
64. Bonamigo, A.; Schimalski, M.B.; Soares, P.R.C.; Liesenberg, V.; de Souza, T.R.; Boesing, T.L.S. Variação Nas Áreas de Preservação Permanente Em Imóveis Rurais Do Planalto Sul Catarinense Segundo as Leis N° 4.771 e 12.651. *Cienc. Rural* **2017**, *47*, 1–6. [CrossRef]
65. Gharari, S.; Fenicia, F.; Hrachowitz, M.; Savenije, H.H.G. Land Classification Based on Hydrological Landscape Units. *Hydrol. Earth Syst. Sci. Discuss* **2011**, *8*, 4381–4425. [CrossRef]
66. Schier, D.T. Avaliação Do Índice Topográfico de Umidade Para Detecção de Zonas Urbanas Inundáveis. Master's Thesis, Universidade do Estado de Santa Catarina, Lages, Brazil, 2020.
67. Biffi, L.J.; Neto, S.L.R. Spatial Behavior of the Agronomic Variables of the Fuji Apple during Two Years in the Planalto Serrano of Santa Catarina State. *Rev. Bras. Frutic* **2008**, 975–980. [CrossRef]
68. Scheidegger, A.E. Horton's Law of Stream Numbers, 3rd ed. *Water Resour. Res.* **1968**, *4*, 655–658. [CrossRef]
69. Tan, Z.; Li, Y.; Zhang, Q.; Liu, X.; Song, Y.; Xue, C.; Lu, J. Assessing Effective Hydrological Connectivity for Floodplains with a Framework Integrating Habitat Suitability and Sediment Suspension Behavior. *Water Res.* **2021**, *201*, 117253. [CrossRef]
70. Sieker, H.; Klein, M. Best Management Practices for Stormwater-Runoff with Alternative Methods in a Large Urban Catchment in Berlin, Germany. *Water Sci. Technol.* **1998**, *38*, 91–97. [CrossRef]
71. Braune, M.J.; Wood, A. Best Management Practices Applied to Urban Runoff Quantity and Quality Control. *Water Sci. Technol.* **1999**, *39*, 117–121. [CrossRef]
72. Pereira, B.W.d.F.; Maciel, M.d.N.M.; Oliveira, F.d.A.; Alves, M.A.M.d.S.; Ribeiro, A.M.; Ferreira, B.M.; Ribeiro, E.G.P. Land Use and Water Quality Degradation in the Peixe-Boi River Watershed. *Ambiente Água* **2016**, *11*, 472–485. [CrossRef]
73. Wroblescki, F.A.; Bertol, I.; Wolschick, N.H.; Bagio, B.; Santos, V.P.d.; Bernardi, L.; Biasiolo, L.A. Assessed Impact of Anthropization on Water and Soil Quality in a Drainage Basin in Southern Brazil. *Rev. De Ciências Agroveterinárias* **2021**, *20*, 74–85. [CrossRef]
74. Neto, S.L.R.; Cordeiro, M.T.A. Análise Do Comportamento de Sistemas Urbanos Por Meio de Componentes de Sistemas Hidrológicos. *GEOUSP Espaço E Tempo* **2015**, *19*, 142–155. [CrossRef]
75. Ice, G. History of Innovative Best Management Practice Development and Its Role in Addressing Water Quality Limited Waterbodies. *J. Environ. Eng.* **2004**, *130*, 684–689. [CrossRef]
76. Tingsanchali, T. Urban Flood Disaster Management. In *Proceedings of the Procedia Engineering*; Elsevier Ltd.: Amsterdam, The Netherlands, 2012; Volume 32, pp. 25–37.
77. Woods Ballard, B. *Construction Industry Research and Information Association. The SuDS Manual*; CIRIA: London, UK, 2016.
78. Maidment, D.; Rajib, A.; Lin, P.; Clark, E.P. *National Water Center Innovators Program Summer Institute Report 2016*; National Water Center: Tuscaloosa, AL, USA, 2016.

International Journal of
Geo-Information

MDPI

Article

Assessment of a Dynamic Physically Based Slope Stability Model to Evaluate Timing and Distribution of Rainfall-Induced Shallow Landslides

Juby Thomas [1], Manika Gupta [1,*], Prashant K. Srivastava [2] and George P. Petropoulos [3]

[1] Department of Geology, University of Delhi, Delhi 110007, India
[2] Remote Sensing Laboratory, Institute of Environment and Sustainable Development, Banaras Hindu University, Varanasi 221005, India
[3] Department of Geography, Harokopio University of Athens, El. Venizelou 70, 17671 Athens, Greece
* Correspondence: manikagup@gmail.com

Abstract: Shallow landslides due to hydro-meteorological factors are one of the most common destructive geological processes, which have become more frequent in recent years due to changes in rainfall frequency and intensity. The present study assessed a dynamic, physically based slope stability model, Transient Rainfall Infiltration and Grid-Based Slope Stability Model (TRIGRS), in Idukki district, Kerala, Western Ghats. The study compared the impact of hydrogeomechanical parameters derived from two different data sets, FAO soil texture and regionally available soil texture, on the simulation of the distribution and timing of shallow landslides. For assessing the landslide distribution, 1913 landslides were compared and true positive rates (TPRs) of 68% and 60% were obtained with a nine-day rainfall period for the FAO- and regional-based data sets, respectively. However, a false positive rate (FPR) of 36% and 31% was also seen, respectively. The timing of occurrence of nine landslide events was assessed, which were triggered in the second week of June 2018. Even though the distribution of eight landslides was accurately simulated, the timing of only three events was found to be accurate. The study concludes that the model simulations using parameters derived from either of the soil texture data sets are able to identify the location of the event. However, there is a need for including a high-spatial-resolution hydrogeomechanical parameter data set to improve the timing of landslide event modeling.

Keywords: rainfall-induced landslides; physically based model; TRIGRS; Western Ghats

Citation: Thomas, J.; Gupta, M.; Srivastava, P.K.; Petropoulos, G.P. Assessment of a Dynamic Physically Based Slope Stability Model to Evaluate Timing and Distribution of Rainfall-Induced Shallow Landslides. *ISPRS Int. J. Geo-Inf.* **2023**, *12*, 105. https://doi.org/10.3390/ijgi12030105

Academic Editors: Walter Chen, Fuan Tsai and Wolfgang Kainz

Received: 24 November 2022
Revised: 23 February 2023
Accepted: 24 February 2023
Published: 2 March 2023

1. Introduction

Shallow landslides due to hydro-meteorological factors are one of the most common destructive geological processes on the Earth's surface and are responsible for the loss of lives, both human and livestock, every year, in mountainous regions. Although landslide occurrences are confined to hillsides and steep terrain, the devastating after-effects are not only limited to their origin but also affect the downstream areas. The cascading effects of frequent landslides, often with intense precipitation, create innumerable adversities for people and ecosystems in both highlands and lowlands [1–3]. Geologically, shallow, rainfall-induced landslides have an important role as an agent of shaping mountainous landscapes. While modifying the topography and renewing the terrain and ecosystems, both in the upstream and downstream areas, landslides cause the degradation of fertile land, damage to agricultural land, traffic disruption, and destruction of infrastructures, especially settlements, and have many other indirect social and economic implications [4–10].

Although landslides are triggered by geological (volcanic eruption, earthquakes, etc.) and anthropogenic (slope or toe excavation, slope loading, drawdown and irrigation, blast-vibrations, etc.) factors, hydro-meteorological factors still remain the major extrinsic factor [6,11]. In tectonically stable areas, e.g., southern peninsular India where this study was carried out, hydro-meteorological (intense rainfall) factors are the significant triggers, in addition to anthropogenic activities.

The current trends and projections of precipitation patterns show an increase in occurrence of high-intensity rainfall events globally, and heavy rainfalls in a short span of time will trigger a greater number of shallow landslides. As observed, a higher number of landslides have been reported globally because of the surge in the occurrence of frequent intense-localized precipitation events. The increased human presence in the mountainous regions makes the landslides and their immediate effects more severe and destructive [12,13]. With the alarming situation of climate change, population expansion, and uncontrolled exploitation of mountainous regions, tropical and subtropical mountainous areas such as Western Ghats, where this study was carried out, will be the worst affected areas by any typical hydro-geotechnical disaster such as landslides [6,14].

Shallow landslides are defined as slope failures that are generally confined to a depth of less than 3 m [15–18], and depth is determined as the distance to the failure plane from the surface. The failure plane, often termed as the slip surface, is between the soil column and bedrock or within the soil column where a sharp contrast in permeability is present [19,20]. The extent of a geographical area directly hit by landslides, especially shallow, rainfall-induced landslides, is often much smaller than the area affected by similar natural hazards such as floods, droughts, volcanic eruptions, and earthquakes. However, landslides are more frequent and recurring than many of the above-mentioned disasters and cause extensive destructions [6,8,21–25]. Although an individual rainfall-induced shallow landslide involves less runout volume, it has the potential to evolve into debris or earth flows and thereby gather more channel sediments and cause significant destruction along its path and in the downstream areas [3,26,27].

In India, landslides are one of the most common geohazards and are often triggered by hydrometeorological factors. India's mountainous regions experience enormous pressure due to increased human activities along with adversities due to climate change. The impacts of geohazards such as rainfall-induced landslides have intensified. Since it is impossible to control the occurrence of rainfall-induced landslides, it is best advised to deploy comprehensive strategies to mitigate the effects. Additionally, it has to be noted that rainfall-induced landslides are site-specific and their destructive power is limited to the area of origin and the channels along the downstream areas. Adequate information on when and where a landslide may occur will help the policymakers and administrators with issuing site- or region-specific early warnings.

A number of methods are used for computing the information on when and where a landslide will occur [28]. Conventionally, the methods utilized for landslide hazard assessments or slope stability analysis to demarcate landslides can be classified into qualitative and quantitative methods [29–31]. The qualitative approaches can be further categorized into inventory-based mapping and heuristic methods. Inventory-based landslide assessments are one of the simplest forms of landslide hazard assessment, in which historical information of individual landslides are directly collected by visiting the affected area or through various remote sensing techniques [32,33]. Heuristic methods provide information on the degree and type of risks based on the knowledge and experience of an expert [34,35]. Because assigning weightage and ranking to causative and triggering parameters are highly subjective and localized, the expert with most experience and knowledge might produce the best results. Since each area is unique and complicated in its own ways, the ranking and weightage for one region might not produce desirable results for another region. The quantitative methods include two major categories: statistical- or stochastic- and physically based models. The former method employs mathematical models to assess slope stability based on the weighting assigned to the causative factors [36,37]. The weightages are relative, and a possibility of generalization is always present in the statistical methods. Statistical methods often work in combination with heuristic methods through bivariate, multivariate, or neural network analysis. Even though physically based models are relatively complicated compared with the rest of the methods, they are most suitable in local-scale and site-specific studies. Landslide hazard is determined using process-based slope stability models that consider the interplay of causal factors [38,39]. The results are often expressed in terms of a Factor of Safety (FoS), where lesser values indicate unstable areas.

Regardless of the complexity and data intensity of computing, physically based models are widely used to forecast rainfall-induced landslides because of the increased control over the hydrological and geotechnical aspects [40,41]. Physically based models in Geographic Information System (GIS) platforms have provided promising results in computing the timing and localization of shallow, rainfall-induced landslides at the regional scale [16,22,25,42–46]. These models can be employed to model slope stability ranging for areas from tens to thousands of kilometers squared. r.rotstab [47], High Resolution Slope Stability Simulator (HIRESS) [48], GEOtop-FS [49], Shallow Landslides Instability Prediction (SLIP) [50], Stability Index Mapping (SINMAP) [51], Shallow Slope Stability Model (SHALSTAB) [52], distributed Shallow Landslide Analysis Model (dSLAM) [53], and Transient Rainfall Infiltration and Grid-Based Regional Slope Stability (TRIGRS) [54] are some of the available physically based models. SINMAP, SHALSTAB, and TRIGRS are widely used in modeling the slope stability of hills. The models have been tested and validated in diverse geological, geomorphological, and climatic conditions globally. SINMAP, a probabilistic steady-state model, and SHALSTAB, a deterministic steady-state model, require geotechnical parameters and altitude information to compute slope stability and combine the Mohr–Coulomb infinite slope stability model with a steady-state hydrological model [55]. The illustration of slope failure mechanism is achieved mainly through the computation of FoS in physically based models. The FoS is the ratio of shear strength (resisting or stabilizing forces) to shear stress (driving or destabilizing forces) parallel to the surface. The SLIP model considers the saturation of part of the soil column as the major factor affecting slope failure. It assumes that the terrain is stable due to shear strength, and water flows easily into the soil column as the rainfall progresses. Large portions of the soil eventually become saturated, and the soil sliding process starts. SLIP assesses the stability of terrain using the FoS computed through the limit equilibrium method.

TRIGRS, a transient model, considers hydrological information in addition to geotechnical parameters and altitude information. Because TRIGRS considers the transient effects of rainfall on infiltration and computes the change in pore water pressure (PWP) and FoS [54], time varying analysis of slope stability is possible with TRIGRS.

TRIGRS has been previously utilized for regional-scale analysis in various studies and has been found to produce reliable results for transient rainfall events varying from hours to days [42,56,57]. In another study, importance was placed on the transient pore water pressure, which may result in the triggering of a slip [58]. In a study, based on flume tests, it was suggested that the initial porosity and saturation have an impact on the prediction capacity of the SLIP model [59]. It has been noted that another model, SHALSTAB, considers a single set of geotechnical values for a whole basin. The study emphasized the spatial discretization of these parameters to improve the predictive nature of the model [60]. One of the major limitations of SHALSTAB and SINMAP is the assumption of uniform thickness of soil, hydraulic conductivity, and steady-state shallow subsurface flow. Additionally, the slide is assumed to be translational, and the hydrological boundaries always follow the topography of the terrain [22]. While SHALSTAB and SINMAP are suggested for primary investigations of terrain stability, TRIGRS is often recommended to model specific events (such as the 2018 monsoon rainfall events in the study area) and early warning systems [45]. Although considering the soil as homogeneous has been found to over-estimate unstable pixels when using TRIGRS, it has been successfully utilized at the regional scale. As stated in one of the studies, the model parameters were set to be constant for the study area, which resulted in false positives [61]. These studies have stated the limitation of the model lies in the accuracy of the hydrogeomechanical parameters, which may need to be obtained through field investigations. These parameters are known to be spatially variable. Additionally, the lack of field data and laboratory analysis at the regional scale makes these parameters scarce. So, most of the studies are based on the inverse optimization of these parameters through the utilization of landslide inventories [56,61–63]. However, most of the areas lack landslide inventories with correct timing of the landslide occurrence, creating uncertainty in the optimization. The present study used the TRIGRS model [54,64,65] to compute the timing and distribution of shallow, rainfall-induced landslides in a catchment region in Western Ghats, India. In this study, the soil hydraulic parameters (SHPs) were based on two different soil texture maps available at different scales, and their influences on the occurrence of a landslide event were compared. Therefore, the impact of hydro-geomechanical parameters on the determination of FoS and PWP was quantified along hill slopes based on two soil texture data sets (FAO-based and regional), which differ in introducing soil heterogeneity to the model.

2. Materials and Methods

A catchment area was delineated to execute the TRIGRS, and the catchment area lies mostly in the district of Idukki, Kerala, India, being on the windward side of the southern part of Western Ghats. Figure 1 shows the geographical location of the study area along with the administrative boundaries and extent of Western Ghats. The present study examined the distribution of 1913 landslides as a result of rainfall events during the devastating 2018 monsoon period (Figure 2) along with timing of 9 selected shallow landslides in the study area (Figure 3).

The study area lies in one of the most landslide-prone areas in India, and both climate and topography play a significant role as causative and triggering factors [66] for the occurrences of landslides in the region. The tropical climate with intense rainfall and scorching summers have resulted in a thick overburden of poorly consolidated soil, with thickness often ranging up to 5 m (depending upon the slope) in the study area [67,68]. The combination of highly weathered bedrock, steep slopes, and heavy precipitation in the monsoon season makes the study area an ideal location for understanding, rainfall-induced shallow landslides.

The study area receives the majority of its rainfall in two monsoon periods (South West monsoon from June to September and North East monsoon from October to November). Additionally, the high-altitude and steep terrain of Western Ghats receives more rainfall than the low-lying areas due to orographic effect of Western Ghats [67,68].

Moreover, the devastating impacts of global climate changes have already been observed in the study area in the form of extreme rainfall, which are expected to increase in the coming years, which will in turn increase the frequency of slope failure in the study area [69–74]. According to the India Meteorological Department (IMD), the entire state of Kerala received 2346.6 mm of rainfall between 1 June 2018 and 19 August 2018 rather than the expected 1649.5 mm, which was about 42% above the normal rainfall rate. Furthermore, the district of Idukki received 3555.5 mm rainfall instead of the normal 1851.7 mm from 1 June 2018 to 22 August 2018 [75]. This is a 92% increase from the normal rainfall, and the IMD has identified this as large-excess deviation from normal rainfall. On 9 August, three stations in the catchment region delineated for this study received 255 mm, 254 mm, and 211 mm of rain. The intense and incessant rainfall received in the steep and undulating terrain of Western Ghats triggered multiple landslides in the study area on 9 and 11 June 2018 [76].

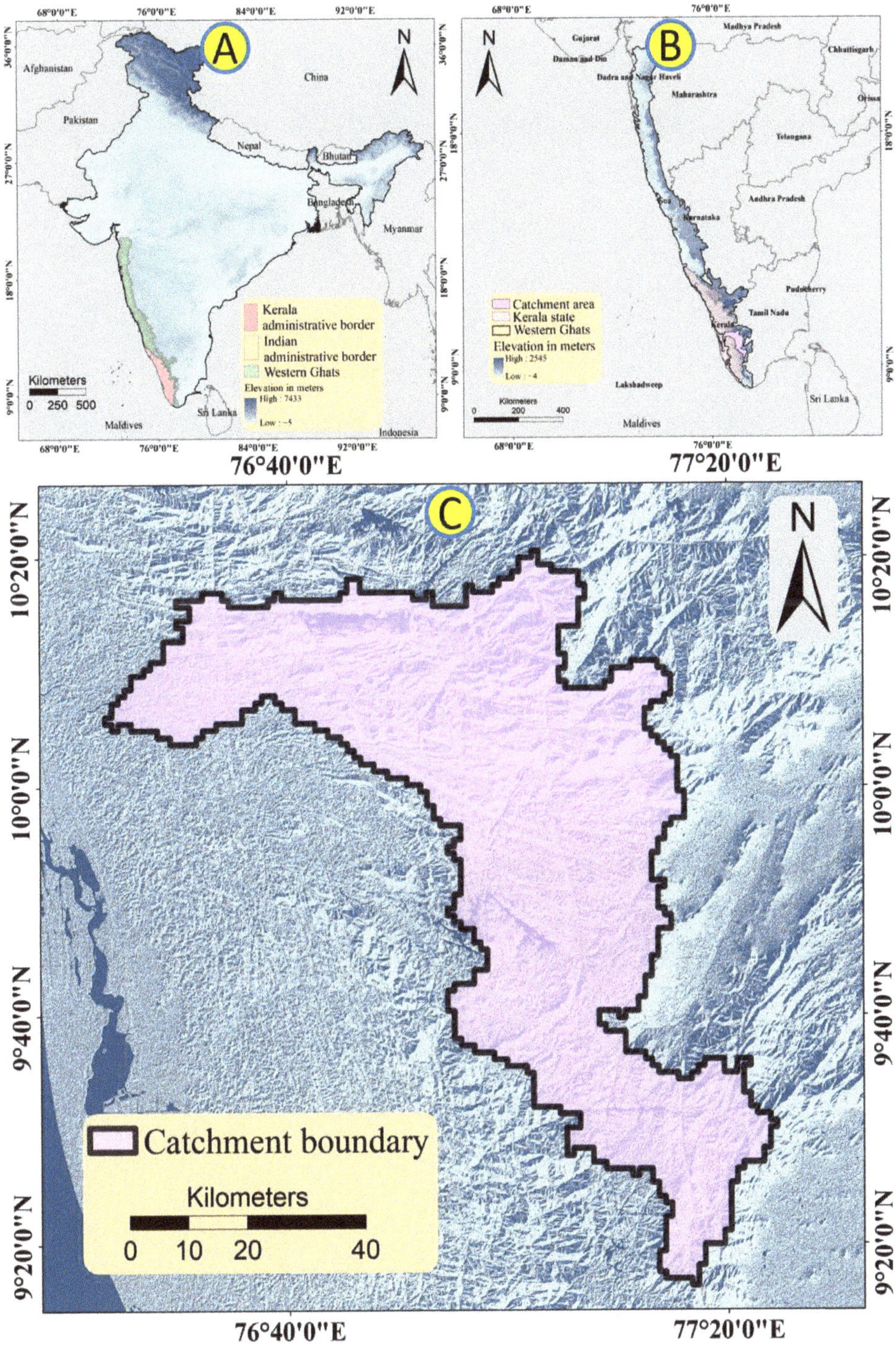

Figure 1. Geographical location of the study area. (**A**) Location of Western Ghats in the political map of India. (**B**) Relative positions of Kerala and the catchment region selected for the study in the Western Ghats. (**C**) Geographical location of the catchment region chosen for the study.

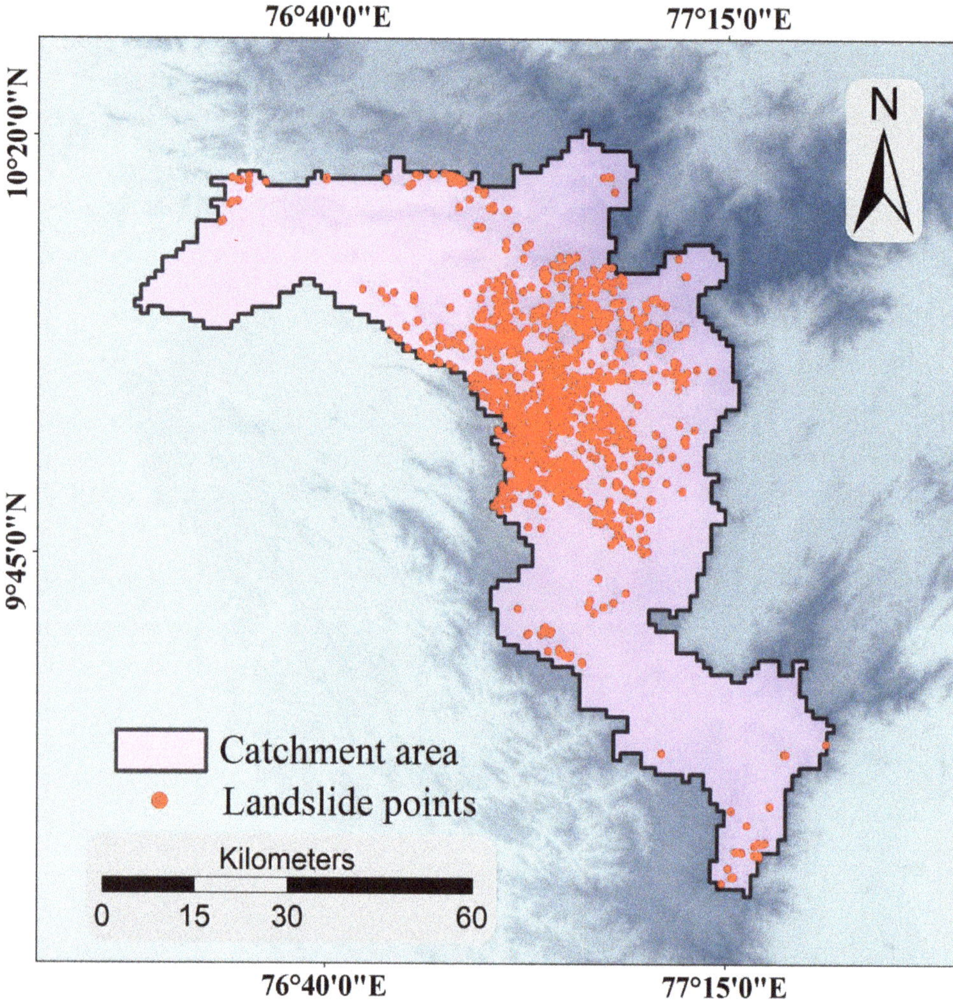

Figure 2. Distribution of landslides in the catchment area. The red dots in the catchment area represent the locations of the landslides occurred during the 2018 monsoon period.

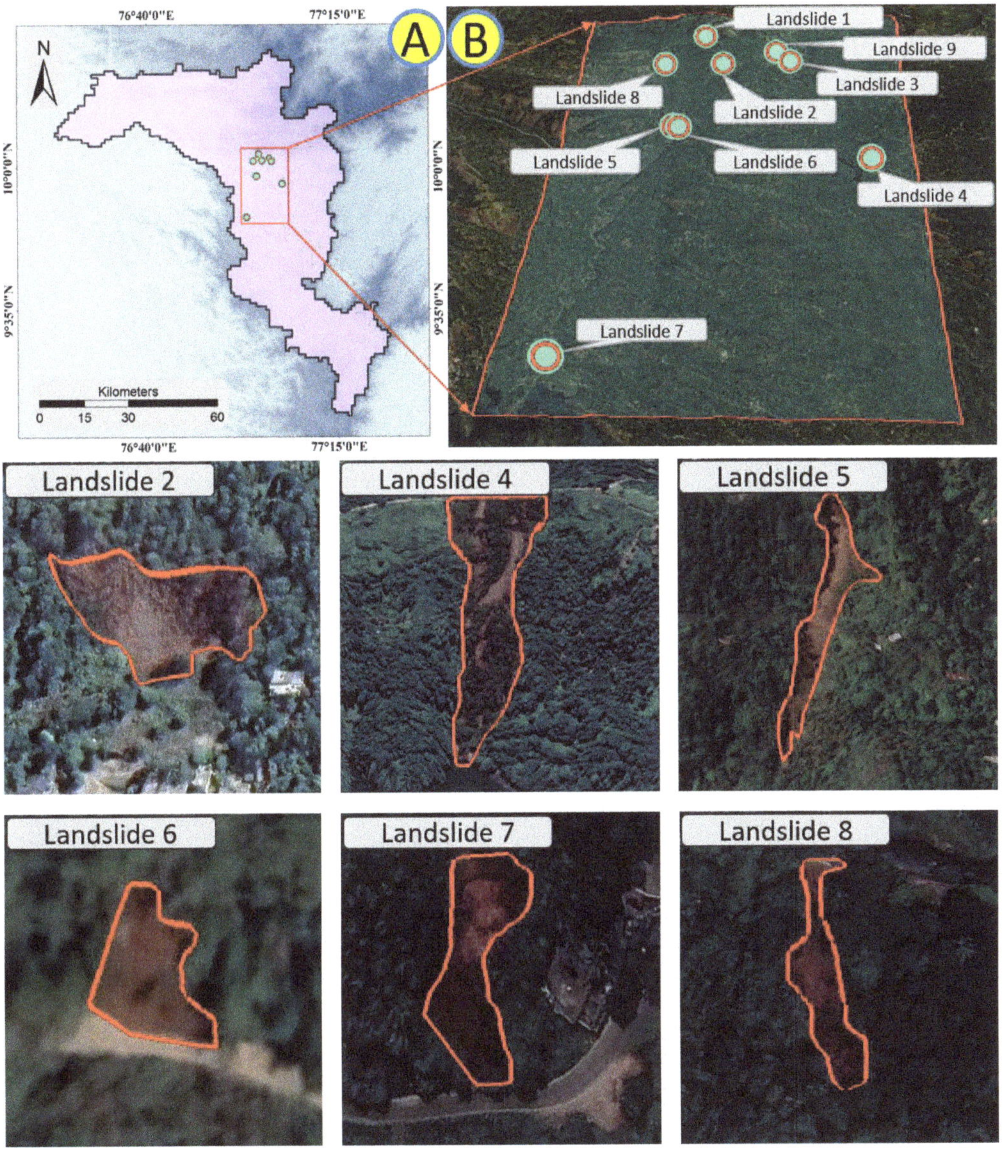

Figure 3. Locations of the 9 events selected for finding the timing of initiation through computing FoS as part of the study in the catchment area. (**A**) Geographical location of the landslide locations. (**B**) Landslide 1 to 9 in Google Earth images. Approximate shapes of Landslides 2, 4, 5, 6, 7, and 8 were available on Google Earth and given as six tiles.

2.1. Data Sets

The determination of landslide occurrences through physically based models involves multiple levels of complexity. The output of models such as TRIGRS are significantly dependent on the resolution and precision of the input data sets. The following data sets, materials, tools, and software were used for understanding the timing and distribution of rainfall-induced shallow landslides in the study area. A detailed description of the data sets used for the present study is provided in the sections below.

2.1.1. Slope Failure Inventory

The entire state of Kerala received one of the most severe rainfalls in the century during June and August 2018, which triggered thousands of landslides, affecting millions of people [77–80]. The severity of the disaster led to studies on understanding the causes of slope failures in the region. The National Remote Sensing Center-Indian Space Research Organization (NRSC-ISRO), Geological Survey of India (GSI), and Kerala State Disaster Management Authority (KSDMA) generated various landslide inventories primarily by analysing pre- and post-disaster high-resolution satellite images. Later, Hao et al. improvised a landslide inventory by adding missing events and eliminating falsely detected landslides [66]. Further details on the elimination of erroneous entries and utilization of object-based image analysis for automated landslide detection can be found in [66]. The present study used the spatial information of the 1913 landslide events that occurred in the study area during 2018 monsoon rainfall based on the above-mentioned landslide inventory.

Along with the spatial information on the distribution of landslides in the study area, 9 slope failures in the study area that were triggered by the very intense rainfall between 1 and 19 June 2018 were validated for the timing of the occurrence of event. The events were documented by Geological Survey of India (GSI) under the program "Post disaster studies in Kerala" (FSP No: M4SI/NC/SR/SU-KRL/2018/21108).

Landslide 1 (Figure 3) happened in the Survey of India (SOI) toposheet number 58F/04 in *Kallar Vettiyar*, and the event happened on 9 June 2018. It had a dimension of 60 m × 30 m × 60 m (length × width × height), and the depth to the failure surface as observed was less than 4 m. The failure can be classified as a rainfall-induced shallow planar landslide. The area was an agricultural land with major crop being cardamom (low height plantation with no deep roots). Landslide 2 (Figure 3) was a shallow planar rainfall-induced landslide in a locality called *Anachal* (SOI toposheet number 58F/04) and occurred in the afternoon, around 2 PM, of 11 June, 2018. The slide had a length of 20 m, width of 30 m, and a height of 25 m. The failure surface was less than 3 m from the top, and the overlying material was very loose coarse-grained in situ soil. The slide was caused by a rise in the PWP as a result of a continuous downpour on the overlying material. A nearby four-story building collapsed as a result of the slide. Landslide 3 (Figure 3) happened on 9 June 2018 and was a shallow planar failure due to incessant rainfall. It happened in a locality called *Eetticity*, which lies in SOI toposheet 58F/04. The failure was a shallow planar rainfall-induced landslide with a 60 m length, 50 m width, and 45 m height. The overlying soil was planted with mixed plantations. On 9 June 2018, another landslide (landslide 4, Figure 3) happened in a locality called *Kallimali* (SOI toposheet number 58G/01). The landslide had a dimension of 45 m × 8 m × 40 m (length × width × height), and the overlying land was used for agriculture (pepper cultivation), and almost 1.5 acres of agricultural land on the hillslope was destroyed. According to the GSI reports, unplanned agricultural practices had caused higher infiltration of rainwater, which in turn caused a rise in the PWP and resulted in the failure. Landslide 5 (Figure 3) happened in *Kallarkutty* village in SOI toposheet 58G/01 on 9 June 2018, and was a shallow rotational failure triggered by rainfall. The failure was relatively small and had a length, width, and height of 8 m, 9 m, and 20 m, respectively. The same village had another landslide (Landslide 6, Figure 3) on the same day early in the morning (2 AM) in a pepper plantation. A shallow planar landslide happened in *Cheruthoni* on 9 June 2018, with a failure depth of 0.5 m deep and a very small run out distance of less than 2.5 m. The area lies in SOI toposheet 58B/16 (Landslide 7, Figure 3). Landslide 8 (Figure 3) occurred on 9 June 2018 in the village of *Anaviratty* (SOI toposheet number 58F/04). The failure was a shallow planar rainfall-induced landslide, and the overburden had thick vegetation. The depth to the failure was almost 1 m and had a very small, less than 5 m, run-out distance. Landslide 9 (Figure 3), chosen in the catchment area for the present study, happened on 9 June 2018, and was a shallow planar landslide less than 1 m deep to the failure plane. The top soil was characterized by moderate vegetation.

2.1.2. Strength and Hydraulic Parameters of the Study Area

The quantification of infiltration, runoff, flow routing, and thereby PWP and FoS was based on the strength and hydraulic parameters of the study area. The present study derived the strength and hydraulic parameters for the area from a soil texture map. The soil texture map of the area was obtained from the NASA Centre for Climate Simulation (NCCS) data portal. The Global Hybrid STATSGO/FAO Soil Texture [81,82] is provided by the NCCS data portal and is a 16-category soil texture map with 30 s resolution. Although the data are provided in 30 s spatial resolution, they were originally remapped data from the Food and Agricultural Organization of United Nations (below) 5 min data. The STATSGO/FAO data's soil texture indices were set to the United States Geological Survey (USGS) index texture values, and further information on the original STATSGO and original FAO soil map can be found on the NCCS data portal. Three major designated soil textures were identified in the study according to the 30 s STATSGO/FAO soil texture map, and Table 1 summarizes the key hydrogeomechanical parameters (hydraulic and strength parameters) with respect to the three major corresponding soil textures of the terrain. A number of authors have examined the sensitivity of hydraulic and mechanical properties of terrain to compute the FoS and have found that physically based models perform better when the input parameters, hydraulic and mechanical, are high resolution [83]. The present study further used a detailed soil texture map from the Department of Soil Survey and Soil Conservation (DSSSC), Thiruvananthapuram, Kerala, India. The DSSSC is the state nodal agency for the conservation as well as management of soil resources for the state of Kerala. The department provides scientific databases to researchers and policymakers for the best practices on soil and land management. The detailed soil texture map was derived from the local soil texture associations provided by the DSSSC. The high-resolution hydrogeomechanical properties derived from the DSSSC soil texture maps were further utilized for deriving high-resolution FoS and then for sensitivity analyses (Sections 2.2 and 4). Figure 4 illustrates the soil texture maps (FAO-STATSGO-derived and DSSSC-derived soil maps) of the catchment area selected for the present study. It can be seen that the FAO-STATSGO database demarcates only three soil types in the region, while eight soil types are identified in the field-based high-resolution DSSSC regional map.

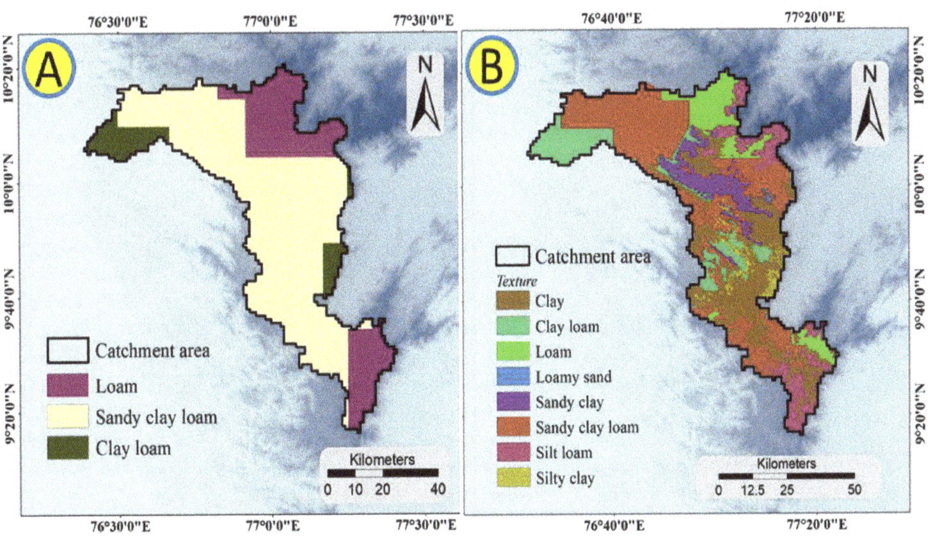

Figure 4. Soil texture map of the study area. (**A**) Soil texture based on the Global Hybrid STATSGO/FAO Soil Texture. (**B**) Field-based soil texture from Department of Soil Survey and Soil Conservation (DSSSC), Thiruvananthapuram, Kerala.

Table 1. Hydrogeomechanical parameters (hydraulic and strength parameters) of the study area for three designated soil textures. C' is cohesion for effective stress, ϕ' is angle of internal friction, γ is unit weight of soil, D_0 is hydraulic diffusivity, and K_s is saturated hydraulic conductivity.

Soil Texture	Loam	Sandy Clay Loam	Clay Loam
FAO soil texture index	6	7	9
C' (KPa)	10	29	35
ϕ' (degree)	22.5	20	20
γ (Nm^{-3})	13,000	15,000	14,000
D_0 (m^{2}s^{-1})	0.0000094	0.0000062	0.000005
K_s (m^{1}s^{-1})	0.00000453	0.00000659	0.00000272

2.1.3. Digital Elevation Model (DEM)

Terrain parameters, such as elevation, slope, and flow direction, have been extensively used in every landslide study at the catchment scale. The present study derived the terrain parameters from the Shuttle Radar Topography Mission (SRTM) Global DEM (Digital Object Identifier number:/10.5066/F7PR7TFT). The void filled elevation data at 1 arc-second (30 m) with a worldwide coverage was used for generating a grid-wise distribution of the altitude, slope, and flow direction. SRTM, hosted by the Endeavour space shuttle, was the first mission to use C-band spaceborne imaging radar and X-band synthetic aperture radar (X-SAR) for the global acquisition of elevation. The void-filled DEM used for the present study is hosted and distributed by the Earth Resources Observation and Science (EROS) Archive. Even though the DEM is tagged as void-filled, there were some anomalously low-altitude pixels, often referred to as depressions or pits or sinks, entirely surrounded by high-altitude pixels. Because sinks significantly interfere with the routing and flow across terrain, they have to be eliminated by filling (increasing the altitude to an extent where it allows draining off to the downhill) or breaching (lowering the edges of the dead-ends or sinks blocking the flow to allow draining off downhill). The present study used a highly efficient utility program called TauDEM, Terrain Analysis Using Digital Elevation Models [84] to remove the sinks and generate a hydrologically conditioned DEM for seamlessly deriving the terrain parameters for topographic indexing and TRIGRS.

2.1.4. Precipitation

Time-varying rainfall intensity is one of the major parameters used as input to compute the FoS and PWP in the TRIGRS model. The present study used a carefully curated synthetic rainfall history for analyzing the distribution and timing of the landslides in the study area. The IMERG and IMD data sets and a number of reports and articles from the State Disaster Management Authority (SDMA) and various authors (Section 1) were used for generating the time-varying rainfall intensity curve for a nine-day study period. The precipitation data sets from the IMD have a spatial resolution of 0.250° × 0.250° and provide daily gridded data over the Indian region [85]. Additionally, the Integrated Multi-satellitE Retrievals for GPM (IMERG) daily "final" precipitation (Level 3, version 6) is a global precipitation data set with high spatial resolution (0.10° × 10°) that is available half-hourly, daily, and monthly [86]. The present study used IMERG's "final" (~3.5 months after the observation) precipitation data sets, which are satellite-gauge products using both forward and backward morphing, and the data sets are available from web portals such as the Global Precipitation Measurement (GPM) portal (https://gpm.nasa.gov/data/sources, accessed on 24 November 2022). Because the routing scheme in TRIGRS is established through the mass balance of rainfall, infiltration, and runoff, the reliability of precipitation data plays a significant role in computing slope stability.

The synthetic rainfall history prepared for the present study is shown in Figure 5. The curve shows a gradual increase in the daily rainfall from the first to the seventh day and a drop for next two days; the highest daily accumulated daily rainfall was 250 mm. Because Central Water Commission (CWC) reported heavy rainfall on 9 August 2018, and three stations in the catchment region received 255 mm, 254 mm, and 211 mm (*Peermade*, *Munnar*, and *Myladumpara* in the Idukki district, respectively), the rainfall history taken for this study is a possible scenario in the catchment region.

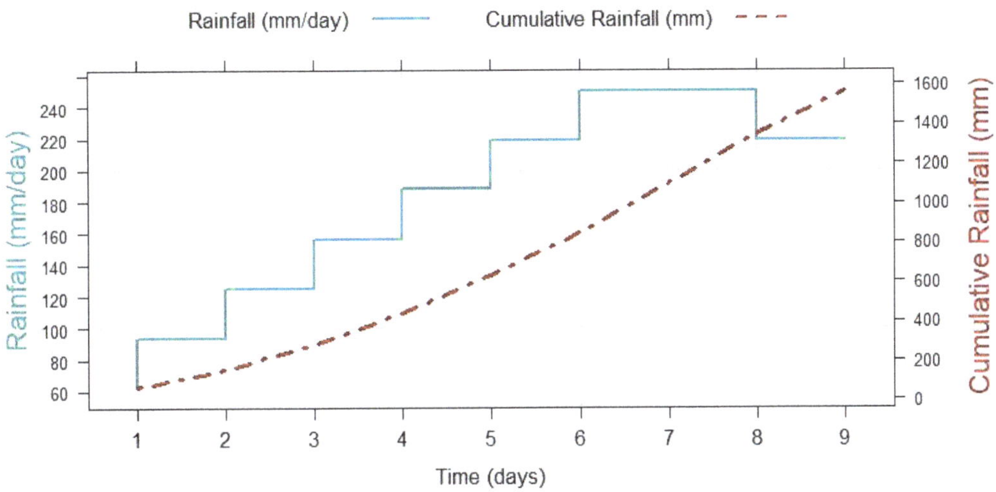

Figure 5. Time-varying rainfall rate prepared for the present study by considering CWC and GPM-IMERG data sets.

2.1.5. Spatial Distribution of Soil Thickness

Soil thickness is one of the significant input parameters in TRIGRS and plays a crucial role in determining the FoS, PWP, depth to slip surface, and hydrological response of the terrain [44,87–89]. Although a uniform soil depth is accepted in the model, it might produce considerably less-realistic FoSs and lead to the erroneous timing and spatial distribution of landslides. The present study obtained the depth to slip surface at 9 nine landslide locations where field observations were available through GSI.

Because the present study was carried out for larger and complex terrain, it was rather difficult to obtain (or interpolate from the available sources) the spatial distribution of soil depth in high resolution. A number of studies have been carried out in different parts of the world, and a firm linear relationship between soil depth and slope has been established [42,44,90,91]. The present study therefore assumed a linear relationship between slope and soil thickness over the study area (Equation (1)) and derived the soil depth for the study area (Figure 6).

$$D = -0.3437s + 3.5 \quad (1)$$

where D is the soil depth at each raster cell with slope s. The slope distribution for the study area was derived from the pit-filled SRTM 30 m DEM. The depth D is in meters, and the slope s is in degrees. Waterbodies in the study area were masked as pixels corresponding to waterbodies (i.e., have the least slope), might have erroneously resulted in thicker soil.

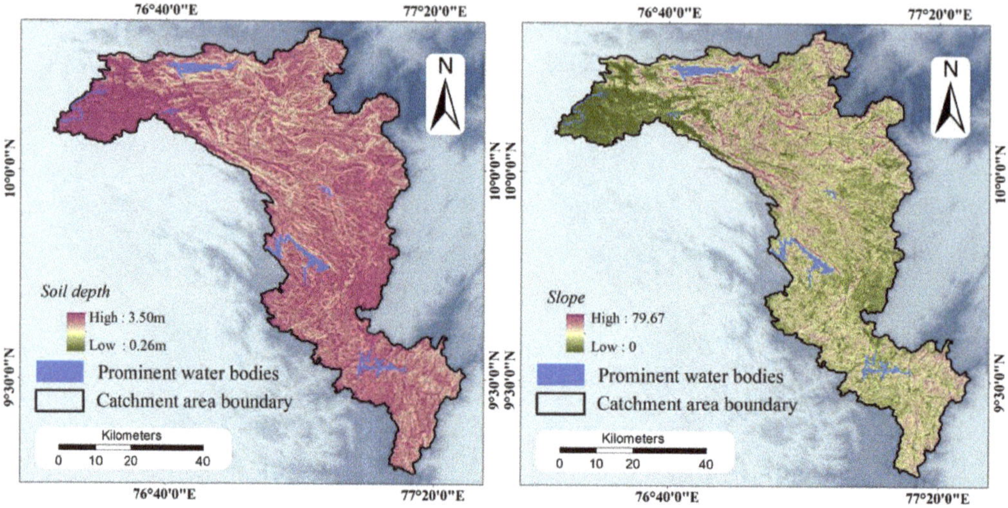

Figure 6. Spatial distribution of soil thickness and slope in the study area.

2.2. Model Description

The dynamic physically based slope stability model used in the present study, Transient Rainfall Infiltration and Grid-based Regional Slope-Stability Model (TRIGRS) is a commonly used slope stability model for computing the timing and distribution of rainfall-induced shallow landslides [54,63,64,92,93]. The TRIGRS model consists of a runoff routing component and a slope stability component. The former computes the infiltration and subsurface flow of storm water, and the latter models the grid-based slope stability over an area of interest. Many studies have been carried out globally for different terrain with varying topographic settings, and a number of researchers have shown that the open-source model is capable of modeling rainfall-induced shallow landslides over a large region. A brief description of infiltration, runoff, flow routing, and slope stability models is given below.

2.2.1. Modeling Infiltration, Runoff, and Flow Routing in TRIGRS

Since the study area is in a low-latitude area and experiences a warm, temperate climate, it can be safely assumed that rainwater is the only form of precipitation in the study area. It is hypothesized in the TRIGRS flow routing model that the rainwater from the upslope cells flows downslope, based on the infiltrability of the cells. The infiltrability or infiltration capacity is a function of the saturated hydraulic conductivity of the medium. Additionally, runoff from a particular cell occurs only when the sum of the direct precipitation received on a cell and the accumulated runoff from the upslope cells exceeds the infiltrability of the cell.

The model computes infiltration (I) at every cell as a sum of precipitation (P) and runoff (R_u) from cells with higher elevation with an assumption that infiltration cannot exceed saturated hydraulic conductivity (K_s).

$$I = P + R_u \text{ [Provided } (I - K_s) \text{ is negative]} \tag{2}$$

Furthermore, whenever the sum of P and R_u exceeds K_s, the excess quantity, $P + R_u - K_s$, is channeled as runoff and routed to adjacent downslope cells, and the runoff diverted to the adjacent downslope cells (R_d) is calculated as,

$$R_d = P + R_u - K_s \text{ [only when } (P + R_u) > K_s\text{]} \tag{3}$$

Additionally, the input water is forced to run off from every cell where the ground water table reaches the surface (implying saturated grids). The saturated cells exfiltrate water received, either in the form of rain or run off from upslope cells, and is modeled to further run off to the downslope cells.

The entire process of flow routing after the computation of run off for each cell is based on the directions of the nearest and steepest confining downslope cell. D8 flow directions [94,95] were used in the present study for flow routing cell-by-cell in the mass balance calculations, and the D8 numbering scheme was set up according to the TopoIndex (Section 2.3.1). D8 flow directions were derived in TauDEM, which output gridded cells with a numbering scheme where 1 to 8 represent east, north east, north, north west, west, south west, south, and south east, respectively. Furthermore, adequate measures were taken to avoid any possible conflicts along the grid cells adjacent to the edges of the hydrologically conditioned DEM and along the flat areas in the basin [96].

2.2.2. Modeling Slope Stability

The slope stability component in the model computes the grid-based slope stability of each cell using an infinite-slope stability analysis. A parameter called FoS, which is the ratio of resisting basal Coulomb friction to gravitationally induced downslope basal driving stress, was employed to determine the failure of infinite slope as result of rainfall. Equation (4) governs the FoS for each grid for the selected time step.

$$FoS = \frac{tan\phi'}{tan\alpha} + \frac{c' - \psi(Z,t)\gamma^{w} tan\phi'}{\gamma_{s} d_{lb} sin\alpha cos\alpha} \tag{4}$$

where ϕ' is the effective angle of internal friction, c' is the effective cohesion, ψ is the pressure head as a function of depth Z and time t, d_{lb} is the depth of the lower impervious boundary, and γ^{w} and γ_{s} are the unit weights of water and soil, respectively. The model was modified to include a layer of an unsaturated zone. The modified equation for obtaining the FoS in this zone is,

$$FoS = \frac{tan\phi'}{tan\alpha} + \frac{c' - \psi(Z,t)\gamma^{w} \chi tan\phi'}{\gamma_{s} d_{lb} sin\alpha cos\alpha} \tag{5}$$

where,

$$\chi = \frac{\theta - \theta_{r}}{\theta_{s} - \theta_{r}} \tag{6}$$

where the χ approximation was given by Vanapalli and Fredlund [97] and is known as the effective stress parameter. θ is the soil moisture content in the unsaturated soil zone, θ_r is the residual soil water content of the soil, and θ_s is the saturated soil water content. Figure 7 (modified from [54]) is a schematic cross-section of the hillslope where the TRIGRS model was applied. The infinite slope or the particular cell is considered stable when FoS is above 1; when the FoS falls below 1, the slope fails. Thus, the depth Z, where FoS first falls below 1, is considered the depth of landslide initiation [92].

The model considers the catchment region as digital terrain, which can be divided into zones, based on the soil texture, and three-dimensional cell grids. The user or modeler can introduce more control by feeding each zone topographic parameters (elevation and slope), hydraulic properties (flow direction, diffusivity, infiltration rate, depth to ground water, saturated water content, residual water content, saturated hydraulic conductivity, and inverse of the height of capillary rise), and strength parameters (cohesion, friction angle, and weight of the soil). Furthermore, control can be achieved by providing elevation, slope, flow direction, depth to ground water, and depth to bedrock grid-wise. The TRIGRS version used for the present study did not have the versatility to provide strength and hydraulic parameters grid-wise. The present study employed version 2.1 [54] of TRIGRS, which is based on version 1.0 [64], but with fewer restrictive hypotheses. The model is capable of performing on saturated as well as unsaturated media and is capable of capturing unstable cells, which are effectively landslide locations, in the study area by considering precipitation and the corresponding changes in the hydrogeomechanical parameters.

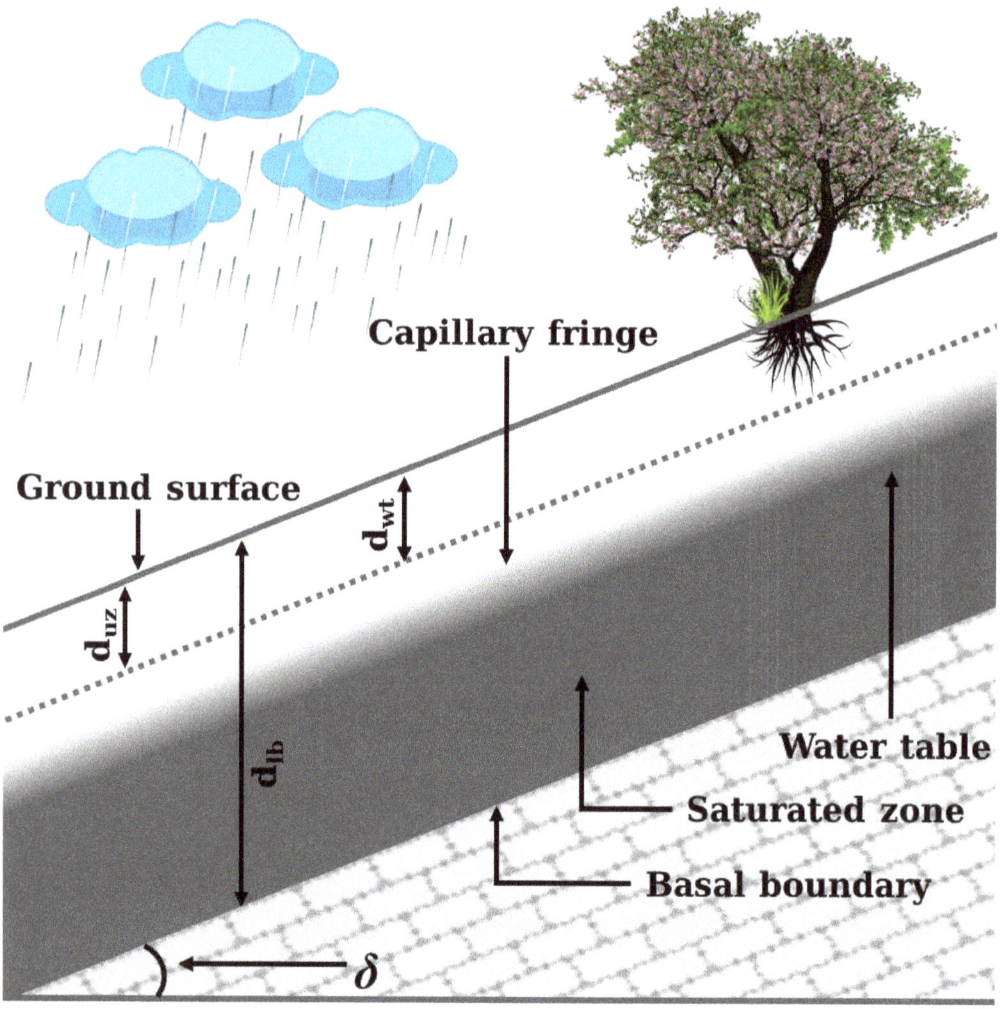

Figure 7. Schematic cross-section showing the shallow ground water condition in the soil. d_{uz} is the unsaturated zone above the water table. d_{wt} is the depth to the ground water table from the surface. d_{lb} is the depth to the impervious lower boundary δ of the slope.

The execution time required for TRIGRS has an exponential negative relationship with the number of pixels, grids, and cells in the input data sets. A larger number of pixels can be expected whenever the model is executed with high-resolution input data sets or over a large area. Because the present study was carried out for a relatively larger catchment area with a higher number of grids in the input data sets, multiple model executions became time consuming. In order to overcome this issue, the study employed TRIGRS MPI [65] for the parallel implementation of TRIGRS by utilizing the maximum computational capacity. The TRIGRS MPI significantly reduced the time consumed for each run, provided more flexibility and freedom while changing the input parameter sets for multiple executions of the model.

2.3. Application of the Model

The present study assessed the distribution of slope failures in the catchment region at 1913 landslide sites along with computing the fall in FoS and rise in PWP at 9 landslide sites in the study area with respect to a 9-day rainfall period. The open-source model, TRIGRS, is capable of considering topographic parameters (elevation and slope), hydraulic properties (flow direction, permeability, diffusivity, conductivity, infiltration rate, and depth to ground water), and strength parameters (cohesion, friction angle, and weight of the soil) while computing the FoS.

A brief description of the significant steps involved in this work is described in the following sections.

2.3.1. Topographic Indexing for Flow Routing

The execution of the model TRIGRS is carried out through two individual programs: TopoIndex and the main TRIGRS program. The former, as the name suggests, is used for topographic indexing; the latter, the main program, is used for flow routing and runoff calculations. TopoIndex prepares a list and a grid of downslope receptor cells (TIdsneiList and TIdscelGrid), a list of index numbers corresponding to each cell number (TIcelindxList), a list of downslope cells for which nonzero weighting factors have been computed (TIdscelList), and a list of weighting factors for downslope receptor cells (TIwfactorList). The outputs of the TopoIndex are generated in ASCII, and text formats and each output's name start with what is given above in the brackets.

A hydrologically conditioned DEM and flow direction are the input files required for generating the essentials data sets for TRIGRS. Section 2.1.3 briefly describes the hydrological conditioning of the DEM. The present study used the D8 algorithm to derive the flow direction through TauDEM (Section 2.1.3). Although TopoIndex uses the D8 flow direction, the numeric coding is different than that of the TauDEM-derived flow direction or ESRI flow direction. Figure 8 shows the designated codes in both the flow direction along with the geographical direction. Although the flow direction for TopoIndex can be generated through with D8 and D-infinity methods, the present study used the TauDEM-derived D8 flow direction scheme for generating the outputs.

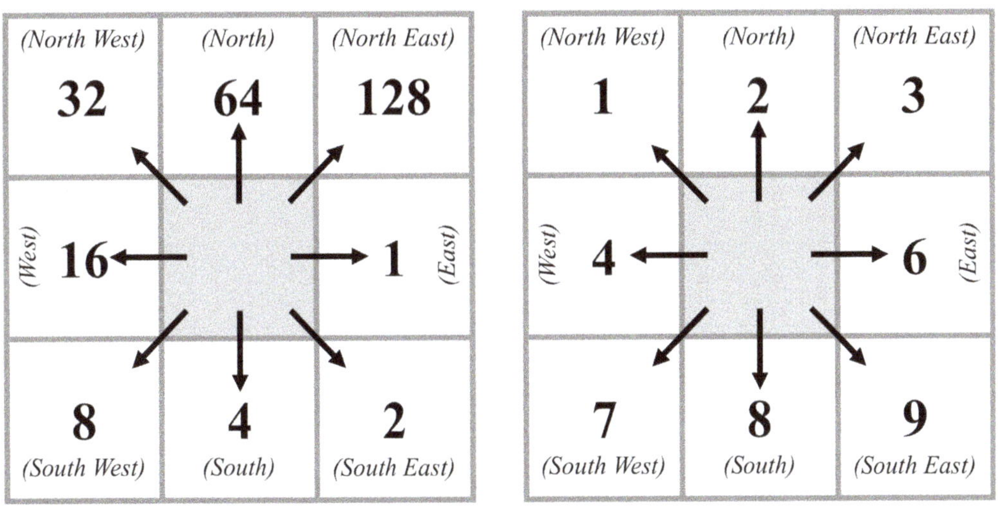

Figure 8. Comparison of D8 ESRI flow direction and D8 TopoIndex flow direction.

Figure 9 shows a schematic diagram of the study, which depicts the derivation of the input data sets, application of the model, and validation of the output (grid-based FoS) from the hydrogeomechanical properties through FAO-STATSGO and high-resolution hydrogeomechanical properties through the field-based DSSSC against the landslide inventory. The landslide inventory is based on the landslides that occurred in the catchment region during the 2008 monsoon season.

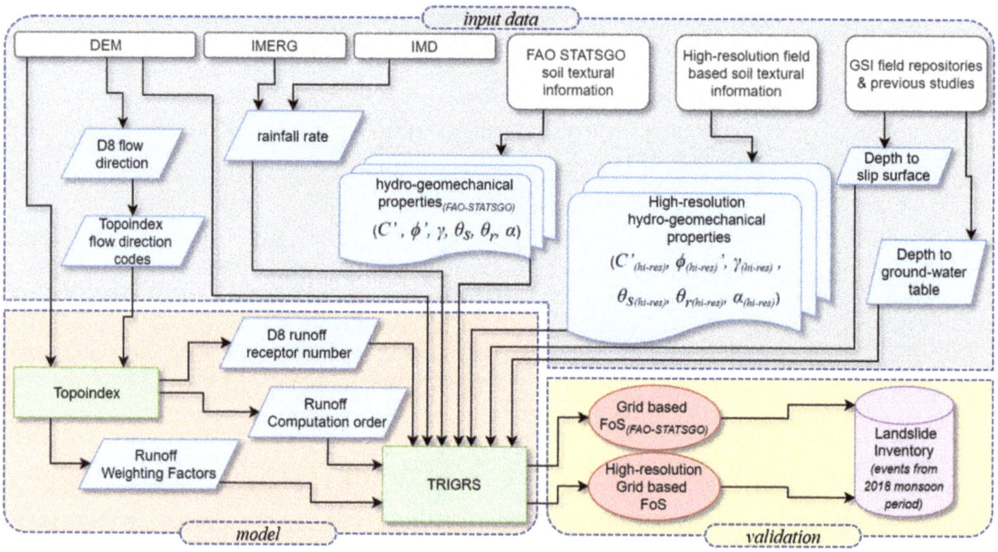

Figure 9. Schematic diagram of the methodology adopted in the present study. The derivation of the input data sets, application of the model, and validation of the output (grid-based FoS) from hydrogeomechanical properties through FAO-STATSGO and high-resolution hydrogeomechanical properties through field-based DSSSC against the landslide inventory are graphically represented in the diagram.

2.3.2. TRIGRS Initialization for Slope Stability

The present study divided the entire study area into three zones based on the FAO soil texture and eight zones based on the regional soil texture map (Figure 4). The hydrological and strength parameters were based on both soil texture maps. The hydrological and strength parameters, such as cohesion (c'), internal friction (ϕ'), unit weight of soil (γ), hydraulic diffusivity (D_0), saturated hydraulic conductivity (K_s), and saturated and residual water content, were derived solely based on the soil texture map. The values of these parameters were obtained from the literature based on the soil type. The present study divided the entire catchment region into 30 m $\times$ 30 m cells, and the FoS was computed for each grid with a time step of 1 day for a total period of nine days. An FoS value of 1 was considered as the threshold for slope failure in the catchment region. Thereby, a value greater than 1 was taken as a measure to classify the pixel as stable (non-landslide) and a value less than 1 was taken as a measure to classify the pixel as unstable (landslide) in the catchment region. The computed FoS for every pixel in the study area for the entire nine-day period was taken as an indicator of the initiation of slope failure.

3. Results

As discussed in Section 2.1.1, the present study examined the landslides in a landslide-prone catchment region in Western Ghats. The shallow landslides were triggered by widespread heavy and incessant rainfall that occurred in the first weeks of June 2018. Details of the hydrogeomechanical parameters and terrain parameters used for computing the FoS and PWP in the physically based model are provided in Sections 2.1.2 and 2.1.3. The rainfall rate fed into the TRIGRS model is described Section 2.1.4. The distribution of the landslide pixels and the changes as the rainfall continued are described in Section 3.2, and the timing of the triggering of the nine landslides as a result of the rainfall in the study are is described in Sections 3.3 and 3.4.

3.1. Impact of Varying the Number of Soil Zones

Even though the number of soil zones in the studied catchment varied for the two runs, being three and eight, the properties were still based on soil texture. It can be seen from Figure 10 that the true identifications of the landslide events in both runs were comparable. The model in both the cases could identify 1524 and 1323 landslides of the 1913 landslides. Additionally, a true positive rate (TPR) of 68% and 60% was obtained with a nine-day rainfall period for the two derived soil texture data sets, respectively. A false positive rate (FPR) of 36% and 31% was also seen, respectively. The total accuracy was 68%. Although previous studies [83] showed that the hydrogeomechanical parameters have an impact on the simulations of FoS and PWP, it was seen here that the properties just based on soil texture did not impact the model result. It was also suggested that these properties may need to be derived from the other sources such as Earth Observation (EO) data sets, which can result in a better simulation of the PWP [98].

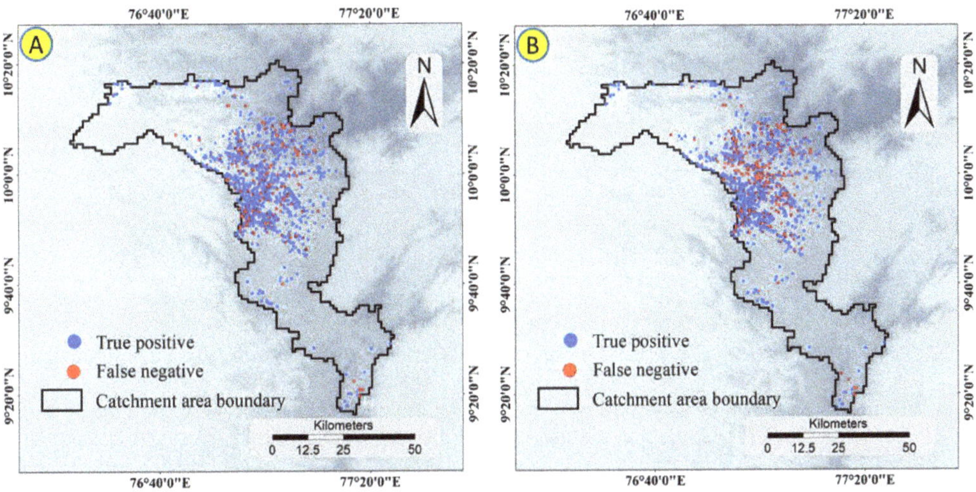

Figure 10. Distribution of false positives and false negatives (**A**) when hydraulic properties were derived from FAO soil texture and (**B**) when hydraulic properties were derived from the soil map of the Soil Survey & Soil Conservation Directorate.

3.2. Change in FoS and PWP in the Catchment Region

In the study area, the FoS corresponding to majority of the grids showed a significant decrease during the nine-day rainfall period. Furthermore, over time, a higher number of grids started showing an FoS below the threshold value of one, and the corresponding time was taken as the initiation of the landslide. Along with the FoS, the PWP was computed for each grid to further illustrate the interconnection among the FoS, landslide initialization, and rainfall history. Figure 11 shows the fall in the FoS and rise in the PWP for the study area for the period of nine days. It is depicted in the figure that the FoS was decreasing for the majority of the pixels, and new areas could be demarcated as landslide pixels as time progressed. The FoS was calculated for each pixel at the end of a 24 h rainfall period for nine days in Figure 11 (left). The same was repeated for the PWP and is plotted in Figure 11 (right). It is clear from the FoS–PWP through the rainfall rate plot that a number of pixels fell below the threshold FoS value and turned into unstable areas.

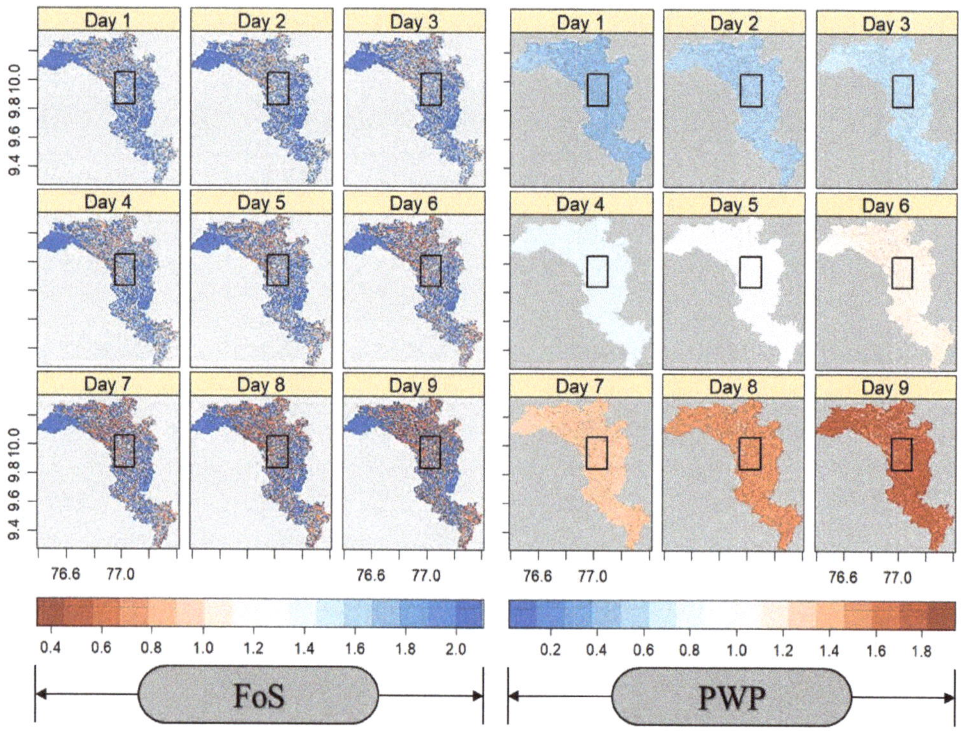

Figure 11. (**Left**) Decrease in factor of safety (FoS) in the catchment region for the 9-day rainfall period. Day 1 (**top left**) is the plot of FoS at each pixel at the end of a 24 h rainfall period, and day 9 is the FoS at each pixel at the end of a 9-day rainfall period. Red region indicates lower-FoS and unstable areas, and blue region indicates higher-FoS and stable area. (**Right**): Rise in pore water pressure (PWP) in the catchment region for the 9-day rainfall period. Day 1 (**top left**) is the plot of PWP at each pixel at the end of a 24 h rainfall period, and day 9 is the PWP at each pixel at the end of a 9-day rainfall period. Red region indicates higher PWP (relatively unstable), and blue region indicates lower PWP (relatively stable). The nine selected landslides are located in the black rectangle (not shown in this plot). Note that the color keys are reversed for easy demarcation of relatively unstable (i.e., red) areas.

Furthermore, Figure 11 shows the inverse relationship between the FoS and PWP in the study area. It can be observed from the figure that when PWP increased as rainfall continued, the FoS significantly decreased. The blue region in the plots represents the relatively stable areas (lower PWP and higher FoS), and the red region represents relatively lesser stable areas (higher PWP and lower FoS). For better understanding, the usage of model with regard to landslide initiation timing, the nine landslides were chosen for the case study. The nine landslides chosen for the study are located in the black rectangle in the plots.

The plot in Figure 8 is a detailed illustration of the FoS–PWP in and around the nine landslide sites chosen in the study area. As discussed above, the rise in the PWP and fall in the FoS were computed and plotted for the nine-day rainfall period. The black dots represent the locations of the nine landslides chosen for the study area. It can be seen from Figures 11 and 12 that there are false positive regions, where the FoS drops below one. This shows that the geohydrological parameters that were governing the hydrological simulation needed to be further improved so that the partition of water among the soil layers could be better simulated. The χ approximation from Equation (6), utilized in

Equation (5) of the FoS, is based on the SHPs, saturated water content, and residual water content. It can be seen that the properties impact both the PWP and FoS. These parameters were literature-based and dependent on the soil texture. One of the limitations is that these properties were not available at a 30 m resolution. Again, this shows that these properties may have to be identified at finer resolution through satellite data that have higher spatial coverage compared with the field analysis, which may be not possible in remote and inaccessible areas. However, the landslide events were identified, which can still act as a warning for susceptible zones.

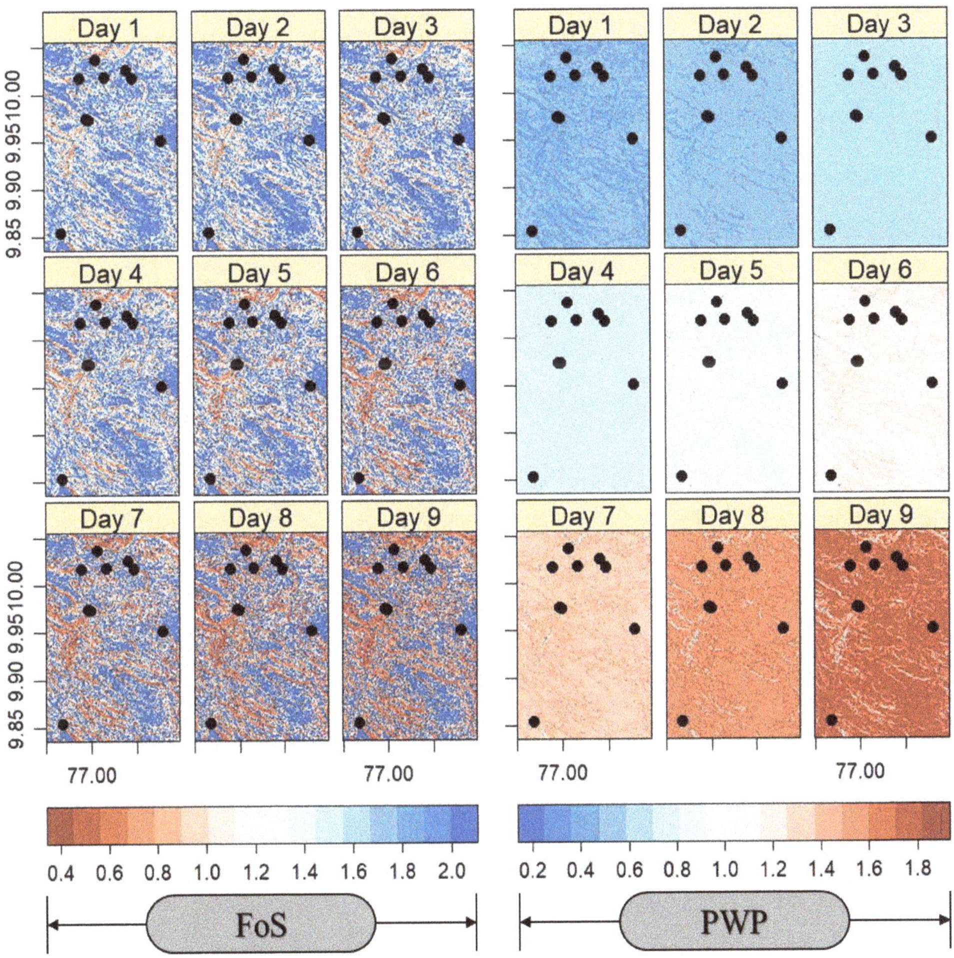

Figure 12. (**Left**) Fall in factor of safety (FoS) in and around the nine landslides for the nine days rainfall period. Day 1 (**top left**) is the plot of FoS at each pixel at the end of a 24-h rainfall period and Day 9 is the FoS at each pixel at the end of a 9-day rainfall period. Red region indicates lower FoS, unstable, areas and blue region indicates higher FoS, stable area. (**Right**): Rise in Pore Water Pressure (PWP) in and around the nine landslides for the nine days rainfall period. Day 1 (**top left**) is the plot of PWP at each pixel at the end of a 24-h rainfall period and Day 9 is the PWP at each pixel at the end of a 9-day rainfall period. Red region indicates higher PWP (relatively unstable), and blue region indicates lower PWP (relatively stable). The black dots indicate the landslide sites. Note that the color keys are reversed for easy demarcation of relatively unstable (~red) areas.

3.3. Fall in FoS

Figures 11 and 12 show the distribution of the landslides in the catchment region and their changes as the rainfall progresses. The 8 landslides were triggered on 9 June 2018 and one landslide was triggered on 11 June 2018. The rainfall event used for computing the eight landslides which were triggered on 9 June 2018 started on 3 June and ended on 11 June 2018. Seventh day of the rainfall event (9 June) was the day on which eight landslides happened. Similarly, a rainfall event started on 5 June and ended on 11 June 2018 was used for computing the FoS at one landslide which was triggered on 11 June 2018.

Figure 13 shows the gradual fall in FoS at nine landslide sites with respect to the nine days rainfall event. A vertical dotted line is added to each plot for easy representation of day on which the landslide occurred (9 June in the case of Landslide 1, 3, 4, 5, 6, 7, and 8 and 11 June 2018 in the case of Landslide 2). As discussed above, the present study considered 1 as a threshold value for FoS and whenever a pixel's FoS fall below 1 is considered as a landslide and the corresponding time is recorded as the landslide initiation time.

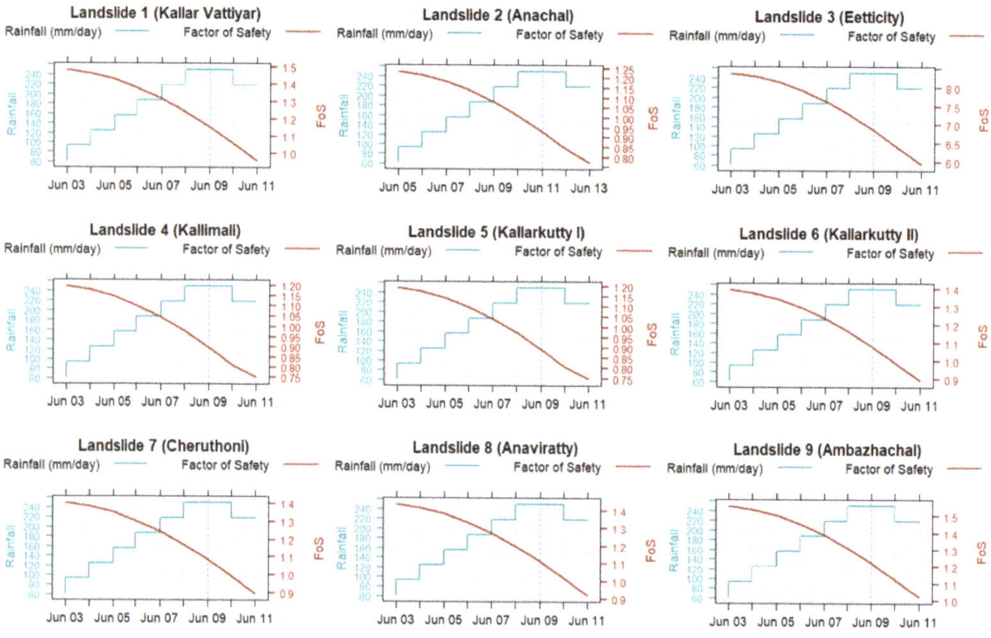

Figure 13. Fall in FoS with respect to nine-day rainfall period at the landslide sites. The vertical dotted line shows the day on which the landslide occurred.

Landslide 1 (*Kallar Vattiyar*), Landslide 3 (*Eetticity*), Landslide 4 (*Kallimali*), Landslide 5 (*Kallarkutty I*), Landslide 6 (*Kallarkutty II*), Landslide 7 (*Cheruthoni*), Landslide 8 (*Anaviratty*) and Landslide 9 (*Ambazhachal*) were triggered on 9 June 2018. The FoS computed at these landslide locations on 9 June are 1.15, 6.87, 0.90, 0.89, 1.08, 1.08, 1.11, and 1.22 (FoS corresponding to dotted line on Figure 13). Additionally, Landslide 2 (*Anachal*), was triggered on 11 June and the FoS computed as per TRIGRS on the same day is 0.93 (FoS corresponding to dotted line on Figure 13). According to the FoS computed Figure 13, the model could indicate the slope failures at three landslide sites out of nine chosen landslides in the study area. Additionally, as the time progresses it can be observed that the model rightly shows the eight out of 9 landslides failures in the study area.

3.4. Rise in PWP

As part of the study, the PWP was calculated for the catchment region. Figure 14 plots the PWP extracted at the nine landslide sites in the study area against the time-varying rainfall rate. The plot shows a near linear rise in the PWP with respect to the rainfall. Furthermore, the PWP and FoS showed an inverse relationship at the nine landslide sites (Figures 13 and 14). As in the case of rainfall-induced shallow landslides anywhere else, an increase in the PWP indeed reduced the FoS at the nine landslide sites in the study area.

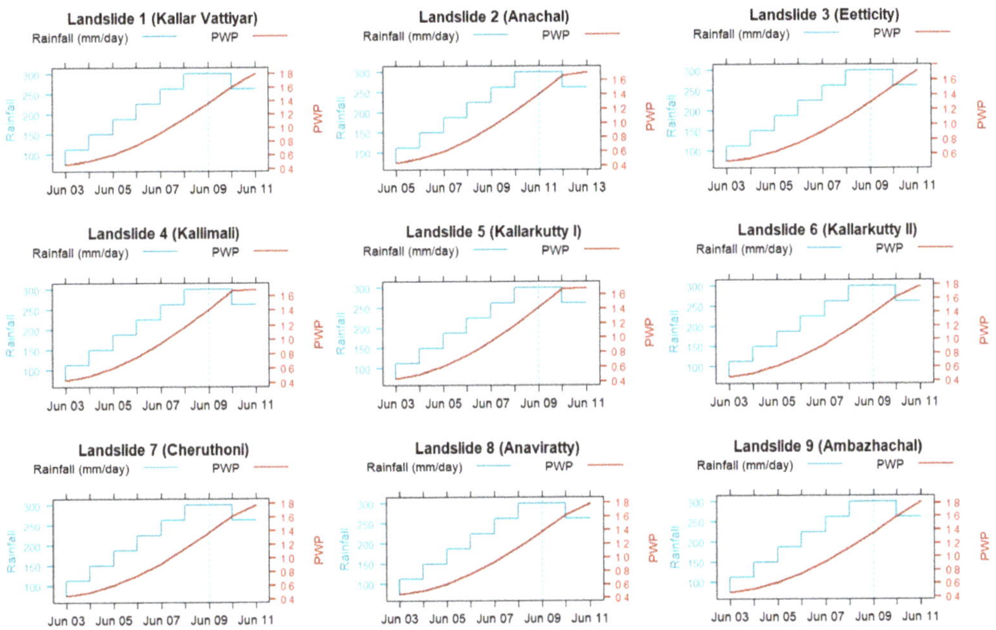

Figure 14. Comparison of the rise in PWP (porewater pressure) with respect to rainfall, from the outputs of nonoptimized and optimized model runs at 9 validation sites. The vertical dotted black line indicates the day on which the landslide was triggered.

4. Discussion

The timing and distribution of shallow landslides were modeled in a catchment region (Figure 1) in Western Ghats through TRIGRS. The hydrogeomechanical properties used in TRIGRS to model the FoS and PWP were derived through two different sources: FAO-STATSGO and a regional soil texture map from DSSSC. Although TRIGRS computes the FoS cellwise (i.e., gridded), the hydrogeomechanical parameters are provided zone-wise. The available versions of TRIGRS do not have the ability to provide cell-wise hydrogeomechanical properties. Because the hydro-geomechanical parameters were derived from the soil texture information, the zones in TRIGRS approximately followed the soil texture map of the study area. The present study incorporated FAO-STATSGO soil information and high-resolution regional soil information from DSSSC, which resulted in three and eight soil types in the study area, respectively (Figure 4). Therefore, the study area was categorized into three and eight zones. Although the spatial heterogeneity was better represented with the DSSSC soil texture, the modeled FoS from both soil textures were comparable (Section 3.1), leading to a lack of improvisation. This may have been due to the use of SHPs based on soil texture. This was further clarified by carrying out a sensitivity analysis to quantify the impact of the SHP on the analysis of slope stability and to further understand the absence of significant changes in modeled FoS when the two soil texture maps were used to derive the SHPs.

Because SHPs are important in determining the stability of hillslopes, saturated water content (θ_s); residual water content (θ_r); and a fitting parameter for soil size distribution (α), which is approximately equivalent to the inverse height of capillary rise [54], were selected for sensitivity analysis. In order to understand the impact of each parameter on the final grid-based FoS, one-parameter-at-a-time (OAT) analyses were carried out. OAT analyses are often used in sensitivity analyses, parameter optimization, and calibrating physically based models [42,99–101].

The current study considered a case of a finite and unsaturated profile. The finite depth was based on Equation (1). Based on Equations (4) and (5), the FoS was calculated for both the saturated and unsaturated zones, where each cell was treated as a unique sliding unit. Thus, failure analysis with pressure head determination provided the FoS for the depth profile for each cell. It is noted that the depth at which the FoS first reduced below one was considered the landslide initiation depth, which could vary depending on the soil properties as well as the rainfall intensity [92,102]. The depth may vary between the water table depth and the basal boundary [103]. The pressure head, utilized in Equations (4) and (5), was calculated based on the below equations based on Richard's equation (Equation (7)).

$$\frac{\partial \theta}{\partial t} = \frac{\partial}{\partial z}\left[K(\psi)\left(\frac{1}{cos^2\delta}\frac{\partial \psi}{\partial Z} - 1\right)\right] \tag{7}$$

where,

$$K(\psi) = K_s \exp(\alpha\psi^*) \tag{8}$$

The pressure head was obtained using Equation (9).

$$\psi(Z,t) = \frac{cos\delta}{\alpha_1} ln\left[\frac{K(Z,t)}{Ks}\right] + \psi_0 \tag{9}$$

A realistic range of θ_s, ranging from 0.280 to 0.450, with a median of 0.365, was chosen to understand the impact of the saturated water content on the stability of terrain and to determine the landslide initiation zone (Figure 15). It can be seen that with the same rainfall intensities, a change in θ_s impacted the FoS of the depth profile (Figure 15) as well as the time taken for the rising of the groundwater table (Table 2). The FoS corresponding to θ_s = 0.365 reached < 1 on the seventh day. At the minimum θ_s, an FoS < 1 was seen on the fifth day, while the cell remained stable for θ_s = 0.450 (Figure 15). Figure 15 further quantifies the change in FoS (%) with respect to changes in θ_s (%), where a 30% increase in θ_s reflects a 15% change in the FoS. The OAT analyses carried out as part of the present study for θ_s showed an approximate linear trend for the study area. Additionally, it can be seen from Table 2 that the landslide was initiated as soon as the groundwater table rose to depth of 0.50 m. The water table rise mechanism was already previously explained (Baum et al., 2010). Thus, as shown in Take et al., 2004, with the rise in pore water pressure to zero, the shear resistance fell. This led to an unstable zone. It is illustrated from both the depth profile in Figure 15 and Table 2 that the zone of weakness increased with the decrease in θ_s, which represents the porosity of the soil. It is deciphered from Figure 15 that θ_s had a significant role in estimating FoS and, thereby, the initiation time and spatial distribution of the landslides in the study area.

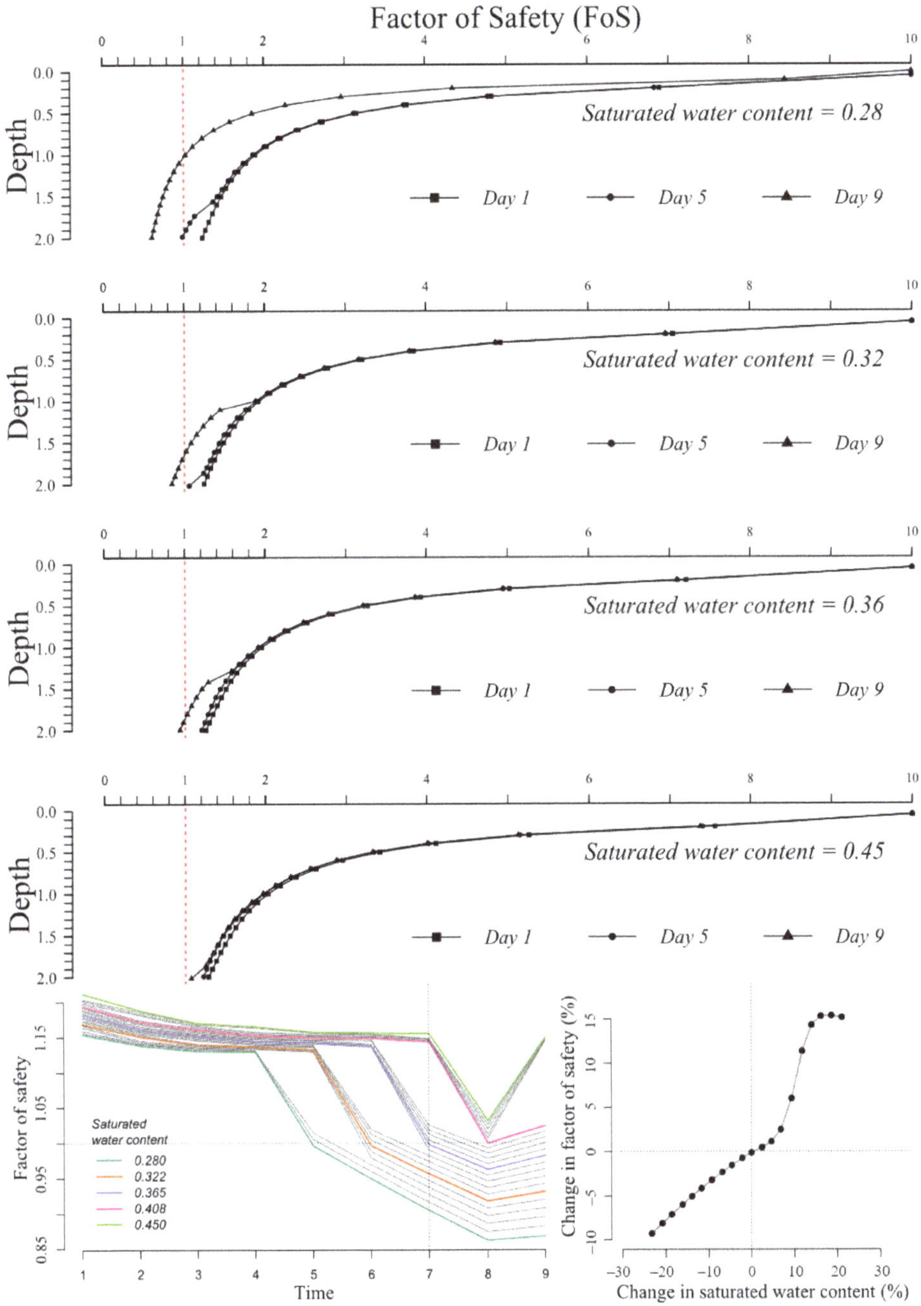

Figure 15. Sensitivity analysis of saturated water content (θ_s): (**Top**) changes in FoS with depth for a 9−day period with respect to the different θ_s values; (**Bottom Left**) changes in FoS and landslide initiation time with change in θ_s values. (**Bottom right**) Percentage change in FoS in accordance with the percentage change θ_s. The vertical dotted line on the bottom left indicates the time at which the median value of the selected parameter crossed the threshold FoS, and the horizontal line indicates the corresponding threshold FoS, which is 1.

Table 2. Comparative analysis for landslide initiation based on variation in θ_s.

Sl. No.	θ_s	FoS < 1 (Landslide Initiation Day)	Depth to Groundwater Table	Depth of Weak Zone
1	0.28	Day 5	0.50 m	1–2 m
2	0.32	Day 6	0.48 m	1.7–2 m
3	0.36	Day 7	0.46 m	1.8–2 m
4	0.45	Day 9 (Stable)	0.76 m	Stable

As described above, an OAT analysis was carried out for θ_r. The θ_r values were considered from 0.01 to 0.1, with a median of 0.55, and the corresponding FoSs were modeled for a 9-day period and are plotted in Figure 16. The median θ_r (0.055) reached the threshold value before the seventh day, and the soils with lower θ_r values needed more time, up to 36 more hours, to become unstable. This change could be attributed to the increase in the water-holding capacity of the soil. On the other hand, higher θ_r values started crossing the threshold FoS of one starting on the fifth day. It can be seen that a 100% change in θ_r produced only a 4% change in the FoS. It might be because of the fact that the θ_r alone might have a lesser impact on the modeled FoS. In reality, landslide is a very complex phenomenon, and the interplay of multiple factors determine the mechanism behind its initiation. Furthermore, in order to understand the impact of θ_r together with θ_s and other factors, many-parameter-at-a-time (MAT) analyses should be carried out by simultaneously varying multiple parameters in the model. Even though MAT analyses will further shed light on the interconnection among various hydrogeomechanical parameters, the present study was limited to OAT because of its simplicity and the significantly less computational effort required.

Along with θ_s and θ_r, α (a fitting parameter for soil size distribution and is approximately equivalent to the inverse height of capillary rise) was analyzed to comprehend its impact on FoS and thereby the initiation of landslides. The α values were varied from 0 to 10, and we modeled the FoS for a 9-day period. The experiments were carried out as in the cases for θ_s and θ_r and are plotted in Figure 16. The fall in FoS for 9 days with the α values 0.35, 1.00, 1.80, 2.00, and 10.00 are highlighted in the plot. It is observed from Figure 16 that α values above two were less significant in determining the FoS than values from zero to two. Because any α value above two had less significance in determining the FoS in the study area, a range of zero to two with a median of one was considered for analyzing the percentage change in the FoS with respect to the percentage change in α (Figure 16). It has to be noted that any α value less than zero acts as a flag in TRIGRS to treat the terrain as tension-saturated.

Two major observations were made as part the present study while selecting the input parameters for modeling the slope stability in complex terrain. Primarily, even though the regional DSSSC soil map had better soil information representing the study area, it still considered a mean θ_s, θ_r, and α for a zone based on the field observations and laboratory measurements of the soil samples. In reality, the SHPs vary for the same soil type based on the porosity, grain size, degree of compaction, etc., and complex terrain is more likely to reflect this trend. The subzonal variation must have strongly affected the modeled distribution and the computed initiation time of the landslides in the study area. Secondly, because the uncertainties present in spatially varying hydrological properties have an impact on the initiation time and spatial distribution in complex terrain, a robust approach has to be implemented to (1) accurately derive the hydrological properties through EO data sets in finer resolution and (2) provide the derived finer-resolution hydrological properties cell-wise rather than zone-wise as in the present versions of TRIGRS. An obvious solution for deriving hydrological properties that have a significant impact on the FoS is through inverse modeling. Although in situ collected soil samples and laboratory analyses are rather more intuitive than modeling approaches, it has to be noted that in situ data collection, sampling in uniform intervals (spatial and temporal), and analyses are very costly and often impossibly laborious for complex and inaccessible terrain.

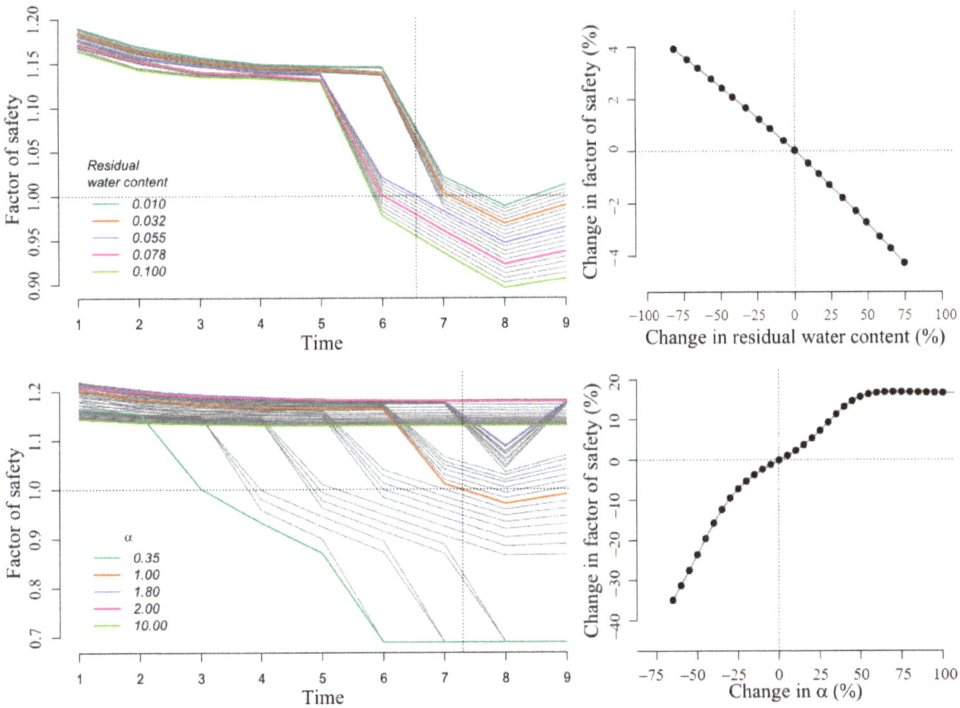

Figure 16. Sensitivity analysis of SHPs, residual water content (θ_r), and a fitting parameter for soil size distribution (α), which is approximately equivalent to the inverse height of capillary rise. (**Left**) Changes in FoS for a 9−day period with respect to the different θ_r and α. (**Right**) Percentage change FoS in accordance with the percentage changes in θ_r, and α. The vertical dotted line on the left side of the figure indicates the time at which the median value of the selected parameter crossed the threshold FoS, and the horizontal line indicates the corresponding threshold FoS, which is 1.

Along with the results and findings of the present study, the significance of the inherent uncertainties present in the hydraulic parameters in determining the model outputs was also put forward in a previous study on a landslide-prone area in Brazil [104]. The study further emphasized the need for developing novel strategies to generate hydraulic parameters that represent spatial variation in finer resolution. A recent review [45] on TRIGRS and its performance compared with that of other physical slope stability models also concluded that refined hydrological parameters can generate more realistic results. Another study carried out in Norway focused on the initial conditions of the terrain and concluded that wetter initial conditions lead to early instability and overestimation of landslide pixels in the study area [105]. A data-sparse region in the northern part of Kerala was studied for landslide hazard zonation using TRIGRS and showed that providing spatially varying SHPs significantly reduces the overestimation of landslide pixels [62]. The study further raised concerns regarding the lack of flexibility to provide local variations in hydrogeomechanical parameters to the current versions of TRIGRS. The study suggested the modification of TRIGRS to receive input parameters with local variations rather than generalizing the properties zone-wise. A number of methods to inversely derive SHPs through EO data sets have been put forward to overcome the limitations of regression functions [106–108]. The high-resolution SHPs derived through inverse modeling [108] can be a possible alternative, especially in areas where field data are difficult to retrieve.

5. Conclusions

The present study modeled the distribution and timing of rainfall-induced shallow landslides in a catchment in Western Ghats. It was shown how a physically based model can be used for understanding the timing and distribution of landslides with minimal ground-based data sets. This study used a set of satellite-derived terrain parameters and hydrogeomechanical parameters with ground-measured rainfall data to compute the FoS and PWP in a landslide-prone area. The model was able to predict the landslide distribution in the spatial region with a TPR of 68%. Additionally, for nine landslides, the model predicted the occurrence of eight landslides, with precise timing of three landslides. In addition, the present study analyzed the impact of the SHPs in computing the FoS and thereby predicting the timing and distribution of landslides in the study area. It could be seen that with the increase in the storage capacity of soil, it became more stable, and a prominent shift was seen in the initiation of the landslide events. Thus, the precise timing of a landslide event was dependent on the SHPs. Although the model could compute the change in the FoS and PWP in accordance with the rainfall received in the study area, it is worth mentioning the challenges and room of improvement when using the model, TRIGRS. Primarily, the lack of high-resolution SHPs was one of the major challenges faced when executing the slope stability model. Although a field-based high-resolution soil texture map from DSSSC was used to generate high-resolution FoS and could reduce the number of false positives, it needs further improvement, especially in computing the timing of landslide initiation. A high-resolution map of SHPs can provide more control and heterogeneity and thereby improve the modeling of the FoS and PWP. Because the hydrologic response of terrain is determined by the spatial distribution of the thickness of soil columns, any physically based slope stability model heavily depends on the resolution and accuracy of the soil depth. In TRIGRS, the depth to slip surface and time taken to reach complete saturation of each cell are computed as a function of soil depth. It further emphasizes the requirement for the high-resolution spatial distribution of soil depth. Although it is extremely difficult to obtain in situ observations of soil thickness for larger and complicated terrain, a higher number of in situ and uniformly distributed soil thickness sampling observations would resolve this problem to a certain extent. Although necessary steps, such as pit removal, were taken prior to the analyses, the present study heavily relied upon SRTM 30 m DEM to derive the terrain and hydrological parameters, such as slope, flow direction, and TopoIndex parameters used in TRIGRS. A high-resolution DEM from airborne surveys, such as drone-based surveys, could have significantly improved TRIGRS' outputs.

Furthermore, the dynamic physically based slope stability model, TRIGRS (version 2.1), does not provide the flexibility to input the hydrogeomechanical parameters grid- or pixel-wise. The grid-wise provision of input data sets with the inclusion of the impact of local geological and anthropogenic features and hydrogeomechanical properties can significantly improve the model outputs. Although the grid-wise provision of the hydrogeomechanical properties will be computationally expensive, it may significantly enhances modeling capabilities. On the basis of the results obtained in the present study and the sensitivity analysis carried out, we further plan to generate SHPs in finer resolution and provide SHPs cell-wise, rather than grid-wise, in TRIGRS.

Despite of the limitations of this study and the unavailability of high-resolution terrain and hydrogeomechanical information, the model could fairly demarcate the distribution and capture the timing of rainfall-induced shallow landslides in the study area. It can be further concluded that the model and method can be used as a measure to assess the landslide vulnerabilities in the Western Ghats area or anywhere else where the input data sets are available. Moreover, the physically based slope stability model used in the present study can be used as a primary approach to comprehensively understand the stability of hillslopes where input data sets are limited or only remotely sensed EO data sets are readily available.

Author Contributions: Conceptualization, Juby Thomas, Manika Gupta, and Prashant K. Srivastava; methodology, Juby Thomas and Manika Gupta; writing—original draft, Juby Thomas; writing—editing and review, Manika Gupta, Prashant K. Srivastava, and George P. Petropoulos; software, Juby Thomas and Manika Gupta; data curation, Juby Thomas; visualization, Juby Thomas; resources, Prashant K. Srivastava and George P. Petropoulos; supervision, Manika Gupta. All authors have read and agreed to the published version of the manuscript.

Funding: This research received no external funding.

Data Availability Statement: The current study used open-source data from various platforms and the web portals that host the EO data sets used in the study. These data sets were from NASA and USGS including GPM Precipitation Data Directory (https://gpm.nasa.gov/data/directory, accessed on 24 November 2022) and Earth Resources Observation and Science (EROS) Archive (https://www.usgs.gov/centers/eros, accessed on 24 November 2022). Precipitation data sets were obtained from India Meteorological Department's web portal (https://www.imdpune.gov.in/cmpg/Griddata/Rainfall_25_NetCDF.html, accessed on 27 January 2023). The landslide events used for the study ere documented by NRSC, GSI, and KSDMA. Upon reasonable request, the data sets used for the present study and the results generated are available from the authors.

Acknowledgments: The authors acknowledge the support and facilities provided by the Department of Geology, University of Delhi, Delhi, India, where this work was carried out. The authors also acknowledge the open-source data sets from National Aeronautics and Space Administration (NASA), United States Geological Survey (USGS), Geological Survey of India (GSI), India Meteorological Department (IMD), and Central Water Commission (CWC).

Conflicts of Interest: The authors declare no conflict of interest.

References

1. Manenti, S.; Amicarelli, A.; Palazzolo, N.; Bordoni, M.; Creaco, E.; Meisina, C. Post-Failure Dynamics of Rainfall-Induced Landslide in Oltrepò Pavese. *Water* **2020**, *12*, 2555. [CrossRef]
2. Saha, S.; Sarkar, R.; Roy, J.; Hembram, T.K.; Acharya, S.; Thapa, G.; Drukpa, D. Measuring Landslide Vulnerability Status of Chukha, Bhutan Using Deep Learning Algorithms. *Sci. Rep.* **2021**, *11*, 16374. [CrossRef] [PubMed]
3. Stancanelli, L.M.; Peres, D.J.; Cancelliere, A.; Foti, E. A Combined Triggering-Propagation Modeling Approach for the Assessment of Rainfall Induced Debris Flow Susceptibility. *J. Hydrol.* **2017**, *550*, 130–143. [CrossRef]
4. Korup, O.; Densmore, A.L.; Schlunegger, F. The Role of Landslides in Mountain Range Evolution. *Geomorphology* **2010**, *120*, 77–90. [CrossRef]
5. von Ruette, J.; Lehmann, P.; Or, D. Rainfall-triggered Shallow Landslides at Catchment Scale: Threshold Mechanics-based Modeling for Abruptness and Localization. *Water Resour. Res.* **2013**, *49*, 6266–6285. [CrossRef]
6. Acharya, K.P.; Bhandary, N.P.; Dahal, R.K.; Yatabe, R. Seepage and Slope Stability Modelling of Rainfall-Induced Slope Failures in Topographic Hollows. *Geomat. Nat. Hazards Risk* **2016**, *7*, 721–746. [CrossRef]
7. Carrión-Mero, P.; Montalván-Burbano, N.; Morante-Carballo, F.; Quesada-Román, A.; Apolo-Masache, B. Worldwide Research Trends in Landslide Science. *Int. J. Environ. Res. Public Health* **2021**, *18*, 9445. [CrossRef]
8. Scaioni, M.; Longoni, L.; Melillo, V.; Papini, M. Remote Sensing for Landslide Investigations: An Overview of Recent Achievements and Perspectives. *Remote Sens.* **2014**, *6*, 9600–9652. [CrossRef]
9. Zhang, J.; van Westen, C.J.; Tanyas, H.; Mavrouli, O.; Ge, Y.; Bajrachary, S.; Gurung, D.R.; Dhital, M.R.; Khanal, N.R. How Size and Trigger Matter: Analyzing Rainfall-and Earthquake-Triggered Landslide Inventories and Their Causal Relation in the Koshi River Basin, Central Himalaya. *Nat. Hazards Earth Syst. Sci.* **2019**, *19*, 1789–1805. [CrossRef]
10. Petley, D. Global Patterns of Loss of Life from Landslides. *Geology* **2012**, *40*, 927–930. [CrossRef]
11. Highland, L.; Bobrowsky, P.T. *The Landslide Handbook: A Guide to Understanding Landslides*; US Geological Survey: Reston, VA, USA, 2008.
12. Park, J.-Y.; Lee, S.-R.; Lee, D.-H.; Kim, Y.-T.; Lee, J.-S. A Regional-Scale Landslide Early Warning Methodology Applying Statistical and Physically Based Approaches in Sequence. *Eng. Geol.* **2019**, *260*, 105193. [CrossRef]
13. Nadim, F.; Kjekstad, O.; Peduzzi, P.; Herold, C.; Jaedicke, C. Global Landslide and Avalanche Hotspots. *Landslides* **2006**, *3*, 159–173. [CrossRef]
14. Ellen, S.D.; Wieczorek, G.F. *Landslides, Floods, and Marine Effects of the Storm of January 3–5, 1982, in the San Francisco Bay Region, California*; US Government Printing Office: Washington, DC, USA, 1902; Volume 1434.
15. Park, D.W.; Nikhil, N.V.; Lee, S.R. Landslide and Debris Flow Susceptibility Zonation Using TRIGRS for the 2011 Seoul Landslide Event. *Nat. Hazards Earth Syst. Sci.* **2013**, *13*, 2833–2849. [CrossRef]
16. Godt, J.W.; Baum, R.L.; Savage, W.Z.; Salciarini, D.; Schulz, W.H.; Harp, E.L. Transient Deterministic Shallow Landslide Modeling: Requirements for Susceptibility and Hazard Assessments in a GIS Framework. *Eng. Geol.* **2008**, *102*, 214–226. [CrossRef]

17. Campbell, R.H. Debris Flows Originating from Soil Slips during Rainstorms in Southern California. *Q. J. Eng. Geol.* **1974**, *7* 339–349. [CrossRef]
18. Crosta, G.B.; Frattini, P. Distributed Modelling of Shallow Landslides Triggered by Intense Rainfall. *Nat. Hazards Earth Syst. Sci* **2003**, *3*, 81–93. [CrossRef]
19. Bordoni, M.; Meisina, C.; Valentino, R.; Lu, N.; Bittelli, M.; Chersich, S. Hydrological Factors Affecting Rainfall-Induced Shallow Landslides: From the Field Monitoring to a Simplified Slope Stability Analysis. *Eng. Geol.* **2015**, *193*, 19–37. [CrossRef]
20. Persichillo, M.G.; Bordoni, M.; Meisina, C.; Bartelletti, C.; Barsanti, M.; Giannecchini, R.; D'Amato Avanzi, G.; Galanti, Y.; Cevasco, A.; Brandolini, P. Shallow Landslides Susceptibility Assessment in Different Environments. *Geomat. Nat. Hazards Risk* **2017**, *8*, 748–771. [CrossRef]
21. Aram, A.; Dalalian, M.R.; Saedi, S.; Rafieyan, O.; Darbandi, S. An Assessment of Data Mining and Bivariate Statistical Methods for Landslide Susceptibility Mapping. *Sci. Iran.* **2022**, *29*, 1077–1094.
22. Zizioli, D.; Meisina, C.; Valentino, R.; Montrasio, L. Comparison between Different Approaches to Modeling Shallow Landslide Susceptibility: A Case History in Oltrepo Pavese, Northern Italy. *Nat. Hazards Earth Syst. Sci.* **2013**, *13*, 559–573. [CrossRef]
23. Mathew, J.; Kundu, S.; Kumar, K.V.; Pant, C.C. Hydrologically Complemented Deterministic Slope Stability Analysis in Part of Indian Lesser Himalaya. *Geomat. Nat. Hazards Risk* **2016**, *7*, 1557–1576. [CrossRef]
24. Kirschbaum, D.B.; Adler, R.; Hong, Y.; Kumar, S.; Peters-Lidard, C.; Lerner-Lam, A. Advances in Landslide Nowcasting: Evaluation of a Global and Regional Modeling Approach. *Environ. Earth Sci.* **2012**, *66*, 1683–1696. [CrossRef]
25. Thomas, J.; Gupta, M.; Prusty, G. Assessing Global Parameters of Slope Stability Model Using Earth Data Observations for Forecasting Rainfall-Induced Shallow Landslides. *J. Appl. Geophys.* 2023; *under review*.
26. Wieczorek, G.F.; Morgan, B.A.; Campbell, R.H. Debris-Flow Hazards in the Blue Ridge of Central Virginia. *Environ. Eng. Geosci.* **2000**, *6*, 3–23. [CrossRef]
27. Crosta, G. Rainfall Threshold Regionalization: An Aid for Landslide Susceptibility Zonation. *Environ. Geol.* **1998**, *35*, 131–145. [CrossRef]
28. Coco, L.; Macrini, D.; Piacentini, T.; Buccolini, M. Landslide Susceptibility Mapping by Comparing GIS-Based Bivariate Methods: A Focus on the Geomorphological Implication of the Statistical Results. *Remote Sens.* **2021**, *13*, 4280. [CrossRef]
29. Chalkias, C.; Ferentinou, M.; Polykretis, C. GIS-Based Landslide Susceptibility Mapping on the Peloponnese Peninsula, Greece. *Geosciences* **2014**, *4*, 176–190. [CrossRef]
30. Bui, D.T.; Lofman, O.; Revhaug, I.; Dick, O. Landslide Susceptibility Analysis in the Hoa Binh Province of Vietnam Using Statistical Index and Logistic Regression. *Nat. Hazards* **2011**, *59*, 1413–1444. [CrossRef]
31. Metternicht, G.; Hurni, L.; Gogu, R. Remote Sensing of Landslides: An Analysis of the Potential Contribution to Geo-Spatial Systems for Hazard Assessment in Mountainous Environments. *Remote Sens. Environ.* **2005**, *98*, 284–303. [CrossRef]
32. Guzzetti, F.; Mondini, A.C.; Cardinali, M.; Fiorucci, F.; Santangelo, M.; Chang, K.-T. Landslide Inventory Maps: New Tools for an Old Problem. *Earth-Sci. Rev.* **2012**, *112*, 42–66. [CrossRef]
33. Casagli, N.; Catani, F.; Puglisi, C.; Delmonaco, G.; Ermini, L.; Margottini, C. An Inventory-Based Approach to Landslide Susceptibility Assessment and Its Application to the Virginio River Basin, Italy. *Environ. Eng. Geosci.* **2004**, *10*, 203–216. [CrossRef]
34. Stanley, T.; Kirschbaum, D.B. A Heuristic Approach to Global Landslide Susceptibility Mapping. *Nat. Hazards* **2017**, *87*, 145–164. [CrossRef]
35. Francipane, A.; Arnone, E.; Lo Conti, F.; Puglisi, C.; Noto, L.V. A Comparison between Heuristic, Statistical, and Data-Driven Methods in Landslide Susceptibility Assessment: An Application to the Briga and Giampilieri Catchments. In Proceedings of the 11th International Conference on Hydroinformatics, New York City, NY, USA, 8 January 2014.
36. Pasang, S.; Kubíček, P. Landslide Susceptibility Mapping Using Statistical Methods along the Asian Highway, Bhutan. *Geosciences* **2020**, *10*, 430. [CrossRef]
37. Zhang, K.; Wu, X.; Niu, R.; Yang, K.; Zhao, L. The Assessment of Landslide Susceptibility Mapping Using Random Forest and Decision Tree Methods in the Three Gorges Reservoir Area, China. *Environ. Earth Sci.* **2017**, *76*, 405. [CrossRef]
38. Marin, R.J.; Mattos, Á.J. Physically-Based Landslide Susceptibility Analysis Using Monte Carlo Simulation in a Tropical Mountain Basin. *Georisk Assess. Manag. Risk Eng. Syst. Geohazards* **2020**, *14*, 192–205. [CrossRef]
39. Raia, S.; Alvioli, M.; Rossi, M.; Baum, R.L.; Godt, J.W.; Guzzetti, F. Improving Predictive Power of Physically Based Rainfall-Induced Shallow Landslide Models: A Probabilistic Approach. *Geosci. Model Dev.* **2014**, *7*, 495–514. [CrossRef]
40. Vieira, B.C.; Fernandes, N.F. Shallow Landslide Prediction in the Serra Do Mar, São Paulo, Brazil. *Nat. Hazards Earth Syst. Sci.* **2010**, *10*, 1829–1837. [CrossRef]
41. Oguz, E.A.; Depina, I.; Thakur, V. Effects of Soil Heterogeneity on Susceptibility of Shallow Landslides. *Landslides* **2022**, *19*, 67–83. [CrossRef]
42. Salciarini, D.; Godt, J.W.; Savage, W.Z.; Conversini, P.; Baum, R.L.; Michael, J.A. Modeling Regional Initiation of Rainfall-Induced Shallow Landslides in the Eastern Umbria Region of Central Italy. *Landslides* **2006**, *3*, 181–194. [CrossRef]
43. Salciarini, D.; Godt, J.W.; Savage, W.Z.; Baum, R.L.; Coversini, P. Modeling the rainfall-induced development of shallow landslides in eastern Umbria, central Italy, using the TRIGRS (transient rainfall infiltration and grid-based slope-stability) approach. In Proceedings of the 1st North American Landslide Conference, Vail, CO, USA, 3–8 June 2007; Turner, A.K., Schuster, R.L., Eds.; Association of Environmental and Engineering Geologists: Lexington, KY, USA, 2007; pp. 294–304.

44. Zhang, S.; Jiang, Q.; Wu, D.; Xu, X.; Tan, Y.; Shi, P. Improved Method of Defining Rainfall Intensity and Duration Thresholds for Shallow Landslides Based on TRIGRS. *Water* **2022**, *14*, 524. [CrossRef]
45. Konig, T.; Hermann, K.U.X.; Corsi, A. Landslide Risk Management Using the Mathematical Model Trigrs. *Geosci. Geociências* **2022**, *41*, 243–254.
46. Zhuang, J.; Peng, J.; Wang, G.; Iqbal, J.; Wang, Y.; Li, W.; Xu, Q.; Zhu, X. Prediction of Rainfall-induced Shallow Landslides in the Loess Plateau, Yan'an, China, Using the TRIGRS Model. *Earth Surf. Process. Landf.* **2017**, *42*, 915–927. [CrossRef]
47. Mergili, M.; Marchesini, I.; Alvioli, M.; Metz, M.; Schneider-Muntau, B.; Rossi, M.; Guzzetti, F. A Strategy for GIS-Based 3-D Slope Stability Modelling over Large Areas. *Geosci. Model Dev.* **2014**, *7*, 2969–2982. [CrossRef]
48. Rossi, G.; Catani, F.; Leoni, L.; Segoni, S.; Tofani, V. HIRESSS: A Physically Based Slope Stability Simulator for HPC Applications. *Nat. Hazards Earth Syst. Sci.* **2013**, *13*, 151–166. [CrossRef]
49. Simoni, S.; Zanotti, F.; Bertoldi, G.; Rigon, R. Modelling the Probability of Occurrence of Shallow Landslides and Channelized Debris Flows Using GEOtop-FS. *Hydrol. Process. Int. J.* **2008**, *22*, 532–545. [CrossRef]
50. Montrasio, L.; Valentino, R. A Model for Triggering Mechanisms of Shallow Landslides. *Nat. Hazards Earth Syst. Sci.* **2008**, *8*, 1149–1159. [CrossRef]
51. Pack, R.T.; Tarboton, D.G.; Goodwin, C.N. SINMAP 2.0—A Stability Index Approach to Terrain Stability Hazard Mapping, User's Manual. 1999. Available online: https://digitalcommons.usu.edu/cee_facpub/16/ (accessed on 18 December 2022).
52. Montgomery, D.R.; Dietrich, W.E. A Physically Based Model for the Topographic Control on Shallow Landsliding. *Water Resour. Res.* **1994**, *30*, 1153–1171. [CrossRef]
53. Wu, W.; Sidle, R.C. A Distributed Slope Stability Model for Steep Forested Basins. *Water Resour. Res.* **1995**, *31*, 2097–2110. [CrossRef]
54. Baum, R.L.; Savage, W.Z.; Godt, J.W. *TRIGRS: A Fortran Program for Transient Rainfall Infiltration and Grid-Based Regional Slope-Stability Analysis, Version 2.0*; US Geological Survey: Reston, VA, USA, 2008.
55. O'loughlin, E.M. Prediction of Surface Saturation Zones in Natural Catchments by Topographic Analysis. *Water Resour. Res.* **1986**, *22*, 794–804. [CrossRef]
56. Dikshit, A.; Satyam, N.; Pradhan, B. Estimation of Rainfall-Induced Landslides Using the TRIGRS Model. *Earth Syst. Environ.* **2019**, *3*, 575–584. [CrossRef]
57. Ciurleo, M.; Ferlisi, S.; Foresta, V.; Mandaglio, M.C.; Moraci, N. Landslide Susceptibility Analysis by Applying TRIGRS to a Reliable Geotechnical Slope Model. *Geosciences* **2021**, *12*, 18. [CrossRef]
58. Take, W.A.; Bolton, M.D.; Wong, P.C.P.; Yeung, F.J. Evaluation of Landslide Triggering Mechanisms in Model Fill Slopes. *Landslides* **2004**, *1*, 173–184. [CrossRef]
59. Montrasio, L.; Schilirò, L.; Terrone, A. Physical and Numerical Modelling of Shallow Landslides. *Landslides* **2016**, *13*, 873–883. [CrossRef]
60. Melo, C.M.; Kobiyama, M.; Michel, G.P.; de Brito, M.M. The Relevance of Geotechnical-Unit Characterization for Landslide-Susceptibility Mapping with SHALSTAB. *GeoHazards* **2021**, *2*, 383–397. [CrossRef]
61. Rana, H.; Babu, G.L. Regional Back Analysis of Landslide Events Using TRIGRS Model and Rainfall Threshold: An Approach to Estimate Landslide Hazard for Kodagu, India. *Bull. Eng. Geol. Environ.* **2022**, *81*, 160. [CrossRef]
62. Weidner, L.; Oommen, T.; Escobar-Wolf, R.; Sajinkumar, K.S.; Samuel, R.A. Regional-Scale Back-Analysis Using TRIGRS: An Approach to Advance Landslide Hazard Modeling and Prediction in Sparse Data Regions. *Landslides* **2018**, *15*, 2343–2356. [CrossRef]
63. Liao, Z.; Hong, Y.; Kirschbaum, D.; Adler, R.F.; Gourley, J.J.; Wooten, R. Evaluation of TRIGRS (Transient Rainfall Infiltration and Grid-Based Regional Slope-Stability Analysis)'s Predictive Skill for Hurricane-Triggered Landslides: A Case Study in Macon County, North Carolina. *Nat. Hazards* **2011**, *58*, 325–339. [CrossRef]
64. Baum, R.L.; Savage, W.Z.; Godt, J.W. *TRIGRS-A Fortran Program for Transient Rainfall Infiltration and Grid-Based Regional Slope-Stability Analysis*; Open-File Rep; US Geological Survey: Reston, VA, USA, 2002; Volume 38, p. 424.
65. Alvioli, M.; Baum, R.L. Parallelization of the TRIGRS Model for Rainfall-Induced Landslides Using the Message Passing Interface. *Environ. Model. Softw.* **2016**, *81*, 122–135. [CrossRef]
66. Hao, L.; Rajaneesh, A.; Van Westen, C.; Sajinkumar, K.S.; Martha, T.R.; Jaiswal, P.; McAdoo, B.G. Constructing a Complete Landslide Inventory Dataset for the 2018 Monsoon Disaster in Kerala, India, for Land Use Change Analysis. *Earth Syst. Sci. Data* **2020**, *12*, 2899–2918. [CrossRef]
67. Sajinkumar, K.S.; Anbazhagan, S.; Pradeepkumar, A.P.; Rani, V.R. Weathering and Landslide Occurrences in Parts of Western Ghats, Kerala. *J. Geol. Soc. India* **2011**, *78*, 249–257. [CrossRef]
68. Kuriakose, S.L.; Sankar, G.; Muraleedharan, C. History of Landslide Susceptibility and a Chorology of Landslide-Prone Areas in the Western Ghats of Kerala, India. *Environ. Geol.* **2009**, *57*, 1553–1568. [CrossRef]
69. Crozier, M.J. Deciphering the Effect of Climate Change on Landslide Activity: A Review. *Geomorphology* **2010**, *124*, 260–267. [CrossRef]
70. Johnston, E.C.; Davenport, F.V.; Wang, L.; Caers, J.K.; Muthukrishnan, S.; Burke, M.; Diffenbaugh, N.S. Quantifying the Effect of Precipitation on Landslide Hazard in Urbanized and Non-Urbanized Areas. *Geophys. Res. Lett.* **2021**, *48*, e2021GL094038. [CrossRef]

71. Seneviratne, S.; Nicholls, N.; Easterling, D.; Goodess, C.; Kanae, S.; Kossin, J.; Luo, Y.; Marengo, J.; McInnes, K.; Rahimi, M. *Changes in Climate Extremes and Their Impacts on the Natural Physical Environment*; Cambridge University Press: Cambridge, UK, 2012.
72. Hunt, K.M.R.; Menon, A. The 2018 Kerala Floods: A Climate Change Perspective. *Clim. Dyn.* **2020**, *54*, 2433–2446. [CrossRef]
73. Rai, P.K.; Singh, G.P.; Dash, S.K. Projected Changes in Extreme Precipitation Events over Various Subdivisions of India Using RegCM4. *Clim. Dyn.* **2020**, *54*, 247–272. [CrossRef]
74. Shashikanth, K.; Ghosh, S.; Karmakar, S. Future Projections of Indian Summer Monsoon Rainfall Extremes over India with Statistical Downscaling and Its Consistency with Observed Characteristics. *Clim. Dyn.* **2018**, *51*, 1–15. [CrossRef]
75. Central Water Commission; Study Report: Kerala Floods of August 2018. Ministry of Water Resources, River Development & Ganga Rejuvenation, Government of India, New Delhi, 2018. Available online: http://cwc.gov.in/main/downloads/KeralaFloodReport/Rev-1.pdf (accessed on 18 December 2022).
76. Sulal, N.L.; Archana, K.G. *Note on Post Disaster Studies for Landslides Occurred in June 2018 At Idukki District, Kerala*; Geological Survey of India: Thiruvananthapuram, India, 2019.
77. Megha, V.; Joshi, V.; Kakde, N.; Jaybhaye, A.; Dhoble, D. Flood Mapping and Analysis Using Sentinel Application Platform (SNAP)—A Case Study of Kerala. *Int. J. Res. Eng. Sci. Manag.* **2019**, *2*, 486–488.
78. Vishnu, C.L.; Sajinkumar, K.S.; Oommen, T.; Coffman, R.A.; Thrivikramji, K.P.; Rani, V.R.; Keerthy, S. Satellite-Based Assessment of the August 2018 Flood in Parts of Kerala, India. *Geomat. Nat. Hazards Risk* **2019**, *10*, 758–767. [CrossRef]
79. Sankar, G. Monsoon Fury in Kerala—A Geo-Environmental Appraisal. *J. Geol. Soc. India* **2018**, *92*, 383–388. [CrossRef]
80. Singh, B.; Singh, P.; Supriya, K.; Singh, M. An Overview on Kerala Floods: Loss of Human Lives as Well as Biodiversity in God's Own Country. *State Gov.* **2018**, *7*, 7.
81. Miller, D.A.; White, R.A. A Conterminous United States Multilayer Soil Characteristics Dataset for Regional Climate and Hydrology Modeling. *Earth Interact.* **1998**, *2*, 1–26. [CrossRef]
82. Tian, Y.; Peters-Lidard, C.D.; Kumar, S.V.; Geiger, J.; Houser, P.R.; Eastman, J.L.; Dirmeyer, P.; Doty, B.; Adams, J. High-Performance Land Surface Modeling with a Linux Cluster. *Comput. Geosci.* **2008**, *34*, 1492–1504. [CrossRef]
83. Marin, R.J.; Mattos, Á.J.; Fernández-Escobar, C.J. Understanding the Sensitivity to the Soil Properties and Rainfall Conditions of Two Physically-Based Slope Stability Models. *Boletín Geol.* **2022**, *44*, 93–109. [CrossRef]
84. Tarboton, D.G. *TauDEM (Terrain Analysis Using Digital Elevation Models)*; Utah State University: Logan, UT, USA, 2004.
85. Pai, D.S.; Rajeevan, M.; Sreejith, O.P.; Mukhopadhyay, B.; Satbha, N.S. Development of a New High Spatial Resolution (0.25 × 0.25) Long Period (1901–2010) Daily Gridded Rainfall Data Set over India and Its Comparison with Existing Data Sets over the Region. *Mausam* **2014**, *65*, 1–18. [CrossRef]
86. Huffman, G.J.; Stocker, E.F.; Bolvin, D.T.; Nelkin, E.J.; Tan, J. GPM 752 IMERG Final Precipitation L3 Half Hourly 0.1 Degree x 0.1 Degree V06, Greenbelt, 753 MD, Goddard Earth Sciences Data and Information Services Center (GES DISC). 2019. Available online: https://disc.gsfc.nasa.gov/datasets/GPM_3IMERGHH_06/summary (accessed on 18 December 2019). [CrossRef]
87. Alvioli, M.; Lee, G.; An, H.U. Three-Dimensional, Time-Dependent Modeling of Rainfall-Induced Landslides over a Digital Landscape: A Case Study. *Landslides* **2018**, *15*, 1071–1084.
88. Tufano, R.; Formetta, G.; Calcaterra, D.; De Vita, P. Hydrological Control of Soil Thickness Spatial Variability on the Initiation of Rainfall-Induced Shallow Landslides Using a Three-Dimensional Model. *Landslides* **2021**, *18*, 3367–3380. [CrossRef]
89. Fusco, F.; Mirus, B.B.; Baum, R.L.; Calcaterra, D.; De Vita, P. Incorporating the Effects of Complex Soil Layering and Thickness Local Variability into Distributed Landslide Susceptibility Assessments. *Water* **2021**, *13*, 713. [CrossRef]
90. Del Soldato, M.; Pazzi, V.; Segoni, S.; De Vita, P.; Tofani, V.; Moretti, S. Spatial Modeling of Pyroclastic Cover Deposit Thickness (Depth to Bedrock) in Peri-volcanic Areas of Campania (Southern Italy). *Earth Surf. Process. Landforms* **2018**, *43*, 1757–1767. [CrossRef]
91. De Vita, P.; Agrello, D.; Ambrosino, F. Landslide Susceptibility Assessment in Ash-Fall Pyroclastic Deposits Surrounding Mount Somma-Vesuvius: Application of Geophysical Surveys for Soil Thickness Mapping. *J. Appl. Geophys.* **2006**, *59*, 126–139. [CrossRef]
92. Baum, R.L.; Godt, J.W.; Savage, W.Z. Estimating the Timing and Location of Shallow Rainfall-inducedlandslides Using a Model for Transient, Unsaturated Infiltration. *J. Geophys. Res.* **2010**, *115*, 3013. [CrossRef]
93. Baum, R.L.; Godt, J.W. Early Warning of Rainfall-Induced Shallow Landslides and Debris Flows in the USA. *Landslides* **2010**, *7*, 259–272. [CrossRef]
94. O'Callaghan, J.F.; Mark, D.M. The Extraction of Drainage Networks from Digital Elevation Data. *Comput. Vis. Graph. Image Process.* **1984**, *28*, 323–344. [CrossRef]
95. Jenson, S.K.; Domingue, J.O. Extracting Topographic Structure from Digital Elevation Data for Geographic Information System Analysis. *Photogramm. Eng. Remote Sens.* **1988**, *54*, 1593–1600.
96. Garbrecht, J.; Martz, L.W. The Assignment of Drainage Direction over Flat Surfaces in Raster Digital Elevation Models. *J. Hydrol.* **1997**, *193*, 204–213. [CrossRef]
97. Vanapalli, S.K.; Fredlund, D.G. Comparison of Different Procedures to Predict Unsaturated Soil Shear Strength. *Adv. Unsaturated Geotech.* **2000**, 195–209. [CrossRef]
98. Thomas, J.; Gupta, M. Prediction of Rainfall-Induced Shallow Landslides through Integration of Hydrological Model with a Slope Stability Model. In Proceedings of the EGU General Assembly, Vienna, Austria, 23–27 May 2022.

99. Gioia, E.; Speranza, G.; Ferretti, M.; Godt, J.W.; Baum, R.L.; Marincioni, F. Application of a Process-Based Shallow Landslide Hazard Model over a Broad Area in Central Italy. *Landslides* **2016**, *13*, 1197–1214. [CrossRef]
100. Zieher, T.; Rutzinger, M.; Schneider-Muntau, B.; Perzl, F.; Leidinger, D.; Formayer, H.; Geitner, C. Sensitivity Analysis and Calibration of a Dynamic Physically Based Slope Stability Model. *Nat. Hazards Earth Syst. Sci.* **2017**, *17*, 971–992. [CrossRef]
101. Hammond, C. *Level I Stability Analysis (LISA) Documentation for Version 2.0*; US Department of Agriculture, Forest Service, Intermountain Research Station: Ogden, Utah, 1992; Volume 285.
102. Iverson, R.M. Landslide Triggering by Rain Infiltration. *Water Resour. Res.* **2000**, *36*, 1897–1910. [CrossRef]
103. Di, B.; Stamatopoulos, C.A.; Stamatopoulos, A.C.; Liu, E.; Balla, L. Proposal, Application and Partial Validation of a Simplified Expression Evaluating the Stability of Sandy Slopes under Rainfall Conditions. *Geomorphology* **2021**, *395*, 107966. [CrossRef]
104. de Lima Neves Seefelder, C.; Koide, S.; Mergili, M. Does Parameterization Influence the Performance of Slope Stability Model Results? A Case Study in Rio de Janeiro, Brazil. *Landslides* **2017**, *14*, 1389–1401. [CrossRef]
105. Schilirò, L.; Cepeda, J.; Devoli, G.; Piciullo, L. Regional Analyses of Rainfall-Induced Landslide Initiation in Upper Gudbrandsdalen (South-Eastern Norway) Using TRIGRS Model. *Geosciences* **2021**, *11*, 35. [CrossRef]
106. Ines, A.V.M.; Droogers, P. Inverse Modelling in Estimating Soil Hydraulic Functions: A Genetic Algorithm Approach. *Hydrol. Earth Syst. Sci.* **2002**, *6*, 49–66. [CrossRef]
107. Mohanty, B.P. Soil Hydraulic Property Estimation Using Remote Sensing: A Review. *Vadose Zone J.* **2013**, *12*, 1–9. [CrossRef]
108. Thomas, J.; Gupta, M.; Srivastava, P.K.; Pandey, D.K.; Bindlish, R. Development of High-Resolution Soil Hydraulic Parameters with Use of Earth Observations for Enhancing Root Zone Soil Moisture Product. *Remote Sens.* **2023**, *15*, 706. [CrossRef]

International Journal of
Geo-Information

Article

A Comparison Study of Landslide Susceptibility Spatial Modeling Using Machine Learning

Nurwatik Nurwatik [1,*], Muhammad Hidayatul Ummah [1], Agung Budi Cahyono [1], Mohammad Rohmaneo Darminto [1] and Jung-Hong Hong [2]

1 Department of Geomatics Engineering, Faculty of Civil, Planning, and Geo Engineering, Institut Teknologi Sepuluh Nopember, Surabaya 60111, Indonesia
2 Department of Geomatics, National Cheng Kung University, Tainan City 701, Taiwan
* Correspondence: nurwatik@its.ac.id; Tel.: +62-81249886362

Abstract: One hundred seventeen landslides occurred in Malang Regency throughout 2021, triggering the need for practical hazard assessments to strengthen the disaster mitigation process. In terms of providing a solution for investigating the location of landslides more precisely, this research aims to compare machine learning algorithms to produce an accurate landslide susceptibility model. This research applies three machine learning algorithms composed of RF (random forest), NB (naïve Bayes), and KNN (k-nearest neighbor) and 12 conditioning factors. The conditioning factors consist of slope, elevation, aspect, NDVI, geological type, soil type, distance from the fault, distance from the river, river density, TWI, land cover, and annual rainfall. This research performs seven models over three ratios between the training and testing dataset encompassing 50:50, 60:40, and 70:30 for KNN and NB algorithms and 70:30 for the RF algorithm. This research measures the performance of each model using eight parameters (ROC, AUC, ACC, SN, SP, BA, GM, CK, and MCC). The results indicate that RF 70:30 generates the best performance, witnessed by the evaluation parameters ACC (0.884), SN (0.765), GM (0.863), BA (0.857), CK (0.749), MCC (0.876), and AUC (0.943). Overall, seven models have reasonably good accuracy, ranging between 0.806 and 0.884. Furthermore, based on the best model, the study area is dominated by high susceptibility with an area coverage of 51%, which occurs in the areas with high slopes. This research is expected to improve the quality of landslide susceptibility maps in the study area as a foundation for mitigation planning. Furthermore, it can provide recommendations for further research in splitting ratio scenarios between training and testing data.

Keywords: landslide susceptibility; machine learning; k-nearest neighbor; naïve Bayes; random forest

Citation: Nurwatik, N.; Ummah, M.H.; Cahyono, A.B.; Darminto, M.R.; Hong, J.-H. A Comparison Study of Landslide Susceptibility Spatial Modeling Using Machine Learning. *ISPRS Int. J. Geo-Inf.* **2022**, *11*, 602. https://doi.org/10.3390/ijgi11120602

Academic Editors: Wolfgang Kainz, Walter Chen and Fuan Tsai

Received: 29 September 2022
Accepted: 21 November 2022
Published: 2 December 2022

1. Introduction

Landslides are the phenomena of downslope movements by soil mass and rock slopes. Landslides occur due to the sliding of a volume above a layer of rock containing clay after the saturation of water acts as a launcher [1]. A landslide is a natural phenomenon controlled by geological factors, rainfall, and land use on the slopes [2]. Indonesia is a country with a high potential for landslides. According to the data from the National Disaster Management Agency of Indonesia (BNPB), throughout 2021, there were 632 incidents reaching 20% of the total disasters in Indonesia throughout 2021.

Malang Regency is situated in East Java Province and is highly vulnerable to landslides. The Malang Regency Regional Disaster Management Agency (BPBD) data accounted for 117 landslides in 2021, reaching 44% of the total disasters in Malang Regency throughout 2021. Geographical conditions render the Malang Regency highly vulnerable to landslides. It is located in a highland area with various slopes, from sloping to very steep, as it is surrounded by the Tengger Mountains, Mounts Kawi and Kelud, and Mounts Arjuna and Welirang.

Landslide susceptibility assessment is a fundamental action for improving the mitigation process. Periodical assessment is necessary, since landslides occur periodically, and the conditioning factors change over time. Implementing various methods to investigate the location of landslides, assess the vulnerability area, and analyze the impacts can be conducted by using a terrestrial survey [3], satellite monitoring [4], or spatial modeling [5]. Spatial modeling has become a prompt solution, along with the growth of technologies and the availability of various data sources. It can integrate various data sources through algorithms to produce maps, such as the machine learning approach. Machine learning (ML) is a branch of computational algorithms developed and designed to imitate human intelligence by learning from environmental data [6]. Machine learning is capable of solving problems regarding predictions and classifications [7]. In terms of landslide susceptibility modeling, a prediction can utilize machine learning using coordinate data of landslide occurrence as training data and landslide conditioning factors as the evaluators [8].

Research trends using the keywords landslide susceptibility and machine learning have grown significantly since 2018 [9]. Research conducted by [10] applied the NB (naïve Bayes) algorithm, the RBF (radial basis function) classifier, and the RBF network for Longhai, China, for analysis of landslide susceptibility modeling. It indicated that the naïve Bayes algorithm showed high performance in predicting landslide susceptibility with an AUC value of 0.872. Moreover, other research conducted by [11] using ANN (artificial neural network) and support vector machine (SVM) algorithms, decision trees (DTs), RF (random forest), and combined models of ANN and SVM was implemented in the Cameron Highlands district located in the state of Pahang, Malaysia. According to this research, the RF algorithm produced the best performance, with an AUC value on the testing data of 0.82. Research conducted by [12] carried out spatial modeling of landslide susceptibility in the Wayanad district in the southern part of India using RF, SVM, and K-NN (k-nearest neighbor) algorithms. The K-NN algorithm has a good predictive ability of landslide susceptibility, with a maximum AUC value of 0.981. The maximum entropy (MAXENT) algorithm was developed for various spatial analyses with good performance results as part of the development of machine learning algorithms for spatial analysis. The Maxent algorithm can perform various spatial analyses, including predictions of urban waterlogging-prone areas, fire hazards, and land subsidence studies [13–15]. A recent study, however, showed that using the maximum-entropy algorithm (MAXENT) in the evaluation of landslide susceptibility produced a lower accuracy than RF [16].

In the study area, research regarding spatial modeling of landslide susceptibility applied scoring and overlay analysis, logistic regression, and spatial multi-criteria evaluation [17–19]. Those methods are subject-oriented and rely on the consistency of various experts in the adjustment process and the time-consuming handling of multiple data sources. In addition, a landslide susceptibility model using the conventional scoring method, multi-criteria evaluation, and expert judgment generates less accuracy [17]. Considering the condition of Malang Regency as a mobility center with a high tourist attraction, a high-accuracy of landslide susceptibility assessment is necessary to mitigate casualties. Therefore, this research applies machine learning algorithms to assess landslide susceptibility in the study area. Using a statistical approach and machine learning techniques can help to reduce the subjectivity of the analysis. The model can be evaluated quantitatively, and producing the contribution level of each variable can be quantitatively based on [20,21]. RF, KNN, and NB are three machine-learning algorithms that have produced accurate models of landslide susceptibility in various case studies.

Therefore, this research will compare the spatial modeling of landslide susceptibility using three machine learning algorithms (RF, NB, and KNN). This research applies three splitting ratios for training and testing data comprising 50:50, 60:40, and 70:30 for NB and KNN. Moreover, the RF only uses 70:30, following the best splitting ratio produced from previous research [22]. Eight evaluation parameters were sequentially used to test the performance of seven models. These parameters were comprised of ROC (receiver operator characteristic), AUC (area under curve), accuracy (ACC), sensitivity (SN), specificity

(SP), balanced accuracy (BA), geometric mean (GM), Cohen's kappa (CK), and Matthew's correlation coefficient (MCC).

2. Materials and Methods

2.1. Study Area

Figure 1 depicts the study area of this research showing the distribution of landslides and non-landslide location. The study area was in Malang Regency, which is located geographically at 112°17′10.9″–112°57′0.0″ E and 7°44′55.11″–8°26′35.45″ S. Malang Regency has 33 sub-districts, 12 urban villages, and 378 villages. Malang Regency is the second largest regency in East Java Province with an area of 334.786 ha. The topography of Malang Regency varies, with elevation values between 0 and 3660 MASL. It has several mountains, including Mount Semeru (4676 MASL), Mount Kelud (1731 MASL), Mount Welirang (3156 MASL), and Mount Arjuno (3339 MASL). Consequently, the slope is varied between 0° and 85.2°. The geological type is dominated by tuff formation with extrusive intermediate pyroclastic composition and derived from volcanic deposits. Malang Regency has a tropical climate with an average surface temperature of 18.25 °C to 31.45 °C.

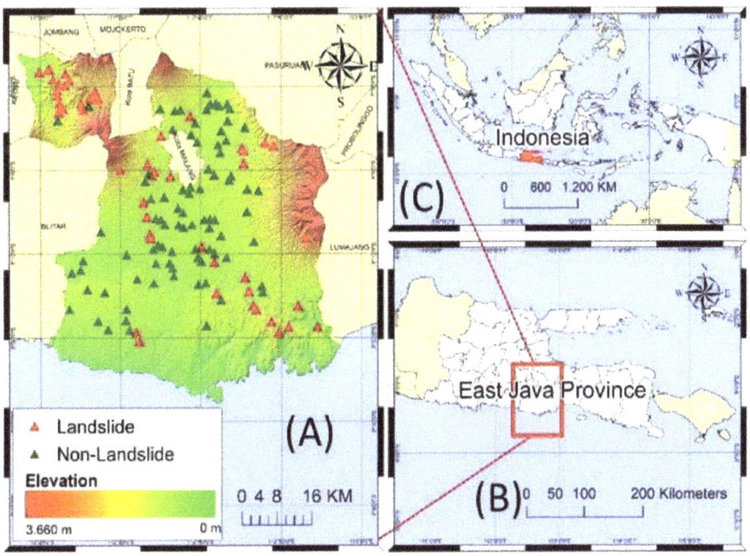

Figure 1. Location of the study area. (**A**) Elevation of the study area and distribution of training points. (**B**) The location of the study area in East Java Province. (**C**) The location of East Java Province in Indonesia.

2.2. Data Sources

2.2.1. Data Training Sample

The training sample consisted of landslides and non-landslide areas [23]. The data type was a point feature acquired from the Malang Regency Disaster Management Agency's daily reports from 2012 to 2021. From the data collected on landslide occurrence during 2012–2021, the number of points was 88. The number of landslides inclined in certain locations from 2012 to 2021. Hence, it was assumed that past events are still actively occurring at some locations. Moreover, the non-slide training sample was obtained by randomly extracting points with a slope of less than 2° [24]. The number of non-landslide training samples, as many as 88 points, was adjusted to the number of landslide points. Eventually, the total number of training sample points was 176.

2.2.2. Spatial Data Landslide Conditioning Factors

The selection of landslide conditioning factors is essential to achieving high modeling accuracy. Standard rules related to the parameters that affect the landslide susceptibility model do not exist [25]. Landslide conditioning factors depend on the characteristics of the case study, the type of occurrence of the landslide, and the scale of analysis [26]. This research proposed 12 landslide conditioning factors to produce landslide susceptibility maps considering study area conditions, literature studies, and data availability. The 12 parameters consist of topography, land cover, and hydrological and trigger factors Topographic factors consist of elevation, slope, and aspect. Moreover, land cover factors include geological type, soil type, distance from faults, and vegetation density. Hydrological factors include TWI (Topographic Wetness Index), distance from the river, and river density, whilst the triggering factor is average annual rainfall in 2012–2021. Figure 2 visualizes the landslide conditioning factors in the study area.

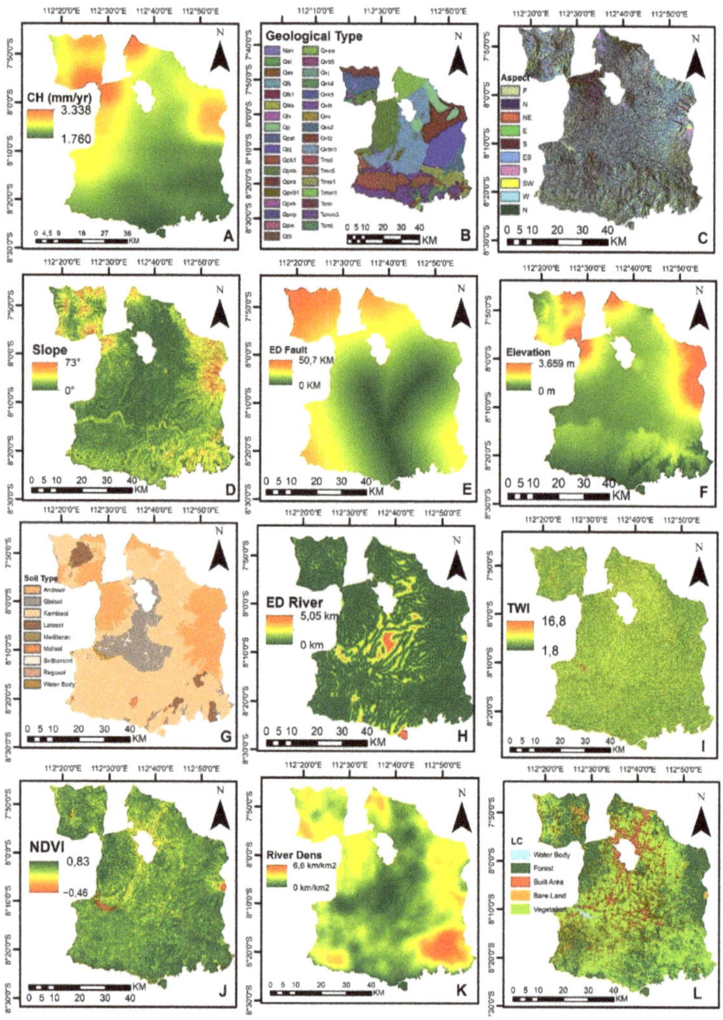

Figure 2. Landslide conditioning factors; (**A**) annual rainfall; (**B**) geological type; (**C**) aspect; (**D**) slope; (**E**) distance to fault; (**F**) elevation; (**G**) soil type; (**H**) distance to river; (**I**) TWI; (**J**) NDVI; (**K**) river density; (**L**) land cover.

Elevation Data

This research used DEMNAS as the elevation data with a resolution of 0.27 arc-second or 8 m, published in 2018 by The Indonesian Geospatial Information Agency (https://tanahair.indonesia.go.id/demnas, accessed on 26 February 2022) [27]. DEMNAS was used to extract elevation, aspect, and slope parameters. Based on the DEMNAS, the study area has an elevation value of 0 to 3660 MASL with a slope ranging from 0° to 73°. Moreover, the aspect distributes from 0 to 360, indicating that the slope angle direction is clockwise. It consists of north, northeast, east, southeast, south, southwest, west, northwest, and flat. Besides extracting topographic factor parameters, elevation data also generated TWI of the study area. The TWI ranged from 1.8 to 16.8. For modeling purposes, this research resampled all the data into 30 m. In addition, the resampling process was carried out to project all datasets into the same coordinate system.

Geological Map Data

Geological map data were acquired from the Geological Agency, Ministry of Energy, and Mineral Resources Indonesia with the scale of 1:100,000. The latest geological map was created in 1992 by the Indonesian ministry of energy and mineral resources. The map was produced from measurements of direct outcrop points in the field, which started in 1921 during the Dutch-Indies period [28]. Geological maps extracted geological type and fault parameter information. Furthermore, this research proceeded with Euclidean distance analysis to calculate the distance from the fault location; moreover, the geological type was converted into a raster format and resampled. According to the geological type, the study area is dominated by a tuff formation with a coverage area of 16%. The formation is a pyroclastic extrusive rock originating from volcanic deposits.

The study area consists of 34 geological unit formations. Table 1 represents the characteristics related to the types of formations, rock formations, and deposits. In general, the rock conditions are composed of rocks brought by volcanic activity consisting of tuff, sandy tuff, volcanic breccia, agglomerates, and lava. Moreover, the distance between the study area and the fault ranges between 0 and 50,000 m. The type of fault which crosses the study area is a local fault with shear, descending, and horizontal faults [29–31]. The local faults pass through the Sub-district of Sumbermanjing, Bantur, Gedangan, Gondang Legi, Turen, Wajak, Poncokusumo, and Dampit. The fault which passes through Sumbermanjing Sub-district is a descending type, while those passing through Sumbermanjing and Bantur Sub-district are shear-type and horizontal, respectively.

Table 1. Geological unit, Malang Regency.

Code	Formation	Rock Formation	Deposit	Area (km^2)
Qvtm1	Malang tuff	E: I: PA	Volcanism: subaerial—Volcanism	633.995
Qpkb	Kawi-butak volcanic rock	E: I: PC	Volcanism: subaerial—Volcanism	446.265
Tomm3	Mandalika formation	E: I: L	Volcanism: subaerial—Volcanism	401.839
Qpj	Jombang formation	ST: CC: CE: B	Volcanism: subaerial—Volcanism:	331.369
Tmn5	Nampol formation	ST: CC: M: S	Sedimentation: transitional—Sed	277.764
Qvt2	Tengger volcanic rock	E: I: PA	Volcanism: subaerial—Volcanism	238.221
Qvaw	Arjuna-Welirang volcanic rock	E: I: PC	Volcanism: subaerial—Volcanism	184.673
Tmw1	Wuni formation	ST: CC: CE: B	—	184.217
Qp	Western volcanic rock	E: I: PC	Volcanism: subaerial—Volcanism	171.113
Qpat	Anjasmara old volcanic rock	E: I: PC	Volcanism: subaerial—Volcanism	160.352
Qvs2	Semeru volcanic deposit	E: I: L	Volcanism: subaerial—Volcanism	96.447
Qpva	Anjasmara young volcano	E: I: PC	Volcanism: subaerial—Volcanism	87.365
Tomt	Tuff member	E: I: PA	Volcanism: subaerial—Volcanism	70.339
Tmcl	Campurdarat formation	ST: CC: LS	Sedimentation: littoral—Sedimen	45.227
Qpvb1	Buring volcanic deposit	E: MC: L	Volcanism: subaerial—Volcanism	39.199
Qas	Swamp and river deposits	S: CC: M: S	Sedimentation: terrestrial: fluv	26.346
Non	Lake	-	-	20.240
Tmwl1	Wonosari formation	ST: R: LS	Sedimentation: littoral: reef—S	15.264
Qvk4	Kelud young volcano	E: I: PC	Volcanism: subaerial—Volcanism:	13.200
Qpvk	Kelud old volcanic rock	E: I: L	Volcanism: subaerial—Volcanism	11.789

Table 1. *Cont.*

Code	Formation	Rock Formation	Deposit	Area (km^2)
Tomi	Rock intrusion	IE: I	Plutonism: sub-volcanic—Plutoni	11.564
Qpvp	Marikeng volcanic rock	IE: I	Plutonism: sub-volcanic—Plutoni	6.937
Qvlh	Lava deposit	E: I: PC	Volcanism: subaerial—Volcanism	5.602
Qvs	Tengger volcanic sand	E: I: PA	Volcanism: subaerial—Volcanism	4.173
Qvk5	Kepolo volcanic deposit	E: I: L	Volcanism: subaerial—Volcanism	3.084
Qpw	Welang formation	ST: CC: M: S	Sedimentation: terrestrial: allu	2.546
Qvj	Jembangan volcanic deposit	E: MC: L	Volcanism: subaerial—Volcanism	2.225
Qt5	Terrace deposit	ST: CC: A	Sedimentation: terrestrial: allu	2.179
Qlk	Katu's peak lava	E: I: L	Volcanism: subaerial—Volcanism	1.829
Qal	Aluvial and coastal deposit	ST: CC: A	Sedimentation: terrestrial: fluv	1.130
Qvb5	Bromo volcanic rock	E: I: PC	Volcanism: subaerial—Volcanism:	0.810
Qlks	Lava Parasite Kepolo Mt. Semeru	E: I: L	Volcanism: subaerial—Volcanism	0.727
Qlk1	Lava andesit parasit	E: I: L	Volcanism: subaerial—Volcanism	0.058
Qlv	Avalanche deposits from volcanoes	E: I: PC	Volcanism: subaerial—Volcanism	0.035
Qpk1	Kalipucang formation	ST: CC: CE: CL	Sedimentation: terrestrial: fluv	0.001

Rock Formation: ST = sediment, CC= clastic, E = extrusive, I = intermediate, L = lava, PC = polymic, A = alluvium, M = medium, PA = pyrocla, R = reef: LS = limestone, S = sands, CE = coarse, B = brecc, MC = mafic.

Soil Type Data

The Indonesian Ministry of Agriculture Indonesia produced soil-type map data with a scale of 1:50,000 in 2014. The rasterization proceeded to convert the data into a raster format. Then, this research resampled the map with 30 m. According to the soil type, cambisol dominates the study area with a coverage area of 60%. Cambisol soil types are rich in mineral matter and vary in drainage, depth, and base saturation [32].

Landsat-8 OLI TIRS Imagery Data

Landsat-8 OLI TIRS Imagery data were acquired from the USGS (United States Geological Survey) directory using the Google Earth engine (https://developers.google.com/earth-engine/datasets, accessed on 24 March 2022). The acquisition time of imagery was 19 August 2021, with a cloud cover of 5.51%. Land-cover analysis and NDVI were chosen in 2021 and on a specific date, as this research tried to utilize the latest and best data specifications with a relatively low cloud cover. Since to produce a good landslide hazard prediction model, the latest land cover and NDVI data are necessary [33]. The imagery has a spatial resolution of 30 m on a multispectral sensor [34]. The Landsat-8 OLI TIRS imagery data were used to extract land cover and triggering factor parameters. The extracted land cover factor was the vegetation index using the $NDVI$ algorithm. $NDVI$ can be used to estimate the level of greenery density in an area of land [35]. The $NDVI$ algorithm can be seen in Equation (1), where NIR is the near infrared band, and R is the red band of the Landsat-8 [36].

$$NDVI = \frac{NIR - R}{NIR + R} \tag{1}$$

This research applied the supervised classification random forest method to generate land cover. It comprised water bodies, forests, vegetation (including agricultural land), built-up land, and bare land. The classification was reasonably well-accepted, with overall accuracy and kappa accuracy values of 0.89 and 0.86, respectively. The classification results indicated that forests cover 38% of the study area. According to the vegetation density, the result showed that the vegetation includes a variety of land cover, namely, water bodies, low-density vegetation, medium-density vegetation, and high-density vegetation, with a density index between −0.46 and 0.83 [37].

Annual Rainfall Data

Annual rainfall data were acquired by calculating daily CHIRPS data with a resolution of 5 km retrieved from 2012 to 2021. This research used CHIRPS data as a database for annual rainfall because the number of rain gauge station points covering the study area is very limited. Consequently, the rain gauge station data were less representative. While

CHIRPS is a terrestrial rainfall database that combining three types of rainfall information (global climatology, satellite-based rainfall estimates, and in-situ rainfall observations) [38]. It can be accessed at https://data.chc.ucsb.edu/products/CHIRPS-2.0/ (accessed on 24 March 2022). Retrieval and processing of this dataset were carried out using the Google Earth engine. Following the process, raster extraction produced 190 rainfall grid points, which were assumed to be rainfall measuring points. Then, this research applied ordinary kriging to generate rainfall value over the study area. Based on the average annual rainfall data, the study area has 1750.56–3338.21 mm/year.

River Net Data

River net data were obtained from a topographic map produced by the Indonesian Geospatial Information Agency. The river net data have a scale of 1:25,000 and were published in 1999. These data are the latest data owned by the Indonesian Geospatial Information Agency. River net data were used to extract the hydrological factor parameters composed of the distance from the river and the density of the river. Euclidean distance analysis was carried out to measure the distance parameter from the river. Based on the distance parameter from the river, the study area has a distance value from the river between 0 and 5055.16 m. A line density analysis proceeded with units of km/km^2 to generate river density. The result demonstrated that the river density has a value of 0–6.58 km/km^2.

2.3. Methods

This research applied three machine learning algorithms composed of random forest, naïve Bayes, and k-nearest neighbor to compare their performance in generating landslide susceptibility analysis. Figure 3 illustrates the workflow of this research. In general, the landslide susceptibility analysis consisted of 3 major steps: (1) conditioning factor parameters preparation, (2) modelling, and (3) model evaluation.

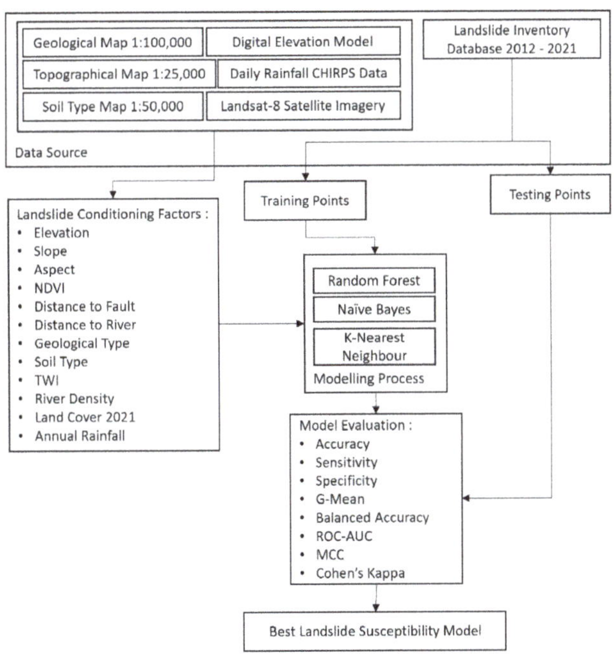

Figure 3. Research workflow of the comparison of landslide susceptibility prediction using RF, NB, and KNN algorithms.

Random forest is an ensemble learning model from a set of decision trees (DTs). Each DT depends on a sample of independent data values, and the distribution of each decision tree is the same [39]. RF is effective for predictions, as it uses the strength of each DT and its correlation and is less sensitive to the problem of over-fitting [40]. It works by performing a majority voting of the overall results of each DT. Equation (2) denotes the RF algorithm, where $\hat{C}_{rf}$ is the class of random forest results, and the hat operator in $\hat{C}$ indicates that the class is the estimated class; x is an input vector; and $\hat{C}_n$ is the predictive class of the nth tree in a random forest [41].

$$\hat{C}_{rf} = majority\ vote\ \{\hat{C}_n(x)\}_{n=1}^{N} \tag{2}$$

The k-nearest neighbor (KNN) is a machine learning algorithm utilizing neighboring techniques in determining the class of a point [42]. A point is classified based on its closest neighbors to the training data. KNN is categorized as a non-parametric ML model because the computational process does not depend on data distribution [12]. The determination of the shortest distance between the new data and the training data commonly utilizes Euclidean distance (Equation (3)), where X_{iv} is the individual characteristic of i; X_{jv} is an individual characteristic of j; p is the number of sample partitions; and v is an individual sample [43].

$$d_{ij} = \sqrt{\sum_{v=1}^{p} (X_{iv} - X_{jv})^2} \tag{3}$$

Naïve Bayes (NB) is a supervised learning method based on statistical measurement for classifying purposes. NB works based on the Bayesian theorem, which is well suited for when the data have a high dimension and is not affected by the distribution of the data [44]. NB is a simple form of a Bayesian network, with all variables considered independent of each other [45]. Equation (4) denotes the NB algorithm for landslide susceptibility modeling, where x is the parameter of the factors causing landslides; y is the classification variable for landslides and non-landslides; $P(y_i)$ is the probability of y_i; and $P(x_i/y_i)$ is a posterior probability that can be calculated by Equation (5) [10].

$$y = \underset{y_i = (landslide,\ non-landslide}{argmax P(y_i)} \prod_{i=1}^{14} P(x_i/y_i) \tag{4}$$

$$P(x_i/y_i) = \frac{1}{\sqrt{2\pi\sigma}} e^{\frac{-(x_i-\mu)^2}{2\sigma^2}} \tag{5}$$

For landslide susceptibility modelling, a stack raster ensures that all parameters are in the exact resolution. Therefore, this research extracts landslide occurrence points for each parameter and conducts a normalization process using the z-score calculation so that all numeric data are in the same dimension (Equation (6)), where X is the value of data, namely, the average value of all the data; and S is the standard deviation of the overall data [46].

$$Z = \frac{X - \overline{X}}{S} \tag{6}$$

Following the normalization process, splitting is performed to separate training and testing data. The training data are used to generate prediction models, while testing is used to evaluate the built models. The ratios between training and testing are 70:30, 60:40, and 50:50 for applying the NB and K-NN algorithms. Moreover, the RF algorithm uses a ratio of 70:30. In general, specific rules in determining the splitting ratio scenario between training and testing data do not exist, since each machine learning algorithm has its optimum splitting ratio to perform the best model. However, some splitting ratio schemes which are commonly used are 50:50, 60:40, and 70:30. The KNN and NB algorithms use these three scenarios to obtain optimum model accuracy [10,22,41,47,48]. Unlike the RF algorithm, previous research with the same physical area characteristics showed that the RF algorithm had maximum accuracy when using a splitting ratio of 70:30 [22]. Therefore, the RF algorithm only used a splitting ratio of 70:30 in this research.

After the modeling process, eight evaluation parameters comprising ROC (receiver operator characteristic), AUC (area under curve), accuracy (ACC), sensitivity (SN), specificity (SP), balanced accuracy (BA), geometric mean (GM), Cohen's kappa (CK), and Matthew's correlation coefficient (MCC) were used to assess the performance of each model. The evaluation values were obtained based on the confusion matrix of four predicted labels, which consisted of tp and fp for the number of positive data samples and tn and fn for the number of negative data samples. Table 2 denotes the equations and objectives of each evaluator.

Table 2. Metric evaluator equation and each objective.

Metric	Equation	Objective
ACC	$\frac{tp+tn}{tp+fp+tn+fn}$	Indicates the ratio of correct prediction to the total number of evaluation samples [49].
SN	$\frac{tp}{tp+fn}$	Measures the fraction of correctly classified positive patterns [49].
SP	$\frac{tn}{tn+fp}$	Measures the fraction of correctly classified negative patterns [49].
GM	$\frac{sn+sp}{2}$	Measures the average sensitivity (sn) obtained under each class [50].
BA	$\sqrt{sn \times sp}$	Measures the roots of the products sn and sp [50].
CK	$\frac{2\times((TP\times TN)-(FP\times FN))}{((TP+FP)\times(FP+TN))+((TP+FN)\times(FN+TN))}$	Consistency value between 2 raters (observation and prediction) [51].
MCC	$\frac{(TP\times TN)-(FP\times FN)}{\sqrt{(TP+FP)\times(TP+FN)\times(TN+FP)\times(TN+FN)}}$	Measures the performance of the classification algorithm through the correlation between observations and predictions [51].
ROC-AUC	$AUC=\frac{S_p-n_p(n_n+1)/2}{n_p n_n}$	The ROC curve is built based on sn (sb-Y) with sp (sb-X), and AUC is an integral ROC [10].

3. Results

3.1. Continuous Data Parameter Normality Characteristics

Some machine learning algorithms assume that the training data are normally distributed, so that identifying the normality characteristics of the data for evaluating the application of machine learning algorithms is necessary. This research uses a non-parametric alternative statistical Kolmogorov–Smirnov test (K-S) to display the normality characteristics. The K-S test uses the cumulative distribution to determine the distribution level of data [52]. Moreover, the K-S test is reliable for various purposes to efficiently establish Goodness of Fit [53]. Table 3 denotes the results of the K-S test from the landslide and non-landslide training dataset.

Table 3. Result of K-S test training dataset.

Parameter	Landslide Training Point			Non-Landslide Training Point		
	D-Value	*p*-Value	Normal Distribution	D-Value	*p*-Value	Normal Distribution
River Density	0.167	2.32×10^{-6}	No	0.096	0.04519	No
Annual Rainfall	0.165	3.47×10^{-6}	No	0.104	0.01946	No
Distance to Fault	0.258	6.76×10^{-16}	No	0.151	3.92×10^{-5}	No
Elevation	0.107	0.01467	No	0.101	0.02766	No
Distance to River	0.192	5.39×10^{-7}	No	0.152	3.03×10^{-5}	No
NDVI	0.175	5.39×10^{-7}	No	0.149	4.74×10^{-5}	No
Slope	0.140	2.04×10^{-4}	No	0.113	7.78×10^{-3}	No
TWI	0.088	9.00×10^{-2}	Yes	0.205	8.66×10^{-10}	No

Hypothesis: H_0 = normally distributed; H1 = not normally distributed.

The error rate in decision making is set to 5% = 0.05, with the decision-making criteria using sig.α or p-value. If sig.$\alpha < \alpha$, then H_0 is rejected [54]. The results of the K-S test shows that all parameters in non-landslide training are not normally distributed. Moreover, in the landslide training data, only the TWI parameter is normally distributed with a p-value of 0.09.

3.2. Landslide Susceptibility Modeling Results

Before performing KNN for the landslide susceptibility model, it is required estimating the value of k to generate the number of nearest neighbors considered from a point. The estimation of the k-value used the cross-validation technique. Cross-validation is performed with three iterations to optimize the accuracy. Figure 4 illustrates the results of the measure of the k-value. Based on the cross-validation results for the estimated k-value for the 50:50 KNN model, the optimum k-value was 3, with a maximum accuracy of 0.814, while for the 60:40 KNN model, the value of k produces a maximum accuracy of 3 with a maximum accuracy of 0.796. The KNN 70:30 model had a maximum accuracy when the k value was 7, with a maximum accuracy value of 0.817.

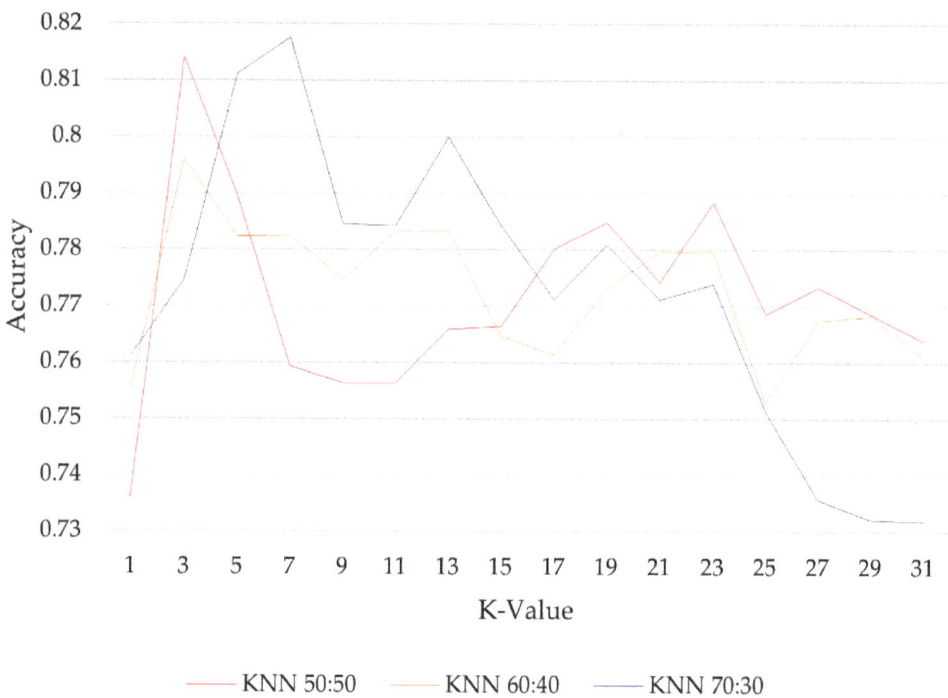

Figure 4. Cross-validation to obtain the best k value for each scenario of the KNN algorithm.

The KNN yields probability values of landslides from 0 to 1. The average probability values of the KNN 50:50, KNN 60:40, and KNN 70:30 models are 0.449, 0.338, and 0.365, respectively. The probability values are then classified into low susceptibility (0–0.3), moderate susceptibility (0.3–0.6), and high susceptibility (0.6–1) [42]. Figure 5 illustrates the result of each scenario, where (A), (B), and (C) demonstrate the results of landslide susceptibility modeling using the KNN. The 50:50 KNN model indicates that high susceptibility dominates the study area with an area of 147,319.29 km^2 (42%) as opposed to the 60:40 KNN model being dominated by moderate susceptibility with an area of 195,318.54 km^2 (56%). For the 70:30 KNN model, the study area was dominated by low susceptible with an area of 180,326.16 km^2 (51%).

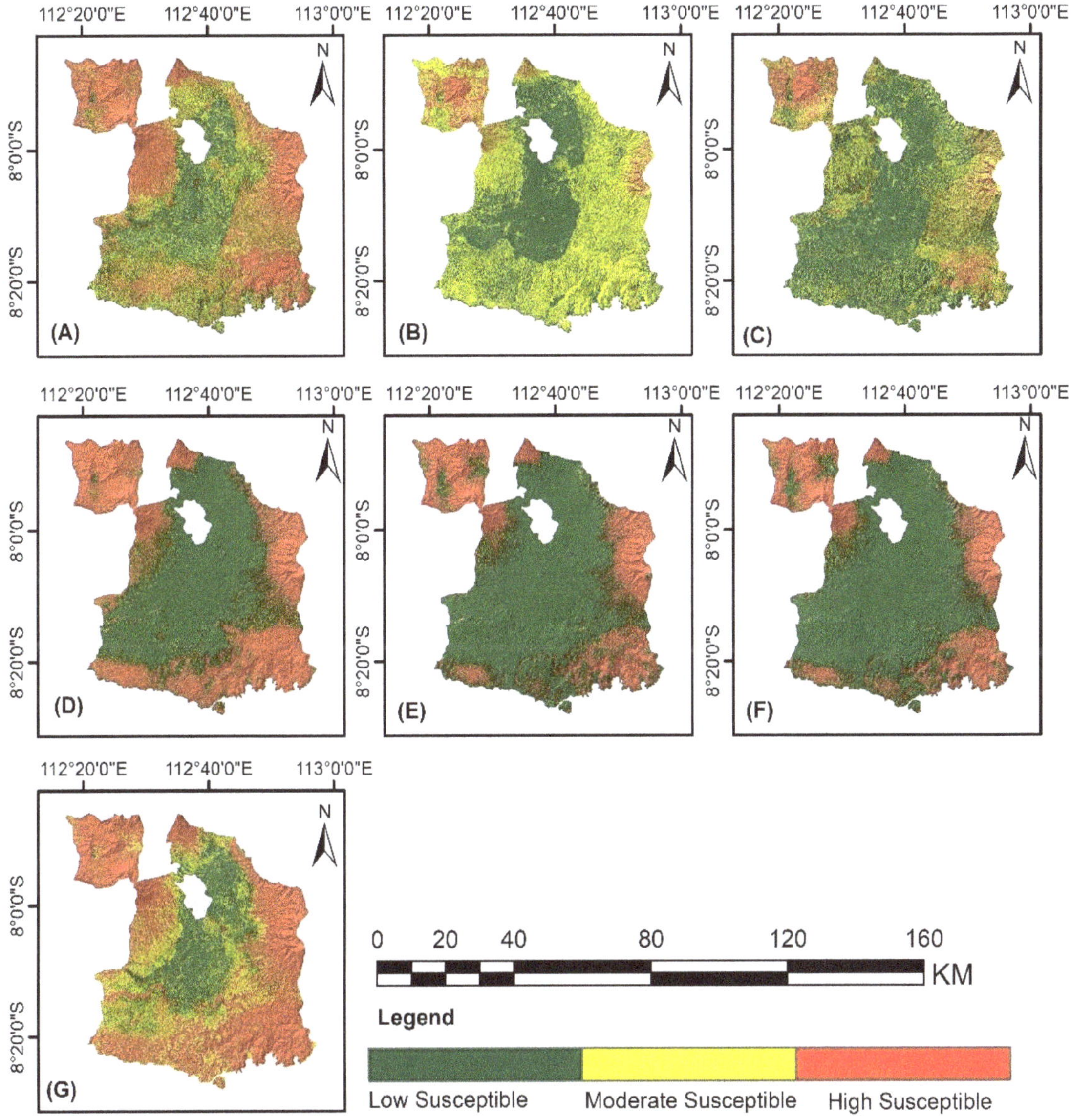

Figure 5. Landslide susceptibility modeling result of 7 models. (**A**) KNN 50:50 model. (**B**) KNN 60:40 model. (**C**) KNN 70:30 model. (**D**) NB 50:50 model. (**E**) NB 60:40 model. (**F**) NB 70:30 model. (**G**) RF 70:30 model.

Likewise, the NB algorithm applies three scenarios between training and testing composed of 50:50, 60:40, and 70:30. The results indicates that the probability values of landslides in the NB 50:50 model has a range of 6.24×10^{-10} to 1, with an average of 0.451. Moreover, the NB 60:40 model generates a probability range between 5.68×10^{-14} and 1, with an average of 0.424. In the NB 70:30 model, the probability values of landslides lay between 5.87×10^{-12} and 1, with an average value of 0.299. In addition, the NB models also classifies the susceptibility. Figure 6 illustrates the proportion of the study area based on the probability classification of landslides. The classification of all scenarios showes that low susceptibility dominated the study area; 51% of the study area was classified as low susceptibility in the NB 50:50 model, with 179,493.57 km^2, as opposed to the NB 60:40 model with 231,354.63 km^2 (66%). Moreover, the area with low susceptibility on the NB 70:30 model was 235.410.39 km^2 (67%). On the contrary, only RF 70:30 generates more than 50% high susceptibility.

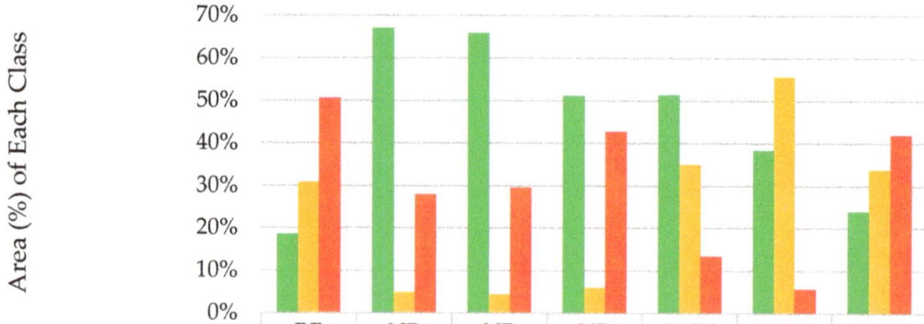

	RF 70:30	NB 70:30	NB 60:40	NB 50:50	KNN 70:30	KNN 60:40	KNN 50:50
Low Susceptible	19%	67%	66%	51%	51%	38%	24%
Moderate Susceptible	31%	5%	5%	6%	35%	56%	34%
High Susceptible	51%	28%	30%	43%	13%	6%	42%

Classes of Landslide Susceptibility Model

Figure 6. Landslide susceptibility classes' percentages for each model.

The RF algorithm only implements a scenario between training and testing (70:30) to produce a landslide susceptibility map. In the RF modelling, it was necessary to estimate the best mtry, which is the number of random variables, before establishing a DT. The best mtry estimation agrees using a cross-validation technique. Figure 7 illustrates the results of the cross-validation.

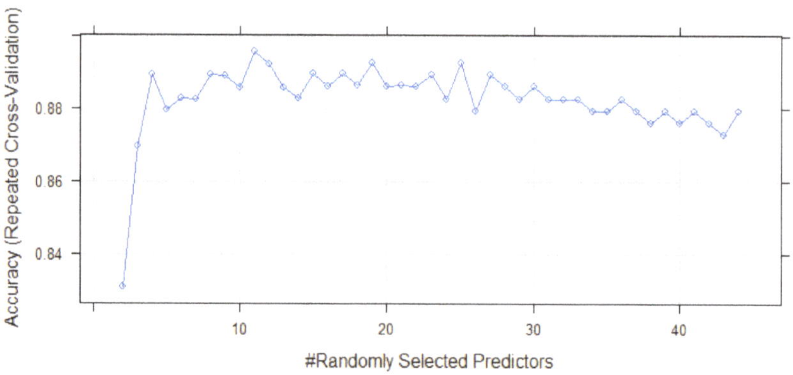

Figure 7. Cross-validation results to obtain the best mtry of the RF model.

According to results, the mtry value which produces the highest accuracy (0.896) of the RF model was 11. The RF 70:30 generates probability values from 0.01 to 1, with an average of 0.595. After generating the RF model, it classified the level of susceptibility to landslides based on the respective value. The result indicates that high susceptibility dominated 51% of the study area, with 177,208.83 km^2 distributed over the edge.

4. Discussion

This research produces seven landslide susceptibility models. All models indicates that high levels of landslide susceptibility located on the edge of the study area, except for the

KNN 60:40 and KNN 70:30 models. The probability values generated by all models ranged from 0 to 1. The probability values of landslides approaching 0 indicated no susceptibility to landslides. On the contrary, once the probability value was close to 1, it refers to an increased susceptibility to landslides [55]. Then the probability value can be classified into three levels of landslide susceptibility, composed of low, moderate, and high when the probability value ranged from 0 to 0.3, from 0.3 to 0.6, and from 0.6 to 1, respectively.

The evaluation is conducted towards training and testing data. For training data, ACC and CK were measured. Figure 8 depicts the evaluation results of each model using these parameters. The RF 70:30 model generates the highest values for ACC and CK, with values of 0.915 and 0.819, respectively. In comparison, the NB60:40 model yields the highest evaluation value for ACC and CK, with values of 0.863 and 0.691, respectively. For the KNN algorithm, the KNN50:50 model produces the highest ACC and CK values among all the scenarios, with ACC and CK values of 0.823 and 0.597, respectively.

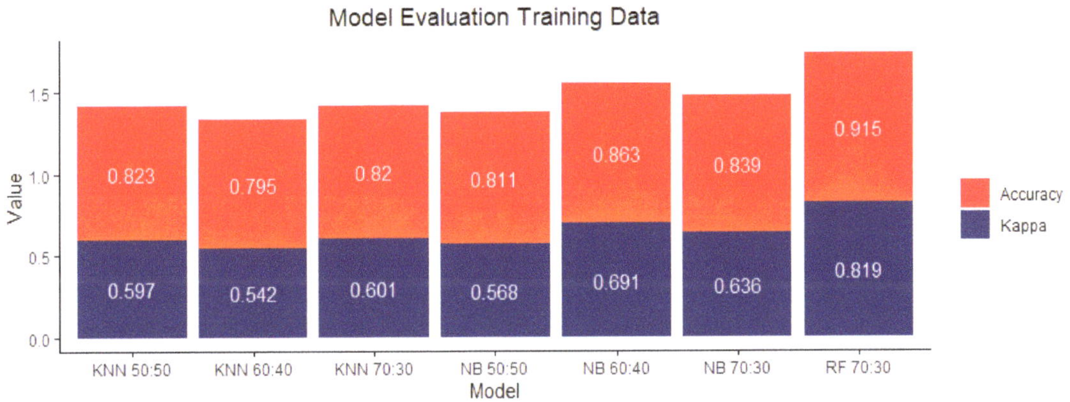

Figure 8. Result of evaluation of each model on training data.

Eight parameters (ACC, SN, SP, BA, GM, MCC, CK, and ROC–AUC) are used to evaluate the performance of each model. Figure 9 depicts the results of the evaluation of each model using these parameters. The RF 70:30 model generates the highest values for six evaluation parameters, namely, ACC, SN, GM, BA, CK, and MCC, with values of 0.884, 0.765, 0.863, 0.857, 0.749, and 0.876, respectively. Moreover, for the SP parameter, the NB 50:50 and KNN50:50 models have the highest value among the other models, namely, 0.977. The NB 50:50 model had the lowest performance with six evaluation parameters, namely, ACC (90.806), SN (0.536), GM (0.757), BA (0.724), CK (0.556), and MCC (0.601). Moreover, the KNN 70:30 model obtains the lowest performance for the SP evaluation parameter (0.846).

ROC–AUC measures the performance of each model for distinguishing landslide and non-landslide as a binary value. ROC–AUC is the relationship curve between SP and SN. Figure 10 illustrates the results of the ROC–AUC of training and testing data. Based on the ROC graph, all models have an AUC of more than 0.7, which indicates that the model had good performance [46]. Compared to testing data of other models, the RF 70:30 generates the highest AUC of 0.943. On the contrary, the model with the lowest AUC value was KNN70:30 (0.852), meaning that the performance of KNN in identifying landslides was low. In line with the AUC value in the testing data, the AUC in the RF algorithm training data and the 70:30 scenario produces the highest AUC compared to other models, with a value of 1. Moreover, the lowest AUC value for training data is obtained with KNN 60:40, with an AUC value of 0.922. In addition, KNN 70:30 produced the lowest values of ACC, SP, CK, and MCC, with respective values of 0.814, 0.846, 0.611, and 0.624. Moreover, the most optimum scenario of the KNN splitting ratio between training and testing was 50:50, which produces the highest values in five of the eight evaluation parameters. It comprises ACC (0.833), SP (0.977), CK (0.625), MCC (0.658), and AUC (0.881).

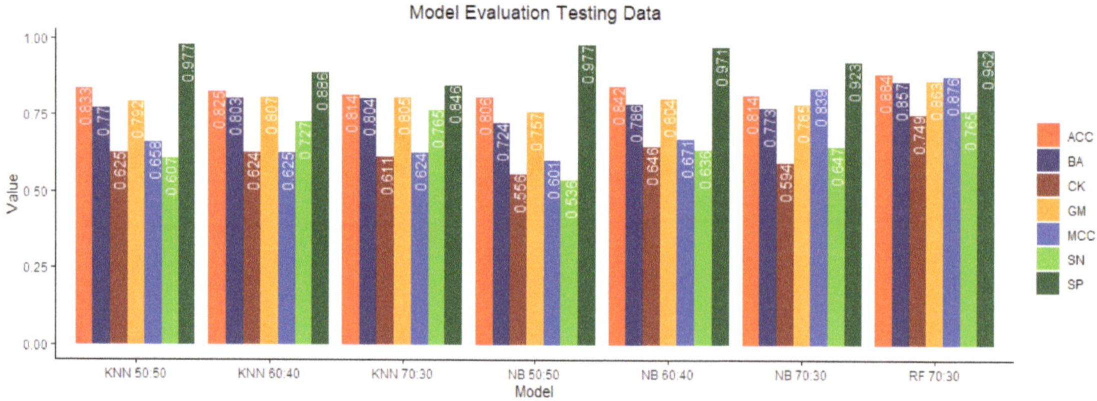

Figure 9. Result of evaluation of each model.

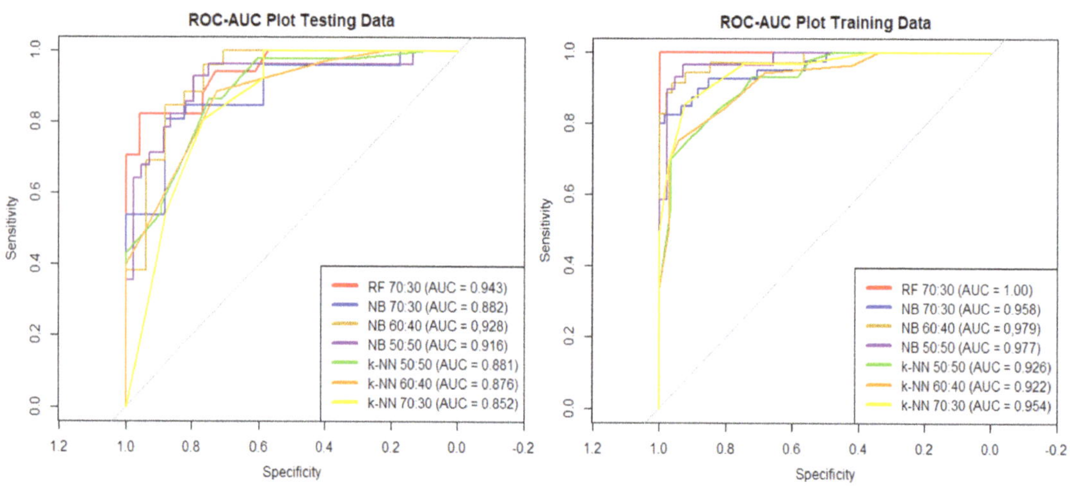

Figure 10. ROC–AUC plot of training and testing data.

Figure 11 depicts the relative variable contribution degree of each model. In general, slope led to the highest relative contribution degree in all models, with a value of 100%. However, each model produces a different sequence of contribution degrees on each parameter. Looking at the lowest contribution degree, the NDVI has the lowest relative contribution degree in the RF 70:30 and NB 70:30 models, with merely 0.44% and 5.28%, respectively. For land use, it possesses the lowest relative contribution degree in the NB 50:50 model (7.31%). Moreover, the proportion of geological type in the NB 60:40 model was just above ten (10.13%). In KNN 50:50 and KNN 60:40 models, the soil type parameter yielded a relative contribution degree of 10.12% for the KNN60:40 model as opposed to the KNN50:50 model (8.64%). In the KNN 70:30 model, the parameter with the lowest relative contribution degree was aspect, with a value of 9.24%.

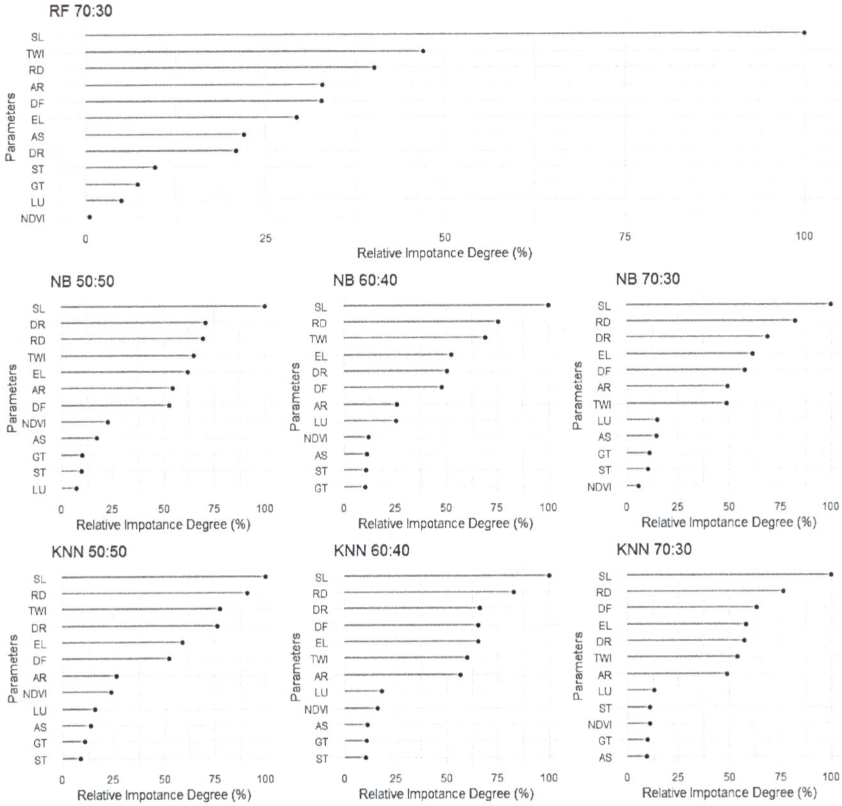

Figure 11. Relative contribution degree of each model (SL = slope; RD = river density; DR = distance to river; EL = elevation; AR = annual rainfall; DF = distance to fault; AS = aspect; GT = geological type; ST = soil type; LU = land use).

Among all models, RF was the appropriate model to discriminate non-landslide areas from landslide areas based on landslide conditioning factors, considering the model evaluation performance and accuracy [56]. The evaluation parameters comprise ROC–AUC, ACC, SP, SN, GM, BA, CK, and MCC. According to the evaluation results, RF 70:30 was the best model with the highest value of seven of the eight evaluation parameters. In the application of the NB algorithm, the optimal ratio between training and testing scenarios is 60:40, as it generated the highest value in five of eight parameters [57]. Moreover, the scenario with the lowest performance wis the 50:50, since it generates the lowest value in six of the eight evaluation parameters. RF performes the best in this research, followed by KNN and NB sequentially. In addition to implementing algorithms using similar conditioning factors, KNN, RF, and NB yielded good performance, with AUC values of 0.8903, 0.8690, and 0.8639, respectively [58]. The excellent performance of these three algorithms in predicting landslides was also approved by additional conditioning factors such as curvature, lithology, road ratios, and forest area ratios [59].

The KNN algorithm shows the lowest performance compared to the best models of the other algorithms. However, compared to the overall splitting ratio scheme, the NB algorithm produces the lowest performance compared to the KNN and RF algorithms. Based on the results of the continuous data normality test in the previous sub-section, the training data do not normally distribute. Otherwise, the NB algorithm assumes that the data does not normally distribute [60]. Therefore, this research applies numerical training normality tests on the data to determine the normality of data distribution using the Kolmogorov–Smirnov (K-S) test. Eventually, the K-S test proves that NB's performance depends on the training data distribution.

The most influential parameter of all models is the slope. In several studies related to landslide susceptibility modeling using machine learning algorithms, the slope parameter dominantly leads to the highest relative contribution as opposed to other parameters [8,11,47]. Other research produces different contribution levels, such as elevation [61] and rainfall [62], while the slope parameter has a contribution level in the fifth order. In this research, the distribution of landslide training data dominantly occurred on slopes between 8° and 30°, which are classified as rather steep slopes. According to the influence of topography on the landslides occurrence, ref. [63] found that landslides tend to occur at slope values between 15° and 25°, as the slope angle controls shear forces and stresses on a slope [64]. The slope angle level affects how much shear stress there is and how low the level of slope stability is [65]. As the slope angle increases, the tangential stress increases in the consolidated soil layer, while the axial stress (shear strength increases on a steeper slope) and the slope stability level decrease accordingly. As a result, slope angle triggers the potential for rock mass increase and ultimately triggers soil movement down the slope [64]. Variations in the slope value affect the magnitude of the stress on the potential shear surface and determine the deformation mechanism [66]. Furthermore, the saturation of the fill slope causes the rock mass to slide down the slope because the high compressibility and mobility of air in the unsaturated void allow the fill slope to initiate undrained failure. The saturation level on the fill slope is determined by the type of soil and the hydrological conditions [67,68].

The lowest relative contribution level is divergent in each model. The NDVI had the lowest relative contribution level in the RF 70:30 and NB 70:30 models. The results of these two models indicates that the model is less associated with NDVI data. The landslide training data tend to occur at NDVI values between 0.24 and 0.787 which is classified as low to high vegetation density [37]. According to the influence of vegetation density in identifying landslides occurrence, it does not significantly contribute [69]. On the other hand, ecological damage, indicated by low vegetation density, will trigger landslides. Therefore, it is necessary to consider ecological restoration as the primary means of preventing and controlling landslides [70]. Vegetation can be an effective measure for mitigating landslides, as it can promote the shear strength of the soil through a series of mechanical and hydrological effects [71].

The land cover parameter is found to be the lowest relative contribution level in the NB 50:50 model. This research plots all the training datasets regardless of the land cover type. However, the locations were mainly in a forest area, and built-up areas, including roads, were non-significant, as the spatial resolution of the imagery is 30 m while the road width usually is less than 30 m. As a result, a misclassification possibly occurred due to the road being covered by vegetation. Hence, land cover is also an essential factor in the assessment of landslide susceptibility [72]. Changes in land cover, such as deforestation, which is used to support various human activities, can increase slope instability, which causes landslides [73].

The soil type parameter has the lowest relative contribution level in the KNN 50:50 and 60:40 models; 57.90% of landslides occurred on Gleisol soil of the study area. Gleisol has a loamy texture, as it is formed in a basin area and is affected by excessive water [74]. Loamy soil increases the potential of landslides because the loose soil is relatively soft after being exposed to water and breaks when the air temperature is too high [75]. There is a relationship between soil type and landslide occurrence regarding geotechnical properties [76]. The geotechnical properties consist of hydraulic conductivity, infiltration rate, runoff and increased pore water pressure on the slope, volume change, and the rate of decrease in shear strength during rain [77]. These geotechnical properties are also related to the type of geology of an area [78]. Other areas that have the potential for landslides are sandy slope areas. When sandy slope areas also have the characteristics of an area with high rainfall, slope instability will increase, and ultimately landslides will occur [79]. In the NB 60:40

model, the geological parameter contributes as the lowest order. Based on the distribution of the training dataset, landslides tend to occur with the characteristics of rocks originating from volcanic deposits. Volcanic deposits are easily weathered rocks, especially tuff, which is highly weathered to wholly weathered. On the other hand, previous studies have proved that geology or lithology contributes relatively significantly [58,69,80].

In the 70:30 KNN model, aspect contributes as the lowest order, which is opposed to other research finding that aspect has a relatively significant contribution level [58,69,80]. In this research, the landslides dominantly occurred on slopes facing northeast, with a percentage of 28%. The direction of the slope is related to the amount of sunlight intensity. In areas continuously exposed to direct sunlight, the organic content of the soil composition in the area is low, which causes the area to be easily dispersed and ultimately triggers landslides. The northern aspect is more susceptible to landslides, where landslides occur in the southern hemisphere, and the southern aspect is more susceptible to landslides in the northern hemisphere and vice versa [81]. In the northern hemisphere, the direction of the slopes facing south has a higher intensity of sunlight than slopes facing north. In areas exposed to direct sunlight continuously, the organic content of the soil constituents in the area is low, which triggers the area to easily disperse, and ultimately causes landslides [81]. However, aspect does not contribute significantly to this research model, since Indonesia is a country situated in the equatorial region. As a result, sunlight intensity is almost the same in all directions [82].

Evaluation of landslide susceptibility is carried out to accurately determine areas that are susceptible to landslides [83]. Mistakes in determining landslide susceptibility can lead to false judgment, resulting in loss of life and property. The landslide susceptibility map becomes fundamental for evaluating sustainable disaster mitigation issues [83]. A machine learning approach can accurately and efficiently predict the level of landslide susceptibility. The application of machine learning to evaluate landslide susceptibility has not been widely implemented in Indonesia. In addition, the landslide susceptibility map in the study area still applies the conventional scoring method with low accuracy. As a result, machine learning has the potential to be implemented. Moreover, machine learning can efficiently update landslide susceptibility maps continuously. Determining the splitting ratio between training and testing data is crucial in determining the model's accuracy. Hence, this research is expected to provide recommendations for further research using the RF, KNN, and NB algorithms. Subsequently, it can save time in the process of determining the splitting ratio between training and testing for landslide susceptibility modelling.

5. Conclusions

This research compares the performance of the RF, KNN, and NB algorithms in producing a spatial model of landslide susceptibility in Malang Regency, East Java Province, Indonesia. According to the results, the RF algorithm dominantly led to the highest value of evaluation parameters, composed of ACC, SN, GM, BA, CK, and MCC, with respective values of 0.884, 0.765, 0.863, 0.857, 0.749, and 0.876. In addition, RF generates the best performance, with an AUC of 0.943. On the other hand, the optimum splitting ratios between the training and testing data for the NB and KNN algorithms in the case study were 60:40 and 50:50, with AUC values of 0.928 and 0.916, respectively. Slope contributes as the highest relative contribution degree for all the models, with the same value of 100%. According to the best model, high susceptibility dominates Malang Regency, which includes 51% of the study area. Thus, the predictive model can assist policymakers in promoting sustainable mitigation for the potential location. However, optimization methods and prior knowledge concerning selecting landslide conditioning factors and landslide occurrence inventories are necessary to improve prediction accuracy. This research recommends utilizing multi-temporal data for more complex analyses in future research.

Author Contributions: Conceptualization, Nurwatik Nurwatik and Muhammad Hidayatul Ummah; methodology, Nurwatik Nurwatik, and Muhammad Hidayatul Ummah; software, Muhammad Hidayatul Ummah; validation, Muhammad Hidayatul Ummah; formal analysis, Nurwatik Nurwatik;

resources, Muhammad Hidayatul Ummah; data curation, Muhammad Hidayatul Ummah; writing—original draft preparation, Muhammad Hidayatul Ummah; writing—review and editing, Nurwatik Nurwatik; visualization, Muhammad Hidayatul Ummah; supervision, Nurwatik Nurwatik and Mohammad Rohmaneo Darminto; project administration, Agung Budi Cahyono; funding acquisition, Agung Budi Cahyono and Jung-Hong Hong. All authors have read and agreed to the published version of the manuscript.

Funding: This research was funded by the Directorate of Research and Community Service (DRPM-ITS), Institut Teknologi Sepuluh Nopember, grant number 1623/PKS/ITS/2022.

Data Availability Statement: The data presented in this study are openly available at https://tanahair.indonesia.go.id/portal-web (accessed on 25 February 2022), https://tanahair.indonesia.go.id/demnas/ (accessed on 26 February 2022), and https://developers.google.com/earth-engine/datasets (accessed on 24 March 2022).

Acknowledgments: The author acknowledges with profound thanks DRPM-ITS for its technical and financial support. The author also acknowledges the tremendous support of the Geoinformatics Laboratory, Department of Geomatics Engineering, and ITS.

Conflicts of Interest: The authors declare no conflict of interest.

References

1. Skempton, A.W.; Hutchinson, J. Stability of natural slopes and embankment foundations. In Proceedings of the 7th International Conference on Soil Mechanics and Foundation Engineering, Mexico City, Mexico, 29 August 1969; pp. 291–340.
2. Muntohar, A. *Tanah Longsor: Analisis-Prediksi-Mitigasi*, 1st ed.; Universitas Muhammadiyah Yogyakarta: Kasihan, Indonesia, 2012.
3. Keefer, D.K. Investigating landslides caused by earthquakes—A historical review. *Surv. Geophys.* **2002**, *23*, 473–510. [CrossRef]
4. Lu, P.; Bai, S.; Casagli, N. Investigating spatial patterns of persistent scatterer interferometry point targets and landslide occurrences in the Arno River basin. *Remote Sens.* **2014**, *6*, 6817–6843. [CrossRef]
5. Hong, H.; Naghibi, S.A.; Pourghasemi, H.R.; Pradhan, B. GIS-based landslide spatial modeling in Ganzhou City, China. *Arab. J. Geosci.* **2016**, *9*, 1–26. [CrossRef]
6. El Naqa, I.; Murphy, M.J. What is machine learning? In *Machine Learning in Radiation Oncology*; Springer: Berlin/Heidelberg, Germany, 2015; pp. 3–11.
7. Ahmad Hania, A. Mengenal Artificial Intelligence, Machine Learning, & Deep Learning. Available online: https://amt-it.com/mengenal-perbedaan-artificial-inteligence-machine-learning-deep-learning/ (accessed on 3 July 2022).
8. Youssef, A.M.; Pourghasemi, H.R. Landslide susceptibility mapping using machine learning algorithms and comparison of their performance at Abha Basin, Asir Region, Saudi Arabia. *Geosci. Front.* **2021**, *12*, 639–655. [CrossRef]
9. Yanbin, M.; Hongrui, L.; Lin, W.; Wengang, Z.; Zhengwei, Z.; Haiqing, Y.; Luqi, W.; Xingzhong, Y. Machine learning algorithms and techniques for landslide susceptibility investigation: A literature review. *J. Civ. Environ. Eng.* **2022**, *44*, 53–67.
10. He, Q.; Shahabi, H.; Shirzadi, A.; Li, S.; Chen, W.; Wang, N.; Chai, H.; Bian, H.; Ma, J.; Chen, Y.; et al. Landslide spatial modelling using novel bivariate statistical based Naïve Bayes, RBF Classifier, and RBF Network machine learning algorithms. *Sci. Total Environ.* **2019**, *663*, 1–15. [CrossRef] [PubMed]
11. Al-Najjar, H.A.H.; Pradhan, B. Spatial landslide susceptibility assessment using machine learning techniques assisted by additional data created with generative adversarial networks. *Geosci. Front.* **2021**, *12*, 625–637. [CrossRef]
12. Abraham, M.T.; Satyam, N.; Lokesh, R.; Pradhan, B.; Alamri, A. Factors Affecting Landslide Susceptibility Mapping: Assessing the Influence of Different Machine Learning Approaches, Sampling Strategies and Data Splitting. *Land* **2021**, *10*, 989. [CrossRef]
13. Adab, H.; Atabati, A.; Oliveira, S.; Moghaddam Gheshlagh, A. Assessing fire hazard potential and its main drivers in Mazandaran province, Iran: A data-driven approach. *Environ. Monit. Assess.* **2018**, *190*, 670. [CrossRef]
14. Lin, J.; He, P.; Yang, L.; He, X.; Lu, S.; Liu, D. Predicting future urban waterlogging-prone areas by coupling the maximum entropy and FLUS model. *Sustain. Cities Soc.* **2022**, *80*, 103812. [CrossRef]
15. Rahmati, O.; Golkarian, A.; Biggs, T.; Keesstra, S.; Mohammadi, F.; Daliakopoulos, I.N. Land subsidence hazard modeling: Machine learning to identify predictors and the role of human activities. *J. Environ. Manage* **2019**, *236*, 466–480. [CrossRef]
16. Shahzad, N.; Ding, X.; Abbas, S. A Comparative Assessment of Machine Learning Models for Landslide Susceptibility Mapping in the Rugged Terrain of Northern Pakistan. *Appl. Sci.* **2022**, *12*, 2280. [CrossRef]
17. Laila Nugraha, A.; Sukmono, A.; Sugistu Firdau, H.S.; Lestari, S. Study of Accuracy in Landslide Mapping Assessment Using GIS and AHP, A Case Study of Semarang Regency. *KnE Eng.* **2019**. [CrossRef]
18. Bachri, S.; Sumarmi; Yudha Irawan, L.; Utaya, S.; Dwitri Nurdiansyah, F.; Erfika Nurjanah, A.; Wahyu Ning Tyas, L.; Amri Adillah, A.; Setia Purnama, D. Landslide Susceptibility Mapping (LSM) in Kelud Volcano Using Spatial Multi-Criteria Evaluation. *IOP Conf. Ser. Earth Environ. Sci.* **2019**, *273*, 012014. [CrossRef]
19. Bachri, S.; Shrestha, R.P.; Yulianto, F.; Sumarmi, S.; Utomo, K.S.B.; Aldianto, Y.E. Mapping landform and landslide susceptibility using remote sensing, gis and field observation in the southern cross road, Malang regency, East Java, Indonesia. *Geosciences* **2021**, *11*, 4. [CrossRef]

20. Ghasemian, B.; Shahabi, H.; Shirzadi, A.; Al-Ansari, N.; Jaafari, A.; Kress, V.R.; Geertsema, M.; Renoud, S.; Ahmad, A. A Robust Deep-Learning Model for Landslide Susceptibility Mapping: A Case Study of Kurdistan Province, Iran. *Sensors* **2022**, *22*, 1573. [CrossRef]
21. Pham, B.T.; Vu, V.D.; Costache, R.; Van Phong, T.; Ngo, T.Q.; Tran, T.-H.; Nguyen, H.D.; Amiri, M.; Tan, M.T.; Trinh, P.T.; et al. Landslide susceptibility mapping using state-of-the-art machine learning ensembles. *Geocarto Int.* **2022**, *37*, 5175–5200. [CrossRef]
22. Darminto, M.R.; Widodob, A.; Alfatinahc, A.; Chuc, H.-J. High-Resolution Landslide Susceptibility Map Generation using Machine Learning (Case Study in Pacitan, Indonesia). *Int. J. Adv. Sci. Eng. Inf. Technol.* **2021**, *11*, 369–379. [CrossRef]
23. Tsangaratos, P.; Ilia, I. Comparison of a logistic regression and Naïve Bayes classifier in landslide susceptibility assessments: The influence of models complexity and training dataset size. *Catena* **2016**, *145*, 164–179. [CrossRef]
24. Xu, C.; Dai, F.; Xu, X.; Lee, Y.H. GIS-based support vector machine modeling of earthquake-triggered landslide susceptibility in the Jianjiang River watershed, China. *Geomorphology* **2012**, *145–146*, 70–80. [CrossRef]
25. Vakhshoori, V.; Pourghasemi, H.R.; Zare, M.; Blaschke, T. Landslide susceptibility mapping using GIS-based data mining algorithms. *Water* **2019**, *11*, 2292. [CrossRef]
26. Tseng, C.M.; Lin, C.W.; Hsieh, W.D. Landslide susceptibility analysis by means of event-based multi-temporal landslide inventories. *Nat. Hazards Earth Syst. Sci. Discuss.* **2015**, 3, 1137–1173.
27. Iswari, M.Y.; Anggraini, K. Demnas: Model Digital Ketinggian Nasional Untuk Aplikasi Kepesisiran. *Oseana* **2018**, *43*. [CrossRef]
28. Ronodirdjo, M.Z. *Buku Ajar Pengantar Geologi*; Duta Pustaka Ilmu: Mataram, Indonesia, 2019.
29. Varianti, E. Geologi daerah Sumberbening dan sekitarnya Kecamatan Bantur Kabupaten Malang Provinsi Jawa Timur. *J. Online Mhs. Bid. Tek. Geol.* **2019**, *1*, 1–10.
30. Wasis, W.; Sunaryo, S.; Susilo, A. Local Fault Line Tracing in Sri Mulyo Village, Dampit Sub District, Malang Regency Based on Geophysical Data. *Nat. B J. Health Environ. Sci.* **2011**, *1*, 41–50.
31. Islami, D.A.L. Al Geologi daerah Klepu dan sekitarnya, Kecamatan Sumbermanjing Wetan Kabupaten Malang, Provinsi Jawa Timur. *J. Online Mhs. Bid. Tek. Geol.* **2017**, *1*, 1–12.
32. Martins, K.G.; Marques, M.C.M.; dos Santos, E.; Marques, R. Effects of soil conditions on the diversity of tropical forests across a successional gradient. *For. Ecol. Manag.* **2015**, *349*, 4–11. [CrossRef]
33. Viet, L.D.; Chi, C.N.; Tien, C.N.; Quoc, D.N. The Effect of the Normalized Difference Vegetation Index to Landslide Susceptibility using Optical Imagery Sentinel 2 and Landsat 8. In Proceedings of the 4th Asia Pacific Meeting on Near Surface Geoscience & Engineering, Online, 30 November–2 December 2021; Volume 2021, pp. 1–5.
34. Yang, I.; Acharya, T.D. Exploring Landsat 8. Available online: https://www.researchgate.net/profile/Tri-Acharya/publication/311901147_Exploring_Landsat_8/links/589c0de6458515e5f4549e58/Exploring-Landsat-8.pdf%0Ahttp://earthobservatory.nasa.gov/IOTD/ (accessed on 20 April 2022).
35. Gessesse, A.A.; Melesse, A.M. Chapter 8—Temporal relationships between time series CHIRPS-rainfall estimation and eMODIS-NDVI satellite images in Amhara Region, Ethiopia. In *Extreme Hydrology and Climate Variability*; Melesse, A.M., Abtew, W., Senay, G.B.T.-E.H., Eds.; Elsevier: Amsterdam, The Netherlands, 2019; pp. 81–92, ISBN 978-0-12-815998-9.
36. Pettorelli, N. *The Normalized Difference Vegetation Index*; Oxford University Press: Oxford, UK, 2013; ISBN 0199693161.
37. Hashim, H.; Abd Latif, Z.; Adnan, N.A. Urban vegetation classification with ndvi threshold value method with very high resolution (vhr) pleiades imagery. *Int. Arch. Photogramm. Remote Sens. Spat. Inf. Sci.-ISPRS Arch.* **2019**, *42*, 237–240. [CrossRef]
38. Funk, C.C.; Peterson, P.J.; Landsfeld, M.F.; Pedreros, D.H.; Verdin, J.P.; Rowland, J.D.; Romero, B.E.; Husak, G.J.; Michaelsen, J.C.; Verdin, A.P. A quasi-global precipitation time series for drought monitoring. *US Geol. Surv. Data Ser.* **2014**, *832*, 1–12.
39. Breiman, L. Random forests. *Mach. Learn.* **2001**, *45*, 5–32. [CrossRef]
40. Liaw, A.; Wiener, M. Classification and regression by randomForest. *R News* **2002**, *2*, 18–22.
41. Rodriguez-Galiano, V.F.; Ghimire, B.; Rogan, J.; Chica-Olmo, M.; Rigol-Sanchez, J.P. An assessment of the effectiveness of a random forest classifier for land-cover classification. *ISPRS J. Photogramm. Remote Sens.* **2012**, *67*, 93–104. [CrossRef]
42. Cunningham, P.; Delany, S.J. K-Nearest Neighbour Classifiers-A Tutorial. *ACM Comput. Surv.* **2021**, *54*, 1–25. [CrossRef]
43. Gonçalves, D.N.S.; Gonçalves, C.D.M.; De Assis, T.F.; Silva, M.A. Da Analysis of the difference between the euclidean distance and the actual road distance in Brazil. *Transp. Res. Procedia* **2014**, 3, 876–885. [CrossRef]
44. Vikramkumar, B.V. Trilochan Bayes and Naive Bayes Classifier. *arXiv* **2014**, arXiv:1404.0933.
45. Zhang, H. The optimality of Naive Bayes. In Proceedings of the Seventeenth International Florida Artificial Intelligence Research Society Conference, Sarasota, FL, USA, 12–14 May 2004; Volume 2, pp. 562–567.
46. Kurniawan, D. *Pengenalan Machine Learning dengan Python*; PT Elex Media Komputindo: Jakarta, Indonesia, 2020.
47. Akinci, H.; Kilicoglu, C. Random Forest-Based Landslide Susceptibility Mapping in Coastal Regions of Artvin, Turkey. *ISPRS Int. J. Geo-Inf.* **2020**, *9*, 553. [CrossRef]
48. Li, X.; Cheng, J.; Yu, D.; Han, Y. Research on Non-Landslide Selection Method for Landslide Hazard Mapping. *Res. Sq.* **2021**, 1–11.
49. Hossin, M.; Sulaiman, M.N. A Review on Evaluation Metrics for Data Classification Evaluations. *Int. J. Data Min. Knowl. Manag. Process* **2015**, *5*, 1–11. [CrossRef]
50. Zhang, H.; Song, Y.; Xu, S.; He, Y.; Li, Z.; Yu, X.; Liang, Y.; Wu, W.; Wang, Y. Combining a class-weighted algorithm and machine learning models in landslide susceptibility mapping: A case study of Wanzhou section of the Three Gorges Reservoir, China. *Comput. Geosci.* **2022**, *158*, 104966. [CrossRef]

51. Chicco, D.; Warrens, M.J.; Jurman, G. The Matthews correlation coefficient (MCC) is more informative than Cohen's Kappa and Brier score in binary classification assessment. *IEEE Access* **2021**, *9*, 78368–78381. [CrossRef]
52. Aslam, M. Introducing Kolmogorov-Smirnov Tests under Uncertainty: An Application to Radioactive Data. *ACS Omega* **2020**, *5*, 914–917. [CrossRef]
53. Massey Jr, F.J. The Kolmogorov-Smirnov test for goodness of fit. *J. Am. Stat. Assoc.* **1951**, *46*, 68–78. [CrossRef]
54. Fleming, T.R.; O'Fallon, J.R.; O'Brien, P.C.; Harrington, D.P. Modified Kolmogorov-Smirnov test procedures with application to arbitrarily right-censored data. *Biometrics* **1980**, *36*, 607–625. [CrossRef]
55. Lee, S.; Choi, J.; Min, K. Landslide susceptibility analysis and verification using the Bayesian probability model. *Environ. Geol.* **2002**, *43*, 120–131. [CrossRef]
56. Hussain, M.A.; Chen, Z.; Zheng, Y.; Shoaib, M.; Shah, S.U.; Ali, N.; Afzal, Z. Landslide Susceptibility Mapping Using Machine Learning Algorithm Validated by Persistent Scatterer In-SAR Technique. *Sensors* **2022**, *22*, 3119. [CrossRef] [PubMed]
57. Bui, D.T.; Tsangaratos, P.; Nguyen, V.T.; Van Liem, N.; Trinh, P.T. Comparing the prediction performance of a Deep Learning Neural Network model with conventional machine learning models in landslide susceptibility assessment. *Catena* **2020**, *188*, 104426. [CrossRef]
58. Abu El-Magd, S.A.; Ali, S.A.; Pham, Q.B. Spatial modeling and susceptibility zonation of landslides using random forest, naïve bayes and K-nearest neighbor in a complicated terrain. *Earth Sci. Inform.* **2021**, *14*, 1227–1243. [CrossRef]
59. Park, S.-J.; Lee, D.-K. Predicting susceptibility to landslides under climate change impacts in metropolitan areas of South Korea using machine learning. *Geomat. Nat. Hazards Risk* **2021**, *12*, 2462–2476. [CrossRef]
60. Soria, D.; Garibaldi, J.M.; Ambrogi, F.; Biganzoli, E.M.; Ellis, I.O. A 'non-parametric' version of the naive Bayes classifier. *Knowl.-Based Syst.* **2011**, *24*, 775–784. [CrossRef]
61. Marjanović, M.; Kovačević, M.; Bajat, B.; Voženílek, V. Landslide susceptibility assessment using SVM machine learning algorithm. *Eng. Geol.* **2011**, *123*, 225–234. [CrossRef]
62. Merghadi, A.; Yunus, A.P.; Dou, J.; Whiteley, J.; ThaiPham, B.; Bui, D.T.; Avtar, R.; Abderrahmane, B. Machine learning methods for landslide susceptibility studies: A comparative overview of algorithm performance. *Earth-Sci. Rev.* **2020**, *207*, 103225. [CrossRef]
63. Nakileza, B.R.; Nedala, S. Topographic influence on landslides characteristics and implication for risk management in upper Manafwa catchment, Mt Elgon Uganda. *Geoenviron. Disasters* **2020**, *7*, 1–13. [CrossRef]
64. Dai, F.C.; Lee, C.F.; Li, J.; Xu, Z.W. Assessment of landslide susceptibility on the natural terrain of Lantau Island, Hong Kong. *Environ. Geol.* **2001**, *40*, 381–391. [CrossRef]
65. Nourani, V.; Pradhan, B.; Ghaffari, H.; Sharifi, S.S. Landslide susceptibility mapping at Zonouz Plain, Iran using genetic programming and comparison with frequency ratio, logistic regression, and artificial neural network models. *Nat. Hazards* **2014**, *71*, 523–547. [CrossRef]
66. Çellek, S. Effect of the slope angle and its classification on landslides. *Himal. Geol.* **2022**, *43*, 85–95.
67. Christian, J.T.; Baecher, G.B. DW Taylor and the foundations of modern soil mechanics. *J. Geotech. Geoenviron. Eng.* **2015**, *141*, 2514001. [CrossRef]
68. Take, W.A.; Bolton, M.D.; Wong, P.C.P.; Yeung, F.J. Evaluation of landslide triggering mechanisms in model fill slopes. *Landslides* **2004**, *1*, 173–184. [CrossRef]
69. Kim, J.-C.; Lee, S.; Jung, H.-S.; Lee, S. Landslide susceptibility mapping using random forest and boosted tree models in Pyeong-Chang, Korea. *Geocarto Int.* **2018**, *33*, 1000–1015. [CrossRef]
70. Gonzalez-Ollauri, A.; Mickovski, S.B. Hydrological effect of vegetation against rainfall-induced landslides. *J. Hydrol.* **2017**, *549*, 374–387. [CrossRef]
71. Norris, J.E.; Stokes, A.; Mickovski, S.B.; Cammeraat, E.; Van Beek, R.; Nicoll, B.C.; Achim, A. *Slope Stability and Erosion Control: Ecotechnological Solutions*; Springer Science & Business Media: Berlin/Heidelberg, Germany, 2008; ISBN 1402066767.
72. Guillard, C.; Zezere, J. Landslide Susceptibility Assessment and Validation in the Framework of Municipal Planning in Portugal: The Case of Loures Municipality. *Environ. Manag.* **2012**, *50*, 721–735. [CrossRef]
73. Karsli, F.; Atasoy, M.; Yalcin, A.; Reis, S.; Demir, O.; Gokceoglu, C. Effects of land-use changes on landslides in a landslide-prone area (Ardesen, Rize, NE Turkey). *Environ. Monit. Assess.* **2009**, *156*, 241–255. [CrossRef]
74. Tufaila, M.; Alam, S. Karakteristik tanah dan evaluasi lahan untuk pengembangan tanaman padi sawah di kecamatan oheo kabupaten konawe utara. *Agriplus* **2014**, *24*, 184–194.
75. Balai, B. Ksda Faktor Penyebab Tanah Longsor. Available online: http://ksdasulsel.menlhk.go.id/post/faktor-penyebab-tanah-longsor#:~:text=Tanahyangkurangpadatdan,longsor%2Cterutamabilaterjadihujan (accessed on 3 July 2022).
76. Mahmood, K.; Kim, J.M.; Ashraf, M. The effect of soil type on matric suction and stability of unsaturated slope under uniform rainfall. *KSCE J. Civ. Eng.* **2016**, *20*, 1294–1299. [CrossRef]
77. Yeh, H.-F.; Lee, C.-C.; Lee, C.-H. A rainfall-infiltration model for unsaturated soil slope stability. *Sustain. Environ. Res.* **2008**, *18*, 271–278.
78. Igwe, O. The geotechnical characteristics of landslides on the sedimentary and metamorphic terrains of South-East Nigeria, West Africa. *Geoenviron. Disasters* **2015**, *2*, 1–14. [CrossRef]
79. Di, B.; Stamatopoulos, C.A.; Stamatopoulos, A.C.; Liu, E.; Balla, L. Proposal, application and partial validation of a simplified expression evaluating the stability of sandy slopes under rainfall conditions. *Geomorphology* **2021**, *395*, 107966. [CrossRef]

80. Chen, W.; Xie, X.; Peng, J.; Wang, J.; Duan, Z.; Hong, H. GIS-based landslide susceptibility modelling: A comparative assessment of kernel logistic regression, Naïve-Bayes tree, and alternating decision tree models. *Geomat. Nat. Hazards Risk* **2017**, *8*, 950–973. [CrossRef]
81. Gilliam, F.S.; Hédl, R.; Chudomelová, M.; McCulley, R.L.; Nelson, J.A. Variation in vegetation and microbial linkages with slope aspect in a montane temperate hardwood forest. *Ecosphere* **2014**, *5*, 1–17. [CrossRef]
82. Singh, S. Understanding the Role of Slope Aspect in Shaping the Vegetation Attributes and Soil Properties in Montane Ecosystems. Available online: www.tropecol.com (accessed on 4 April 2022).
83. van Westen, C. Landslide Risk Assessments for Decision-Making. In Proceedings of the 2012 UR Forum, Cape Town, South Africa, 2–6 July 2012; pp. 67–71.

International Journal of *Geo-Information*

MDPI

Article

Classification of Floods in Europe and North America with Focus on Compound Events

Steven Brazda [1,2], Mojca Šraj [2] and Nejc Bezak [2,*]

[1] BGC Engineering Inc., Vancouver, BC V6Z 0C8, Canada
[2] Faculty of Civil and Geodetic Engineering, University of Ljubljana, 1000 Ljubljana, Slovenia
* Correspondence: nejc.bezak@fgg.uni-lj.si

Abstract: Compound events occur when multiple drivers or hazards occur in the same region or on the same time scale, hence amplifying their impacts. Compound events can cause large economic damage or endanger human lives. Thus, a better understanding of the characteristics of these events is needed in order to protect human lives. This study investigates the drivers and characteristics of floods in Europe and North America from the compound event perspective. More than 100 catchments across Europe and North America were selected as case study examples in order to investigate characteristics of floods during a 1979–2019 period. Air temperature, precipitation, snow thickness, snow liquid water equivalent, wind speed, vapour pressure, and soil moisture content were used as potential drivers. Annual maximum floods were classified into several flood types. Predefined flood types were snowmelt floods, rain-on-snow floods, short precipitation floods and long precipitation floods that were further classified into two sub-categories (i.e., wet and dry initial conditions). The results of this study show that snowmelt floods were often the dominant flood type in the selected catchments, especially at higher latitudes. Moreover, snow-related floods were slightly less frequent for high altitude catchments compared to low- and medium-elevation catchments. These high-altitude areas often experience intense summer rainstorms that generate the highest annual discharges. On the other hand, snowmelt-driven floods were the predominant flood type for the lower elevation catchments. Moreover, wet initial conditions were more frequent than the dry initial conditions, indicating the importance of the soil moisture for flood generation. Hence, these findings can be used for flood risk management and modelling.

Keywords: floods; compound events; flood typologies; precipitation; catchment characteristics

Citation: Brazda, S.; Šraj, M.; Bezak, N. Classification of Floods in Europe and North America with Focus on Compound Events. *ISPRS Int. J. Geo-Inf.* **2022**, *11*, 580. https://doi.org/10.3390/ijgi11120580

Academic Editor: Wolfgang Kainz

Received: 6 September 2022
Accepted: 20 November 2022
Published: 22 November 2022

1. Introduction

Floods are a natural hazard that can cause large economic damage and endanger human lives [1–3]. In order to protect human lives and property, either effective early warning systems or comprehensive flood-risk management are needed. In order to implement flood risk management measures such as hybrid flood protection infrastructure, understanding of flood mechanisms across different spatial scales, climates, elevations and other catchment-related characteristics is required [1]. This also applies to the snowmelt-related floods [2,3], which are the focus of this study.

Snow cover and snowmelt can affect the occurrence of floods in different ways. For example, a combination of snowmelt and intense precipitation with higher air temperature can generate so-called rain-on-snow floods, which can cause significant flood damage due to the compound effect. In recent years, special attention has been given to various compound events. Definitions and typologies for compound events were recently presented by Zscheischler et al. (2020) [1]. According to Zscheischler et al. (2020) [1], compound events can be classified into four main types, namely preconditioned events, multivariate events, temporally compounding events and spatially compounding events. Preconditioned events are hazards created or exacerbated by a pre-existing condition, as in the case of rain-on-snow floods [2,3]. Multivariate events are caused by multiple drivers and/or hazards occurring in the same geographic region within a temporal boundary, such as a combination of fluvial and coastal floods or a combination of drought and heat waves [4,5]. Temporally compounding events are a sequence of hazards occurring in a spatially bounded region, such as a series of large rainstorms causing flooding [6–8]. Spatially compounding events experience single or multiple hazards within a given time period [9,10], such as the large floods that occurred in Germany, Belgium and the Netherlands in 2021 [11].

This study focuses on compound events, where snowmelt is one of the driving forces of flooding. These type of events occur most frequently in northern regions in the northern hemisphere and conversely in the southern hemisphere, and in alpine areas [3,12,13]. This type of hazard becomes a potential threat for society when snow depth increases during the winter and then melts rapidly as a result of a sudden temperature rise or precipitation event. The significant melting of snow can saturate the soil and consequently lead to excessive surface runoff that can cause flooding, especially in cases when the ground is frozen. Many rivers around the globe experience this type of flooding each year. If it rains at the same time as the snow melts, even more severe flooding can occur, known as rain-on-snow floods [2,3]. Not many studies have been conducted that focus on examining the characteristics of rain-on-snow floods at large spatial scales. Most studies have been conducted focusing on smaller spatial scales. For example, Sikorska et al. (2015) [14] classified the most frequent flood types in Switzerland into six categories: snowmelt, rain-on-snow, flash, glacier-melt, short-rainfall and long-rainfall floods. Floods were classified using decision tree and the fuzzy method [14]. This study demonstrated that the predominant flood types in Switzerland are long-rainfall and short-rainfall floods. The potential drivers of flood events were also examined by Merz and Blöschl (2003) [15], who investigated floods in Austria. They found that there are significant regional differences between different climatic and terrain zones in Austria. Furthermore, they also analyzed the seasonality of flooding, which indicates the time of the year when the catchments are most likely to be flooded. Additionally, the authors reported that long-term rainfall events are the main cause of flooding in Austria. Recently, Berghuijs et al. (2019) [16] examined the potential drivers of the most extreme floods across Europe. The main drivers considered were snowmelt, extreme precipitation, and high antecedent soil moisture. Extreme precipitation (i.e., maximum annual discharge is a result of the largest precipitation event) was found to be the least dominant driver in generating floods in Europe. On the other hand, this mechanism was the most pronounced in the mountainous regions of the Alps and the Carpathians. Furthermore, the melting of snow was the second-most important flood generation mechanism in Europe, with this factor dominating in Eastern Europe and Scandinavia. However, [16] showed that the most important flood generation mechanism across Europe was high antecedent moisture. It should be noted that [16] focused primarily on the flood dates and did not consider the complete flood hydrographs.

There are still many open questions that need to be addressed to improve the understanding and prediction of floods, e.g., the seasonal characteristics of snow-related events, which climate factors are the main drivers of floods, etc. Therefore, the main aim of this paper is to classify floods into different categories according to their causes and to identify which flood types are most common in different parts of Europe and North America. Additionally, this study also focuses on analyzing the relationships between flood types and elevation, climate zone, and catchment area.

2. Data

The research includes 107 catchments throughout Europe and North America. The selected catchments are shown in Figures 1 and 2 for North America and Europe, respectively. A detailed list of selected catchments with their main characteristics is presented in the Supplement prepared (Table S1) based on Brazda (2021) [17]. These 107 catchments were manually selected to include catchments in different climate zones, elevations, etc., and with the most complete discharge dataset. The focus of this study was Europe and North

America and catchments located between 40° and 70° latitude were selected. Hence, all these catchments are located in the mid-latitudes, meaning that all four seasons (autumn, winter, spring, summer) define the climate characteristics. Additionally, nested catchments were not taken into consideration. Hence, the idea was to have roughly uniform distribution of catchments in these two continents based on the above limitations and discharge data availability. For all catchments, the AM sample was visually checked in order to detect possible significant changes in the sample size due to human impact (e.g., dam construction). We argue that these 107 catchments are a valid representation of a typical catchment for the selected study area.

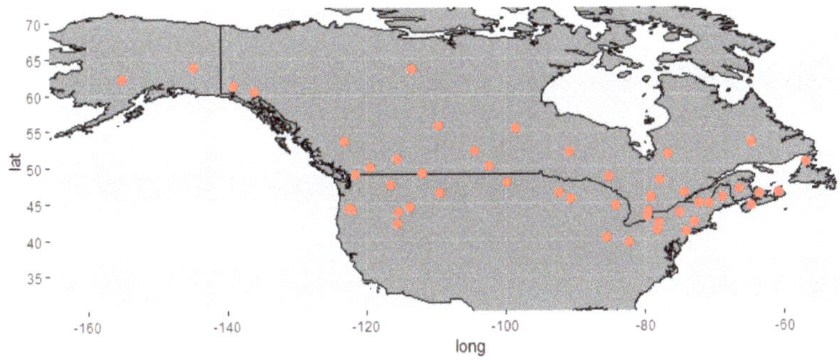

Figure 1. Gauging stations locations for catchments that were selected in North America.

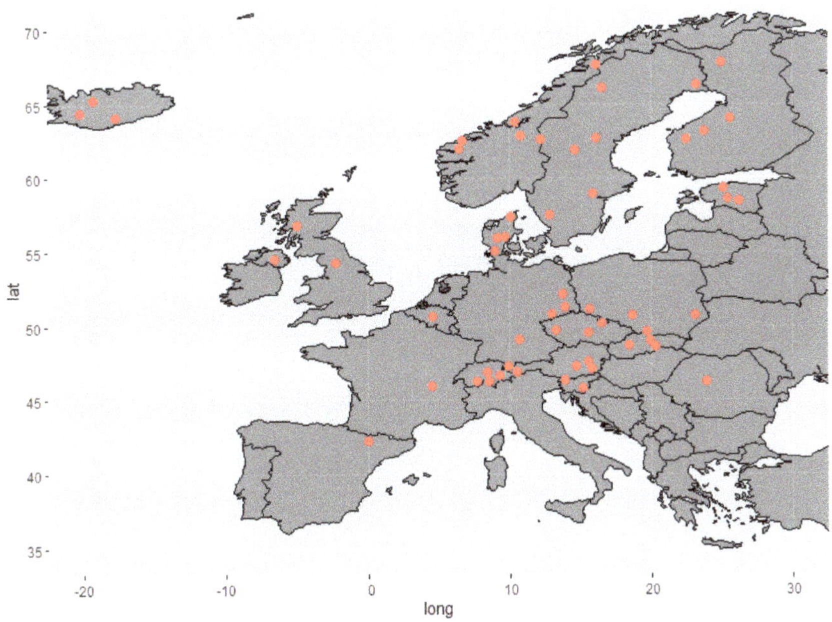

Figure 2. Gauging stations locations for catchments that were selected in Europe.

Daily discharge data for the selected catchments were obtained from the Global Runoff Data Centre [18]. Daily mean discharge from 1979 to 2019 was used in the study. Additionally, the catchment boundaries were also obtained from the GRDC [18]. Figure 3 shows an example of the daily discharge time series for the Penobscot River catchment in the USA, which was one of the selected catchments in North America.

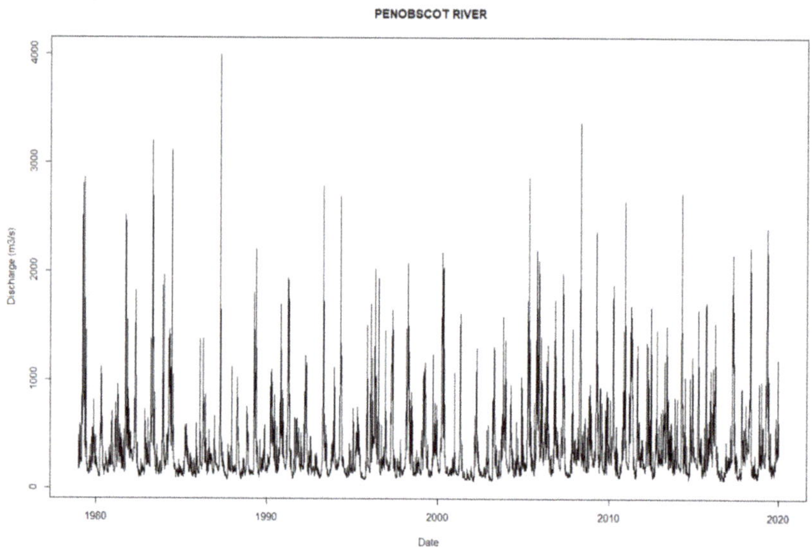

Figure 3. Example of the river discharge time series for one of the investigated catchments for the period 1979–2019. River discharge data were obtained from the GRDC. The Penobscot river is located in the USA and has a catchment area of 17,317 km^2, and most of the catchment area is located in the Warm Summer Humid Continental climate zone.

The Köppen–Geiger system [19] was used to identify the corresponding climate zone of each of the selected catchments. The Climate Change and Infectious Diseases Group [19] was used to obtain the climate zone data. Table 1 shows the climate zones that were considered. The distribution of catchments per climate zone is shown in the Supplement (Figure S1).

Table 1. The climate zones of the considered catchments based on the Köppen–Geiger system.

Acronym	Climate Zone
BSK	Cold Semi-Arid
CFA	Humid Subtropical
CFB	Temperate Oceanic
CFC	Subpolar Oceanic
CSB	Warm Summer Mediterranean
DFA	Hot Summer Humid Continental
DFB	Warm Summer Humid Continental
DFC	Subarctic
DSB	Mediterranean-Influenced Warm Summer Humid Continental
DSC	Mediterranean-Influenced Subarctic
ET	Tundra

Global elevation data were downloaded from EarthEnv [20] to determine the catchment mean elevation. The elevation data used in this study were a gridded dataset with a spatial resolution of 1 km. Figure 4 shows the elevation data and the catchment boundaries. The distribution of catchments per elevation zone is shown in the Supplement (Figure S3).

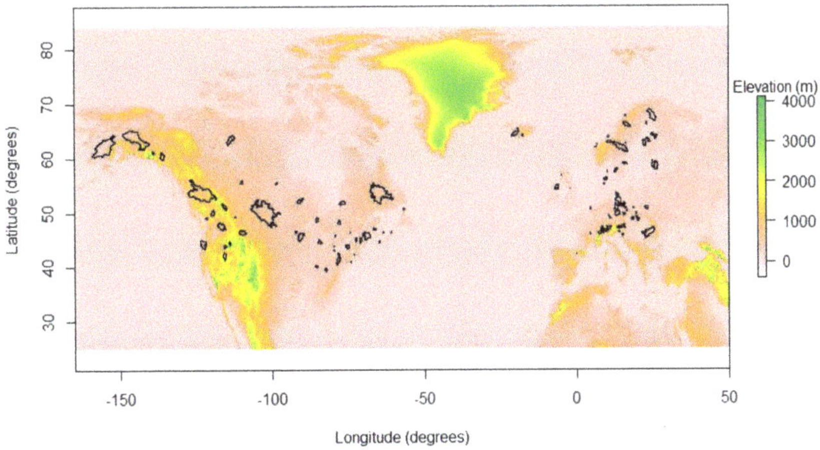

Figure 4. The elevation (m.a.s.l.) data with catchment boundaries (black polygons).

All climate data, except soil moisture data, were obtained from the Copernicus Agrometeorological Indicators Data Store [21]. The period used was 1979–2019. The grids had a spatial resolution of 0.1°. Table 2 displays all the climate information downloaded and used in the scope of this study.

Table 2. List of climate variables that were taken into consideration in the scope of this study.

Variable	Description	Unit
Temperature	Mean 24 h air temperature at a 2 m height	K
Precipitation	Total volume of water fallen per unit area over the 24 h period	mm/day
Snow Thickness	Mean depth of snow cover over the 24 h period	cm
Snow Thickness Liquid Water Equivalent (LWE)	Mean depth of liquid over the 24 h period assuming all snow melts and there is no runoff, soil penetration or evaporation	cm
Vapour Pressure	Mean water vapour pressure measured over the 24 h period	hPa
Wind Speed	Mean wind speed at 10 m height	m/s
Soil Moisture	Volume of water in the top soil layer (0–7 cm depth)	m^3/m^3

Soil moisture data were obtained from the Copernicus Data Store [22]. Hourly ERA5 data at individual levels from 1979 to the present was used. The hourly value at 12:00 was used for further analysis. ERA5 is one of the products that is frequently used in many different fields [4,23–25], providing a reanalysis of global climate and weather that combines model data with observations into a globally complete and consistent dataset using data assimilation technique.

3. Methods

3.1. Flood Hydrograph Seperation

R software was used to conduct the data analysis (i.e., data import, clipping of the gridded data, etc.) [26]. The Annual Maximum (AM) method [27–30] was used to determine the flood events. In the scope of this study, we did not focus only on peak discharge values but we extracted the entire flood hydrographs [29], which was not the case in some previous studies [16]. We decided to extract the climate data for the entire duration of the hydrograph rather than just extracting data only on the day of the maximum peak discharge. This is because in many catchments there is a lag between precipitation and runoff, and the flood-driving climate factors often occur on the days before the peak. Baseflow separation was used to determine the start and the end of the hydrograph. It is a frequently used method for hydrograph definition [29,31,32]. Baseflow is often considered as part of the stream water that originates from groundwater [32]. The surface runoff hydrograph starts when overland flow exceeds baseflow and ends when there is no more water belonging to the overland runoff. To determine the shape of the hydrograph, as well as the start and the end date, the Baseflow Index Method (BFI) from the "lfstat" package (Koffler et al., 2016) [33] in R was used. A detailed explanation of the BFI method can be found in the report on Low-flow Estimation and Prediction [34]. Hence, for each AM peak discharge value, the corresponding flood hydrograph was extracted from the daily discharge time series for all 107 stations.

3.2. Flood Typology

In order to analyze the compound flooding, the hydrographs were first divided into classes that included multivariate compound events and pre-conditioned compound events. The methodology used in this study is relatively similar to one implemented by Sikorska et al. (2015) [14], as some similar flood types were used. However, we decided to additionally distinguish between dry and wet event conditions (Table 3). This resulted in the eight flood types presented in Table 3.

Sikorska et al. (2015) [14] indicated that when snow cover exceeded 5% of the catchment area, the flood can be considered as influenced by snow, and the flood type in this case is either a snowmelt (SMF) flood or a rain-on-snow (ROS) flood. In case rain falls on top of the existing snow cover, then this is considered a ROS event. The threshold for precipitation used in this study to identify the ROS flood was 12 mm, which causes more than 1 mm of snowmelt (Table 3). The amount of snowmelt was determined by subtracting the snow thickness liquid water equivalent (LWE) from the previous day's LWE. The total thickness of the solid snow was not considered in this calculation because snow thickness can decrease, which changes the density of the snow without causing snowmelt runoff [35]. A SMF flood occurs when snowmelt exceeds 1 mm and less than 12 mm of precipitation falls (Table 3). To determine if the snow cover was in an initial wet or dry condition, snow density characteristics were investigated. Kuusisto (1984) [35] studied snow density during melting periods. Snow density depends on many factors, including snow thickness, air and snow temperature, precipitation, etc. Moreover, snow density can also vary regionally, and thus a direct comparison of snow densities between the selected catchments would not be an optimal solution. Therefore, the percentage of snow density increase during the melting phase can be compared. Kuusisto (1984) [35] wrote that snow density increased by more than 20% during the final phase of melting. This value (i.e., 20%) was used in the study as a threshold to determine whether the conditions were initially wet or dry. If snow density increased by more than 20% from the beginning of the hydrograph to the day of maximum snowmelt, it can be assumed that the snow was not in the process of melting, and the event (i.e., hydrograph) can be classified as an initial dry condition. If the increase in density was less than 20%, it can be assumed that the snow was already close to melting. Hence, the hydrograph can be classified as having initial wet conditions. Furthermore, the day of maximum snowmelt was used to determine the change in density, since there may be situations where, on the day of the peak hydrograph, the snow cover and thickness is close to zero, meaning that all the snow has already been melted. To obtain the snow density, the LWE was compared to the total solid snow thickness.

In cases when snow cover is less than 5% of the catchment area, it can be assumed that snowmelt has a minor impact on flood generation [14]. In this case, the main driver of the flood event is precipitation. The precipitation floods in the study were divided into short-precipitation floods (SPF) and long-precipitation floods (LPF). A SPF event occurs when the rainfall duration does not exceed 1 day and rainfall amount is greater than 12 mm. A LPF occurs when the rainfall duration is from 2 to 4 days and the rainfall amount exceeds 25 mm [14,36]. In the case that both conditions are fulfilled, the event was classified as LPF.

Moreover, multiple peaks in the hydrograph were assumed to be the LPF event [36]. The R package "pracma" was used to determine the number of peaks in the hydrograph [37].

Soil moisture data were used to determine the initial conditions (i.e., wet or dry) of the event. In order to calculate the percent saturation, data for each catchment were examined, and the largest daily volumetric water content for each catchment was defined. This maximum water content was assumed to be 100% saturation. All other values were then selected as numerators above this maximum value, resulting in the daily percent saturation content. The threshold was set at 75% (Table 3). This threshold was selected after some preliminary investigations and it was found to be a reasonable threshold to be used in relation to defining the antecedent conditions. By applying these constraints to each of the identified flood hydrographs, they were classified into eight categories. If an individual flood hydrograph did not meet any of the eight pre-determined categories (Table 3), it was classified in the "other" category. Figure 5 shows the classification process.

Table 3. Flood typology used in this study. References that were used to define the threshold values are presented in the square brackets.

Flood Type	Precipitation [14]	Snow Cover [14]	Snowmelt [14]	Antecedent Moisture Condition	Other [36]	Abbreviation
Rain-on-Snow Flood with Dry Conditions	>12 mm	>5%	>1 mm	>20% Increase in snow density [35]		ROS-D
Rain-on-Snow Flood with Wet Conditions	>12 mm	>5%	>1 mm	<20% Increase in snow density [35]		ROS-W
Snowmelt Flood with Dry Conditions	<12 mm	>5%	>1 mm	>20% Increase in snow density [35]		SMF-D
Snowmelt Flood with Wet Conditions	<12 mm	>5%	>1 mm	<20% Increase in snow density [35]		SMF-W
Long-Precipitation Floods with Dry Conditions	>25 mm over 4 days	<5%	<1 mm	<75% soil saturation at start of hydrograph	Multiple Peaks	LPF-D
Long-Precipitation Floods with Wet Conditions	>25 mm over 4 days	<5%	<1 mm	>75% soil saturation at start of hydrograph	Multiple Peaks	LPF-W
Short-Precipitation Floods with Dry Conditions	>12 mm in 1 day	<5%	<1 mm	<75% soil saturation at start of hydrograph		SPF-D
Short-Precipitation Floods with Wet Conditions	>12 mm in 1 day	<5%	<1 mm	>75% soil saturation at start of hydrograph		SPF-W

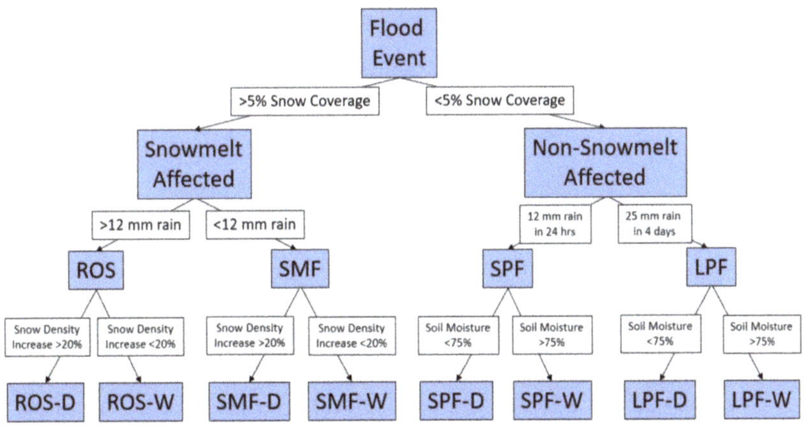

Figure 5. Visual representation of the flood classification process shown in Table 3.

Additionally, the differences in climate zones, catchment areas, and altitudes were also considered in analysis. The climate zone covering most of the catchment area was selected as the dominant zone of the individual catchment (see Supplement Figure S1). Four dominant climate zones were identified in the study, namely warm summer humid continental climate, temperate oceanic climate, tundra, and subarctic climate zone (see Supplement Figure S1). Other climate zones were only relevant to 1–3 catchments and were not considered in the classification by climate zone. The catchments were also divided into three categories according to their size (see Supplement Figure S2): large (>10,000 km^2), medium (between 200 and 10,000 km^2), and small (<200 km^2). Furthermore, the catchments were also divided into three categories by catchment mean elevation: high (>1000 m.a.s.l.), medium (between 500 and 1000 m.a.s.l.), and low (<500 m.a.s.l.) (see Supplement Figure S3).

4. Results and Discussion

4.1. Flood Typology Classification for All Catchments

Following the presented methodology for the extracted AM events (i.e., 41 events were extracted for each catchment), main climate characteristics during these events were examined (see Supplement Figures S4–S7). Some relatively large variability in the snow thickness, soil moisture and other variables during these events can be seen across Europe and North America (see Supplement Figures S4–S7). Hence, the percentage of snowmelt-related floods differed among the selected 107 catchments (see Supplement Figures S8 and S9). We argue that selected catchments represent a variety of conditions between the 40° and 70° latitude. Figures 6 and 7 show the results of the flood classification methodology for the selected European and North American catchments using heat maps. Several conclusions can be drawn from the heatmaps shown in Figures 6 and 7. Firstly, the majority of the AM floods that occurred in the investigated catchments (107 in total) were classified as snowmelt-driven floods (SMF). The snowmelt-driven floods (i.e., SMF-W and SMF-D types) represent 39% of all floods considered on both continents (Figures 6 and 7). Secondly, for all categories (i.e., ROS, SMF, SPF, and LPF), the wet conditions (-W) were always more prevalent than the dry conditions (-D) (Figures 6 and 7). The dry initial conditions accounted for 45%, 36%, 20%, and 19% of ROS, SMF, SPF, and LPF, respectively (Figures 6 and 7). These results are in accordance with what was reported in some previous studies that pointed to the importance of soil moisture on flood generation [16,38]. It should be noted that the percentage of dry initial conditions (-D) was higher for the snow-influenced floods (i.e., ROS and SMF) compared to SPF and LPF. Hence, high soil moisture values frequently occur with SPF and LPF events, and compound occurrence frequently results in flooding. It should be noted that the dry initial conditions for the snow-influenced floods were determined based on snow density. More specifically, the ROS floods made up 55% of the floods in the wet initial condition (Figures 6 and 7). The main reason for this is that rain quickly increases snow density prior to melting [39]. Furthermore, less than 20% of SPF and LPF events were seen where the maximum annual flood started at less than 75% soil saturation. Based on this fact, it is reasonable to assume that antecedent soil moisture is an important driver in flood generation in Europe and North America. These results are consistent with findings from previous studies [16,38]. It can also be seen that the percentage of SMF-W in general decreased from north to south (Figures 6 and 7). For flood hydrographs that we could not classify into any of the predefined categories for the North America catchments (i.e., Other type), an opposite situation is evident (Figure 6). For European catchments, the percentage of SPF-W and LPF-W generally increases from south to north (Figure 7). It should be noted that the selected thresholds (Table 3) could have an impact on the percentage of the occurrence of different flood types. For example, increasing the snow cover threshold (i.e., from 5% to 10 or 15%) would reduce the number of snow-related flood events. Additionally, increasing the 12 mm precipitation threshold related to the SPF events would decrease the number of these events. However, we argue that smaller changes in the selected thresholds would have a relatively minor impact on the presented results.

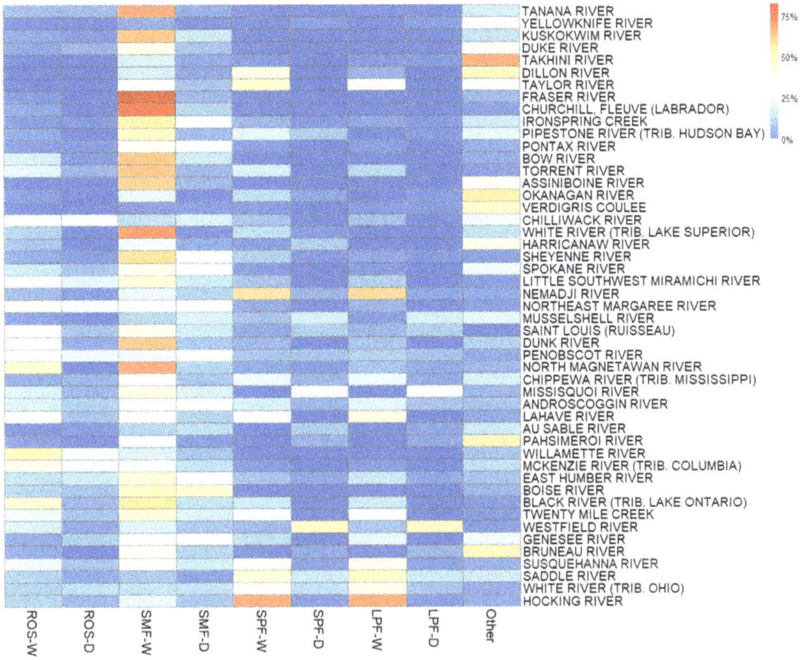

Figure 6. Heatmap for the North American catchment distribution of annual maximum (AM) flood typologies. Catchments are sorted from the south to the north.

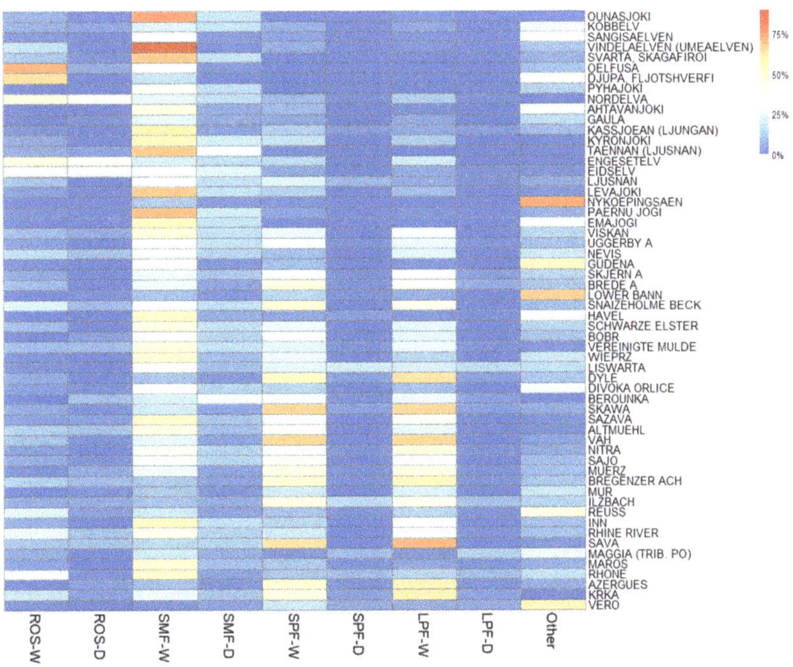

Figure 7. Heatmap for the European catchment distribution of annual maximum (AM) flood typologies. Catchments are sorted from south to north.

4.2. Flood Typology Classification Based on Climate Zone

The flood types were also classified based on the predefined climate zone of each catchment. For each climate zone, the mean distribution of flood typologies was considered. Figure 8 shows that the CFB and DFA zones are more evenly distributed across the different typologies, with the SPF and LPF being nearly equal to the SMF. The DFC zone has the highest magnitude of floods in the SMF-W category (Figure 8). The ET climate zone is dominated by the two dominant flood types, namely SMF-W and ROS-W (Figure 8). ET is the only one of the four climate zones (i.e., CFB, DFA, DFC, and ET) that includes more than three catchments, where ROS-W occured as the dominant flood type (Figure 8). This is likely due to the large amounts of rainfall in many mountainous regions where the ET climate is found [14].

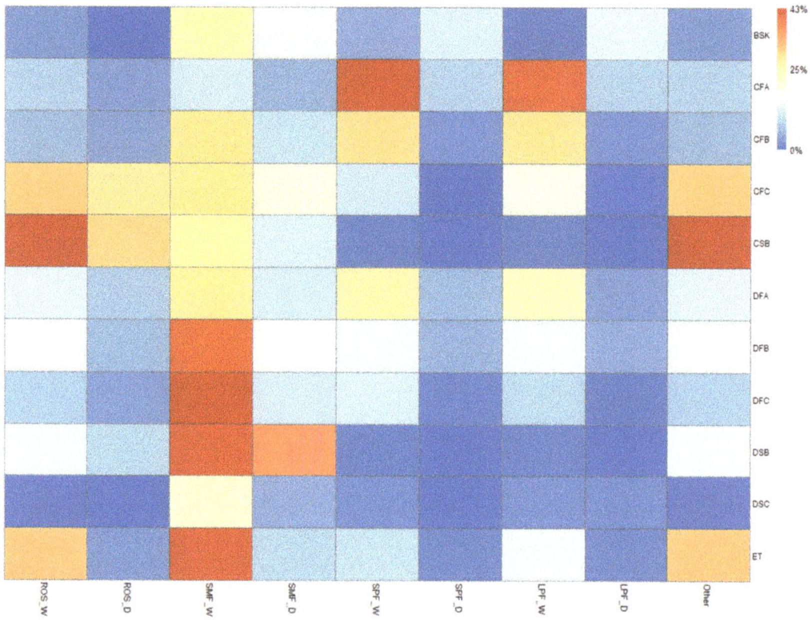

Figure 8. Heatmap for the annual maximum (AM) flood typology classification by climate zone, showing results for the four climate zones with the highest number of catchments.

4.3. Flood Typology Classification Based on Catchment Area

Figure 9 shows the flood types classified based on the size of the catchments. The SMF floods dominate in the large catchments, whereas the SPF and LPF are more common in the other two categories (Figure 9). There are several reasons for these results, one of which is the threshold related to the snow cover used to classify hydrographs (Table 3). The threshold for the snow cover, above which the flood was considered to be influenced by snow, was 5% (Table 3). In case of very large catchments, it is more likely that a part of the catchment is covered with snow, exceeding the threshold for a flood to be influenced by snowmelt. On the other hand, small catchments may be completely without snow coverage since small catchments do not cover large geographical areas. Additionally, several larger catchments in North America are located at higher latitudes. Another possible reason for these results is the soil moisture concept discussed by Harpold et al. (2015) [39]. Harpold et al. (2015) [39] argued that the highest soil moisture is reached within 5 days after the snowpack has completely melted. Moreover, in the case of large catchments, a rainfall storm would need to have a large spatial extent in order that catchment reaches the soil moisture required for an annual maxima flood generation. Hence, if a spatially extensive snowpack melts throughout the watershed, the entire watershed can reach the high soil moisture required. The number of LPFs is relatively small for large catchments (Figure 9), which is to some extent an unexpected result, as LPFs generally have the ability to saturate large catchments, while on the other hand, the spatial extent of most extreme and short-duration precipitation events (i.e., SPFs) may be spatially limited [9]. Moreover, the results indicate that the medium and small catchments are dominated by SPFs and LPFs (Figure 9). This is likely due to the fact that the spatial extent of the storm required for these catchments to become saturated is generally smaller. Very extreme rainfall events (e.g., summer thunderstorms) tend to have a smaller spatial extent, which is why SPFs are more common in small- and medium-sized catchments (Figure 9) [15]. Additionally, small- and medium-sized catchments often have a shorter time of concentration compared to large catchments [15]. Thus, SPF and LPF can more easily saturate the entire small and medium catchments, resulting in high peak discharge values.

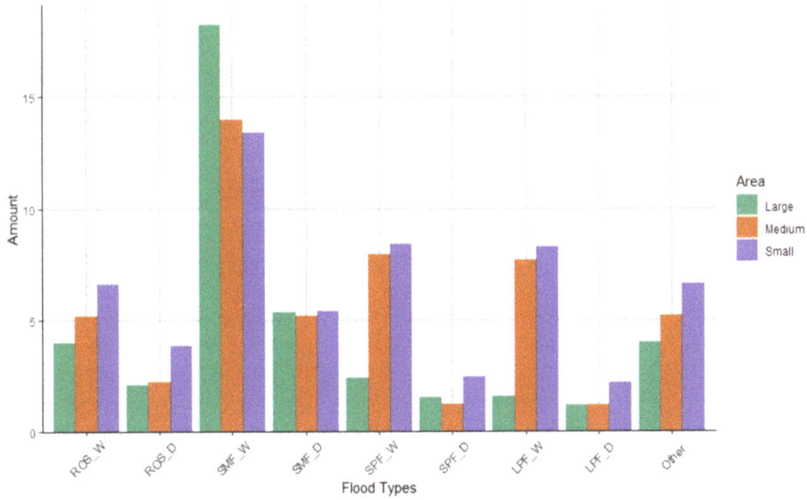

Figure 9. Flood typology classification based on the catchment area size.

4.4. Flood Typology Classification Based on Elevation

Figure 10 shows the distribution of floods in each flood type based on the mean elevation of the 107 catchments considered. It can be seen that the catchments at higher elevations have a slightly lower number of SMF and ROS floods than catchments at the medium and low elevations (Figure 10). A similar conclusion was also reached by Sikorska et al. (2015) [14]. Sikorska et al. (2015) [14] used two different methods to classify the floods. When investigating high-elevation catchments, they found that precipitation floods were the dominant flood type when using the crisp decision tree method [14], which is similar to the findings of this study (Figure 10). However, when they applied the fuzzy method, they found that although the dominant flood type remained the precipitation-driven flood, many of the floods were also classified as SMF or ROS floods [14]. However, the snow-related aspects do not exceed the thresholds to classify the flood as a snowmelt-affected flood. Sikorska et al. (2015) [14] hypothesized that this is due to a large amount of rainfall that falls in mountainous regions. Flash floods also often occur in mountainous catchments [14], which is not a flood type in the assessment shown in this study (Table 3). Moreover, Berghuijs et al. (2019) [16] found that extreme precipitation is the only predominant driver in mountainous regions [16]. Despite the large amounts of snow in the mountainous regions, rainfall can form the dominant flood type, which is caused by rainfall rather than snowmelt.

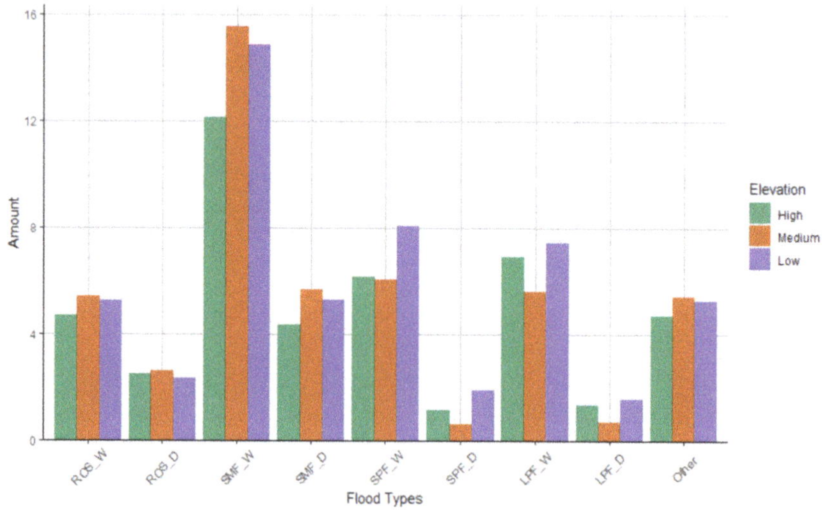

Figure 10. Flood typology classification based on the catchment elevation.

5. Conclusions

Based on the conducted analyses that were made based on the 107 selected catchments, several important conclusions can be made. It should be noted that the selection of 107 catchments that are relatively uniformly distributed between 40° and 70° latitude in Europe and North America represent a subset of all catchments and that different selection of catchments could yield different results. However, we argue that 107 catchments are representative for the selected study area. The results presented in this study indicate that snowmelt floods (especially SMF-W events) are often the dominant flood type in the catchments considered, especially for the catchments located at higher latitudes.

When comparing the relationships between the flood types (Table 3) and the climate zones, catchment elevation, and size, further conclusions could be made. Firstly, the primary flood type in the DFC and ET zones was SMF-W (Figure 8). Secondly, it was also shown that the large catchments had a slightly higher proportion of SMFs (Figure 9), while the medium and small catchments had slightly larger numbers of SPFs and LPFs (Figure 9). Thirdly, the occurrence of some specific flood types was found to change with latitude (e.g., SMF-W slightly decreased from south to north in Europe and North America). Finally, floods with wet soil initial condition (-W) occurred much more frequently than floods with the dry soil initial conditions (-D) in our analysis (Figures 6–10). The high frequency of the floods with wet soil initial conditions (-W) indicates that this may be a type of compound event—where a high antecedent moisture condition would qualify as a pre-existing event. In summary, these findings could be useful in the flood forecasting process, where special focus could be given to situations where soil moisture is high and a medium-precipitation event is expected in the following days. Additionally, these results could be used in the process of optimizing the flood risk management in relation to specific catchment characteristics (i.e., size, elevation, location climate). Moreover, finding could also be used for design of flood protection measures such as hybrid infrastructure. In the future, a similar study could be conducted, taking into account additional influencing factors (e.g., soil temperature as a proxy of ground frozenness) using an even larger number of catchments and testing different thresholds, which could be supplemented with additional statistical analysis.

Supplementary Materials: The following supporting information can be downloaded at: https://www.mdpi.com/article/10.3390/ijgi11120580/s1, Figure S1: Distribution of the selected 107 catchments per climate zone; Figure S2: Number of catchments based on the size of the catchment area. Small catchments have less than 200 km^2, large catchments are greater than 10,000 km^2 and medium catchments are between 200 and 10,000 km^2; Figure S3: Number of catchments based on the elevation. High elevation catchments are greater than 1000 m.a.s.l. Low elevation catchments are less than 500 m.a.s.l. Medium elevation catchments are between 500 and 1000 m.a.s.l; Figure S4: Mean snow thickness in the selected North American catchments during annual maximum events in the period 1979–2019; Figure S5: Mean snow thickness in the selected European catchments during annual maximum events in the period 1979–2019; Figure S6: Mean soil moisture in the selected North American catchments during annual maximum events in the period 1979–2019; Figure S7: Mean soil moisture in the selected European catchments during annual maximum events in the period 1979–2019; Figure S8: Percentage of annual maximum events that were affected by snowmelt in North America in the period 1979–2019; Figure S9: Percentage of annual maximum events that were affected by snowmelt in Europe in the period 1979–2019; Table S1: A list of 107 selected catchments with their main characteristics.

Author Contributions: Steven Brazda: methodology, investigation, formal analysis, data curation, writing—original draft; Mojca Šraj: writing—review and editing, conceptualization, supervision; Nejc Bezak: conceptualization, methodology, writing—review and editing, supervision. All authors have read and agreed to the published version of the manuscript.

Funding: The authors would like to acknowledge the support of the Slovenian Research Agency (ARRS) through grants P2-0180, V2-2137 and J6-4628 and support from the UNESCO Chair on Water-related Disaster Risk Reduction.

Data Availability Statement: Data can be obtained upon request.

Conflicts of Interest: The authors declare no conflict of interest.

References

1. Zscheischler, J.; Martius, O.; Westra, S.; Bevacqua, E.; Raymond, C.; Horton, R.M.; van den Hurk, B.; AghaKouchak, A.; Jézéquel, A.; Mahecha, M.D.; et al. A typology of compound weather and climate events. *Nat. Rev. Earth Environ.* **2020**, *1*, 333–347. [CrossRef]
2. Wachowicz, L.J.; Mote, T.L.; Henderson, G.R. A rain on snow climatology and temporal analysis for the eastern United States. *Phys. Geogr.* **2020**, *41*, 54–69. [CrossRef]
3. Sezen, C.; Šraj, M.; Medved, A.; Bezak, N. Investigation of rain-on-snow floods under climate change. *Appl. Sci.* **2020**, *10*, 1242. [CrossRef]
4. Sutanto, S.J.; Vitolo, C.; Di Napoli, C.; D'Andrea, M.; Van Lanen, H.A.J. Heatwaves, droughts, and fires: Exploring compound and cascading dry hazards at the pan-European scale. *Environ. Int.* **2020**, *134*, 105276. [CrossRef] [PubMed]
5. Bezak, N.; Mikoš, M. Changes in the compound drought and extreme heat occurrence in the 1961–2018 period at the european scale. *Water* **2020**, *12*, 3543. [CrossRef]
6. Mikoš, M.; Četina, M.; Brilly, M. Hydrologic conditions responsible for triggering the Stože landslide, Slovenia. *Eng. Geol.* **2004**, *73*, 193–213. [CrossRef]
7. Mikoš, M. After 2000 Stože landslide: Part II—Development in landslide disaster risk reduction policy in Slovenia—Po zemeljskem plazu Stože leta 2000: Del II—Razvoj politike zmanjševanja tveganja nesreč zaradi zemeljskih plazov v Sloveniji. *Acta Hydrotech.* **2021**, *34*, 39–59. [CrossRef]
8. Mikoš, M. After 2000 Stože landslide: Part I—Development in landslide research in Slovenia—Po zemeljskem plazu Stože leta 2000: Del I—Razvoj raziskovanja zemeljskih plazov v Sloveniji. *Acta Hydrotech.* **2020**, *33*, 129–153. [CrossRef]
9. Berghuijs, W.R.; Allen, S.T.; Harrigan, S.; Kirchner, J.W. Growing Spatial Scales of Synchronous River Flooding in Europe. *Geophys. Res. Lett.* **2019**, *46*, 1423–1428. [CrossRef]
10. Bezak, N.; Borrelli, P.; Panagos, P. A first assessment of rainfall erosivity synchrony scale at pan-European scale. *Catena* **2021**, *198*, 105060. [CrossRef]
11. Dietze, M.; Bell, R.; Ozturk, U.; Cook, K.L.; Andermann, C.; Beer, A.R.; Damm, B.; Lucia, A.; Fauer, F.S.; Nissen, K.M.; et al. More than heavy rain turning into fast-flowing water—A landscape perspective on the 2021 Eifel floods. *Nat. Hazards Earth Syst. Sci.* **2022**, *22*, 1845–1856. [CrossRef]
12. Brunner, M.I.; Viviroli, D.; Sikorska, A.E.; Vannier, O.; Favre, A.-C.; Seibert, J. Flood type specific construction of synthetic design hydrographs. *Water Resour. Res.* **2017**, *53*, 1390–1406. [CrossRef]
13. Poschlod, B.; Zscheischler, J.; Sillmann, J.; Wood, R.R.; Ludwig, R. Climate change effects on hydrometeorological compound events over southern Norway. *Weather Clim. Extrem.* **2020**, *28*, 100253. [CrossRef]

14. Sikorska, A.E.; Viviroli, D.; Seibert, J. Flood-type classification in mountainous catchments using crisp and fuzzy decision trees. *Water Resour. Res.* **2015**, *51*, 7959–7976. [CrossRef]
15. Merz, R.; Blöschl, G. A process typology of regional floods. *Water Resour. Res.* **2003**, *39*, 1–20. [CrossRef]
16. Berghuijs, W.R.; Harrigan, S.; Molnar, P.; Slater, L.J.; Kirchner, J.W. The Relative Importance of Different Flood-Generating Mechanisms Across Europe. *Water Resour. Res.* **2019**, *55*, 4582–4593. [CrossRef]
17. Brazda, S. *Snowmelt Floods in Relation to Compound Drivers in North Americ and Europe*; University of Ljubljana: Ljubljana, Slovenia, 2021.
18. GRDC. Global Runoff Data Centre (GRDC). Available online: https://www.bafg.de/GRDC/EN/Home/homepage_node.html (accessed on 1 March 2021).
19. CCID the Climate of the European ALPS: Shift of Very High Resolution Köppen-Geiger Climate Zones 1800–2100. Available online: http://koeppen-geiger.vu-wien.ac.at/alps.htm (accessed on 1 March 2021).
20. Amatulli, G.; Domisch, S.; Tuanmu, M.; Parmentier, B. Data Descriptor: A suite of global, cross-scale topographic variables for environmental and biodiversity modeling. *Sci. Data* **2018**, *5*, 180040. [CrossRef]
21. Copernicus Agrometeorological Indicators from 1979 to Present Derived from Reanalysis. Available online: https://cds.climate.copernicus.eu/cdsapp#!/dataset/10.24381/cds.6c68c9bb?tab=form (accessed on 15 March 2021).
22. Copernicus ERA5 Hourly Data on Single Levels from 1979 to Present. Available online: https://cds.climate.copernicus.eu/cdsapp#!/dataset/reanalysis-era5-single-levels?tab=overview (accessed on 20 March 2021).
23. Ozturk, U.; Saito, H.; Matsushi, Y.; Crisologo, I.; Schwanghart, W. Can global rainfall estimates (satellite and reanalysis) aid landslide hindcasting? *Landslides* **2021**, *18*, 3119–3133. [CrossRef]
24. Beck, H.E.; Pan, M.; Roy, T.; Weedon, G.P.; Pappenberger, F.; Van Dijk, A.I.J.M.; Huffman, G.J.; Adler, R.F.; Wood, E.F. Daily evaluation of 26 precipitation datasets using Stage-IV gauge-radar data for the CONUS. *Hydrol. Earth Syst. Sci.* **2019**, *23*, 207–224. [CrossRef]
25. Reder, A.; Rianna, G. Exploring ERA5 reanalysis potentialities for supporting landslide investigations: A test case from Campania Region (Southern Italy). *Landslides* **2021**, *18*, 1909–1924. [CrossRef]
26. R Core Team. *A Language and Environment for Statistical Computing*; R Foundation for Statistical Computing: Vienna, Austria, 2020.
27. Bezak, N.; Brilly, M.; Šraj, M. Comparison between the peaks-over-threshold method and the annual maximum method for flood frequency analysis | Comparaison entre les méthodes de dépassement de seuil et du maximum annuel pour les analyses de fréquence des crues. *Hydrol. Sci. J.* **2014**, *59*, 831174. [CrossRef]
28. Xiao, Y.; Guo, S.; Liu, P.; Yan, B.; Chen, L. Design flood hydrograph based on multicharacteristic synthesis index method. *J. Hydrol. Eng.* **2009**, *14*, 1359–1364. [CrossRef]
29. Bezak, N.; Horvat, A.; Šraj, M. Analysis of flood events in Slovenian streams. *J. Hydrol. Hydromech.* **2015**, *63*, 134–144. [CrossRef]
30. De Luca, P.; Hillier, J.K.; Wilby, R.L.; Quinn, N.W.; Harrigan, S. Extreme multi-basin flooding linked with extra-tropical cyclones. *Environ. Res. Lett.* **2017**, *12*, 114009. [CrossRef]
31. Eckhardt, K. How to construct recursive digital filters for baseflow separation. *Hydrol. Process.* **2005**, *19*, 507–515. [CrossRef]
32. Stoelzle, M.; Schuetz, T.; Weiler, M.; Stahl, K.; Tallaksen, L.M. Beyond binary baseflow separation: A delayed-flow index for multiple streamflow contributions. *Hydrol. Earth Syst. Sci.* **2020**, *24*, 849–867. [CrossRef]
33. Koffler, D.; Gauster, T.; Laaha, G. Package "Lfstat". 2016. Available online: https://cran.r-project.org/web/packages/lfstat/index.html (accessed on 15 September 2022).
34. Gustard, A.; Demuth, S. *Manual on Low-Flow Estimation and Prediction*; German National Committee for the International Hydrological Programme (IHP) of UNESCO and the Hydrology and Water Resources Programme (HWRP) of WMO Koblenz: Koblenz, Germany, 2009.
35. Kuusisto, E. *Snow Accumultation and Snowmelt in Finland*; National Board of Waters: Helsinki, Finland, 1984.
36. Fischer, S.; Schumann, A.; Bühler, P. Timescale-based flood typing to estimate temporal changes in flood frequencies. *Hydrol. Sci. J.* **2019**, *64*, 1867–1892. [CrossRef]
37. Borchers, H. Package "Pracma". 2021. Available online: https://cran.r-project.org/web/packages/pracma/index.html (accessed on 15 September 2022).
38. Blöschl, G.; Hall, J.; Viglione, A.; Perdigão, R.A.P.; Parajka, J.; Merz, B.; Lun, D.; Arheimer, B.; Aronica, G.T.; Bilibashi, A.; et al. Changing climate both increases and decreases European river floods. *Nature* **2019**, *573*, 108–111. [CrossRef]
39. Harpold, A.A.; Molotch, N.P.; Musselman, K.N.; Bales, R.C.; Kirchner, P.B.; Litvak, M.; Brooks, P.D. Soil moisture response to snowmelt timing in mixed-conifer subalpine forests. *Hydrol. Process.* **2015**, *29*, 2782–2798. [CrossRef]

International Journal of *Geo-Information*

Article

Evaluation of Coastal Erosion in the Watersheds of Municipality of Buenaventura, Colombia: Using Geospatial Techniques and the Composite Vulnerability Index

Jose Eduardo Fuentes [1,2,3,*], Robin Alexis Olaya [4] and Cesar Edwin Garcia [4]

1 AGESAT Research Group, Geography Department, Universidad del Valle, Cali 760031, Colombia
2 TERRITORIOS Research Group, Geography Department, Universidad del Valle, Cali 760031, Colombia
3 INCIMAR Institute of Marine Sciences and Limnology, Universidad del Valle, Cali 760031, Colombia
4 School of Civil and Geomatics Engineering, Universidad del Valle, Cali 760031, Colombia
* Correspondence: jose.fuentes@correounivalle.edu.co

Abstract: Buenaventura on the Colombian Pacific coast has experienced a wide range of threats, mainly due to the effects of coastal erosion and flooding. Globally, millions of people will experience increased vulnerability in the coming decades due to climate change. The change in the coastline (1986–2020) over time was analyzed with remote sensors and the Digital Shoreline Analysis System (DSAS) in conjunction with GIS. A total of 16 indicators were selected to quantitatively evaluate exposure, sensitivity, and adaptive capacity to construct a composite vulnerability index (COVI). The endpoint rate (EPR) of the change in the coastline was estimated. The results showed that 35% of the study area was stable, 18% of the coastline experienced erosion processes, and 47% experienced accretion. The COVI analysis revealed that coastal watersheds show great spatial heterogeneity; 31.4% of the area had moderate vulnerability levels, 26.5% had low vulnerability levels, and 41.9% had high vulnerability levels. This analysis revealed that the watersheds located in the northern (Málaga Bay) and central (Anchicaya, Cajambre, and Rapposo basins) parts of the coastal zone were more vulnerable than the other areas.

Keywords: coastal vulnerability index; coastal erosion; shoreline change; GIS; remote sensing; coastal watersheds

Citation: Fuentes, J.E.; Olaya, R.A.; Garcia, C.E. Evaluation of Coastal Erosion in the Watersheds of Municipality of Buenaventura, Colombia: Using Geospatial Techniques and the Composite Vulnerability Index. *ISPRS Int. J. Geo-Inf.* **2022**, *11*, 568. https://doi.org/10.3390/ijgi11110568

Academic Editors: Walter Chen, Fuan Tsai and Wolfgang Kainz

Received: 27 October 2022
Accepted: 14 November 2022
Published: 15 November 2022

1. Introduction

The coastal areas of Colombia cover less than 7% of the land surface of the country and support a population of 6 million inhabitants [1]. In recent years, the Colombian Pacific coast has experienced a wide range of catastrophic threats to its ecosystems, population, and infrastructure, mainly due to the effects of coastal erosion and flooding [2]. The destruction of ecosystems, climate change, population growth, and human activities, such as deforestation and mining, will increase vulnerability even more in the coming decades [3–6]. Globally, approximately 10 million people experience negative effects from tropical storms, coastal erosion, floods, and storm surges each year, which is expected to increase to 50 million by 2080 due to climate change and high sociodemographic pressure [7]. Coastal flooding and sea level are expected to increase significantly by the middle of the century [8]. How vulnerability should be assessed to generate adaptation and resilience strategies in the face of potentially disastrous events in the coastal zone is a global concern of scientific communities. [8,9]. However, the intensity and severity of hazardous events vary spatially, and they often become disasters when combined with the vulnerable socioeconomic environment of the human population [10]. Vulnerability is the degree to which a system is susceptible to natural hazards and social changes; it is a concept with multiple dimensions, encompassing the economic, political, physical, social, and environmental dimensions [11]. Vulnerability to any event can be explained as a function of exposure, sensitivity, and the ability to adapt or cope [12]. The definition of vulnerability implemented by the Intergovernmental Panel on Climate Change (IPCC) is one of the most widespread concepts in the world for conducting vulnerability assessments of multiple hazards [7]. To address this problem in Colombia, it has been proposed to include coastal erosion in disaster risk management as a public policy approach [13,14].

In recent years, evaluation of coastal vulnerability with an emphasis on geomorphological and physical factors has focused on the use of the coastal vulnerability index (CVI). This index was designed to estimate areas of risk caused by environmental and socioeconomic hazards and is widely used to implement decision-making within the framework of risk reduction. The CVI approach was initially developed by Gornitz [15,16] to study the vulnerability of the east coast of the United States of America due to sea level rise. The index allows to relate six physical variables in a quantifiable way and produces numerical data that cannot be directly equated with particular physical effects, but it does highlight the regions where the various effects of sea level rise may be greater [17]. Subsequently, the CVI was used to assess vulnerability along the Atlantic coast by the United States Geological Survey (USGS) in the study of Thieler and Hammar-Klose [18]. In the analysis of coastline change, some studies focused on analyzing the geomorphological and physical factors of the CVI but also included socioeconomic variables to develop resilience to the threats of climate change [19–24]. The state of vulnerability can be determined based on a group of conditions and processes resulting from physical, environmental, and socioeconomic factors that increase the susceptibility of people living in coastal areas to natural hazards, including their ability to adapt and respond to disasters [25,26].

Based on the CVI, other indices have been developed that focus more extensively on the conceptual structure of the vulnerability index using a process of analytical hierarchy, and this index is called the composite vulnerability index (COVI). Recently, several researchers have used this method to evaluate coastal vulnerability by incorporating different factors that indicate different dimensions (physical, ecological, social, and economic), including parameters such as biophysical exposure, sensitivity, and adaptive capacity or resilience to evaluate multiple hazards. For example, Zhang et al. [27] evaluated the coastal vulnerability to climate change of Bohai in China considering fifteen factors related to ecological, physical, and socioeconomic conditions in a COVI. Ghosh and Mistri [28] evaluated coastal vulnerability as a function of multiple factors with the composite vulnerability index in the lower delta of the Sundarban, India considering 22 indicators, mainly physical, climatic, and socioeconomic variables. Sahana and Sajjad [29] evaluated floods focusing on storm surge with a vulnerability index composed of remote sensing information in the Sundarban Biosphere Reserve, India considering seventeen factors. Finally, Furlan et al. [30] developed a multidimensional CVI to evaluate vulnerability to flood scenarios along the Italian coast considering multiple indicators. Although there are studies that evaluate the general vulnerability of the coasts of Buenaventura [13,21,22,31,32], these have focused mainly on geomorphological and physical dimensions. Numerous studies have been conducted around the world to examine different aspects of coastal vulnerability with a geospatial approach using the CVI and COVI (Table 1).

The objective of this work is to analyze coastal erosion at the watershed level using remote sensors in conjunction with GIS to build a COVI. Selection of indicators and the weighting assigned to each indicator are important parts of the study. Incorporation of physical, environmental, and socioeconomic variables to evaluate various indices using a weighting method allows a comprehensive view of spatial vulnerability considering that coastal watershed is the most appropriate scale to assess vulnerability to natural and anthropogenic changes.

Table 1. Studies conducted worldwide with a geospatial approach that uses coastal vulnerability indices.

Index	Tools *	Year	Country	Author
Coastal Vulnerability Index (CVI)	GIS, RS, DSAS	2021	India	Bera and Maiti, 2021 [19]
	GIS, RS, DSAS	2021	India	Pramanik et al., 2021 [23]
	GIS, RS, DSAS	2021	Egypt	Abdelaty, 2021 [33]
	GIS, RS, DSAS	2021	Greece	Boumboulis et al., 2021 [34]
	GIS, RS, DSAS	2020	Italy	Sekovskia et al., 2020 [35]
	GIS	2019	Brazil	Serafim et al., 2019 [24]
	GIS	2019	Spain	Koroglua et al.,2019 [36]
	GIS	2019	Malaysia	Mohda et al.,2019 [37]
	GIS, RS, DSAS	2019	Bangladesh	Hoquea et al., 2019 [38]
	GIS	2019	Colombia	Coca and Ricaute, 2019 [21]
	GIS, RS, DSAS	2019	Colombia	Gallego and Selvaraj, 2019 [22]
Composite Vulnerability Index (COVI)	GIS, RS, DSAS	2021	Tunisia	Hzami et al., 2021 [39]
	GIS	2021	China	Zhang et al., 2021 [27]
	GIS	2021	India	Ghosh and Mistri 2021 [28]
	GIS	2021	Italy	Furlan et al., 2021 [30]
	GIS	2020	India	Rehman et al., 2020 [40]
	GIS	2020	India	Sahana and Sajjad,2019 [29]
	GIS, RS, DSAS	2019	Bangladesh	Mullick et al., 2019 [41]

* Geographic information system (GIS), remote sensing (RS), and Digital Shoreline Analysis System (DSAS).

2. Materials and Methods

2.1. Study Area

Buenaventura is located in the Valle del Cauca in the central zone of the Colombian Pacific in one of the four Colombian departments on the coast of the Pacific Ocean. It encompasses the extensive area of the municipality and special port district of Buenaventura. It is bounded by the San Juan River to the north and by the Naya River to the south, semienclosed by two bays: Bay of Buenaventura and Bay of Malaga. Geographically, the coastal area has a total coastline of approximately 686 km and extends over the coordinates 4°2.23′82″ N and 77°26′18.87″ W, at 3°13′33.21″ N and 77°32′41.63″ W, as shown in Figure 1. The coastline is composed of barrier islands, intertidal zones, rocky cliffs, rocky platforms, alluvial and intertidal plains, estuaries, sandy beaches, and salt marshes [42]. The beaches of Buenaventura are of natural origin according to the sedimentological description from the granulometric analysis made by the Institute of Marine and Coastal Research of Colombia (INVEMAR) on beach samples for the sectors of Punta Soldado, La Bocana, and Piangüita in the department of Valle del Cauca. The average grain size distribution in different sampling campaigns in the years 2012, 2014, and 2015, indicates that, in this area of the municipality of Buenaventura, sediments showed a tendency mostly to a fine grain size [1]. The Chocó Biogeografico (biodiversity hotspot) includes the Pacific coastline between Darién in Panama and northwestern Ecuador, passing through the entire coastal strip of Colombia.

Within this hotspot, mangroves are one of the most important ecosystems in Valle del Cauca, covering 140 km^2 [43]. The tides of the Colombian Pacific coast are regular semidiurnal, that is, with two high tides and two low tides per day with a period of approximately 12.25 h, and their tidal range can reach slightly more than 4 m [44]. Precipitation generally shows monomodal behavior, with an annual average between 6821 mm and 7673 mm, and there are approximately 228 days with rain. The average annual temperature for the Pacific is 25.7 °C [45]. Structurally, Buenaventura is characterized by a flat morphology in the south and cliff formations to the north. There are three levels of terraces present in the river courses that seem to indicate recent tectonic activity of uplift and subsidence, formed by Quaternary deposits [44]. Economically, the port area of Buenaventura consists of several maritime terminals that provide port and logistics services in the most important port of the Colombian Pacific through which a large part of Colombia's foreign trade occurs. This port moves 47% of Colombian exports and imports, including those related to mining, oil and its derivatives [46]. Its population for 2020 was approximately 311,827 inhabitants, demographically composed of African descendants and mixed-ancestry and indigenous communities [47,48]. A percentage of the population lives in stilt houses located at or close to the shoreline exposed to waves, increasing their degree of exposure to the tidal regime [45]. Insufficient resources, multidimensional poverty, and remoteness are great challenges for the community. Almost 33.3% of the population lives below the poverty level [47]. Regarding studies of the coastal zone of Buenaventura related to vulnerability, the study by Ricaurte et al., 2021 [11] stands out, where the dominance of each component of the threat in the Colombian Pacific region was analyzed and it was established that it is determined by fragility, mainly social, economic, and institutional. Coca and Ricaurte 2019 [21] studied the town of La Barra since 2013, when a process of avulsion towards the sea began; associated with this event, an accelerated coastal erosion process could be measured, where the vulnerability of the population was evaluated. Gallego and Selvaraj 2019 [22] applied the coastal vulnerability index (CVI) using eight variables, three physical/hydrodynamic, three geological/geomorphological, and two socioeconomic variables. The coastline was classified into five relative vulnerability ranges. Cifuentes et al., 2017 [31] focused on studying the magnitude of shoreline change north of Buenaventura District over a 30-year period. On average, they found a rate of change of −0.2 m per year in the coastline, reflecting its erosional trend, with maximum EPR values of 26.9 m of accretion and −21 m of coastal erosion. Uribe et al., 2020 [32] explored the degree of vulnerability of ecosystem services in the northern area of Buenaventura to natural and anthropogenic hazards. Sea level rise and coastal erosion are the most likely threats to ecosystem services. One of the most significant dangers that threatens the study area is coastal erosion. To evaluate vulnerability, the analyses were grouped using the division of coastal watersheds (Figure 1).

2.2. Shoreline Change

The change in a coastline is an important parameter that can have a natural or anthropogenic origin and indicates the pattern of accretion/erosion in conjunction with different processes, such as waves, tides, sea levels, and topographic shape [49]. The coastline represents the boundary between the sea and the landmass. Evaluating coastal erosion is essential for planning future management strategies, land use planning, and risk management [50]. Historical photographs and high-resolution satellite data were used to monitor coastline changes during a period of 34 years (from 1986 to 2020). Initially, orthomosaics were created for 1986 based on data acquired from official datasets (aerial photos) of the Agustín Codazzi Geographical Institute (IGAC). Two sets of airborne synthetic aperture radar images (synthetic-aperture radar SAR) were used, the first for 2009 and the second of 2015, being the most accurate public use datasets available for the terrain of Buenaventura (Table 2). In addition, 19 high-resolution orthorectified images of the PlanetScope satellite from 2020 were acquired. The constellation of PlanetScope satellites consists of groups of individual high-resolution satellites; each satellite has a 3U CubeSat format (10 cm by 10 cm by 30 cm). The complete constellation of PlanetScope is approximately 130 satellites and is capable of taking images of the entire Earth's surface with four spectral bands (blue, green, red, and near infrared (NIR)); it has a spatial resolution of 3 m and a high temporal resolution (24 h) [51]. All sensors were used to extract the multitemporal coastline (Table 2).

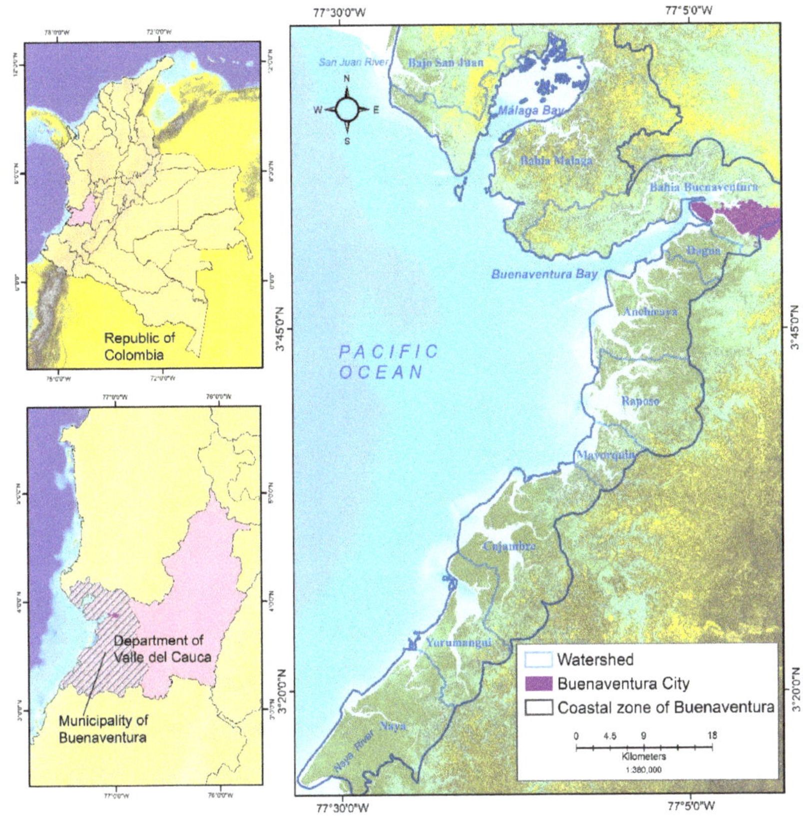

Figure 1. Map of the municipality of Buenaventura and its coastal watersheds.

Table 2. Images and products of the remote sensors used for coastline data extraction.

Year	Sensor	Product *	Spatial Resolution	Source
1986	Aerial photography	Orthomosaic	3 m	Geographic Institute of Colombia (IGAC)
2009	Synthetic-Aperture Radar Image—Airborne	Orthomosaic, DSM, DTM	3 m	Geographic Institute of Colombia (IGAC)
2015	Synthetic-Aperture Radar Image—Airborne	Orthomosaic, DSM, DTM	3 m	Regional Autonomous Corporation of Valle del Cauca (CVC)
2020	Satellite PlanetScope	Orthomosaic	3 m	This project

* Digital surface model (DSM) and digital terrain model (DTM).

To reduce uncertainty during the vectorization of the coastline from remote sensing data at the time of image capture, the tidal fluctuation error was taken into account [52]. The tidal errors were considered low since the acquired images showed the tides oscillating between 0.4 and ±2.3 m based on the data from the port of Buenaventura tide gauge, obtained from tide tables for the study area [53]. Taking into account the spatial resolution of 3 m, the possible displacement of the coastline by the tide was within the spatial uncertainty of the data. A visual interpretation of the coastline was performed using Catalyst software (v 2022, PCI Geomatics, Ontario, Canada) and ArcGIS (v 10.8, ESRI, Redlands, California, USA), in conjunction with Digital Shoreline Analysis System (DSAS) software, which complements ArcGIS and was developed by the Coastal Change Hazards project of the US Geological Survey. The DSAS allows the user to calculate rate of change statistics from multiple historical coastal positions; the rate of shoreline disposal/change was estimated through the software by calculating the end point rate (EPR) statistical parameter. The EPR is obtained by dividing the distance of coastline movement with the elapsed time between the oldest and youngest coastline position [54]. From the digitized coastlines for the four examined years (Figure 2), the date, uncertainty value, and type of coastline were standardized as required by the DSAS format. To create a uniform baseline, an interval of 25 m was used to create transects perpendicular to coastlines.

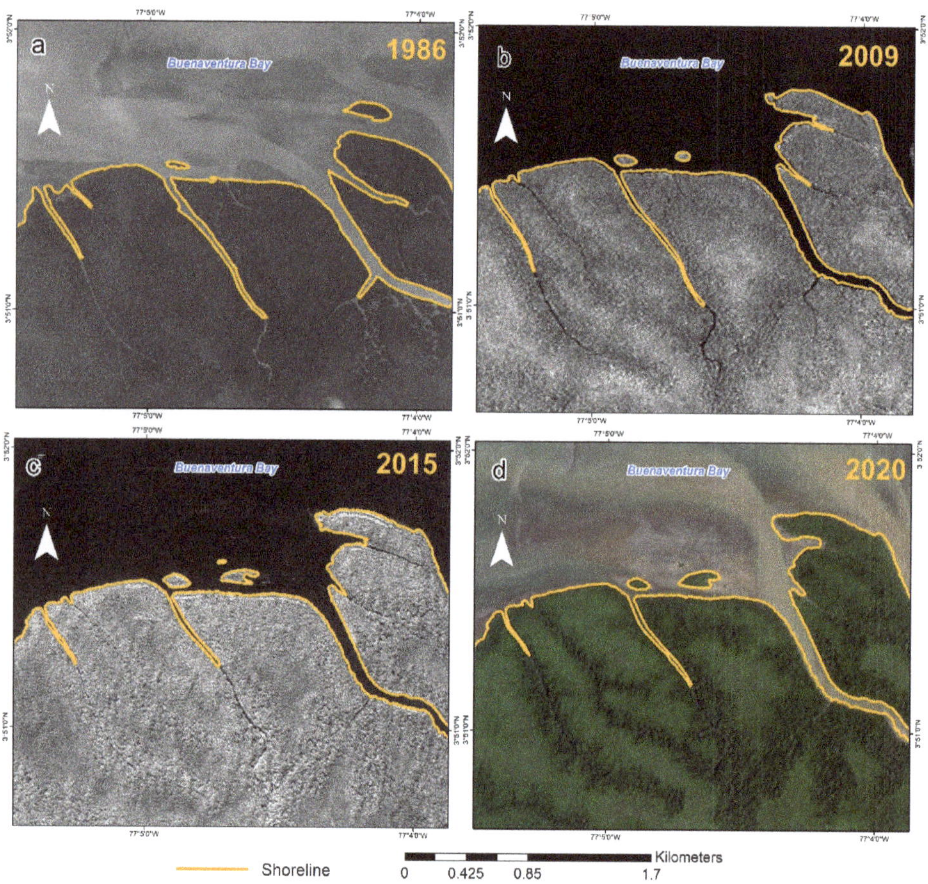

Figure 2. Example of digitization of the coastline in Buenaventura Bay in different years: (**a**) aerial photography mosaic from 1986, (**b**) airborne radar image from 2009, (**c**) airborne radar image from 2015, (**d**) PlanetScope 2020 satellite image.

2.3. Justification of the Indicators

A methodology was implemented to estimate vulnerability as a function of exposure, sensitivity, and adaptive capacity in conjunction with an analysis of coastal erosion at the basin level to understand how to mitigate and adapt to the risks from disasters in the coastal zone of Buenaventura in the Colombian Pacific. According to IPCC [26], vulnerability is explained in terms of exposure, sensitivity, and adaptive capacity. In this work, the COVI refers to the extent to which coastal systems are susceptible to the impacts of coastal erosion and global change. The COVI was developed based on the exposure index (EI), the sensitivity index (SI), and the adaptive capacity index (ACI) modified from the Sahana and Sajjad methodology [29], where resilience is replaced by adaptive capacity. The index allowed the relationship of physical variables in a quantifiable way with the decision-making approach using spatial analysis, giving equal weight to the indicators. Exposure and sensitivity together have a potential impact on coastal systems and are positively correlated with vulnerability based on the propensity of populations and coastal properties to be negatively affected by natural hazards [25]. In contrast, adaptive capacity helps to generate resilience against the adverse consequences of hazards, and this is negatively correlated with vulnerability [26,55].

Based on bibliographic research (Table 1), a total of 16 indicators were selected for the quantitative evaluation of the vulnerability indices, the exposure index, the sensitivity index, and the adaptability index. A detailed description of the selected indicators and their functional relationships with vulnerability are shown in Table 3. After the establishment of the index system, values were assigned and weighted using the appropriate formulas. Each variable was rated from 1 (very low) to 5 (very high) in qualitative ranges. To evaluate coastal vulnerability in the context of environmental hazards, the multicriteria spatial analysis (MCSA) approach and the simple average method (SAM) were used; these quantitative methods are widely used to evaluate vulnerability in the framework of coastal risk reduction [27–30,40,41].

Table 3. Selected indicators to construct the composite vulnerability index.

Components	Indicators	Class	Range
Exposure	Geomorphology	Barrier island, Flood plain, Intertidal flat without vegetation, Beach	5—Very High
		vegetated intertidal flat	4—High
		Alluvial valley	3—Moderate
		Island	2—Low
		Water body, Coastal lagoon, Hillocks and hills, Continental shelf, Marine terrace	1—Very Low
	Slope (degrees)	0–18°	5—Very High
		18–25°	4—High
		25–75°	3—Moderate
		75–80°	2—Low
		80–88°	1—Very Low
	Shoreline change rate (m/year)	−96.30 to −3.0	5—Very High
		−2.99 to 0.5	4—High
		−0.49 to 0.5	3—Moderate
		0.51 to 3.0	2—Low
		3.1 to 95.9	1—Very Low
	Sea level rise rate (mm/year)	>9	5—Very High
		6 to 9	4—High
		3.9 to 6	3—Moderate
		0 to 3.9	2—Low
		<0	1—Very Low
	Mean tidal range (m)	3.0 to 3.74	5—Very High
		2.25 to 2.99	4—High
		1.26 to 2.24	3—Moderate
		0.38 to 0.75	2—Low
		0.26 to 0.38	1—Very Low
	Significant wave height (m)	>6	5—Very High
		4 to 6	4—High
		2 to 4	3—Moderate
		1 to 2	2—Low
		<1.0	1—Very Low
	Flood inundation risk	ENSO floods	5—Very High
		Hydrometeorological flooding	4—High

Table 3. *Cont.*

Components	Indicators	Class	Range
Sensitivity	Roughness of terrain	0.131 (Very rough) 0.128 (Rough) 0.047 (Roughly open) 0.020 (Open) 0.001 (Smooth)	5—Very High 4—High 3—Moderate 2—Low 1—Very Low
	Multidimensional poverty	70.1% to 98.5% 50.1% to 70% 40.1% to 50% 30.1% to 40% 4.15% to 30%	5—Very High 4—High 3—Moderate 2—Low 1—Very Low
	Settlements	Urbanized area Villages Rural No settlement	5—Very high 4—High 3—Medium 2—Low
	Land Use and Land Cover	Urban zones, Artificial surfaces Cultivation areas, Banana, Coconut palm, Miscellaneous Shrubland, Guandal forest, Mangrove Forest, Mixed Forest, Natural grassland, Island, Cultivated grassland, Secondary vegetation Temporary flooded areas, Natural areas, Other marshy areas Shallows and intertidal flats, Littoral barriers, Artificial ponds, Ocean, Beaches, Rivers	5—Very High 4—High 3—Moderate 2—Low 1—Very Low
	Population (inhabitant/km^2)	80 to 20,656 50 to 80 15 to 50 5 to15 1 to 5	5—Very High 4—High 3—Moderate 2—Low 1—Very Low
Adaptive capacity	Economic activities	Industrial fishing Artisanal fishing Ecotourism Landscape Recreation—beaches	5—Very High 4—High 3—Moderate 2—Low 1—Very Low
	Medical services (Health care provided)	0 to 56 57 to179 180 to 327 328 to 628 629 to 1186	5—Very High 4—High 3—Moderate 2—Low 1—Very Low
	Distance to roads	2000 m 1000 m 500 m 250 m 100 m	5—Very High 4—High 3—Moderate 2—Low 1—Very Low
	Literacy rate	<67% 67% to 73% 73% to 81% 81% to 86% >86%	5—Very High 4—High 3—Moderate 2—Low 1—Very Low

2.4. *Exposure Index (EI)*

An EI includes the eight factors that trigger the risk of biophysical exposure, and these factors were compiled from an extensive review of previous studies and expert opinions. The shape of a coastline is fundamental in analyzing vulnerability due to the degree of relative resistance that a coastal geoform can have against erosion [56]. To determine the geomorphology in the coastal zone of Buenaventura, the geomorphological maps developed by the Institute of Marine and Coastal Research of Colombia (INVEMAR) [45] and the geomorphological maps of the Center for Oceanographic and Hydrographic Research of Colombia (CIOH) [57] were used. The geomorphological units were reinterpreted and adjusted to a finer scale (1:10,000) using a geomorphometric analysis [58] with the digital terrain model (DTM) derived from the 2015 radar data. The degree of topographic variation influences the processes through which hydrometeorological events can expose a coast to floods and coastline retreat [26,59]. To determine the slope in the coastal zone, the 2015 radar DTM was used. The rate of coastline change indicated change characteristics that were largely due to erosion and accretion, and the rate of change was calculated with the DSAS software to determine EPR.

One of the aspects that can generate the greatest vulnerability in coastal areas is sea level rise [59]. To analyze this aspect, raster data of monthly and annual sea level averages from satellite observations for the global ocean from 1993 to 2020 were used, and they were provided by the Copernicus Marine Environment Monitoring Service (CMEMS) [60]. This dataset provides global estimates of sea level based on satellite altimetry measurements, and these estimates are calculated with respect to an average reference period of twenty years (1993–2012) using updated altimetry standards. The data are provided in NetCDF format with a horizontal resolution of $0.25° \times 0.25°$. To define the mean tidal range (MTR), the values obtained for the municipality of Buenaventura from the analysis of Gallego and Selvaraj [20] were used. The MTR was obtained using the mean high tide difference (MHW), a record of at least 19 years, and the mean low tide (MLW) from the recorded data of the Buenaventura tide gauge provided by the University of Hawaii Sea Level Center (UHSLC) [22]. Significant wave height is a representation of wave energy that is related to the movement and transport of coastal sediments [22]. For the study area, the significant wave height data were provided by the CMEMS information system. This information was compiled with data from the European Center for Medium-Range Weather Forecasts (ECMWF) [61] in the ERA5, an analysis of historical data. The analysis combines monthly averages of model data with observations from around the world in a globally complete and consistent dataset. Three points were selected from the global grid ERA5 at a relative depth of 15 m located between coordinates 3°49′44.25″ N, 77°23′32.05″ W, and 3°49′58.67″ N, 77°7′46.89″ W. The data were recalculated in a regular latitude and longitude grid with a horizontal resolution of $0.25° \times 0.25°$ for 1959 to 2020. Significant wave height dataset has been extracted for the Colombian Pacific area in NetCDF format.

The risk of flooding due to hydrometeorological processes was determined using hazard maps at a scale of 1:100,000 from the Institute of Hydrology, Meteorology and Environmental Studies (IDEAM). These maps showed flooding from 1 m to 6 m as a result of storm surge, heavy rains due to historical accumulation, and flooding due to the El Niño Southern Oscillation (ENSO). The percentage of area under the different flood risk levels was calculated to evaluate the average flood risk. Terrain roughness can be defined generally as a characteristic related to the irregularity or topography of the terrain [62]. This represents the resistance of the land surface to water intrusion. The greater the roughness and sinuosity of land surfaces, the lower the vulnerability is [63]. From the 2015 digital elevation model, the roughness values were obtained.

The EI was determined by Equation (1) as:

$$\text{Exposure index} = (\text{EX1} + \text{EX2} + \text{EX3} + \text{EX4} + \text{EX5} + \text{EX6} + \text{EX7} + \text{EX8})/8 \quad (1)$$

where the EI is a function of the ranges of geomorphology (EX1), slope (EX2), shoreline change rate (EX3), sea level rise rate (EX4), mean tidal range (EX5), significant wave height (EX6), flood inundation risk (EX7), and terrain roughness (EX8).

2.5. Sensitivity Index (SI)

Sensitivity is expressed as the elements at risk where a potential danger could be triggered. A very important aspect of sensitivity is population density since it plays a very influential role in the vulnerability of the coastal zone. The population density was calculated from the last National Population and Housing Census, which was in 2018 [48]. The multidimensional poverty factor was obtained from official census data. In Colombia, the index for measuring multidimensional poverty was designed by the National Planning Department (DNP) based on an adaptation of the methodology of Alkire and Foster (2011) [64]. For Colombia, the direct method evaluates the results of satisfaction (or no deprivation) that an individual has with respect to certain demographic characteristics that are considered vital, such as health, education, and employment [65]. In the study area, the coastal settlements were located within 6 km of the coastline, and their coastal vulnerability increases as the settlements are located closer to the coast. These data were obtained from the National Administrative Department of Statistics (DANE) from the national geostatistical framework (NGF). This layer contains the political-administrative divisions of Colombia: departments and municipalities, population centers, and other geostatistical areas where populations appear. Finally, this study used land cover and land use data at municipal level that were developed by the Regional Autonomous Corporation of Valle del Cauca (CVC) using the Corine Land Cover methodology adapted for Colombia [66]. Determining the land use and land cover in different human settlements is critical when assessing coastal vulnerability due to the possible socioeconomic impacts and their impacts on communities.

The SI was determined by Equation (2) as follows:

$$\text{Sensitivity index} = (\text{SE1} + \text{SE2} + \text{SE3} + \text{SE4})/4 \quad (2)$$

where the sensitivity index is the function of the ranges of multidimensional poverty (SE1), settlements (SE2), land use and land cover (SE3), and population (SE4).

2.6. *Adaptive Capacity Index (ACI)*

Adaptive capacity is the potential ability of a system to resist the adverse impact of hazards using available resources, skills, and technology [26]. Therefore, it is a crucial factor in determining the impacts of climate change. The distance to road networks was used to determine transportation services by generating a map of proximity to the road network using a specific distance, and this area was divided into five buffer zones. Generally, an increase in the distance to the roads increases exposure due to the difficulty of accessing or exiting an area. The vulnerable population based on access to medical services was quantified from the geostatistical data of the 2018 National Population and Housing Census [48]. The availability of medical services in rural areas helps to strengthen health services in case of hazardous events. The economic activity factor was calculated as an aggregation of five factors that can be affected by coastal hazards, and they were industrial fishing, artisanal fishing, ecotourism, landscapes, and recreation/beaches, which are vulnerable to varying degrees.

These indicators have a negative correlation with vulnerability and help increase the resistance of coastal communities to natural disasters. The ACI was determined by Equation (3) as follows:

$$\text{Adaptive capacity index} = (\text{AC1} + \text{AC2} + \text{AC3} + \text{AC4})/4 \quad (3)$$

where the ACI is a function of the ranges of economic activities (AC1), medical services (AC2), distances to roads (AC3), and literacy rates (AC4).

2.7. *Composite Vulnerability Index (COVI)*

The composite vulnerability index (COVI) was calculated as positively correlated with the EI and SI but negatively correlated with the ACI. Using the three indices, the vulnerability was determined by Equation (4) as follows:

$$\text{COVI} = (\text{Exposure x Sensitivity}) - \text{Adaptive capacity} \quad (4)$$

It was calculated by a simple average of normalized scores (Equation (4)). The value lies between 0 and 1, denoting very low to very high coastal vulnerability. There are several studies [27,28,40] that, in recent years, have validated Equation (4); in the case of India, Sahana and Sajjad 2019 [29] used a variation that replaces the term adaptive capacity with resilience.

3. Results

3.1. Rate of Coastline Displacement

Coastline movements with respect to erosion and accretion are a direct indicator of risk. The EPR statistical parameter of the multitemporal coastlines (1986 to 2020) was estimated after calculating 7818 transects with the DSAS tool; the coastline of Buenaventura was 406 km long. Further, 35% of the transects on the analyzed coastline in the study area were identified as stable. However, 18% of the coastline reflected erosion processes, and 47% reflected accretion processes (Figure 3a).

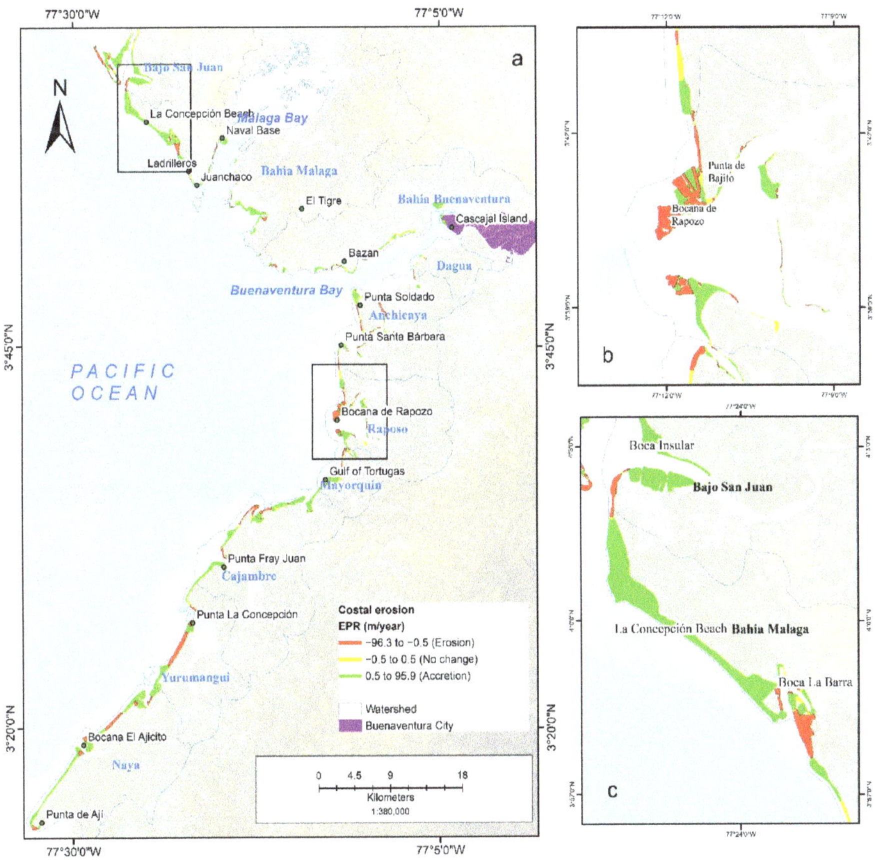

Figure 3. Results of the EPR parameter analysis showing the zones of erosion and coastal accretion. Locations b and c show places with areas of great change.

Grouping the results of the watershed analysis, the watersheds and the places with the greatest erosion and accretion are identified in Table 4. The greatest change in coastline (erosion–accretion) according to the EPR analysis was −96.23 m/year adjacent to the Raposo watershed in the Bocana de Raposo zone (Figure 3b) and 95.83 m/year adjacent to the Málaga Bay watershed in Playa La Concepción (Figure 3c). The general trend of the coastline was accretion according to the average EPR (2.84 m/year).

Table 4. Watersheds and locations where erosion/accretion occurred along the coastline.

State	Watershed	km	Location	km
Erosion	Anchicaya	19.96	Punta Soldado, Punta Santa Barbara	18.27
	Cajambre	12.22	Punta Bonita	12.22
	Yurumangui	9.52	Punta La Concepción	8.31
	Bajo San Juan	9.16	El Choncho	6.87
	Naya	7.50	El Ajicito beach	4.7
	Malaga Bay	7.27	The Bar	6.05
Accretion	Naya	35.35	Ajicito and Ají beach	32.09
	Malaga Bay	32.83	La Concepción Beach	20.1
	Bajo San Juan	27.27	Boca de Bajo San Juan	20.08
	Cajambre	21.57	Punta Fray Juan	20.73
	Raposo	15.65	Raposo mouth	15.29

3.2. Exposure to Coastal Erosion

Approximately 31.7% of the watersheds, concentrated in the central and southern regions of the municipality of Buenaventura, had low to very low exposure scores, while the remaining areas had medium exposure scores (23.2%). Of the total watersheds (10), at least four watersheds were highly exposed to coastal erosion (44.9%) in the high to very high ranges (Figure 4a). The EI values varied with geographic heterogeneity among the coastal watersheds, with Málaga Bay registering the highest values. This is mainly attributed to the relatively higher scores for slope, average wave height, coastal erosion, and risk of flooding over the area of the watershed directly exposed to the sea (Figure 4a). The places in the outermost part of the bay are defined as estuarine areas and directly affected by intertidal processes, and they contain sandy beaches with sand of continental origin, transported by coastal rivers, and a moderate drainage density. In contrast, Málaga Bay has the lowest biophysical exposure (Figure 4b); this scenario also occurs in Buenaventura Bay, the area with the second lowest exposure value. The interior area of the bays is characterized by large areas of mangroves, small beaches, and cliffs, and the lower scores for this area were attributed mainly to the risk factors related to flooding, coastal erosion, and average wave height (Figure 4c).

3.3. Sensitivity to Coastal Erosion

The analysis of the SI revealed that approximately 37% of the areas were classified as having very low to low sensitivity scores, while the remaining areas were classified as having medium scores (32%); additionally, some watersheds were highly to very sensitive (29.4%), as shown in Figure 5a. Of all the coastal areas, the highest SI was calculated for the city of Buenaventura and the Dagua River watershed (Figure 5b) due to these areas having the highest scores for multidimensional poverty and population density and being the only large, urbanized areas. In contrast, the lowest SI was calculated for the interior of Málaga Bay, which was mainly due to the lower scores for population density and multidimensional poverty and its lack of urbanized areas (Figure 5c). Most coastal watersheds with mangrove forests and natural vegetation showed low sensitivity to coastal erosion.

3.4. Adaptive Capacity to Coastal Erosion

The adaptive capacity of the Buenaventura watersheds was low since more than 85% of the total areas were characterized by low to very low resilience; places that generally have little or no infrastructure development have no connection to land by roads, and these areas depend on coastal resources and connections through maritime routes. The remaining watersheds had medium adaptive capacity scores of 11.2% and high to very high adaptive capacity scores of 3.1%. The ACI values showed great spatial heterogeneity between the coastal watersheds (Figure 6a), with the highest value occurring in Buenaventura Bay (Figure 6b) and the lowest value occurring in Málaga Bay (Figure 6c). The factor that most contributed to the high adaptation capacity of the Bay of Buenaventura was the existence of better facilities and greater access to public services, transportation routes, trade and tourism services, education, and medical services, while Málaga Bay's adaptive capacity was reduced mainly by its relatively lower scores for these factors; in addition, its low literacy rate resulted in this area having little or no adaptive capacity.

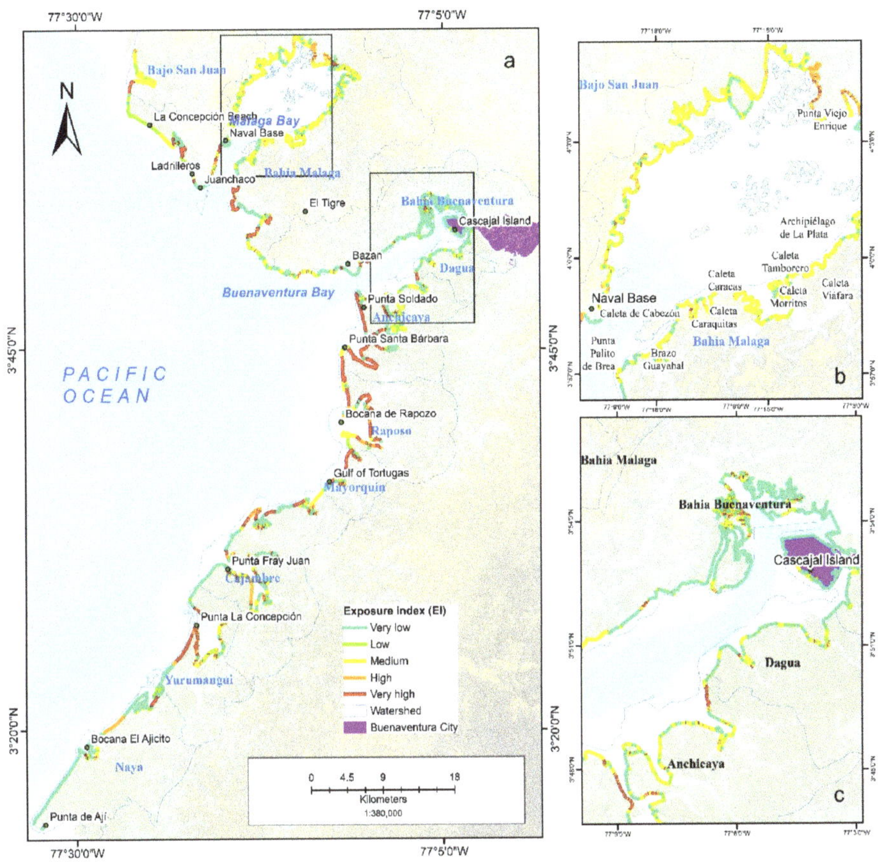

Figure 4. Exposure to coastal erosion in the study area.

3.5. Results of the Composite Vulnerability Index

The COVI analysis revealed that the coastal watersheds showed great spatial heterogeneity in their vulnerability levels (Figure 7). In the coastal zone of Buenaventura, 31.4% of the area had moderate vulnerability values, 22% had low vulnerability values, and 4.5% had very low vulnerability values. In contrast, 16.18% of the area had high vulnerability values, followed by a 25.8% with very high vulnerability values, indicating that 41.9% of the study area had the highest combined vulnerability values. The marine areas exposed to the ocean in the Málaga Bay watershed (Figure 8a,b), such as the Anchicayá, Raposo, and Cajambre watersheds (Figure 8a,c), are very vulnerable to coastal erosion as they are located in the southern–central zone, where hydrodynamic forcing increases due to coastal relief change.

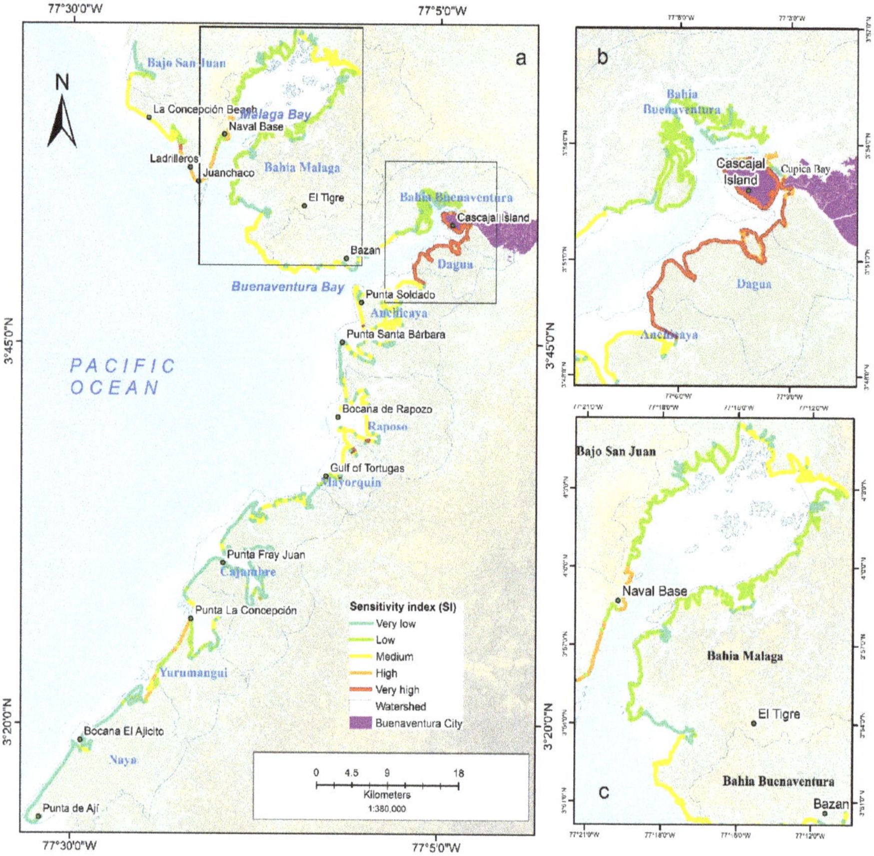

Figure 5. Sensitivity to coastal erosion in the study area.

The causes of the high vulnerability levels of these watersheds were their lack of economic development and resilience, with greater rates of multidimensional poverty, illiteracy rates, coastal erosion, higher average wave height, and risk of flooding over the area of the basin directly exposed to the sea. Within these watersheds, there are places with high vulnerability levels, such as La Boca de La Barra, Ladrilleros, Juanchaco, La Base Naval, and the sectors between Punta Domingo and Punta Culo de Barco (Figure 8b). In the Málaga Bay watershed, the interior of the bay had moderate vulnerability levels. This watershed, although located in a remote, difficult-to-access area, has a high capacity for recovery of its biophysical environment since a national protected area (Parque Nacional Natural Uramba Málaga Bay) and a regional protected area (Parque Natural Regional La Sierpe) have been developed for environmental tourism development and ecosystem protection. This area also has a low population density and minimal effects from coastal erosion, but it is highly vulnerable in terms of its ability to adapt due to the low literacy rate of the population, its lack of public services, and minimal access to medical services. Watersheds with low to very low vulnerability levels occur in the central (Buenaventura) and southern (Naya) parts of the coastal zone.

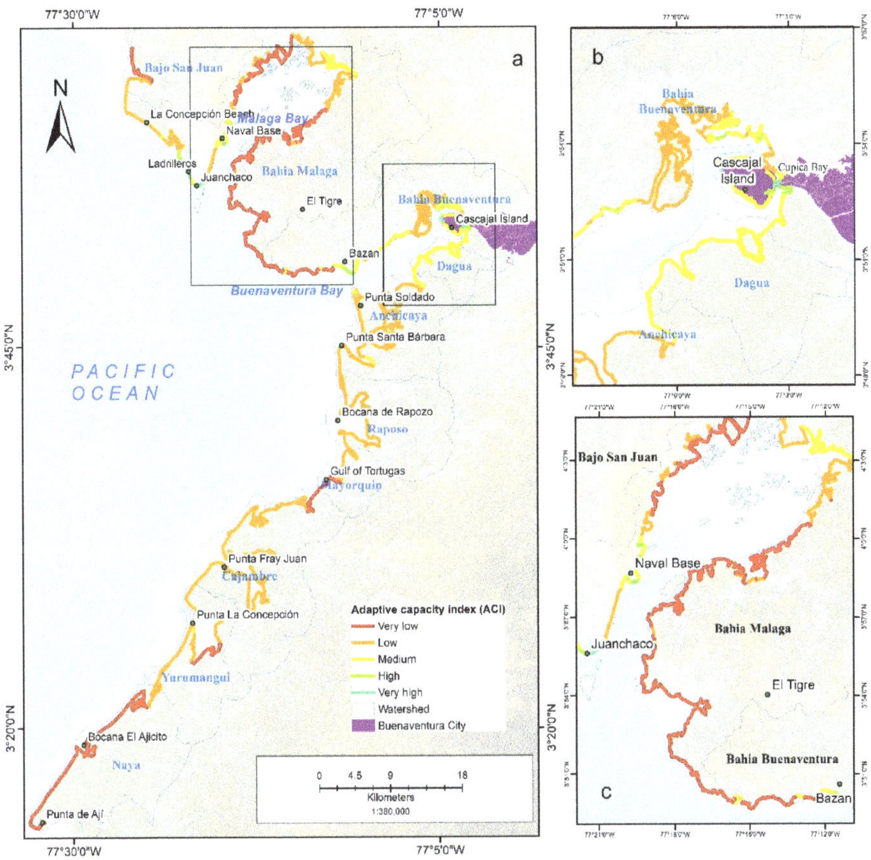

Figure 6. Adaptive capacity in the study area.

Buenaventura has a network of urban roads and the most important port in Colombia, where most of the commercial activity is generated, whereas, around the port, infrastructure and services have been developed. In addition, this area has greater access to services, transportation, education, and medical services. This is associated with its relatively higher adaptive capacity and its lower biophysical exposure, which was mainly attributed to the lower vulnerability scores for flooding, coastal erosion, and average wave height. On the other hand, the community in this area had a very high sensitivity level due to its high multidimensional poverty level and highest population density of the entire coastal area. The Naya watershed has a low vulnerability level mainly due to minimal coastal erosion, whereas it has the highest accretion rate, low population density, and low multidimensional poverty. The COVI helped to estimate the most vulnerable coastal population as a function of coastal erosion. Currently, 82,008 people, representing 26.2% of the total coastal population of Buenaventura, live in the Málaga Bay watershed (Figure 8b) and the Anchicayá and Raposo basins, which are the most prone to erosion and have a greater number of people in areas with very high vulnerability levels compared to those in the other watersheds (Figure 8c). With the 2100 sea level rise, it is estimated that the number of highly threatened people will increase with respect to the current scenario.

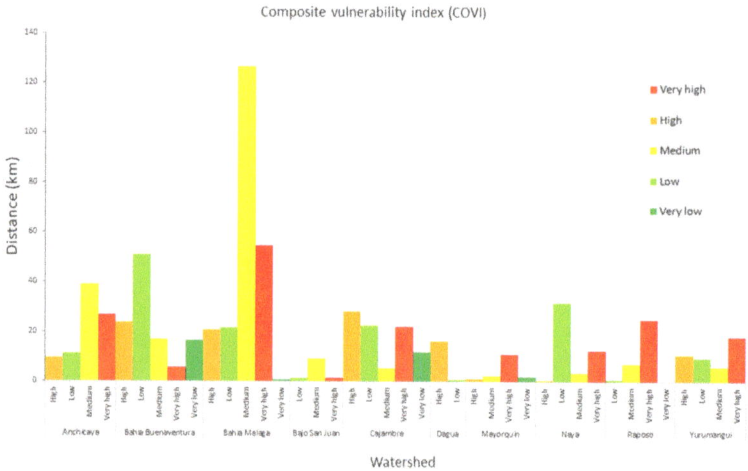

Figure 7. Distribution of vulnerability by watershed in the study area.

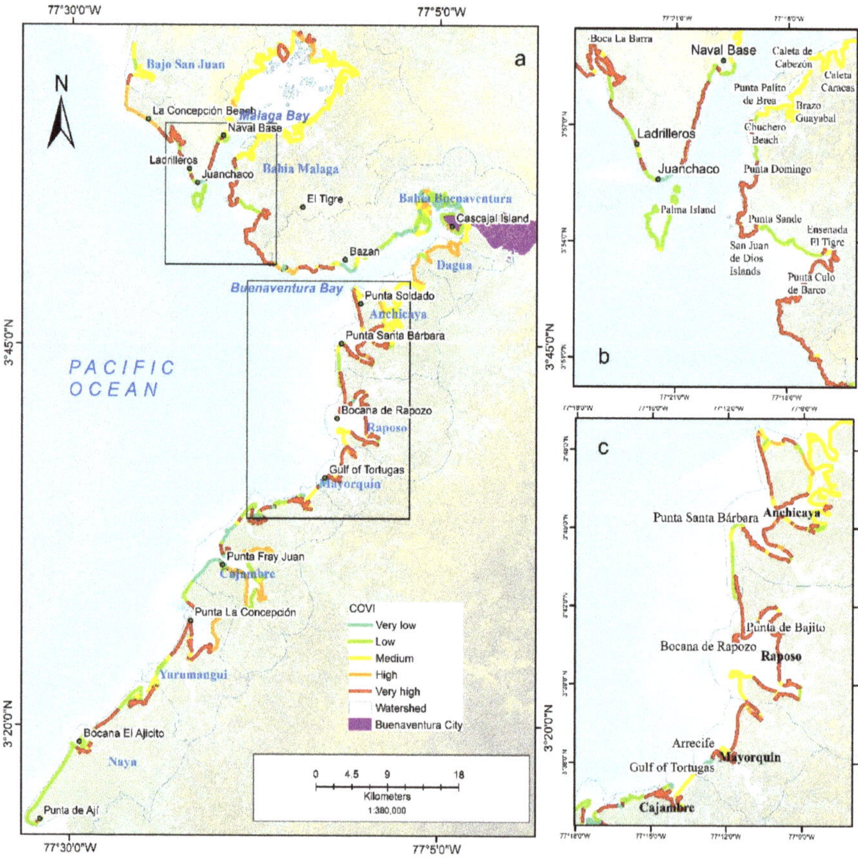

Figure 8. Composite vulnerability index in the study area.

4. Discussion

A coastal vulnerability assessment is important for spatial planning and disaster risk reduction [7,59]. In this study, remote sensing technology and GIS were integrated to study the change in the coastline in the long-term using active and passive sensors with reasonable precision. In the analysis with the DSAS tool from 1989 to 2020, the coastline was shown to experience the greatest change in erosion according to the EPRs, which were recorded as −96.2 m/year and 95.8 m/year per erosion and accretion event, respectively. There was a general trend in coastline accretion according to the average EPR of 2.8 m/year, with a total accretion of 48%. However, the Cifuentes [31] study from 1986 to 2015 showed that, on average, a change rate of −0.2 m/year occurred, reflecting an erosion trend with maximum EPR values of 26.9 m/year of accretion and −21 m/year of coastal erosion. The marked differences in the values of the previous study with those of this study may be explained from the higher temporal coverage used in this work, mainly in the La Concepción beach area. This scenario also explains why the study of Cifuentes [31] used data with a low spatial resolution (30 m). This study used the mean tidal range parameter from the work of Gallego and Selvaraj 2019 [22]; the data were taken from this study and integrated as a layer of analysis. We emphasize that, in our work, a higher spatial and temporal resolution was used in all physical and socioeconomic variables. Regarding the limitations of the study, oceanographic information was requested from the Colombian Center for Oceanographic Data, from which it was not possible to obtain a dataset that covered the entire period of analysis, so the global scale data from ERA5 were used.

This study used the COVI to estimate spatial vulnerability to coastal erosion through an analysis of coastal watersheds. The vulnerability in the Buenaventura area has been characterized by previous studies [13,20–22,31,32]. However, an evaluation of its vulnerability to multiple risks and physical, environmental, and socioeconomic parameters, such as the COVI, has received minimal attention in studies on the coastal zone of Buenaventura. The delineation of the COVI as a function of three indices (EI, SI, and ACI) indicates comprehensively the critical aspects of coastal vulnerability.

The general analysis of the indices showed that the Málaga Bay basin is highly exposed according to the EI of the oceanic portion of the coastal zone, mainly due to the weight of physical factors (slope, average wave height, coastal erosion, and risk of flooding). Consistent with the results of Uribe et al. [32], the most frequent natural and anthropogenic hazards occur in the external region of Málaga Bay, and hazards are mainly caused by water-related erosion, bioerosion, and landslides. Uribe et al. [32] stated that the beaches of Juanchaco and Ladrilleros in the Málaga Bay area are permanently exposed to strong waves and high-energy processes. By contrast, in the internal part of the same watershed, the lowest biophysical exposure occurs because it is protected by the bay, and this internal area has the lowest SI. Ricaurte et al. [13] stated that the factor determining the erosion threat in the Pacific region is vulnerability, mainly social, economic, and institutional vulnerability, and we agree with this conclusion. In this study, the influence that socioeconomic factors (lack of medical services, low literacy rates, and minimal economic activity) have on vulnerability scores was verified, and this impact results in a low ACI, hindering the ability to face coastal hazards. The watershed with the greatest sensitivity to erosion according to the SI is Buenaventura, mainly due to the high score for the multidimensional poverty factor and its highest population density. However, this watershed also shows the highest value of adaptive capacity (ACI) because it has the best facilities and greater access to all types of services as it is the most urbanized area. The EI showed low biophysical exposure within the area of the bay because its exposure to biophysical factors is reduced, a concept consistent with that in Gallego and Selvaraj [22], who affirmed that the biophysical system of Buenaventura Bay, due to its geomorphological characteristics, steep slopes, and high vegetation cover, provides it a lower degree of relative vulnerability.

The results of the study revealed that the high exposure (EI) and especially the lack of adaptive capacity (ACI) directly affected the high vulnerability of the study area, as shown by the COVI. Coastal watersheds, such as Málaga Bay and Anchicayá, Cajambre, and

Raposo basins, are the most vulnerable to coastal hazards physically and socioeconomically. Socioeconomic factors are the main internal element that increases vulnerability. Further, 85% of the watersheds have low adaptive capacity, and households do not have access to drinking water and medical resources. Most households in the coastal area of Buenaventura live below the poverty level and have very limited access to economic resources, so they cannot cope with any type of extreme event. This work analyzed the socioeconomic, biophysical, and ecological aspects together to provide valuable information on the factors that are critical to vulnerability at the spatial level. Those responsible for decision-making will likely be able to act in the most vulnerable areas, using the results as an analytical basis to develop adaptation strategies in the face of vulnerability and climate change.

5. Conclusions

The use of the COVI provided a broad view of the coastal area of Buenaventura at a detailed scale; 41.9% of the coastal area has a high level of vulnerability to erosion, with approximately 82,000 people at risk. This analysis revealed that the watersheds located in the northern (Málaga Bay) and central (Anchicayá, Cajambre, and Raposo) parts of the coastal zone were more vulnerable. If this situation continues, then, clearly, the population will be extremely affected by the increase in potentially disastrous events, such as the rise in sea level and ENSO events.

Therefore, the vulnerability of the region to coastal erosion will increase in the near future, further affecting socioeconomic and ecological systems and making it necessary to plan disaster management actions. Structural and nonstructural measures are needed to improve the adaptive capacity of the inhabitants, improve the physical resilience of the environment, and improve the socioeconomic activities of the community. The results of this type of study can assist decision-makers in highlighting actions to prioritize coastal areas in the development of management plans and land use planning to improve the adaptability of coasts to climate change.

This vulnerability assessment can be refined by incorporating other factors into the indices, such as a three-dimensional hydraulic flood model, a valuation of ecosystem services, and broader sociodemographic aspects, for which more in situ data and complementary physical monitoring data are needed. This work used data available from optical sensors and radar; however, in the future, data from high-resolution active sensors, such as light detection and ranging (LIDAR), could be used as they can generate more accurate results in terms of modeling the terrain and coastline, providing a more accurate measurement of spatial vulnerability.

Author Contributions: Conceptualization, Jose Eduardo Fuentes and Cesar Edwin Garcia; methodology, Jose Eduardo Fuentes; software, Jose Eduardo Fuentes; validation, Jose Eduardo Fuentes and Cesar Edwin Garcia; formal analysis, Jose Eduardo Fuentes and Cesar Edwin Garcia; investigation, Jose Eduardo Fuentes and Robin Alexis Olaya; resources, Jose Eduardo Fuentes and Robin Alexis Olaya; data curation, Robin Alexis Olaya; writing—original draft preparation, Jose Eduardo Fuentes and Robin Alexis Olaya; writing—review and editing, Jose Eduardo Fuentes and Robin Alexis Olaya; visualization, Jose Eduardo Fuentes; supervision, Jose Eduardo Fuentes and Cesar Edwin Garcia; project administration, Jose Eduardo Fuentes; and funding acquisition, Jose Eduardo Fuentes and Robin Alexis Olaya. All authors have read and agreed to the published version of the manuscript.

Funding: This work was supported in part by the Vice-Rectory of Research of the Universidad del Valle under grant CI 4406—"Internal Call 124-2020" associated with the project "Multitemporal analysis of vulnerability by coastal erosion in the Coastal Environmental Unit Malaga Buenaventura".

Institutional Review Board Statement: Not applicable.

Informed Consent Statement: Not applicable.

Data Availability Statement: Not applicable.

Conflicts of Interest: The authors declare no conflict of interest.

References

1. INVEMAR. *Informe del Estado de los Ambientes y Recursos Marinos y Costeros en Colombia, 2019*; Invemar: Santa Marta, Colombia, 2020; p. 183.
2. Ricaurte Villota, C.; Gonzélez Arteaga, M.E.; Coca Dominguez, O.; Bejarano Espinosa, H.M.; Morales Giraldo, D.F.; Correa Rojas, C.X.; Briceño Zuluaga, F.; Legarda, G.; Arteaga, M. *Amenaza y Vulnerabilidad por Erosión Costera en Colombia: Enfoque Regional para la Gestión del Riesgo*; INVEMAR, Ed.; Instituto de Investigaciones Marinas y Costeras "José Benito Vives De Andréis": Santa Marta, Colombia, 2018; Volume 33, p. 268.
3. Gabler, C.A.; Osland, M.J.; Grace, J.B.; Stagg, C.L.; Day, R.H.; Hartley, S.B.; Enwright, N.M.; From, A.S.; McCoy, M.L.; McLeod, J.L. Macroclimatic change expected to transform coastal wetland ecosystems this century. *Nat. Clim. Change* **2017**, *7*, 142–147. [CrossRef]
4. Richard, J.T.K.; Nicholls, R.J.; Sachooda, R.; Michele, C.; James, A.; Buckley, E.N. Technological Options for Adaptation to Climate Change in Coastal Zones. *J. Coast. Res.* **2001**, *17*, 531–543.
5. Hermelin, M. Geomorphological Landscapes and Landforms of Colombia. In *Landscapes and Landforms of Colombia*; Hermelin, M., Ed.; Springer International Publishing: Cham, Switzerland, 2016; pp. 1–21. [CrossRef]
6. Lichter, M.; Zviely, D.; Klein, M.; Sivan, D. Sea-Level Changes in the Mediterranean: Past, Present, and Future—A Review. In *Seaweeds and their Role in Globally Changing Environments*; Seckbach, J., Einav, R., Israel, A., Eds.; Springer: Dordrecht, The Netherlands, 2010; pp. 3–17. [CrossRef]
7. Adger, W.N. Vulnerability. *Glob. Environ. Change* **2006**, *16*, 268–281. [CrossRef]
8. Church, J.A.; Clark, P.U.; Cazenave, A.; Gregory, J.M.; Jevrejeva, S.; Levermann, A.; Merrifield, M.A.; Milne, G.A.; Nerem, R.S.; Nunn, P.D. *Sea Level Change*; PM Cambridge University Press: Cambridge, UK, 2013.
9. Nicholls, R.J. Analysis of global impacts of sea-level rise: A case study of flooding. *Phys. Chem. Earth Parts A/B/C* **2002**, *27*, 1455–1466. [CrossRef]
10. Pandey, B.W. *Geoenvironmental Hazards in Himalaya: Assessment and Mapping (the Upper Beas Basin)*; Mittal Publications: New Delhi, India, 2002.
11. Vogel, C.; O'Brien, K.J. Vulnerability and global environmental change: Rhetoric and reality. *Aviso Inf. Bull. Glob. Environ. Change Hum. Secur.* **2004**, *13*, 1–8.
12. McCarthy, J.J.; Canziani, O.F.; Leary, N.A.; Dokken, D.J.; White, K.S. *Climate Change 2001: Impacts, Adaptation, and Vulnerability: Contribution of Working Group II to the Third Assessment Report of the Intergovernmental Panel on Climate Change*; Cambridge University Press: Cambridge, UK, 2001; Volume 2.
13. Ricaurte-Villota, C.; Santamaría-del-Ángel, E.; Coca-Domínguez, O.; Morales Giraldo, D.; González-Arteaga, M. Determining factors of the hazard and vulnerability by coastal erosion in Colombia. *Rev. Geográfica Chile Terra Aust.* **2021**, *1*, 129–139. [CrossRef]
14. Rozemeijer, M.J.C. *Marine and Coastal Ecological Potential for the Economic Development of Colombia*; IMARES: Yerseke, The Netherlands, 2013.
15. Gornitz, V. Vulnerability of the East Coast, U.S.A. to Future Sea Level Rise. *J. Coast. Res.* **1990**, *9*, 201–237.
16. Gornitz, V.; White, T.W.; Cushman, R.M. *Vulnerability of the US to Future Sea Level Rise*; Oak Ridge National Lab.: Oak Ridge, TN, USA, 1991.
17. Gornitz, V.M.; Daniels, R.C.; White, T.W.; Birdwell, K.R. The Development of a Coastal Risk Assessment Database: Vulnerability to Sea-Level Rise in the U.S. Southeast. *J. Coast. Res.* **1994**, *12*, 327–338.
18. Thieler, E.R.; Hammar-Klose, E.S. *National Assessment of Coastal Vulnerability to Sea-Level Rise: Preliminary Results for the US Atlantic Coast*; US Geological Survey: Reston, VA, USA, 1999.
19. Bera, R.; Maiti, R. An assessment of coastal vulnerability using geospatial techniques. *Environ. Earth Sci.* **2021**, *80*, 306. [CrossRef]
20. Coca-Domínguez, O.; Ricaurte-Villota, C. Validation of the Hazard and Vulnerability Analysis of Coastal Erosion in the Caribbean and Pacific Coast of Colombia. *J. Mar. Sci. Eng.* **2019**, *7*, 260. [CrossRef]
21. Coca-Domínguez, O.; Ricaurte-Villota, C. Análisis de la evolución litoral y respuesta de las comunidades afro-descendientes asentadas en la zona costera: Caso de estudio La Barra, Buenaventura, Pacífico Colombiano. *Entorno Geográfico* **2019**, *17*, 7–26. [CrossRef]
22. Gallego Perez, B.E.; Selvaraj, J.J. Evaluation of coastal vulnerability for the District of Buenaventura, Colombia: A geospatial approach. *Remote Sens. Appl. Soc. Environ.* **2019**, *16*, 100263. [CrossRef]
23. Pramanik, M.K.; Dash, P.; Behal, D. Improving outcomes for socioeconomic variables with coastal vulnerability index under significant sea-level rise: An approach from Mumbai coasts. *Environ. Dev. Sustain.* **2021**, *23*, 13819–13853. [CrossRef]
24. Serafim, M.B.; Siegle, E.; Corsi, A.C.; Bonetti, J. Coastal vulnerability to wave impacts using a multi-criteria index: Santa Catarina (Brazil). *J. Environ. Manag.* **2019**, *230*, 21–32. [CrossRef]
25. Arkema, K.K.; Guannel, G.; Verutes, G.; Wood, S.A.; Guerry, A.; Ruckelshaus, M.; Kareiva, P.; Lacayo, M.; Silver, J.M. Coastal habitats shield people and property from sea-level rise and storms. *Nat. Clim. Change* **2013**, *3*, 913–918. [CrossRef]
26. Klein, R.J.; Midglev, G.; Preston, B.; Alam, M.; Berkhout, F.; Dow, K.; Shaw, M. *Climate Change 2014: Impacts, Adaptation, and Vulnerability*; IPCC Fifth Assessment Report; IPCC: Stockholm, Sweden, 2014.
27. Zhang, Y.; Wu, T.; Arkema, K.K.; Han, B.; Lu, F.; Ruckelshaus, M.; Ouyang, Z. Coastal vulnerability to climate change in China's Bohai Economic Rim. *Environ. Int.* **2021**, *147*, 106359. [CrossRef]

28. Ghosh, S.; Mistri, B. Assessing coastal vulnerability to environmental hazards of Indian Sundarban delta using multi-criteria decision-making approaches. *Ocean Coast. Manag.* **2021**, *209*, 105641. [CrossRef]
29. Sahana, M.; Sajjad, H. Vulnerability to storm surge flood using remote sensing and GIS techniques: A study on Sundarban Biosphere Reserve, India. *Remote Sens. Appl. Soc. Environ.* **2019**, *13*, 106–120. [CrossRef]
30. Furlan, E.; Pozza, P.D.; Michetti, M.; Torresan, S.; Critto, A.; Marcomini, A. Development of a Multi-Dimensional Coastal Vulnerability Index: Assessing vulnerability to inundation scenarios in the Italian coast. *Sci. Total Environ.* **2021**, *772*, 144650. [CrossRef]
31. Cifuentes-Ossa, M.A.; Rosero-Henao, L.V.; Josephraj-Selvaraj, J. Detección de cambios de la línea costera al norte del distrito de Buenaventura mediante el uso de sensores remotos. *Boletín Investig. Mar. Costeras—INVEMAR* **2017**, *46*, 137–152. [CrossRef]
32. Uribe-Castañeda, N.; Satizabal, C.A.; Herrera-Orozco, L.; Cantera Kintz, J.R. Vulnerabilidad de los servicios ecosistémicos del área marina protegida Uramba. *Boletín Investig. Mar. Costeras* **2020**, *49*, 95–118. [CrossRef]
33. Abdelaty, E.F. Coastal Erosion Assessment of the Nile Delta Coast using Remote Sensing, GIS, and Modified Coastal Vulnerability Index. *Alex. Sci. Exch. J.* **2021**, *42*, 645–655. [CrossRef]
34. Boumboulis, V.; Apostolopoulos, D.; Depountis, N.; Nikolakopoulos, K. The Importance of Geotechnical Evaluation and Shoreline Evolution in Coastal Vulnerability Index Calculations. *J. Mar. Sci. Eng.* **2021**, *9*, 0423. [CrossRef]
35. Sekovski, I.; Del Río, L.; Armaroli, C. Development of a coastal vulnerability index using analytical hierarchy process and application to Ravenna province (Italy). *Ocean Coast. Manag.* **2020**, *183*, 104982. [CrossRef]
36. Koroglu, A.; Ranasinghe, R.; Jiménez, J.A.; Dastgheib, A. Comparison of Coastal Vulnerability Index applications for Barcelona Province. *Ocean Coast. Manag.* **2019**, *178*, 104799. [CrossRef]
37. Mohd, F.A.; Abdul Maulud, K.N.; Karim, O.A.; Begum, R.A.; Awang, N.A.; Ahmad, A.; Wan Mohamed Azhary, W.A.H.; Kamarudin, M.K.A.; Jaafar, M.; Wan Mohtar, W.H.M. Comprehensive coastal vulnerability assessment and adaptation for Cherating-Pekan coast, Pahang, Malaysia. *Ocean Coast. Manag.* **2019**, *182*, 104948. [CrossRef]
38. Hoque, M.A.-A.; Ahmed, N.; Pradhan, B.; Roy, S. Assessment of coastal vulnerability to multi-hazardous events using geospatial techniques along the eastern coast of Bangladesh. *Ocean Coast. Manag.* **2019**, *181*, 104898. [CrossRef]
39. Hzami, A.; Heggy, E.; Amrouni, O.; Mahé, G.; Maanan, M.; Abdeljaouad, S. Alarming coastal vulnerability of the deltaic and sandy beaches of North Africa. *Sci. Rep.* **2021**, *11*, 2320. [CrossRef]
40. Rehman, S.; Sahana, M.; Kumar, P.; Ahmed, R.; Sajjad, H. Assessing hazards induced vulnerability in coastal districts of India using site-specific indicators: An integrated approach. *GeoJournal* **2021**, *86*, 2245–2266. [CrossRef]
41. Mullick, M.R.A.; Tanim, A.H.; Islam, S.M.S. Coastal vulnerability analysis of Bangladesh coast using fuzzy logic based geospatial techniques. *Ocean Coast. Manag.* **2019**, *174*, 154–169. [CrossRef]
42. Casanova Rosero, R.F.; Zambrano Ortiz, M.M.; Velasco Vinasco, E.; Rodríguez Cuitiva, D.E.; Escobar Olaya, G.A.; Narváez Flórez, S.; Bautista Duarte, P.; Betancourt Portela, J.M.; Parra, J.P. *Panorama de la Contaminación del Marina del Pacífico Colombiano, 2005–2010*; Dirección General Marítima: Bogota, Colombia, 2012.
43. Castellanos-Galindo, G.A.; Krumme, U.; Rubio, E.A.; Saint-Paul, U. Spatial variability of mangrove fish assemblage composition in the tropical eastern Pacific Ocean. *Rev. Fish Biol. Fish.* **2013**, *23*, 69–86. [CrossRef]
44. INVEMAR. *Informe del Estado de los Ambientes y Recursos Marinos y Costeros en Colombia, 2011*; Invemar: Santa Marta, Colombia, 2012; Volume 8, p. 203.
45. Posada Posada, B.O.; Henao Pineda, W.; Guzmán Ospitia, G. *Diagnóstico de la Erosión y Sedimentación en la Zona Costera del Pacífico Colombiano*; Instituto de Investigaciones Marinas y Costeras—INVEMAR: Santa Marta, Colombia, 2009.
46. Vega, L.; Cantillo, V.; Arellana, J. Assessing the impact of major infrastructure projects on port choice decision: The Colombian case. *Transp. Res. Part A Policy Pract.* **2019**, *120*, 132–148. [CrossRef]
47. Comovamos, B. *Informe de Calidad de Vida de Buenaventura 2019–2020*; Buenaventura Comovamos: Buenaventura, Colombia, 2021; p. 84.
48. DANE. Censo Nacional de Población y Vivienda (CNPV). 2018. Available online: https://www.dane.gov.co/index.php/estadisticas-por-tema/demografia-y-poblacion/censo-nacional-de-poblacion-y-vivenda-2018 (accessed on 24 January 2021).
49. Fletcher, C.; Rooney, J.; Barbee, M.; Lim, S.-C.; Richmond, B. Mapping Shoreline Change Using Digital Orthophotogrammetry on Maui, Hawaii. *J. Coast. Res.* **2003**, *38*, 106–124.
50. Rangel-Buitrago, N.; Neal, W.J.; de Jonge, V.N. Risk assessment as tool for coastal erosion management. *Ocean Coast. Manag.* **2020**, *186*, 105099. [CrossRef]
51. Inc, P.L. *Specification, Planet Imagery Product*; Planet Labs Inc.: San Francisco, CA, USA, 2020; p. 97.
52. White, K.; El Asmar, H.M. Monitoring changing position of coastlines using Thematic Mapper imagery, an example from the Nile Delta. *Geomorphology* **1999**, *29*, 93–105. [CrossRef]
53. Mareas, T.d. Tabla de Mareas, Valle del Cauca—Buenaventura. Available online: https://tablademareas.com/co/valle-del-cauca/buenaventura (accessed on 10 July 2021).
54. Oyedotun, T.D. Shoreline geometry: DSAS as a tool for historical trend analysis. *Geomorphol. Tech.* **2014**, *3*, 1–12.
55. Hahn, M.B.; Riederer, A.M.; Foster, S.O. The Livelihood Vulnerability Index: A pragmatic approach to assessing risks from climate variability and change—A case study in Mozambique. *Glob. Environ. Change* **2009**, *19*, 74–88. [CrossRef]
56. López Royo, M.; Ranasinghe, R.; Jiménez, J.A. A Rapid, Low-Cost Approach to Coastal Vulnerability Assessment at a National Level. *J. Coast. Res.* **2016**, *32*, 932–945. [CrossRef]

57. Álvarez, M.C.; Bermúdez-Rivas, C.; Niño, D.C. Caracterización de la geomorfología costera y sus coberturas vegetales asociadas, a través de sensores remotos en la Bahía de Buenaventura, Valle del Cauca. *Boletín Científico CIOH* **2016**, *34*, 49–63. [CrossRef]
58. Reuter, H.I.; Hengl, T.; Gessler, P.; Soille, P. Chapter 4 Preparation of DEMs for Geomorphometric Analysis. In *Developments in Soil Science*; Hengl, T., Reuter, H.I., Eds.; Elsevier: Amsterdam, The Netherlands, 2009; Volume 33, pp. 87–120.
59. Ashraful Islam, M.; Mitra, D.; Dewan, A.; Akhter, S.H. Coastal multi-hazard vulnerability assessment along the Ganges deltaic coast of Bangladesh—A geospatial approach. *Ocean Coast. Manag.* **2016**, *127*, 1–15. [CrossRef]
60. CMEMS. Sea Level Daily Gridded Data from Satellite Observations for the Global Ocean from 1993 to Present. Available online: https://cds.climate.copernicus.eu/cdsapp#!/dataset/satellite-sea-level-global?tab=overview (accessed on 5 May 2021).
61. ECMWF. ERA5 Monthly Averaged Data on Single Levels from 1959 to Present. Available online: https://cds.climate.copernicus.eu/cdsapp#!/dataset/reanalysis-era5-single-levels-monthly-means?tab=overview (accessed on 5 May 2021).
62. Gisbert, F.J.G.; Martí, I.C. Rugosidad del terreno: Una característica del paisaje poco estudiada. *Doc. Trab.* **2010**, *10*, 1.
63. Saaty, T.L. The Analytic Hierarchy and Analytic Network Processes for the Measurement of Intangible Criteria and for Decision-Making. In *Multiple Criteria Decision Analysis: State of the Art Surveys*; Greco, S., Ehrgott, M., Figueira, J.R., Eds.; Springer: New York, NY, USA, 2016; pp. 363–419. [CrossRef]
64. Alkire, S.; Foster, J. Counting and multidimensional poverty measurement. *J. Public Econ.* **2011**, *95*, 476–487. [CrossRef]
65. Angulo Salazar, R.C.; Díaz Cuervo, Y.; Pardo Pinzón, R. Índice de pobreza multidimensional para Colombia (IPM-Colombia). *Arch. Econ.* **2011**, *382*, 1997–2010.
66. IDEAM. *Leyenda Nacional de Coberturas de la Tierra: Metodología CORINE Land Cover Adaptada para Colombia: Escala 1:100.000*; Ideam: Bogota, Colombia, 2010.

International Journal of *Geo-Information*

MDPI

Article

Use of a MODIS Satellite-Based Aridity Index to Monitor Drought Conditions in the Pearl River Basin from 2001 to 2021

Kunlong Niu [1], Junliang Qiu [2], Shirong Cai [1], Wenxin Zhang [1], Xiaolin Mu [1], Edward Park [3,4,5] and Xiankun Yang [1,6,*]

1 School of Geography and Remote Sensing, Guangzhou University, Guangzhou 510006, China
2 Department of Land, Environment, Agriculture and Forestry, University of Padova, Agripolis, Viale dell'Università 16, 35020 Legnaro, Italy
3 National Institute of Education, Nanyang Technological University, Singapore 637616, Singapore
4 Asian School of the Environment, Nanyang Technological University, Singapore 639798, Singapore
5 Earth Observatory of Singapore, Nanyang Technological University, Singapore 639798, Singapore
6 Rural Non-Point Source Pollution Comprehensive Management Technology Center of Guangdong Province, Guangzhou University, Guangzhou 510006, China
* Correspondence: yangxk@gzhu.edu.cn

Abstract: In recent decades, global climate change has made natural hazards increasingly prevalent. Droughts, as a common natural hazard, have been a hot study topic for years. Most studies conducted drought monitoring in arid and semi-arid regions. In humid and sub-humid regions, due to climate change, seasonal droughts and seasonal water shortages were often observed too, but have not been well studied. This study, using a MODIS satellite-based aridity index (SbAI), investigated spatiotemporal changes in drought conditions in the subtropical Pearl River Basin. The study results indicated that the inter-annual SbAI exhibited a significant decreasing trend, illustrating a wetter trend observed in the basin in the past two decades. The decreasing trend in the SbAI was statistically significant in the dry season, but not in the monsoon season. The drought conditions displayed an insignificant expansion in the monsoon season, but exhibited statistically significant shrinking in the dry season. The Pearl River Basin has become wetter over past two decades, probably due to the results of natural impacts and human activities. The areas with increased drought conditions are more likely impacted by human activities such as water withdrawal for irrigation and industrial uses, and fast urbanization and increased impervious surfaces and resultant reduction in water storage capacity. This study provided a valuable reference for drought assessment across the Pearl River Basin.

Keywords: drought monitoring; Pearl River Basin; MODIS satellite; SbAI; Google Earth Engine

Citation: Niu, K.; Qiu, J.; Cai, S.; Zhang, W.; Mu, X.; Park, E.; Yang, X. Use of a MODIS Satellite-Based Aridity Index to Monitor Drought Conditions in the Pearl River Basin from 2001 to 2021. *ISPRS Int. J. Geo-Inf.* **2022**, *11*, 541. https://doi.org/10.3390/ijgi11110541

Academic Editors: Walter Chen and Wolfgang Kainz

Received: 29 August 2022
Accepted: 25 October 2022
Published: 28 October 2022

Publisher's Note: MDPI stays neutral with regard to jurisdictional claims in published maps and institutional affiliations.

1. Introduction

Natural hazards are a huge threat to humans and social development. Global warming has increased the likelihood of the occurrence of natural hazards such as droughts, heat waves, wildfires, and floods [1,2]. As one of the most serious natural hazards affecting humans, droughts cause huge casualties and economic losses worldwide every year [3]. For example, more than half of the U.S. states are affected by droughts, leading to an annual economic loss of approximately USD 6–8 billion [4]. However, currently, humans still know very little about droughts and need effective ways to combat droughts. Thus, drought monitoring is an important prerequisite for drought migration.

Previous studies have proposed different drought indices for drought monitoring, most of which are based on the observations from meteorological stations. Such indices include the PDSI (Palmer Drought Severity Index) [5], the SPI (Standard Precipitation Index) [6], and the SPEI (Standardized Precipitation Evapotranspiration Index) [7]. Based on the observations from meteorological stations, ambient drought conditions can be accurately characterized. However, due to the stations' low density and uneven distribution, it is difficult to achieve large-scale drought monitoring, especially in remote rural areas. Therefore, these indices cannot represent the general trends and changes in drought conditions for large areas. Satellites can provide spatiotemporal images with continuous coverage for large areas, and thus have unique advantages for drought monitoring, especially in remote areas with few meteorological stations. Based on satellite images, many scholars have proposed new drought monitoring indices such as the VCI (Vegetation Condition Index) [8], the TCI (Temperature Condition Index) [8], the SDCI (Scaled Drought Condition Index) [9], the MIDI (Microwave Integrated Drought Index) [10], the OVDI (Optimized Vegetation Drought Index) [11] and the OMDI (Optimized Meteorological Drought Index) [11] These drought indices can provide better spatial coverage for drought conditions, and also combine multiple bands to build an integrated drought monitoring index to better represent drought conditions.

Therefore, with the development in remote sensing techniques, satellite images have been widely used in drought detection and have made great contributions to drought monitoring at the global, continental and regional scales. Dehghan et al. [12] used the PDSI for drought assessment in Fars Province, Iran. Their results proved the strong correlation between the results simulated from site-based precipitation ($R^2 > 0.63$) and temperature ($R^2 > 0.95$) observations. At a global scale, Trenberth et al., using the PDSI, concluded that increased heating due to global warming may not lead to droughts but it is expected that, when droughts occur, they are likely to occur quicker as well as be more intense [13]. Gidey et al. [14] used the SPI to analyze the temporal and spatial variation of drought in northern Ethiopia, predicting that due to insufficient precipitation, the region will continue to be dry in the future. In facts, the SPI based on historical observations also indicated that the increase in drought frequency, duration, and severity will be significant in Africa, the Mediterranean region, Eastern Asia, and Southern Australia [15]. The SPI can work stably even with the difference of precipitation in different regions. It can also derive various results for different time scales. This is very meaningful for generating diurnal, monthly, seasonal and annual variations of droughts across a large area. On the basis of the SPI, considering evapotranspiration, Hernandez et al. [16] proposed the SPEI and used the metric for drought assessment in southern Texas, USA. The results illustrated that due to the increase in temperature and decrease in precipitation in the future, the region will be much drier. Additionally, based on the SPEI and the SPI, Chiang et al. found that that the presence of anthropogenic forcing has increased the drought frequency, maximum drought duration, and maximum drought intensity experienced in large parts of the Americas, Africa, and Asia [17]. In addition, the VCI is also a new drought monitoring method based on remote sensing technology. Dutta et al. [18] used the VCI to monitor agricultural drought in Rajasthan (India), and compared this with the results derived from the SPI. They found that the VCI and SPI results were consistent. The VCI is also a useful tool for assessing agricultural drought. Zhang et al. [10] proposed the MIDI based on precipitation, soil moisture and surface temperature, and used the index for drought monitoring in northern China. Based on validation with the results of the multi-scale SPI, they concluded that it can be applied well to monitoring of short-term droughts.

As discussed above, most studies conducted drought monitoring in arid and semi-arid regions. In fact, in humid and sub-humid areas, due to the influence of climate change, seasonal droughts were often observed too, and seasonal water shortages also reported from time to time [19–23]. However, few studies have been performed on drought monitoring in humid and sub-humid areas. This context necessitates studies on seasonal droughts or water shortages in humid and sub-humid regions. In addition, in the context of global climate change, drought monitoring is also an important tool to evaluate regional long-term dry and wet changes. It is also an important indicator for regional environmental changes.

The Pearl River Basin (PRB) is a typical humid and sub-humid area, which is an appropriate study area for such aforementioned studies. The PRB is located in the subtropical zone, with a mild and rainy climate. However, in recent years, seasonal water shortages were often reported in the PRB [24,25] but the changes in wet and dry environmental conditions and their long-term changes are still unknown. In this study, using MODIS surface reflectance and surface temperature data, by calculating the metric of the SbAI (Satellite-Based Aridity Index) and comparing it with the precipitation data, the major objectives of this study on the spatiotemporal changes in drought conditions in the PRB are as follows: (1) the inter-annual and monthly variation in the SbAI and the corresponding changes in drought conditions for the period 2001–2021 for the PRB; (2) spatial distribution of droughts across the PRB; (3) spatiotemporal variation in drought conditions in the basin's representative sub-regions. The study results will improve the understanding of the seasonal drought characteristics and the long-term changes in wet and dry environmental conditions in the PRB. It can provide valuable references for drought prevention and environmental management in the PRB.

2. Materials and Methods

2.1. Study Area

The Pearl River Basin (PRB) is located in southern China (Figure 1a), from 21°31′ N to 26°49′ N and 102°14′ E to 115°53′ E, and administratively covers 6 provinces (Yunnan, Guizhou, Guangxi, Guangdong, Hunan, and Jiangxi) and 2 special administrative regions (Hongkong and Macau).

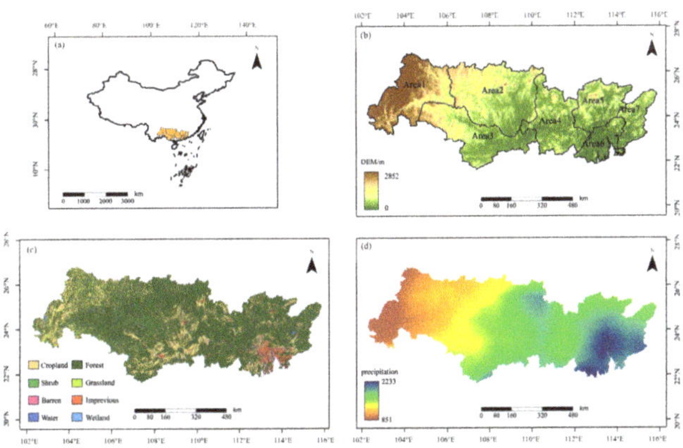

Figure 1. Geographical configuration of the PRB and its natural characteristics: (**a**) rough location of the PRB; (**b**) topographical characteristics of the PRB; (**c**) land use configuration in the PRB in 2020; (**d**) distribution of the annual precipitation in the PRB from 2001 to 2019.

Controlled by a subtropical climate, the basin, with a size of 442,100 km^2, has an annual average temperature from 14 to 22 °C and an average precipitation from 1200 to 2200 mm [26]. Similar to its high northwest and low southeast topography, its precipitation varies widely in spatial and temporal distribution: the precipitation in spatial distribution gradually decreases from west to east. The monsoon season is from April to September, when 70~85% of the annual precipitation occurs.

With its wide span from west to east across the Yunnan–Guizhou Plateau, the mountains in Guangdong and Guangxi provinces, and the Pearl River Delta Plain, we spilt the PRB into 7 sub-regions for detailed analysis (Figure 1b): the headwater region (including the Nanpanjiang River and Beipanjiang River basins) (Area 1); the Hongshui River and Liujiang River basins (Area 2); the Zuojiang River, Youjiang River and Yujiang River basins (Area 3); the Heguijiang River, Qianxunjiang River and Xijiang mainstem river basins (Area 4); the Beijiang River basin (Area 5); the Pearl River Delta (Area 6); the Dongjiang River basin (Area 7).

According to previous studies [27–30], from the 1960s to 2019, precipitation in the PRB showed a slight increasing trend, with the most pronounced declines occurring in the middle and upper reaches, and an increase in the lower reaches. Inter-annual precipitation in spring and winter gradually increased in volume, and reduced in summer and autumn.

In addition to the changes in precipitation, studies also displayed that land use also experienced a great change in the PRB [31,32]. In the last 30 years, the area of waterbodies and forest land in the PRB has shrunk due to the expansion of urbanization, and there has been a dramatic increase in built-up areas. The decrease in grassland mainly occurred in the western part, which was transformed into forest land. The decrease in cropland was concentrated in the central basin. However, the increase in built-up areas primarily occurred in the southeast, especially in the Pearl River Delta, contributing to approximately 6.3% of the delta area, which was mainly converted from crop land and forest land. Nowadays, the predominant land use in the PRB is forest land, accounting for approximately 67.3% of the basin; followed by crop land (approximately 25.6%) mainly in the western and central PRB. After that is built-up areas (approximately 2.5%) in the southeastern PRB; and grassland (approximately 2.2%) in the western PRB.

2.2. *Data*

2.2.1. MODIS Satellite Data

Two datasets were used to calculate the SbAI—the surface reflectance dataset and the surface temperature dataset—both of which were provided by MODIS satellite series. Compared with other satellite products, MODIS satellites provide better-quality and longer continuous datasets. Their performance and applicability have been proved in various studies. However, due to the low spatial resolution, MODIS products can only be used in large-scale regions.

The MODIS reflectance product is the most common data used for albedo production. It has two datasets—MOD09GA and MYD09GA—corresponding to products generated from the Terra and Aqua satellites, respectively. In this study, the MOD09GA reflectance dataset produced by the Terra satellite was used. The MOD09GA product has a temporal resolution of 1 day, a spatial resolution of 1 km, and a reflectance spatial resolution of 500 m. The 500 m reflectance dataset also provided reflectance for Bands 1–7, data quality, data extent, observation number, and 250 m scanning information. The 1 km datasets provided descriptive information including observation frequency, data quality, sensor azimuth and zenith, solar altitude and azimuth, and orbit information. The reflectance of 7 bands from the 500 m datasets and the solar zenith from the 1 km dataset were used for data preprocessing and analysis. To reduce the error, we firstly made a cloud removal operation for the MOD09GA dataset. The dataset was retrieved from the Google Earth Engine (GEE) image library (https://developers.google.com/earth-engine/datasets/catalog/MODIS_006_MOD09GA, accessed on 24 June 2022) for image processing and analysis on the GEE platform using JavaScript.

The surface temperature dataset was acquired from the MOD11A1 dataset produced by the Terra satellite. A global daily average surface temperature production framework was established by the coupling of the annual temperature cycle (ATC) model and the diurnal temperature cycle (DTC) model. Then, the global 1 km average daily surface temperature product was generated based on the framework. The MOD11A1 dataset has a temporal resolution of 1 day and a spatial resolution of 1 km. To reduce the error, we also only used good-quality pixels in the MOD11A1 dataset. The dataset was acquired on the GEE platform (https://developers.google.com/earth-engine/datasets/catalog/MODIS_006_MOD11A1, accessed on 24 June 2022). Additionally, it was also processed on the GEE platform using JavaScript.

2.2.2. Precipitation Data

Precipitation data were used to verify the analysis results of the SbAI. The Tropical Rainfall Measuring Mission (TRMM) and the Global Precipitation Measurement mission (GPM) are global typical representatives of high-precision historical precipitation data. The TRMM satellite provided precipitation data with a temporal resolution of 3 h and a spatial resolution of 0.25° in the latitude range of 50° S–50° N in the period 1997–2015. Since 2014, the National Aeronautics and Space Administration (NASA) and the Japan Aerospace Exploration Agency (JAXA) began to provide the Global Precipitation Measurement mission (GPM) dataset. As a continuation of the TRMM, the GPM provides higher-precision precipitation data with a temporal resolution of 0.5 h and a spatial resolution of 0.1°, using multiple sensors, multiple satellites, and multiple algorithms combined with satellite networks and rain gauge inversion. The most used precipitation dataset in this study was the GPM dataset. This dataset has a monthly dataset on the GEE platform, which was acquired and processed on the GEE platform (https://developers.google.com/earth-engine/datasets/catalog/NASA_GPM_L3_IMERG_MONTHLY_V06, accessed on 24 June 2022).

2.2.3. Other Datasets

To support the results of this study, we also obtained the 2001–2020 Water Resources Bulletin for the PRB from the Pearl River Water Resources Commission (PRWRC) (http://www.pearlwater.gov.cn/zwgkcs/lygb/szygb/, accessed on 28 August 2022). The statistical results of the annual precipitation in the PRB were compared with the GPM precipitation data, and the overall trend obtained in this study was verified. Most of these datasets came from the government's official reports. Thus, the data quality can be guaranteed, meaning that they were able to support the relevant analysis in this study. The land use data were obtained from the 30 m annual land cover datasets in China from 1990 to 2021 (https://zenodo.org/record/5816591#.Yvj5-3ZBxPY, accessed on 14 May 2022). DEM data came from the Geospatial Data Cloud (http://www.gscloud.cn, accessed on 14 May 2022). These datasets were mainly used to investigate the causes for changes in drought conditions.

2.3. Methods

2.3.1. The Satellite-Based Aridity Index (SbAI)

The SbAI [33] is the ratio of the land surface temperature (LST) difference between day and night to the absorbed solar radiation. The physical meaning of the SbAI is the opposite of heat capacity determined by land surface wetness, which can be derived as follows:

$$SbAI = \frac{\Delta T_s}{R_s} \tag{1}$$

where

$$\Delta T_s = LST_{day} - LST_{night} \tag{2}$$

$$R_s = (1 - r)S_0 \cos\theta_c \tag{3}$$

where r in Formula 3 can be computed as:

$$r = 0.160r_1 + 0.291r_2 + 0.243r_3 + 0.116r_4 + 0.112r_5 + 0.081r_7 \tag{4}$$

ΔT_s is the LST difference between the day and night values, and R_s is the absorbed solar radiation, calculated from the broadband albedo r, the solar constant S_0 (1367 W·m^{-2}) and the solar zenith angle at the Sun's apex θ_c, where r is obtained from the experimental formula [34]. For dry surfaces, the SbAI is larger because of its larger heat capacity, resulting in a larger ΔT_s, and vice versa.

2.3.2. Trend Analysis

The Mann–Kendall (M–K) test [35] was used to analyze the significance of the SbAI's changing trends. The M–K test is a non-parametric method, and it does not require the samples to follow a random distribution. The test's results are not affected by missing data and a few outliers, giving it strong applicability. The test steps are as follows:

$$S = \sum_{j=1}^{n-1} \sum_{i=j+1}^{n} sgn(x_i - x_j) \tag{5}$$

where x_i and x_j are observations:

$$sgn(x_i - x_j) = \begin{cases} -1\ x_i - x_j < 0 \\ 0\ x_i - x_j = 0 \\ 1\ x_i - x_j > 0 \end{cases} \tag{6}$$

The calculation of the variance *Var*(*S*) is as follows:

$$Var(S) = \frac{n(n-1)(2n+5)}{18} \tag{7}$$

Z and *p* are obtained by the following equations:

$$Z = \begin{cases} \frac{S-1}{\sqrt{Var(s)}}\ S < 0 \\ 0\ S = 0 \\ \frac{S+1}{\sqrt{Var(s)}}\ S > 0 \end{cases} \tag{8}$$

$$p = 2(1 - cdf(|Z|)) \tag{9}$$

For *Z*, if $|Z| \geq 1.96$, there is a significant change at the confidence level of 0.05.

Theil Sen Median slope estimation [36] was used to calculate the intensity of the trend, which is calculated by Equation (10).

$$TS_{Slope} = median\left(\frac{x_j - x_i}{j - i}\right) \tag{10}$$

The Pettitt test [37] was used to detect the abrupt change year, and the test steps are as follows:

$$U_{t,N} = U_{t-1,N} + \sum_{J=1}^{N} sgn(x_t - x_j) for\ t = 2, 3, \ldots, N \tag{11}$$

where x_t and x_j are observations, and $U_{t,N}$ indicates whether two sample sets are from the same population.

The test statistics $K_{(t)}$ and p are given as follows:

$$K_{(t)} = max|U_{t,N}| \tag{12}$$

$$p = 2e^{\frac{-6(K_N)^2}{N^3+N^2}} for\ t = 2, 3, \ldots, N \tag{13}$$

3. Results

3.1. Temporal Trends of the SbAI

To quantify the temporal trends of drought conditions in the PRB during the period 2001–2021, this study calculated the monthly SbAI (Figure 2a–l) and the yearly SbAI (Figure 2m) on the basis of daily MOD09GA and daily MOD11A1 products. According to seasonal division in China [38], this study assumed that spring is March, April, and May; summer is June, July, and August; autumn is September, October, and November; winter is December, January, and February. According to the precipitation in different months in the PRB [27,39], this study defined the monsoon season in the PRB as from May to October, and the dry season as the remaining months. The seasonal SbAI, the monsoon seasonal SbAI, and the dry seasonal SbAI were derived from the monthly SbAI.

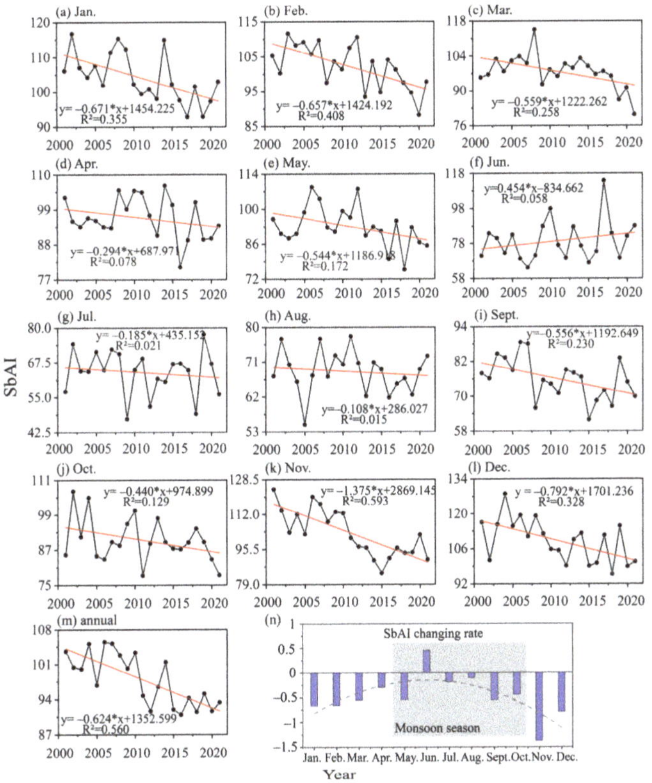

Figure 2. Annual and monthly changing trends of the SbAI in the PRB: (**a–l**) Monthly SbAI trend; (**m**) Annual SbAI trend; (**n**) Monthly SbAI changing rate. Note: * represents multiplication sign; Jan, Feb, Mar, etc. is the abbreviation of each month.

3.1.1. Annual Trends of the SbAI

As shown in Figure 2 and Table 1, the SbAI significantly decreased ($p < 0.01$) during the period 2001–2021, with a decreasing trend of −0.624/yr, indicating that the PRB is becoming wetter. From 2001 to 2021, the maximum, minimum, and mean SbAI was 104.983 in 2006, 90.428 in 2016, and 97.41, respectively; there are 13 years with an SbAI higher than the multi-year average, and 9 years with an SbAI lower than the multi-year average, mostly occurring after 2010, indicating that the PRB became wetter mainly after 2010. In the past 20 years, North China, even arid and semi-arid Northwest China, and the Qinghai–Tibet Plateau also became wetter [40–42], revealing an increasingly humid trend in most parts of China.

Table 1. Trend analysis for the monthly and annual trends of the SbAI in the PRB during the period 2001–2021.

	PRB			Area 1			Area 2			Area 3		
	z	Slope	Year	z	Slope	Year	z	Slope	Year	z	Slope	Year
Annual	−3.29 ***	−0.60	2005	−3.35 ***	−0.80	2013	−3.23 ***	−0.70	2010	−2.75 ***	−0.65	2003
January	−2.81 ***	−0.69	2010	−2.45 **	−0.94	2014	−1.06	−0.50	2014	−1.24	−0.59	2015
February	−2.75 ***	−0.69	2017	−2.26 ***	−0.72	2014	−1.18	−0.54	2008	−1.54	−0.37	2013
March	−1.96 **	−0.45	2008	−1.12	−0.42	2010	−0.94	−0.60	2008	−2.26 **	−1.17	2008
April	−1.06	−0.30	2009	−3.11 ***	−1.08	2013	0.00	0.00	2015	−1.72 *	−0.49	2010
May	−1.66 *	−0.51	2011	−2.14 **	−1.23	2012	−1.60	−0.78	2008	−3.35 ***	−1.26	2012
June	0.75	0.34	2008	0.69	0.17	2017	0.33	0.34	2012	0.09	0.10	2012
July	−0.51	−0.29	2008	−1.12	−0.45	2008	−1.54	−0.38	2011	−0.45	−0.21	2006
August	−0.57	−0.11	2012	−0.21	−0.12	2018	−0.45	−0.14	2012	−1.00	−0.34	2012
September	−2.08 **	−0.57	2007	−1.90 *	−0.74	2005	−2.08 **	−0.94	2007	−1.00	−0.38	2007
October	−1.30	−0.34	2011	−0.39	−0.42	2006	−0.94	−0.35	2011	0.00	−0.01	2005
November	−3.77 ***	−1.35	2011	−2.81 ***	−0.87	2011	−3.05 ***	−1.40	2010	−2.87 ***	−1.47	2010
December	−2.81 ***	−0.88	2011	−2.51 **	−0.69	2009	−2.51 **	−0.94	2009	−1.90 *	−0.98	2014

	Area 4			Area 5			Area 6			Area 7		
	z	Slope	Year	z	Slope	Year	z	Slope	Year	z	Slope	Year
Annual	−2.51 **	−0.47	2010	−2.45 **	−0.42	2007	−2.14 **	−0.42	2010	−3.47 ***	−0.76	2007
January	−0.27	−0.15	2014	−0.09	−0.04	2015	−1.54	−0.41	2009	−1.66 **	−0.82	2015
February	−1.72 *	−0.67	2005	−0.94	−0.24	2013	−2.02 **	−0.79	2004	−1.66 *	−0.58	2012
March	−0.15	−0.09	2006	−0.39	−0.30	2011	−1.66 *	−0.67	2007	−2.08 **	−0.76	2005
April	0.27	0.26	2006	0.63	0.26	2017	−0.94	−0.50	2008	0.33	0.12	2017
May	−0.88	−0.46	2013	−1.78 *	−1.04	2011	−1.42	−0.63	2013	−2.14 **	−0.97	2013
June	0.27	0.15	2009	0.82	0.65	2010	−0.63	−0.44	2006	−0.39	−0.22	2010
July	−1.90 *	−0.76	2016	−1.06	−0.48	2008	−1.36	−0.94	2016	−0.51	−0.13	2008
August	−0.33	−0.09	2017	0.75	0.23	2005	−1.00	−0.43	2008	0.09	0.02	2019
September	−0.75	−0.26	2007	0.33	0.07	2007	−0.03	−0.03	2007	−1.66 *	−0.49	2006
October	−2.02 **	−0.66	2015	−1.66 *	−0.57	2008	−1.12	−0.40	2018	−2.26 **	−0.88	2015
November	−2.87 ***	−1.63	2010	−3.17 ***	−1.48	2012	−4.14 ***	−1.65	2008	−4.14 ***	−1.72	2010
December	−1.78 *	−0.70	2014	−2.26 **	−0.73	2008	−1.90 *	−0.54	2017	−2.02 **	−0.89	2014

Note: trends were calculated by the M–K test (z), the magnitude by the Sen test (slope), and the starting trend by the Pettitt test (year); *** means significance with $p < 0.01$; ** means significance with $p < 0.05$; * means significance with $p < 0.1$.

3.1.2. Monthly Trends of the SbAI

Collectively, during the period 2001–2021, except for June, the SbAI showed a decreasing trend in all other months, but the downward trends were more significant in the dry season. The fastest decrease in the SbAI was observed in November, with a rate of −1.375/yr. The slowest decrease in the SbAI was in August, with a rate of −0.108/yr. However, June was the only month with an increase in the SbAI, although without statistical significance. According to previous studies [27–30], from the 1960s to 2019, precipitation in the Pearl River Basin showed a slight increasing trend, with an increase in the lower reaches, and the most pronounced declines in the middle and upper reaches. Inter-annual precipitation in spring and winter showed a gradual increase, and displayed a reduction in summer and autumn. This revealed that the changes in the SbAI in the PRB are consistent with the trends of precipitation changes.

3.1.3. Drought Events Detected by SbAI

We used the monthly SbAI time series from 2001 to 2021 for monthly drought analysis (Table 2). The Water Resources Bulletin for the PRB from the PRWRC shows that drought occurs almost every year in the PRB in the monsoon season, so we chose the median SbAI in the monsoon season over 21 years to calculate a threshold. Then, the threshold was used to analyze the drought duration, drought severity, and drought intensity. Most of the droughts occurred in November, December, January and February. Drought duration is generally 2–5 months. Although some places experienced droughts for a long time, they are small in size and are primarily located in the upper reaches of the Pearl River, thus the consequences were not typical.

Table 2. Drought events detected by the SbAI.

Start Year	Start Month	End Year	End Month	Duration	Severity	Intensity
2001	1	2001	2	2	211.42	105.71
2001	11	2002	1	3	356.69	118.9
2002	10	2003	3	6	642.96	107.16
2003	11	2004	2	4	431.15	107.79
2004	10	2005	3	6	662.99	110.5
2005	11	2006	3	5	528.28	105.66
2006	11	2007	3	5	560.9	112.18
2007	11	2008	1	3	342.58	114.19
2008	3	2008	4	2	219.19	109.59
2008	11	2009	2	4	443.07	110.77
2009	11	2010	2	4	428.2	107.05
2010	11	2010	12	2	217.74	108.87
2011	11	2012	2	4	416.95	104.24
2013	12	2014	4	5	536.69	107.34
2014	12	2015	1	2	213.85	106.92
2017	12	2018	1	2	212.31	106.16

3.2. Spatial Patterns of the SbAI

To quantify the spatial patterns of the SbAI in the PRB during the period 2001–2021, we calculated the monthly and annual SbAI at the pixel scale. As shown in Figure 3a, the west part of Area 1, the headwater region of the PRB, was the driest region in the PRB. The southeast part of Area 2 and the east part of Area 3 were relatively dry regions in the PRB. Collectively, the monthly and annual SbAI exhibited similar spatial patterns, showing that Areas 1–3 were relatively dry regions in the PRB. Obviously, the SbAI in the dry season was higher than that in the monsoon season, indicating that the dry season is much drier than the monsoon season in the PRB due to low precipitation in the dry season, which has the potential to cause seasonal water shortages in the PRB.

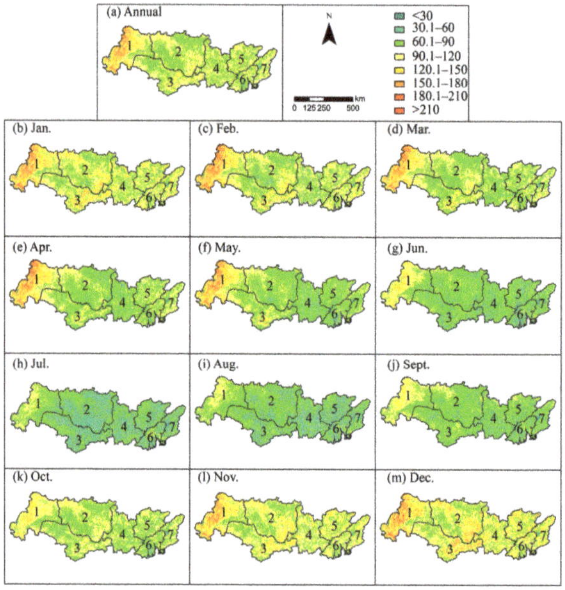

Figure 3. Annual and monthly spatial patterns of the SbAI in the PRB: (**a**) Spatial patterns of annual SbAI, 2001–2021; (**b–m**) spatial patterns of the monthly SbAI, 2001–2021. Note: numbers 1–7 represent 7 areas described in Section 2.1; Jan, Feb, Mar, etc. is the abbreviation of each month, respectively.

From June to September, almost all of the PRB was humid, with the SbAI mostly lower than 90. Even in the relatively dry region of Area 1, the overall SbAI was also lower than 120, indicating that the region also experienced relative humidity. Areas 4–7 experienced high humidity in the monsoon season. In the dry season, for instance in December, parts of wettest region (Areas 4–7) in the PRB became dry, with the SbAI ranging from 120 to 180. From March to June, the switch from the dry season to the monsoon season, the wet area significantly expanded in the PRB. In July and August, when the peak SbAI occurs (Figures 3 and 4), the overall SbAI across the entire PRB is lower than 90; the SbAI for areas 1–7 was similar, and all areas were in the wet stage.

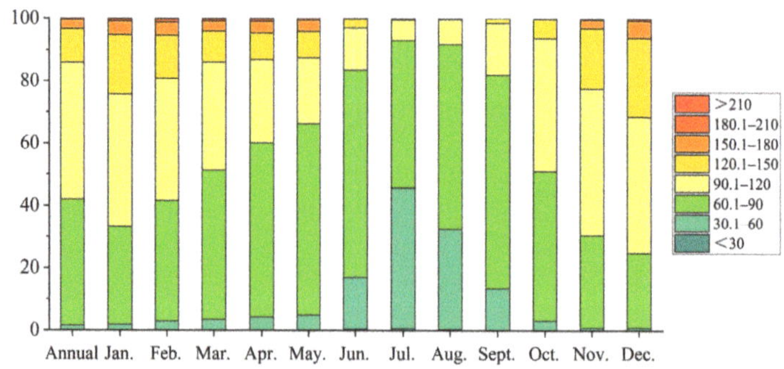

Figure 4. Monthly and annual statistics on the areal proportion of the SbAI in different ranges. Note: Jan, Feb, Mar, etc. is the abbreviation of each month, respectively.Jan etc. is the abbreviation of each month.

3.3. *The M–K Test for the Spatiotemporal Trends of the SbAI*

Figure 5 shows the significance level of SbAI trends assessed using the Mann–Kendall test. As shown in Figure 5a, at the annual scale, areas that were statistically significantly drier were mainly distributed in the central area of Area 6 and some small patches in Areas 1–3. At the monthly scale (Figure 5b–m), the drier trend gradually weakened from the center to the east and west directions. For April to September, many areas became drier, especially in Areas 2–5. Areas with a wetter trend were mostly in Areas 1 and 7 in east and west PRB, respectively. The fastest trend of becoming drier occurred in June, while the fastest change in becoming wetter occurred in November.

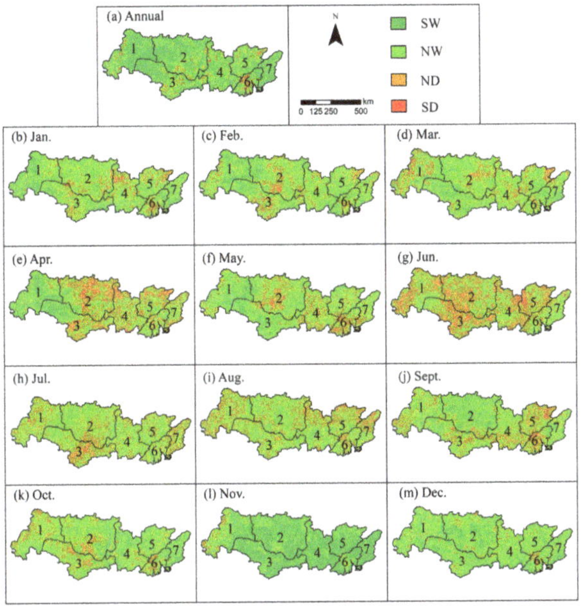

Figure 5. Monthly and annual trends of the SbAI in the PRB during the period 2001–2021 based on the Mann–Kendall test: (**a**) Annual trends of the SbAI during the period 2001–2021; (**b–m**) Monthly trends of the SbAI, 2001–2021. Note: numbers 1–7 represent 7 areas described in Section 2.1; Jan, Feb, Mar, etc. is the abbreviation of each month, respectively. (SD: significantly drier trend; ND: not significantly drier trend; NW: not significantly wetter trend; SW: significantly wetter trend).

In accordance with Figure 6, excluding June, areas with a wetter trend were predominant in the PRB. However, one interesting result should be highlighted: as shown in Figure 3, Area 6 is among the wettest areas in the PRB, but a large part of Area 6 tended to be drier according to the M–K test. Conversely, as one of the driest regions in the PRB, Area 1 has become wetter according to the M–K test.

On the basis of Figure 5b–m with Figure 6, in the dry season, wetter areas were predominant in the PRB, especially in November (almost the driest month according to Figure 4), as the SbAI significantly decreased in almost the entire PRB in the dry season. However, in the monsoon season, areas with a drier trend were much larger than drier areas in the dry season. Therefore, despite high precipitation in the monsoon season, the drier areas that occurred in the monsoon season were much larger than those that occurred in the dry season. In summary, for the temporal trends of the SbAI in the PRB, the PRB became wetter at the inter-annual scale, but the wetter trends in the dry season were more significant than the drier trend in the monsoon season.

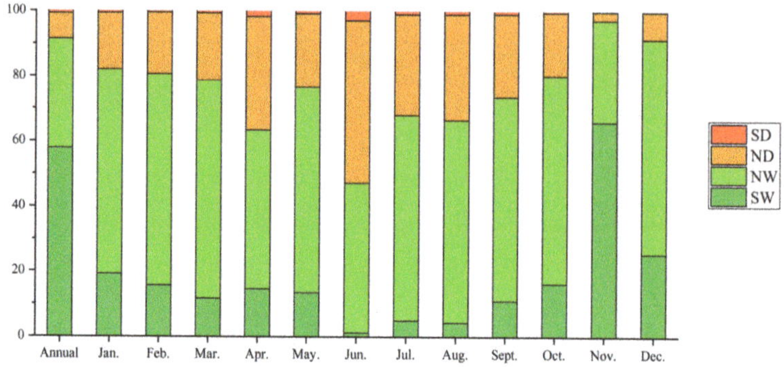

Figure 6. The areal proportion of different trends in the PRB. Note: Jan etc. is the abbreviation of each month. (SD: significantly drier trend; ND: not significantly drier trend; NW: not significantly wetter trend; SW: significantly wetter trend).

3.4. *Temporal Trends of the SbAI in the Representative Regions of Areas 1 and 6*

Area 1 is the remote headwater region of the PRB, with a natural environment, relatively less precipitation, and less impact from anthropogenic activities. Area 6 is the most economically developed region and under most impact from anthropogenic activities, with relatively higher precipitation and the highest and densest population. Thus, Areas 1 and 6 can be regarded as representatives of the most natural region and the most urbanized region in the PRB. As mentioned in Section 3.2, as the driest region in the PRB, Area 1 was becoming wetter. Conversely, as one of the wettest regions in the PRB, Area 6 was becoming drier. Thus, Areas 1 and 6 can be regarded as representative regions for the further analysis of the trends of drought conditions in the PRB.

3.4.1. Temporal Trends of the SbAI in Areas 1 and 6

As shown in Table 1, trends of the SbAI in Area 1 showed similar patterns to that of the entire PRB—a significantly wetter trend in the dry season, and an overall insignificantly drier trend in the monsoon season (except for June). According to Table 3, in Area 1, at the monthly scale, the SbAI decreased fastest in May, with a decreasing rate of −1.10/yr; the SbAI decreased at the slowest rate in August, with a decreasing rate of only −0.07/yr. The SbAI in Area 1 increased only in June, at a rate of 0.38/yr. The mean SbAI in Area 1 in the dry season was obviously higher than that in the monsoon season. From January to December, the SbAI firstly decreased and then increased, with a maximum value of 131.59 in February, and a minimum value of 81.33 in July.

Table 3. Monthly mean SbAI and changing rates for Area 1 and Area 6, 2001–2021.

		Annual	January	February	March	April	May	June	July	August	September	October	November	December
Area 1	mean	124.30	130.21	131.59	127.53	130.67	128.19	99.13	81.33	84.11	92.87	105.62	122.58	128.33
	slope	−0.77	−0.93	−0.69	−0.33	−0.96	−1.10	0.38	−0.46	−0.07	−0.75	−0.20	−0.88	−0.83
Area 6	mean	87.86	96.77	94.40	85.36	80.65	67.05	55.27	49.92	58.11	64.57	78.22	93.88	98.97
	slope	−0.40	−0.51	−0.79	−0.77	−0.43	−0.75	−0.02	−0.69	−0.44	−0.14	−0.45	−1.56	−0.60

In Area 6, at the monthly and annual scales, the SbAI had a decreasing trend, with a significant decrease in the dry season, in which the fastest decrease of −1.56/yr was observed in November. An insignificant decrease was observed in the monsoon season, with a slowest rate of −0.02/yr in June. From January to December, similar to the monthly pattern in Area 1, the SbAI in Area 6 also firstly decreased and then increased, with a maximum value of 98.97 in December and a minimum value of 19.92 in July.

3.4.2. Monthly Fluctuations of the SbAI in Areas 1 and 6

According to Figure 3, at all the time scales, the west part of Area 1 is much drier than the east part due to lower precipitation in the west. With the switch from the dry season to the monsoon season, the difference in the SbAI between western and eastern Area 1 shrunk. As shown in Figure 7, the drier part of Area 1 rapidly decreased in proportion after May. However, the proportion of drier areas in Area 1 is still much higher than across the entire PRB, as the absolute precipitation in Area 1 is still relatively low in the PRB [28].

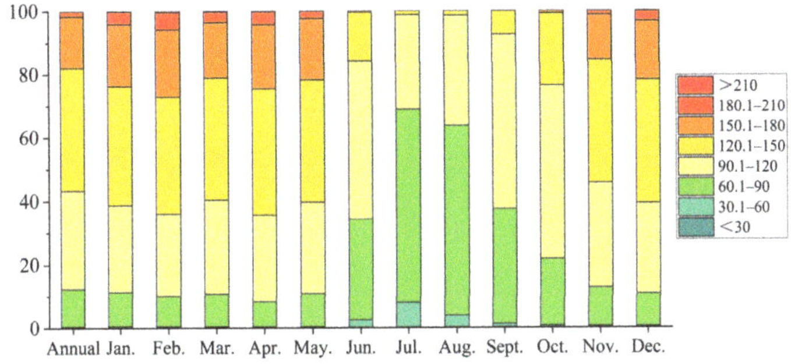

Figure 7. Areal proportion of parts in different ranges of the SbAI in Area 1. Note: Jan etc. is the abbreviation of each month.

According to Figure 8, at the monthly and annual scales, wetter areas were predominant in Area 1, indicating that Area 1 has become wetter in the past 20 years. Excluding November, the proportion of NW areas are in Area 1 was higher than that of SD, ND and SW areas (Figure 8). In November, the proportion of SW areas was greatest Area 1. In May, however, the proportion of ND areas was greatest in Area 1. Collectively, in Area 1, the wetter trend in the dry season was more significant than the drier trend in the monsoon season, showing an overall inter-annual wetter trend in Area 1.

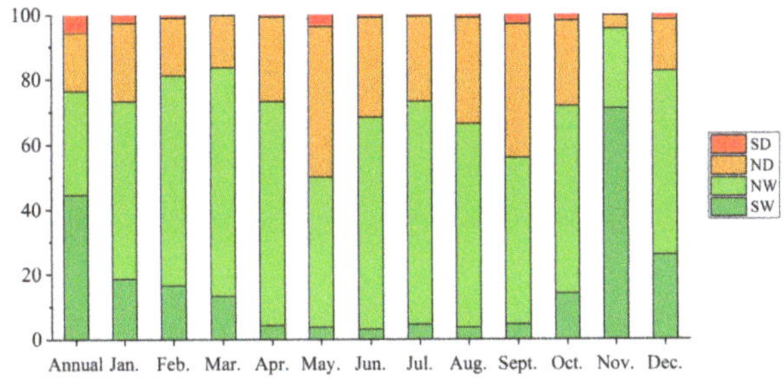

Figure 8. The areal proportions for different trends in Area 1. Note: Jan etc. is the abbreviation of each month. (SD: significantly drier trend; ND: not significantly drier trend; NW: not significantly wetter trend; SW: significantly wetter trend).

In Figure 3, Area 6 is generally humid. In the dry season, from November to April, from the center Area 6 outward, the SbAI gradually increased. After the switch from the dry season to the monsoon season, the SbAI difference in different parts of Area 6 rapidly narrowed. As shown in Figure 9, similar to the entire PRB, the proportion of drier areas decreased after April.

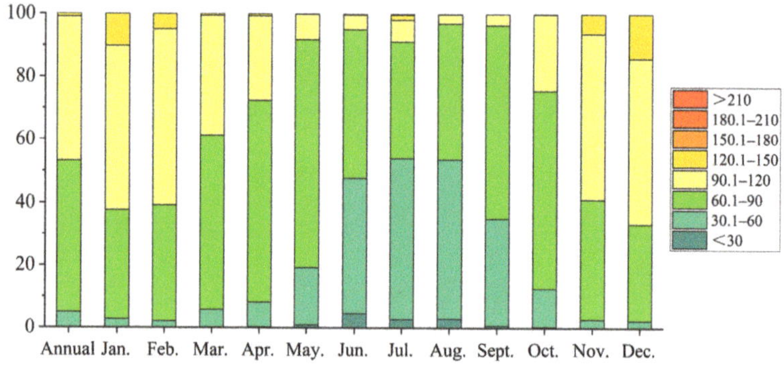

Figure 9. Proportion of areas in different SbAI ranges in Area 6. Note: Jan etc. is the abbreviation of each month.

According to Figure 10, in addition to June, most parts of Area 6 exhibited a wetter trend at the monthly and annual scales. At the annual scale, more than 70% of Area 6 significantly became wetter. However, in June, in the monsoon season, more than 50% of Area 6 became drier. The ratio is much higher than that in other sub-regions. Collectively, in Area 6, the proportion of wetter areas in the dry season was slightly larger than that of drier parts in the monsoon season. It should be pointed out that, Area 6 is among the areas with the highest precipitation in the PRB; however, it had a higher proportion of drier areas in the monsoon season. The possible causes were that Area 6 experienced more significant impacts from human activities (i.e., fast industrialization and urbanization), resulting in a decrease in water storage capacity and aggravating drought conditions.

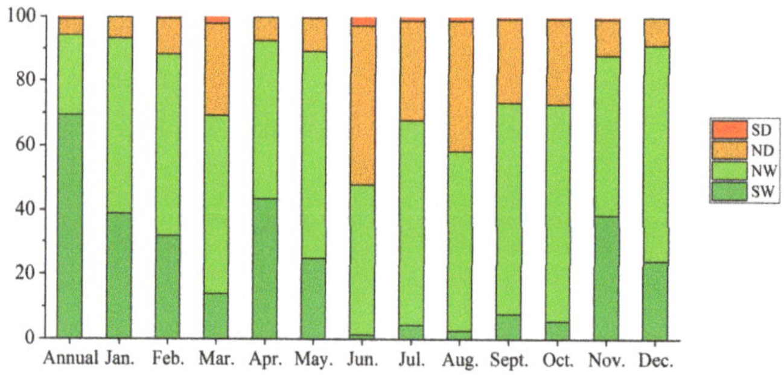

Figure 10. The areal proportion of different trends in Area 6. Note: Jan etc. is the abbreviation of each month. (SD: significantly drier trend; ND: not significantly drier trend; NW: not significantly wetter trend; SW: significantly wetter trend).

4. Discussion

4.1. Accuracy Assessemnt

The SbAI was initially proposed by Kimura and was first used to determine surface dryness in Northeast Asia [33]. Later, it was also used to determine land drought and land degradation levels [43,44]. Using the SbAI to classify land drought levels was more accurate than using a traditional drought index. The SbAI was also used to monitor drought in Mongolia from 2001 to 2013 [45]. The results showed that the SbAI was able to accurately identify drought using satellite images. The SPI is sometimes more promising than the SBAI. However, the SPI relies heavily on observation data from meteorological stations. The SPI may not be ideal for areas with few meteorological stations. In the PRB (with a basin size of 440,000 km^2), there are only approximately 80 meteorological stations available (many county-level stations do not release their data publicly), especially in the upper reaches of the Pearl River which have fewer stations. This is why we used the SBAI for this study.

Some studies have evaluated the accuracy of GPM data in the Pearl River Basin [46]. The assessments displayed that the accuracy of GPM data is satisfactory in the Pearl River Basin. This study also made a simple verification of the GPM data through a comparison with the precipitation data collected by the Pearl River Water Conservancy Commission (PRWCC). As shown in Figure 11, the GPM data and the precipitation data of the PRWCC show great correlation (R^2 is greater than 0.9) across the entire basin and the seven sub-regions. Therefore, we can verify our results using the drought trends obtained from the GPM data.

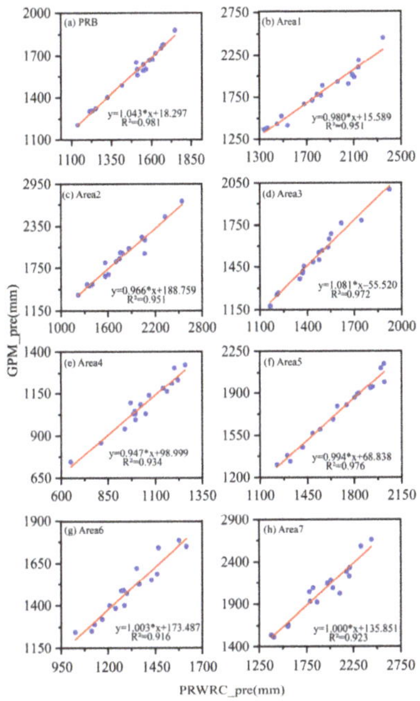

Figure 11. (**a**) Correlation analysis between GPM data and precipitation data provided by the PRWCC in PRB; (**b–h**) Correlation analysis between GPM data and precipitation data provided by the PRWCC in Areas 1–7. Note: * represents multiplication sign.

We checked the accuracy of the SbAI using GPM data and the results are shown in Figure 12. In this validation step, we compare the monthly GPM data from 2001 to 2019 and the mean SbAI value for the entire Pearl River Basin and its seven sub-regions. Figure 12 shows the relationship between the two tends consistent. Across the entire Pearl River Basin and seven sub-regions, R^2 values are all greater than 0.3; half of the R^2 values are even greater than 0.4. The maximum R^2 value is 0.563, and the minimum R^2 value is 0.310. Figure 12 shows that the SbAI and the GPM displayed a significant negative correlation. The increase in rainfall can weaken the drought condition, thereby reducing the SbAI, and vice versa. This trend is pronounced in all of the seven sub-regions. However, it should be pointed out that GPM data and the SbAI data are not be completely consistent, because the GPM data are the result of natural climate change, but the surface drought condition monitored by the SbAI is the combined result of human activities and natural climate change. Therefore, Figure 12 shows that the SbAI is directly affected by natural rainfall, but, at the same time, is not completely regulated by natural rainfall. This shows the incomplete consistency of the two. Therefore, the two trends are inconsistent in some sub-regions, mainly because these areas are disturbed by human activities. For example, humans can increase water storage capacity by afforestation, thereby reducing the drought conditions. Conversely, humans can also increase the impervious area through rapid urbanization, resulting in a decrease in water storage capacity and aggravating drought conditions.

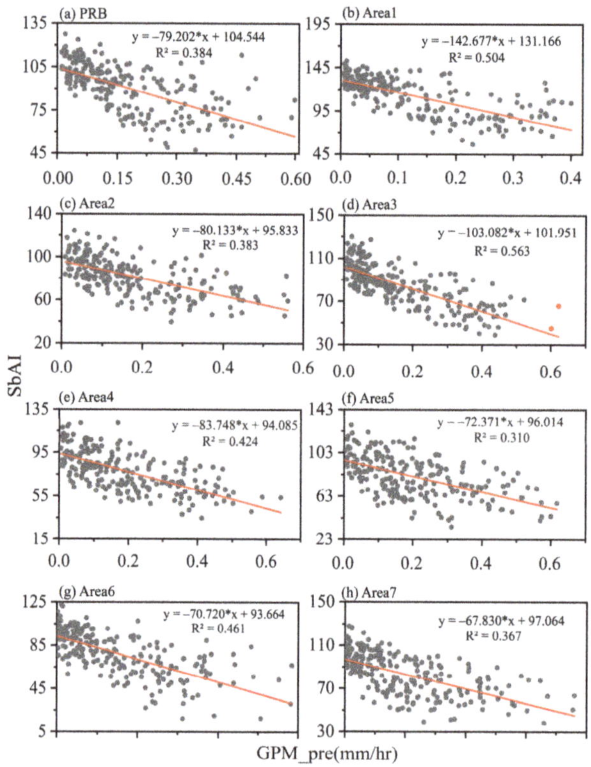

Figure 12. (**a**) Correlation analysis between SbAI and GPM data series in the PRB; (**b–h**) Correlation analysis between SbAI and GPM data series in Areas 1–7. Note: * represents multiplication sign.

4.2. Comparisons with Previous Studies

Studies have shown that most of the arid and semi-arid regions in the world are becoming drier, such as the west of America and the Middle East region [47–49]. Other areas, for example the northwest and southwest regions of China, are becoming wetter [50,51]. Compared with most arid and semi-arid regions, the PRB has a wetter trend, and drought in arid and semi-arid regions mainly occurs in summer, while drought in the PRB mainly occurs in winter and spring.

He et al. [52] found that the NDVI in the Pearl River Basin showed a fluctuating increasing trend ($p < 0.01$); 81.44% of the Pearl River Basin showed an increase in vegetation coverage, mainly in the upper and middle reaches of the Pearl River Basin. The areas with medium and low vegetation coverage are mainly located in the Pearl River Delta. Our results are in good agreement with those of He et al. (Figure 13). In this study, the SbAI in the Pearl River Basin showed a significant downward trend ($p < 0.05$). The SbAI decreased in most areas of the Pearl River Basin. Areas with an SbAI are mainly distributed in the Pearl River Delta. The study by He et al. [52] only analyzed the annual NDVI data for the years before 2015. We obtained NDVI data for recent years and performed a correlation analysis with the SbAI. The results are still consistent.

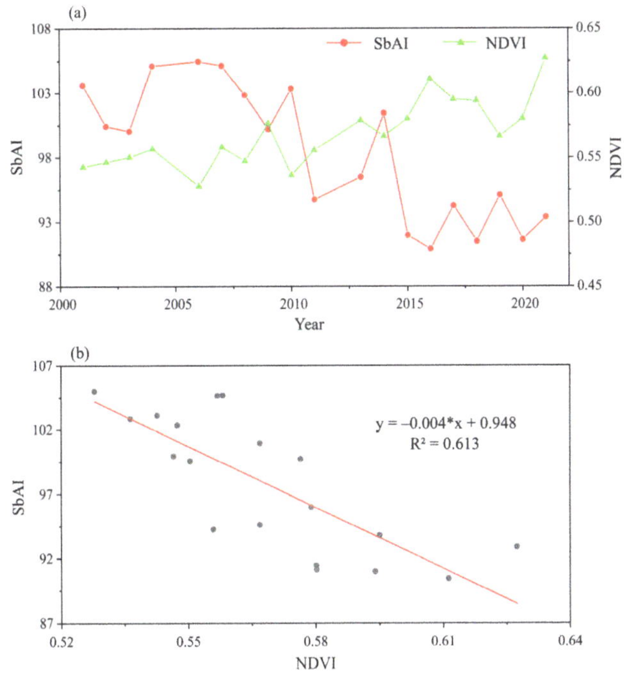

Figure 13. Comparisons of SbAI and NDVI results: (**a**) Comparison of temporal trends; (**b**) Correlation analysis of the two datasets. Note: * represents multiplication sign.

In addition, Zhang et al. [39] found that the Pearl River Basin has become drier in the monsoon season and wetter in the dry season. Areas with relatively low rainfall are distributed in the western part of the basin, while humid areas with high rainfall are distributed in the southeastern part of the basin. The results of this study show that, in the monsoon season, the basin became drier (the area with severer drought condition became larger), while in the dry season, the basin became wetter (the area with a wetter condition became larger). These conclusions are similar to Zhang's findings.

Our results indicated that the Pearl River Basin presents a wetter trend over the past two decades. As discussed in Section 4.1, the wetter trend presented by the SbAI is affected by natural precipitation and human activities. For example, human activities can change land use, among which afforestation and the resulting water storage capacity are paramount [53]. In the Pearl River Delta, due to fast urbanization, a large number of cultivated land and forests have been transformed into impervious areas. Studies [54] have shown that the cultivated land, forests and waters in the Pearl River Delta decreased by 15.64%, 2.9% and 7.45%, respectively, from 2000 to 2015, while the impervious surface increased by 42.1%, of which the transfer ratios of the above three land use types are 23.09%, 11.29% and 7.04%, respectively. The studies by Zhang et al. and Chen et al. have shown [55,56] that the Pearl River Delta is a high-risk area for decreasing water storage capacity. In this study, the Pearl River Delta region was reported to gradually be drier. This drought is not due to a reduction in natural rainfall, but rather a reduction in water storage capacity caused by human activities.

4.3. Implications for Sustainable River Basin Management

In August 2022, as we wrote this paper, the world was experiencing the worst droughts in recent decades. The droughts across the Northern Hemisphere—expanding from the USA to Europe and China—are further adding pressure to global large river ecosystems already under stress. In recent years, climate change has caused droughts to occur more frequently [57,58]. Based on the SbAI and satellite images, this study analyzed the spatiotemporal changes in drought conditions over the past 20 years, providing a valuable reference drought assessment across the Pearl River Basin. Our research results also provide consultative information for watershed management, such as agricultural irrigation, water resource management, and urbanization. Based on our findings, the headwaters of the Pearl River have become wetter and may sustain more vegetation and agricultural activities. Crops planted in the dry season should be prioritized. However, most areas of the Pearl River Basin have displayed a slight trend towards becoming drier in the monsoon season, which may be related to the rise in monsoon temperatures caused by climate change [38]. In the monsoon season, natural vegetation and human activities have a high demand for water resources. A more scientific management of water resources is therefore required. Our study also indicates that that uncontrolled human activities such as excessive water withdrawal for irrigation and industrial use and fast urbanization to increase impervious surfaces and reduce water storage capacity can worsen drought conditions. In the Pearl River Delta region, with rapid urbanization and industrialization, more scientific management for urban development and more appropriate allocation of water resources are required for sustainable development.

5. Conclusions

In this study, we calculated the daily SbAI for the whole Pearl River Basin for the period 2001–2021 using MODIS surface reflectance and surface temperature data, and further derived the annual and monthly SbAI. We investigated the spatiotemporal changes in drought conditions in the Pearl River Basin. Based on the study results, we obtained the following findings:

(1) The inter-annual SbAI in the Pearl River Basin exhibited a significant downward trend. The decreasing trend in the SbAI was statistically significant in the dry season, and the monsoon season also showed a decreasing except for an insignificant increase in June.

(2) In the dry season, areas with droughts are mainly located in sub-regions of Areas 1, 2, and 3; as the flood season arrives, the basin receives more water and gradually becomes humid, and the total area with droughts decreases rapidly.

(3) In the past 20 years, most parts of the Pearl River Basin have become wetter. However, the drought conditions illustrated an insignificant increase in the monsoon season, corresponding to a more statistically significant shrinking in the dry season.

(4) Overall, the Pearl River Basin has become wetter over the past two decades, which may be the result of natural and human factors (i.e., increased precipitation and vegetation coverage); areas with increased drought conditions were likely impacted by human activities such as water withdrawal for irrigation and industrial uses, fast urbanization and increased impervious surfaces and the resulting reduction in water storage capacity.

Author Contributions: Conceptualization, Kunlong Niu and Xiankun Yang; formal analysis, Kunlong Niu and Xiankun Yang; data curation, Kunlong Niu; writing—original draft preparation, Kunlong Niu; writing—review and editing, Xiankun Yang, Junliang Qiu, Edward Park, Shirong Cai, Wenxin Zhang and Xiaolin Mu; supervision, Xiankun Yang. All authors have read and agreed to the published version of the manuscript.

Funding: The National Natural Science Foundation of China (Grant No.: 41871017) and the Natural Science Foundation of Guangdong Province (Grant No.: 2021A1515011533) funded this research.

Data Availability Statement: The data presented in this study are available on request from the corresponding author.

Acknowledgments: The authors are very grateful to the NASA for providing the MODIS products. Thanks also to the NASA and the JAXA for providing the GPM precipitation dataset.

Conflicts of Interest: The authors declare no conflict of interest.

References

1. West, H.; Quinn, N.; Horswell, M. Remote sensing for drought monitoring & impact assessment: Progress, past challenges and future opportunities. *Remote Sens. Environ.* **2019**, *232*, 111291.
2. Tabari, H. Climate change impact on flood and extreme precipitation increases with water availability. *Sci. Rep.* **2020**, *10*, 13768. [CrossRef] [PubMed]
3. Cook, B.I.; Mankin, J.S.; Anchukaitis, K.J. Climate change and drought: From past to future. *Curr. Clim. Chang. Rep.* **2018**, *4*, 164–179. [CrossRef]
4. Wilhite, D.A. Drought as a natural hazard: Concepts and definitions. In *Drought: A Global Assessment*; University of Nebraska-Lincoln: Lincoln, NE, USA, 2000.
5. Palmer, W.C. *Meteorological Drought*; US Department of Commerce, Weather Bureau: Washington, DC, USA, 1965; Volume 30.
6. McKee, T.B.; Doesken, N.J.; Kleist, J. The relationship of drought frequency and duration to time scales. In Proceedings of the 8th Conference on Applied Climatology, Anaheim, CA, USA, 17–22 January 1993; pp. 179–183.
7. Vicente-Serrano, S.M.; Beguería, S.; López-Moreno, J.I. A multiscalar drought index sensitive to global warming: The standardized precipitation evapotranspiration index. *J. Clim.* **2010**, *23*, 1696–1718. [CrossRef]
8. Kogan, F.N. Application of vegetation index and brightness temperature for drought detection. *Adv. Space Res.* **1995**, *15*, 91–100. [CrossRef]
9. Rhee, J.; Im, J.; Carbone, G.J. Monitoring agricultural drought for arid and humid regions using multi-sensor remote sensing data. *Remote Sens. Environ.* **2010**, *114*, 2875–2887. [CrossRef]
10. Zhang, A.; Jia, G. Monitoring meteorological drought in semiarid regions using multi-sensor microwave remote sensing data. *Remote Sens. Environ.* **2013**, *134*, 12–23. [CrossRef]
11. Fayne, A. Current Environment of the Offshore Voluntary Disclosure Initiative (OVDI) Opt-out and Civil Exam Cases. *J. Tax Pract. Proced.* **2013**, *15*, 21.
12. Dehghan, S.; Salehnia, N.; Sayari, N.; Bakhtiari, B. Prediction of meteorological drought in arid and semi-arid regions using PDSI and SDSM: A case study in Fars Province, Iran. *J. Arid Land* **2020**, *12*, 318–330. [CrossRef]
13. Trenberth, K.E.; Dai, A.; Van Der Schrier, G.; Jones, P.D.; Barichivich, J.; Briffa, K.R.; Sheffield, J. Global warming and changes in drought. *Nat. Clim. Chang.* **2014**, *4*, 17–22. [CrossRef]
14. Gidey, E.; Dikinya, O.; Sebego, R.; Segosebe, E.; Zenebe, A. Modeling the spatio-temporal meteorological drought characteristics using the standardized precipitation index (SPI) in raya and its environs, northern Ethiopia. *Earth Syst. Environ.* **2018**, *2*, 281–292. [CrossRef]
15. Spinoni, J.; Naumann, G.; Carrao, H.; Barbosa, P.; Vogt, J. World drought frequency, duration, and severity for 1951–2010. *Int. J. Climatol.* **2014**, *34*, 2792–2804. [CrossRef]
16. Hernandez, E.A.; Uddameri, V. Standardized precipitation evaporation index (SPEI)-based drought assessment in semi-arid south Texas. *Environ. Earth Sci.* **2014**, *71*, 2491–2501. [CrossRef]
17. Chiang, F.; Mazdiyasni, O.; AghaKouchak, A. Evidence of anthropogenic impacts on global drought frequency, duration, and intensity. *Nat. Commun.* **2021**, *12*, 2754. [CrossRef]
18. Dutta, D.; Kundu, A.; Patel, N.; Saha, S.; Siddiqui, A. Assessment of agricultural drought in Rajasthan (India) using remote sensing derived Vegetation Condition Index (VCI) and Standardized Precipitation Index (SPI). *Egypt. J. Remote Sens. Space Sci.* **2015**, *18*, 53–63. [CrossRef]
19. Huang, W.; Yang, X.; Li, M.; Zhang, X.; Wang, M.; Dai, S.; Ma, J. Evolution characteristics of seasonal drought in the south of China during the past 58 years based on standardized precipitation index. *Trans. Chin. Soc. Agric. Eng.* **2010**, *26*, 50–59.
20. Hänsel, S.; Ustrnul, Z.; Łupikasza, E.; Skalak, P. Assessing seasonal drought variations and trends over Central Europe. *Adv. Water Resour.* **2019**, *127*, 53–75. [CrossRef]
21. Dong, Z.; Liu, H.; Baiyinbaoligao; Hu, H.; Khan, M.Y.A.; Wen, J.; Chen, L.; Tian, F. Future projection of seasonal drought characteristics using CMIP6 in the Lancang-Mekong River Basin. *J. Hydrol.* **2022**, *610*, 127815. [CrossRef]
22. Jiménez-Muñoz, J.C.; Mattar, C.; Barichivich, J.; Santamaría-Artigas, A.; Takahashi, K.; Malhi, Y.; Sobrino, J.A.; van der Schrier, G. Record-breaking warming and extreme drought in the Amazon rainforest during the course of El Niño 2015–2016. *Sci. Rep.* **2016**, *6*, 33130. [CrossRef]
23. Pramudya, Y.; Onishi, T.; Senge, M.; Hiramatsu, K.; Nur, P.M. Evaluation of recent drought conditions by standardized precipitation index and potential evapotranspiration over Indonesia. *Paddy Water Environ.* **2019**, *17*, 331–338. [CrossRef]
24. He, Z.; Xu, S.; Xie, X.; Chen, H. Brief introduction of flood and drought disaster prevention in the Pearl River Basin in 2019. *China Flood Drought Manag.* **2020**, *30*, 24–26. [CrossRef]
25. Wu, S.; Huang, F. Summary of flood and drought disaster prevention in the Pearl River Basin in 2021. *China Flood Drought Manag.* **2021**, *31*, 25–26, 43. [CrossRef]
26. Qiu, J.; Cao, B.; Park, E.; Yang, X.; Zhang, W.; Tarolli, P. Flood monitoring in rural areas of the Pearl River Basin (China) using Sentinel-1 SAR. *Remote Sens.* **2021**, *13*, 1384. [CrossRef]
27. Jing, J.-l.; Xu, Y.; Wang, Y.-f.; Dou, S.-q.; Yin, M. Characteristics of multi-scale drought and flood in the Pearl River Basin from 1960 to 2019. *Res. Agric. Mod.* **2021**, *42*, 557–569. [CrossRef]
28. Deng, S.; Chen, T.; Yang, N.; Qu, L.; Li, M.; Chen, D. Spatial and temporal distribution of rainfall and drought characteristics across the Pearl River basin. *Sci. Total Environ.* **2018**, *619–620*, 28–41. [CrossRef]

29. Zhang, Q.; Singh, V.P.; Peng, J.; Chen, Y.D.; Li, J. Spatial–temporal changes of precipitation structure across the Pearl River basin, China. *J. Hydrol.* **2012**, *440–441*, 113–122. [CrossRef]
30. Liu, B.; Chen, J.; Lu, W.; Chen, X.; Lian, Y. Spatiotemporal characteristics of precipitation changes in the Pearl River Basin, China. *Theor. Appl. Climatol.* **2016**, *123*, 537–550. [CrossRef]
31. Hu, M.; Li, Z.; Wang, Y.; Jiao, M.; Li, M.; Xia, B. Spatio-temporal changes in ecosystem service value in response to land-use/cover changes in the Pearl River Delta. *Resour. Conserv. Recycl.* **2019**, *149*, 106–114. [CrossRef]
32. Zhang, S.-X.; Zhang, J.-M.; Zhang, W.-K.; Zhang, D.-N.; Fu, J.-Y.; Zang, C.-F. Spatiotemporal variability characteristics and driving forces of land use in the Pan-Pearl River Basin, China. *J. Appl. Ecol.* **2020**, *31*, 573–580.
33. Kimura, R.; Moriyama, M. Application of a satellite-based aridity index in dust source regions of northeast Asia. *J. Arid Environ.* **2014**, *109*, 31–38. [CrossRef]
34. Liang, S. *Quantitative Remote Sensing of Land Surfaces*; John Wiley & Sons: Hoboken, NJ, USA, 2005.
35. Mann, H.B. Nonparametric tests against trend. *Econometrica* **1945**, *13*, 245–259. [CrossRef]
36. Sen, P.K. Estimates of the regression coefficient based on Kendall's tau. *J. Am. Stat. Assoc.* **1968**, *63*, 1379–1389. [CrossRef]
37. Pettit, A. Anon-parametric approach to the change-point detection. *Appl. Stat.* **1979**, *28*, 126–135. [CrossRef]
38. Qiu, J.; Yang, X.; Cao, B.; Chen, Z.; Li, Y. Effects of urbanization on regional extreme-temperature changes in China, 1960–2016. *Sustainability* **2020**, *12*, 6560. [CrossRef]
39. Zhang, Q.; Xu, C.-Y.; Zhang, Z. Observed changes of drought/wetness episodes in the Pearl River basin, China, using the standardized precipitation index and aridity index. *Theor. Appl. Climatol.* **2009**, *98*, 89–99. [CrossRef]
40. Peng, D.; Zhou, T. Why was the arid and semiarid northwest China getting wetter in the recent decades? *J. Geophys. Res. Atmos.* **2017**, *122*, 9060–9075. [CrossRef]
41. Liang, C.; Chen, T.; Dolman, H.; Shi, T.; Wei, X.; Xu, J.; Hagan, D.F.T. Drying and wetting trends and vegetation covariations in the drylands of China. *Water* **2020**, *12*, 933. [CrossRef]
42. Li, H.; Liu, F.; Zhang, S.; Zhang, C.; Zhang, C.; Ma, W.; Luo, J. Drying–Wetting Changes of Surface Soil Moisture and the Influencing Factors in Permafrost Regions of the Qinghai-Tibet Plateau, China. *Remote Sens.* **2022**, *14*, 2915. [CrossRef]
43. Kimura, R.; Moriyama, M. Recent trends of annual aridity indices and classification of arid regions with satellite-based aridity indices. *Remote Sens. Earth Syst. Sci.* **2019**, *2*, 88–95. [CrossRef]
44. Kimura, R.; Moriyama, M. Determination by MODIS satellite-based methods of recent global trends in land surface aridity and degradation. *J. Agric. Meteorol.* **2019**, *75*, 153–159. [CrossRef]
45. Kimura, R.; Moriyama, M. Use of a modis Satellite-based aridity index to monitor drought conditions in Mongolia from 2001 to 2013. *Remote Sens.* **2021**, *13*, 2561. [CrossRef]
46. Xia, X.; Liu, Y.; Jing, W.; Yao, L. Assessment of Four Satellite-Based Precipitation Products Over the Pearl River Basin, China. *IEEE Access* **2021**, *9*, 97729–97746. [CrossRef]
47. Kousari, M.; Ekhtesasi, M.; Malekinezhad, H. Investigation of long term drought trend in semi-arid, arid and hyper-arid regions of the world. *Desert Manag.* **2017**, *4*, 36–53.
48. Zarei, A.R. Evaluation of drought condition in arid and semi-arid regions, using RDI index. *Water Resour. Manag.* **2018**, *32*, 1689–1711. [CrossRef]
49. Mohammed, R.; Scholz, M. Climate variability impact on the spatiotemporal characteristics of drought and Aridityin arid and semi-arid regions. *Water Resour. Manag.* **2019**, *33*, 5015–5033. [CrossRef]
50. Dong, B.; Yu, Y.; Pereira, P. Non-growing season drought legacy effects on vegetation growth in southwestern China. *Sci. Total Environ.* **2022**, *846*, 157334. [CrossRef] [PubMed]
51. Wei, W.; Zhang, H.; Zhou, J.; Zhou, L.; Xie, B.; Li, C. Drought monitoring in arid and semi-arid region based on multi-satellite datasets in northwest, China. *Environ. Sci. Pollut. Res.* **2021**, *28*, 51556–51574. [CrossRef] [PubMed]
52. He, C.; Ma, B.; Jing, J.; Xu, Y.; Dou, S. Spatial-Temporal Variation and Future Changing Trend of NDVI in the Pearl River Basin from 1982 to 2015. In Proceedings of the International Conference on Human Centered Computing, Virtual Event, 14–15 December 2020; pp. 209–219.
53. Liu, Y.; Tian, J.; Liu, R.; Ding, L. Influences of Climate Change and Human Activities on NDVI Changes in China. *Remote Sens.* **2021**, *13*, 4326. [CrossRef]
54. Hu, P.; Li, F.; Sun, X.; Liu, Y.; Chen, X.; Hu, D. Assessment of Land-Use/Cover Changes and Its Ecological Effect in Rapidly Urbanized Areas—Taking Pearl River Delta Urban Agglomeration as a Case. *Sustainability* **2021**, *13*, 5075. [CrossRef]
55. Zhang, Q.; Xiao, M.; Singh, V.P.; Li, J. Regionalization and spatial changing properties of droughts across the Pearl River basin, China. *J. Hydrol.* **2012**, *472–473*, 355–366. [CrossRef]
56. Chen, Y.D.; Zhang, Q.; Xiao, M.; Singh, V.P. Transition probability behaviors of drought events in the Pearl River basin, China. *Stoch. Environ. Res. Risk Assess.* **2017**, *31*, 159–170. [CrossRef]
57. Liu, S.; Shi, H.; Niu, J.; Chen, J.; Kuang, X. Assessing future socioeconomic drought events under a changing climate over the Pearl River basin in South China. *J. Hydrol. Reg. Stud.* **2020**, *30*, 100700. [CrossRef]
58. Duan, R.; Huang, G.; Li, Y.; Zhou, X.; Ren, J.; Tian, C. Stepwise clustering future meteorological drought projection and multi-level factorial analysis under climate change: A case study of the Pearl River Basin, China. *Environ. Res.* **2021**, *196*, 110368. [CrossRef] [PubMed]

International Journal of *Geo-Information*

MDPI

Article

Dynamics of the Burlan and Pomacochas Lakes Using SAR Data in GEE, Machine Learning Classifiers, and Regression Methods

Darwin Gómez Fernández *, Rolando Salas López, Nilton B. Rojas Briceño, Jhonsy O. Silva López and Manuel Oliva

Instituto de Investigación para el Desarrollo Sustentable de Ceja de Selva (INDES-CES), National University Toribio Rodríguez de Mendoza (UNTRM), Chachapoyas 01001, Peru

* Correspondence: darwin.gomez@untrm.edu.pe; Tel.: +51-996-916-717

Abstract: Amazonas is a mountain region in Peru with high cloud cover, so using optical data in the analysis of surface changes of water bodies (such as the Burlan and Pomacochas lakes in Peru) is difficult, on the other hand, SAR images are suitable for the extraction of water bodies and delineation of contours. Therefore, in this research, to determine the surface changes of Burlan and Pomacochas lakes, we used Sentinel-1 A/B products to analyse the dynamics from 2014 to 2020, in addition to evaluating the procedure we performed a photogrammetric flight and compared the shapes and geometric attributes from each lake. For this, in Google Earth Engine (GEE), we processed 517 SAR images for each lake using the following algorithms: a classification and regression tree (CART), Random Forest (RF) and support vector machine (SVM).) 2021-02-10, then; the same value was validated by comparing the area and perimeter values obtained from a photogrammetric flight, and the classification of a SAR image of the same date. During the first months of the year, there were slight increases in the area and perimeter of each lake, influenced by the increase in rainfall in the area. CART and Random Forest obtained better results for image classification, and for regression analysis, Support Vector Regression (SVR) and Random Forest Regression (RFR) were a better fit to the data (higher R^2), for Burlan and Pomacochas lakes, respectively. The shape of the lakes obtained by classification was similar to that of the photogrammetric flight. For 2021-02-10, for Burlan Lake, all 3 classifiers had area values between 42.48 and 43.53, RFR 44.47 and RPAS 45.63 hectares. For Pomacohas Lake, the 3 classifiers had area values between 414.23 and 434.89, SVR 411.89 and RPAS 429.09 hectares. Ultimately, we seek to provide a rapid methodology to classify SAR images into two categories and thus obtain the shape of water bodies and analyze their changes over short periods. A methodological scheme is also provided to perform a regression analysis in GC using five methods that can be replicated in different thematic areas.

Keywords: changes; Google Earth Engine; sentinel; random forest; SVM; CART; Colaboratory; Amazonas region

Citation: Gómez Fernández, D.; Salas López, R.; Rojas Briceño, N.B.; Silva López, J.O.; Oliva, M. Dynamics of the Burlan and Pomacochas Lakes Using SAR Data in GEE, Machine Learning Classifiers, and Regression Methods. *ISPRS Int. J. Geo-Inf.* **2022**, *11*, 534. https://doi.org/10.3390/ijgi11110534

Academic Editors: Walter Chen, Fuan Tsai and Wolfgang Kainz

Received: 2 September 2022
Accepted: 16 October 2022
Published: 24 October 2022

1. Introduction

Only 2.5% of the planet's water is fresh, of which only 1.2% is surface water, and much of the latter is found in glaciers and 20.9% is found in lakes [1]. There are more than 1.43 million lakes and reservoirs [2,3]. This type of coastal and continental ecosystem is important, being a source of nutritional resources for animals and humans, in addition to providing various ecosystem services [4].

Surface water resources also play important roles in economic development, the balance of terrestrial and aquatic ecosystems, agriculture, and the environment [5]. Therefore, it is crucial to monitor the dynamics of the area and water storage of a lake to evaluate the impacts of climate change and to predict future scenarios [6]. In addition, monitoring the extension of surface water supports the management of water resources and climate modelling, among other functions [7].

Detecting bodies of water near urban centres is also necessary for the delimitation of flood zones and therefore water accumulation, which become possible sources of outbreaks of water-borne diseases [8].

In recent years, with the increasing availability of free synthetic aperture radar (SAR) data, research on water resources has increased, for example, for the monitoring of the flooded surfaces of lakes in wet and dry seasons, especially small lakes [9], surface water quality monitoring [10], humidity mapping [11], river mapping [12], and the analysis of the spatiotemporal variation in the water surface of lakes [13].

In Jiangxi (China), the changes in the area of the water surface of Poyan Lake were analysed during 2014–2016 using 33 SAR images of Sentinel-1 and were processed in the Sentinel Application Platform (SNAP) [14]. In turn, Dongting Lake in China was monitored using SAR images from Environmental Satellite (Envisat) during 2002–2009 [15]. In Latin America, RADARSAT level 1 and 7 images, Japanese Earth Resources Satellite (JERS)-1 images, and aquatic vegetation were combined to calculate the area of the swamps of southern Brazil [16]. The lakes of northern Alaska were also mapped in the winter season of 2009 using European Remote Sensing satellite (ERS)-2 images to quantify the availability of water in winter and summer [17].

The classification of satellite images through classification and regression trees (CART), random forests (RF), and support vector machines (SVM) has achieved efficient and accurate results [18]. The image classification process mainly involves the assignment of pixels to a class based on spectral signatures, indices, contextual information, etc. [19]. For this, two known methods of joint learning are boosting and bagging [20]. In boosting, successive trees give extra weight to the points incorrectly predicted by previous predictors, and then a weighted vote is taken for the prediction [20,21]. In bagging, successive trees do not depend on previous trees, and each tree is constructed independently using an initial sample of the dataset. Then, a simple majority vote is carried out for the prediction [20,22]. These processes were optimized with the launch of Google Earth Engine (GEE), allowing the parallel processing of geospatial data at a global scale using the Google cloud [23,24].

Statistical models are a simplification of reality expressed in a mathematical language, so to achieve such simplification assumptions must be made, such is the case of this research that we simplify the behaviour model of the lakes based on different dates from 2014 to 2020. The regression tries to predict a quantity or an expected value, unlike classification which tries to predict a category or class [25]. The main regression algorithms include simple linear regression (SLR), polynomial regression (PR), random forest regression (RFR), support vector regression (SVR), and decision tree regression (DTR), which can be quickly executed in Google Colaboratory (GC).

We analysed the dynamics of the water surface of two lakes in the Amazonas region of Peru. For this, *(i)* we processed 517 Sentinel-1 images for the period 2014–2020, using the GEE platform, *(ii)* with the area and perimeter values of each lake we applied five regression methods executed in Google Collaboratory, *(iii)* we calculated area and perimeter by classifying a SAR image from 2021-02-10 and compared with the value predicted by the best regressor and *(iv)* finally we compared the values calculated in *iii* with a photogrammetric flight performed on the same date (2021-02-10). In effect, this research sought to show the dynamics of the water surfaces of two lakes approximately 50 km apart, with different climatic conditions, geographic, and socioeconomic conditions, relying on the continuity of SAR image data from Sentinel-1.

In contrast to other studies, we calculated the water mask by classifying SAR images in Google Earth Engine using Classification and Regression Trees, Random Forest and Support Vector Machine, and compared them with a high-resolution orthomosaic obtained by a Remote Pilot Aircraft System. We also show the flexibility of performing a regression analysis in Google Colaboratory using Simple Linear Regression, Polynomial Regression, Support Vector Regression, Decision Trees Regression and Random Forest Regression methods, and the same regression methods can be applied to different thematic areas.

2. Materials and Methods

2.1. Study Area

Burlan and Pomacochas are two of the main lakes in the Amazonas region (NW Peru). Next, Figure 1 shows the geographic location of Burlan and Pomacochas Lake, in Utcubamba and Bongará provinces, respectively, in Amazonas region, Peru.

At Burlan Lake, the climate is warm, with an average temperature of 24.9 °C and an altitude of 450 m.a.s.l. [26]. Pomacochas Lake is in a warm and temperate climate, with an average annual temperature of 15 °C and an altitude of 2220 m.a.s.l. [27].

Both lakes have socioeconomic and environmental importance in terms of tourism, fishing and landscape services, water for agricultural activities, water resource regulation, and biodiversity.

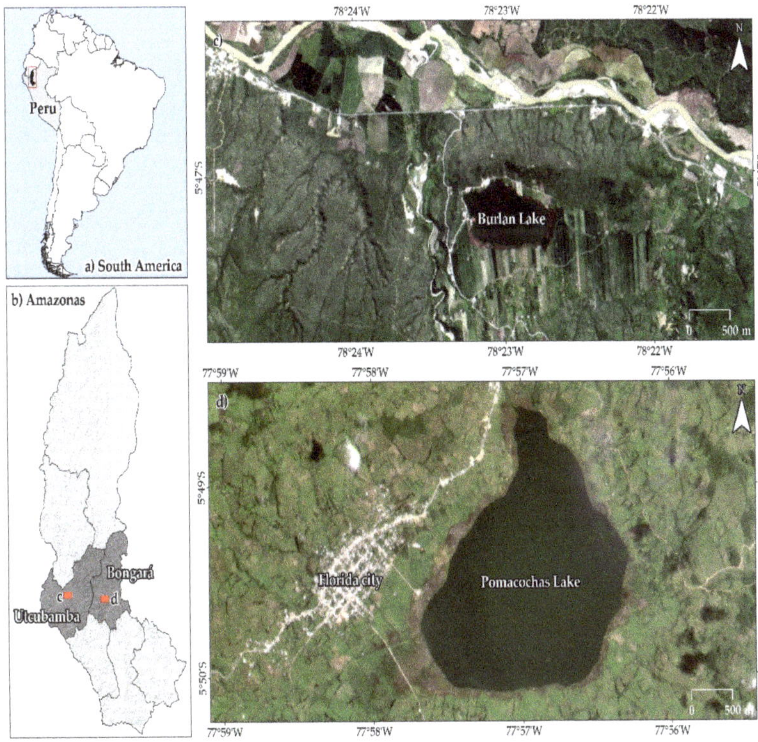

Figure 1. Geographic location of the study lakes in NW Peru.

2.2. Methodological Scheme

Figure 2 summarizes the procedure for analyzing the water surface dynamics of the Burlan and Pomacochas lakes during 2014–2020 using images from the Sentinel-1 mission in GEE and five regression methods: SLR, PR, SVR, DTR and RFR. For this, initially, the speckle of the SAR images was reduced, for a subsequent classification using the CART, RF and SVM algorithms, the classified images were processed in QGIS 3.10. Subsequently, in Google Collaboratory through five regression methods and the area and perimeter values calculated in QGis, the area and perimeter were predicted for 2021-02-10, calculating the R^2 of each regression method. Finally, to validate the calculations performed in GEE and GC, the area and perimeter of each lake were measured in the field using a remotely piloted aircraft system (RPAS) for comparison to the area and perimeter obtained by the extraction in GEE and regression estimation of the method of greater R^2.

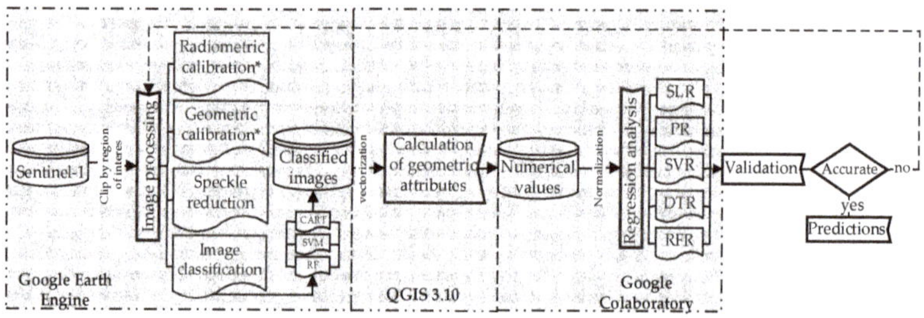

Figure 2. Methodological design for analysing the dynamics of the water surface of Burlan and Pomacochas Lakes during 2014–2020 using SAR images. * This procedure was performed internally by GEE.

2.3. *SAR Dataset and Training Points*

Sentinel-1 A/B images (COPERNICUS/S1_GRD) available in GEE were used [28] with a temporal resolution of 6 days. The data used were level 1 in the ground range detection (GRD), interferometric wideband (IW) format (Beam Mode), with a 10 m resolution, using the ascending and descending Flight Direction, in addition, VH and VV cross-polarized scenes [29].

In the supervised classification of all SAR images, 23 and 12 training points were used for Pomacochas and Burlan Lakes, respectively. The points were categorized as water (1) and land (0), those labelled 1 were distributed in the center of the lake because the previous inspection of images is an area where water is always present, on the other hand, the points labelled 0 were distributed to the edges of the lakes, generally higher parts where there is no water concurrence. For more details on the training points, check file 09 of the web repository Available online: https://github.com/dargofer/SAR_image_classification (accessed on 15 October 2022).

2.4. *SAR Image Processing*

The processing of the SAR images was carried out in the GEE platform [23]. For this, a code was developed (check file 01 of the web repository), that included the import of Sentinel-1 images speckle reduction, classification, and export of SAR images. In addition, according to the availability of data and the objective of the research, water masks were generated in four combinations. For this, Flight Direction and the polarization of the images were combined. These combinations were Descending-VH (DVH), Ascending-VH (AVH), Descending-VV (DVV), and Ascending-VV (AVV) from 2014 to 2020.

For a correct analysis of the SAR images, they must be corrected radiometrically and geometrically, in addition, depending on the objective of the study, the speckle of the images is reduced [30]. In our case, we use the Sentinel-1 images available in the GEE data catalogue, as mentioned in the GEE processing guide for Sentinel-1 images. (Available online: https://developers.google.com/earth-engine/guides/sentinel1 (accessed on 15 January 2021)) these images were already radiometrically and geometrically corrected [29], so we only reduced the speckle of the images using ee.Image.focal_median [31].

A variable was created that contained the filtered collection and a band with the details of each of the four combinations. Then, we performed supervised classification with three machine learning algorithms [32], RF [33], CART [34], and SVM [35,36] algorithms, and 23 and 12 training points for Pomacochas and Burlan Lakes, respectively. Additionally, to evaluate the accuracy of the classification, we calculated the confusion matrix and kappa index [37]. Finally, the images classified in GeoTIFF format and the EPSG coordinate reference system were exported: 32,717 and 32,718 for Burlan and Pomacochas Lakes, respectively.

2.5. *Calculation of the Geometric Attributes*

The geometric attributes were calculated in the QGIS 3.10 LTR software, where the classified images were vectorized using the raster *polygonize* tool executed in batches. The classified images were dissolved according to their coding to avoid calculation errors because, in some images, separate polygons were generated with the same coding. Finally, the geometric values of the area and perimeter were added for each group of images.

2.6. Regression Analysis

With the values obtained for the area and perimeter of each lake and each combination and classifier, five regression methods were applied to estimate the area value of a lake at a specific subsequent time. Simple Linear Regression, Polynomial Regression, Support Vector Regression, Decisions Trees Regression and Random Forest Regression were executed in scripts with Python coding language in Google Colaboratory.

Figure 3 shows the methodological flow chart used in the five regression methods, initially, the Python libraries were imported to input the database, in all 5 methods the database was split into training and evaluation, finally, a feature scaling and execution of the regression method script was performed.

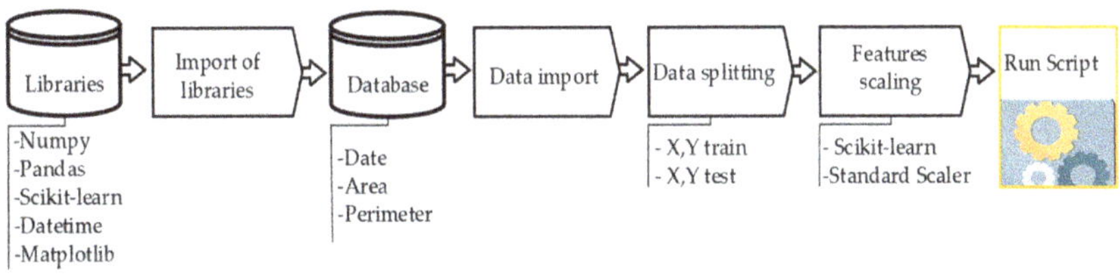

Figure 3. Methodological flow for the regression analysis.

The dependent variables were the area and perimeter (separately), and as an independent variable, the date of acquisition of the SAR image was transformed to an ordinal integer because, in the regressions, chains generate problems in the prediction. The main library used was Scikit-learn [38], which contains all the regression methods used in this research. Next, the procedure followed in each regression method is described.

For Simple Linear Regression, the Numpy, Pandas, Matplotlib, and Scikit-learn libraries were used. The fundamental equation of SLR was determined by the intercept (b_0), the slope (b_1), the independent variable (X) and the Random error term (e_i) (Equation (1)); since the goal of linear regression is to fit a straight line through the data that predicts Y based on X, the calculation of b_0 and b_1 is usually estimated by the ordinary least squares method (Equation (2)) [39,40]. The *LinearRegression* function was used as a regressor [41], imported from the linear models module of the Scikit-learn library.

$$Y_i = b_0 + b_1 X_i + e_i \tag{1}$$

$$\Sigma\,(y_i - \hat{y}_i)^2 \tag{2}$$

To build the polynomial regression we mainly used the "PolynomialFeatures" function [42], which belongs to the scikit-learn library, for which, we used a simple linear regression equation, which was transformed to a second degree using the above-mentioned function.

For Support Vector Regression, the imported data were standardized using the Standard Scaler [43]. Then, to apply the principles of the theory of Vapnik Chervonenkis [44], in which at least the epsilon insensitive tube width and kernel function are required, the SVR function [45], from the Sklearn.svm module was imported. In addition, to complete the regressor, we used the Gaussian Radial Basis Function (RBF) as a kernel function for the Support Vector Regression [46], and 0.1 as the epsilon value.

To build the Random Forest Regression, we imported *RandomForestRegressor* [47], from the *Sklearn.ensemble* module and we considered the default number of trees (*n_estimator = 10*) and 0 as the state of randomness (*random_state*). Finally, to apply Decision Tree Regression to the data, *DecisionTreeRegressor* was imported [48] as a regressor from the *Sklearn.tree* module and the state of randomness was given the value of 0.

2.7. Field Data and Validation

The validation of the area and perimeter of each lake was carried out using images from photogrammetric flights performed on 2021-02-10 with a Phantom 4 RTK in post-processed kinematic mode (PPK) and ground control points (GCPs) collected with a Trimble R10 GNSS. For Pomacochas Lake, 2065 images with 4.57 cm average Ground Sampling Distance (GSD) were obtained, and for Burlan Lake, 729 images with 4.01 cm average GSD were obtained. All images were processed in PIX4D Mapper v 4.6.4 using 9 GCPs for each lake, then to uniformize the images, the orthomosaics were exported at resolutions of 50 cm/pixel.

The measurement of the tie point errors was performed by calculating the root mean square (RMS) error, because the RMS considers the mean error and the variance. Therefore, for a given direction (X, Y, or Z) the RMS is defined as:

$$\mathrm{RMS} = \sqrt{\Sigma\, \frac{{e_i}^2}{N}} \tag{3}$$

where, $\boldsymbol{e_i}$ is the error of each point for the given direction, and $\boldsymbol{N}$ is the number of GCPs.

Finally, for each lake, a SAR image from 2021-02-10 was classified and overlaid with the orthomosaics obtained by the RPAS.

On the other hand, the five regression methods were applied to each group with the area and perimeter data according to each classifier, from which the coefficient of determination (R^2) available in Scikit-learn was calculated [49] to indicate the fit of the data. The R^2 values range from $-\infty$ to 1, the best possible score is 1, and negative values refer to the model can be arbitrarily worse. Therefore, if $\hat{y}_i$ is the predicted value of the ***i-th*** sample, and $\boldsymbol{y_i}$ is the corresponding true value for the total of $\boldsymbol{n}$ samples, the $\mathbf{R^2}$ is defined as:

$$\mathrm{R}^2\,(y,\hat{y}) = 1 - \frac{\sum_{i=1}^{n}(y_i-\hat{y}_i)^2}{\sum_{i=1}^{n}(y_i-\bar{y}_i)^2} \text{ and } \bar{y} = \frac{1}{n}\sum_{i=1}^{n} y_i,\ \sum_{i=1}^{n}(y_i-\hat{y}_i)^2 = \sum_{i=1}^{n}\epsilon_i^2 \tag{4}$$

3. Results

3.1. Distribution and Availability of SAR Data

Figure 4 shows the distribution and monthly availability of the SAR images used for the analysis of the dynamics of Burlan and Pomacochas Lakes from 2014–2021.

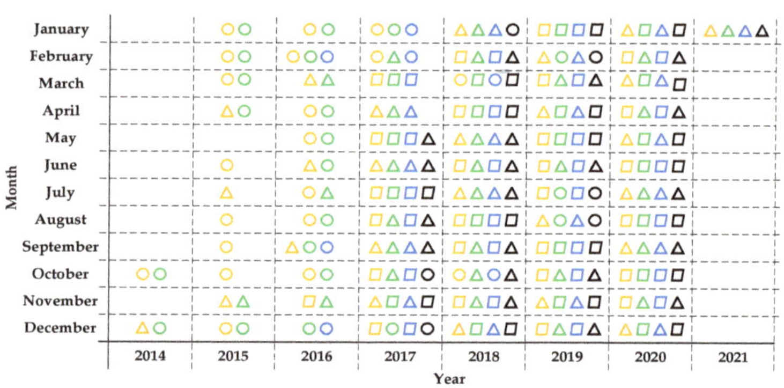

Figure 4. Distribution and monthly availability of the Sentinel-1 images used for the analysis of the dynamics of Burlan and Pomacochas Lakes from 2014 to 2021. The geometric figures represent the number of images available in a month, where circles, triangles, and parallelograms represent 1, 2, and 3 images, respectively. In addition, the colour of each represents the combination of the direction of passage and polarization, where orange, green, blue, and black represent the combinations of DVV, AVV, DVH, and AVH, respectively.

A total of 517 Sentinel-1 images were analysed for each study lake from 2014/10/06 to 2021/01/29 (Table 1). In addition, due to the classification using CART, RF, and SVM, 3 products were obtained per image, generating a total of 3102 water masks for both lakes.

Table 1. Number of SAR images used to generate the water masks of Burlan and Pomacochas Lakes using CART, RF, and SVM.

Lake	SAR Images Available					Water Masks Analysed			
	DVV 2014/10/15–2021/01/29	AVV 2014/10/06–2021/01/20	DVH 2016/02/07–2021/01/29	AVH 2017/05/17–2021/01/20	Total	CART	RF	SVM	Total
Burlan	153	137	123	104	517	517	517	517	1551
Pomacochas	153	137	123	104	517	517	517	517	1551
		Total			1034				3102

Table S1 shows the attributes of all the images used to obtain the water masks of Burlan and Pomacochas Lakes from 2014–2021. We worked with the same scene because we used IW products (250 km for each sweep), and the distance between the lakes was approximately 50 km.

3.2. Obtaining the Geometric Attributes

Figure 5 shows the variation of area and perimeter for Burlan Lake (left) and Pomacochas (right). For Burlan Lake, the maximum values differ according to the flight direction and polarization, for example, 2018 and 2019 show maximum values for VH, while for VV, the maximum values are shown in 2016, 2017 and 2018. On the other hand, Pomacochas Lake presents a homogenous trend, for example, VH presents maximum values in 2018 and 2019, while VV presents a homogeneous trend with maximum values in 2019.

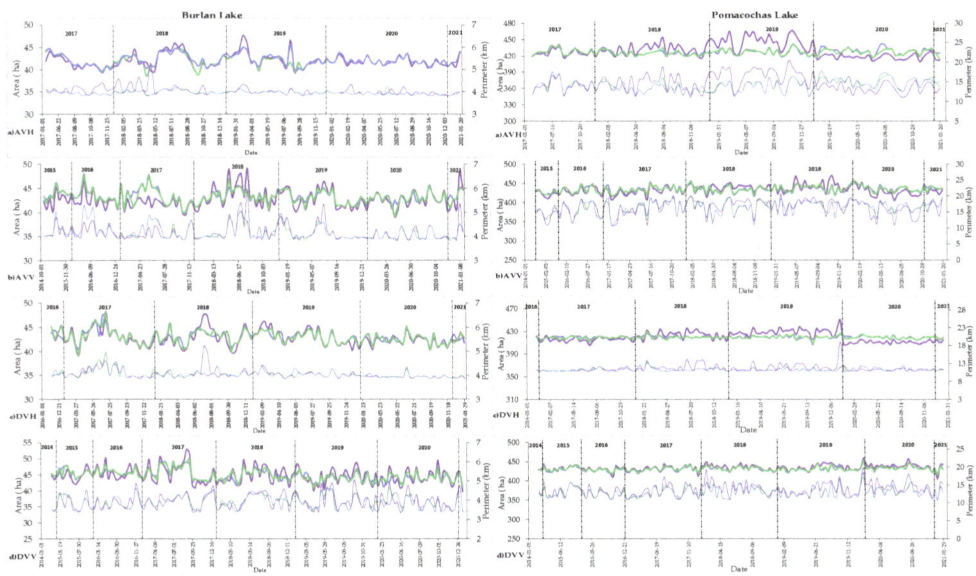

Figure 5. Variation in the area and perimeter of the Burlan and Pomacochas lakes using CART, RF, and SVM as classifiers of the SAR images. The thick lines represent the area (ha), and the thin lines represent the perimeter (km). In addition, the purple, green and blue lines represent the values obtained by SVM, CART and RF, respectively. In addition, AVH, AVV, DVH and DVV, represents the dataset obtained by: flight directions Ascending (A) and Descending (D), polarizations transmitted and received vertically (VV) and transmitted vertically and received horizontally (VH).

Table 2 shows the minimum, maximum and average values obtained for the area (A) and perimeter (P) of Burlan and Pomacochas Lakes calculated according to the classification of SAR images using CART, RF, and SVM. The behaviour of the values obtained by CART and RF was similar for both lakes, while the SVM values were much higher, due to the algorithm used in the classification.

Table 2. Minimum, maximum, and average values of the area and perimeter of the Burlan and Pomacochas lakes obtained by classification of SAR images in the 2014–2020 period.

Classifier	Geometric Attribute		Burlan Lake				Pomacochas Lake			
			AVH	AVV	DVH	DVV	AVH	AVV	DVH	DVV
Classification and regression tree(CART)	Area (ha)	Minimum	38.6	39	39.2	40.9	414	417.8	408.3	415.6
		Maximum	45	48	48.1	50.2	441.4	455.8	430.1	452.2
		Average	42.1	43.3	43	44.9	426.8	434.8	419.3	431.4
	Perimeter (km)	Minimum	3.31	3.34	3.36	3.42	11.06	10.94	10.9	10.89
		Maximum	3.72	4.8	4.46	4.72	17.59	20.06	13.54	17.26
		Average	3.46	3.67	3.55	3.93	14.16	16.52	11.36	13.03
Random Forest(RF)	Area (ha)	Minimum	39.3	40	40.3	41.1	416	416.2	414.8	415.6
		Maximum	45.6	48	47.6	49.3	441.4	455.8	426.5	456.5
		Average	42.2	43.3	43	44.8	427.2	435.1	419.7	431.3
	Perimeter (km)	Minimum	3.33	3.37	3.36	3.43	11.06	10.92	10.92	10.89
		Maximum	3.67	4.8	4.12	4.72	17.79	19.79	13.2	18.52
		Average	3.46	3.68	3.54	3.91	14.2	16.59	11.38	13.02
Support Vector Machine(SVM)	Area (ha)	Minimum	38.9	39	39.2	39.8	409.2	405.4	405.5	405.4
		Maximum	47.7	49.2	48.3	53	466.8	470.8	450.6	458
		Average	42.1	42.8	43	44.9	430.5	433.5	420.1	434.3
	Perimeter (km)	Minimum	3.33	3.34	3.36	3.37	11.14	10.87	10.88	10.8
		Maximum	4.81	5.73	4.72	5.05	20.52	20.58	19.17	19.92
		Average	3.55	3.66	3.57	3.95	14.72	16.43	11.65	13.86

Figure 6 compares the averaged values of the area (ha) and perimeter (km) of the Pomacochas and Burlan lakes for the combinations AVH, AVV, DVH, and DVV resulting from the classification of SAR images using CART, RF and SVM.

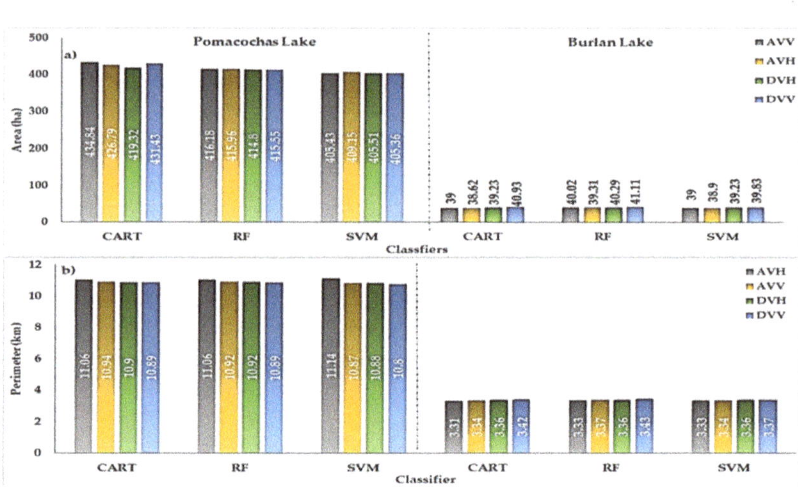

Figure 6. Average values of the (**a**) area and (**b**) perimeter of Pomacochas and Burlan Lakes.

3.3. Data Analysis and Prediction

3.3.1. Data Normalization

For each combination, graphs of letter values (Boxenplots) were created in Google Colaboratory [50] because each batch of data was less than 200 elements [51].

Figure 7 shows the data distribution for each combination (AVV, AVH, DVV, and DVH), where the subfigures of a—l, m—x of each lake represent the data distributions of the area and perimeter, respectively.

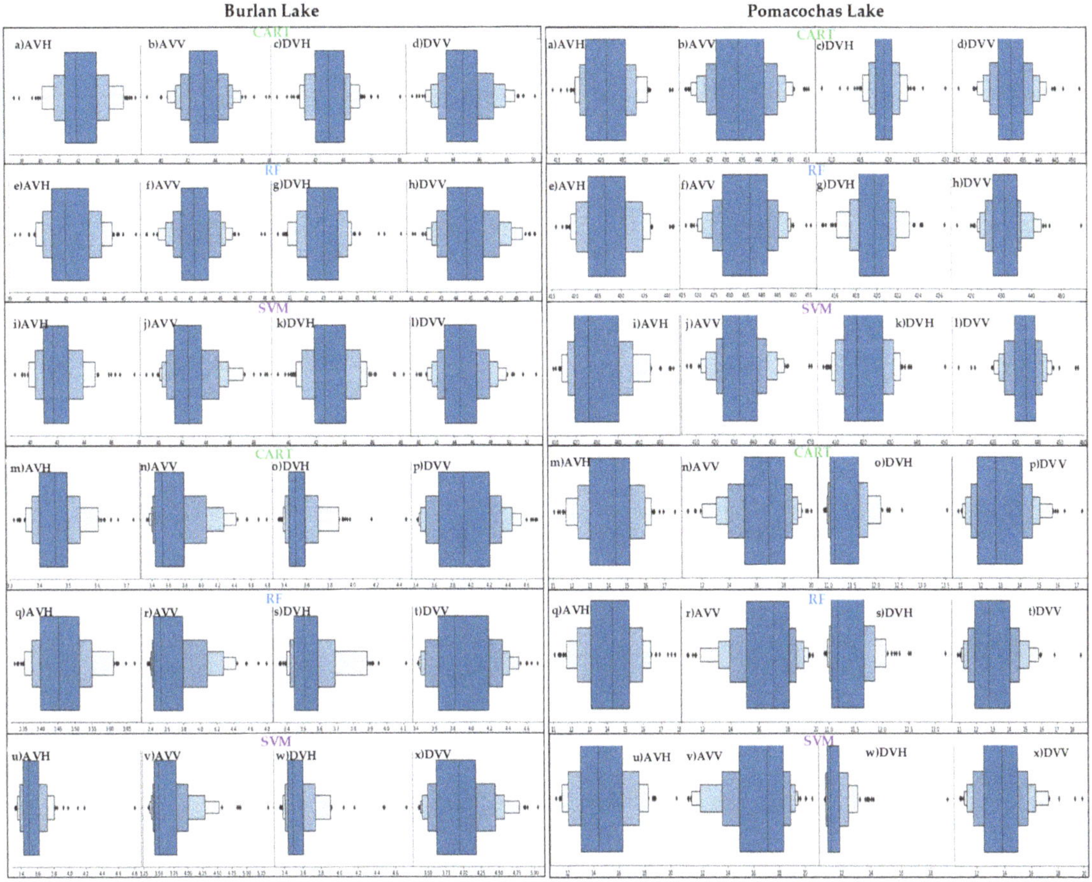

Figure 7. Dispersion of the area and perimeter data for Burlan and Pomacochas Lakes, where the black points represent outliers, and the rectangles represent the highest clustering of data according to quantiles.

As shown in Figure 7, there are outliers for each lake dataset. For example, in sub-figure (o) DVH of Burlan Lake, which represents the distribution of perimeter data obtained in Sentinel-1 descending pass and VH polarization, the majority of data are grouped from 3.4 to 3.6 km, but there are outliers that exceed 4 km (4.2 and 4.4). For its part, the data for Pomacochas Lake was also dispersed, for example, in sub-figure (i) AVH, most of the area data were grouped from 420 to 440 ha, but there are also values that exceed 460 ha and there are also values smaller than 415 ha. Therefore, to perform a correct regression analysis without the inclusion of outliers that can negatively impact the regression models, we proceeded to delete those values.

3.3.2. Regression Methods

Table 3 shows the values of the area, perimeter, and coefficients of determination (R^2) that had the highest degree of fit estimated according to SLR, PR, SVR, DTR, and RFR for 2021-02-10. Table S2 of the supplementary material shows all the R^2 calculated in the present investigation.

Table 3. Area (ha) and perimeter (km) estimated with SLR, PR, SVR, DTR and RFR of greater R^2.

		SLR	PR	SVR	DTR	RFR
Burlan Lake	Area	42.46	42.3	42.43	45.2	**44.47**
	R^2	0.12	0.15	0.22	0.37	**0.46**
	Combination	DVH	DVH	DVH	AVV	**AVV**
	Perimeter	3.43	3.41	3.41	3.43	**3.82**
	R^2	0.15	0.2	0.29	0.23	**0.43**
	Combination	AVH	AVH	DVH	DVV	**DVV**
Pomacochas Lake	Area	417.8	408	**411.42**	414	413.1
	R^2	-0.004	0.38	**0.41**	0.13	0.15
	Combination	DVH	DVH	**DVH**	DVH	DVH
	Perimeter	13.28	16.5	**15.14**	17.1	17.46
	R^2	0.095	0.24	**0.42**	0.16	0.26
	Combination	DVV	AVV	**AVH**	AVV	AVV

To complement Table 3, Figures 8 and 9 show the best fit of the model to the area and perimeter data. For Burlan Lake, SLR, PR, and SVR were better fit to the area data of the DVH combination classified by SVM, while DTR and RFR were better fit to the AVV combination classified by CART. For the perimeter, SLR and PR was a better fit to the AVH combination classified by SVM, and SVR was a better fit to the DVH combination classified by SVM. Finally, DTR and RFR were a better fit for the DVV combination classified by CART. For Pomacochas Lake, all regression models were better fit to the area data of the DVH combination classified by SVM; for the perimeter, SLR was better fit to the DVV combination classified by CART, PR to AVV classified by SVM, SVR to AVH classified by RF, and finally, DTR and RFR to DVV classified by CART.

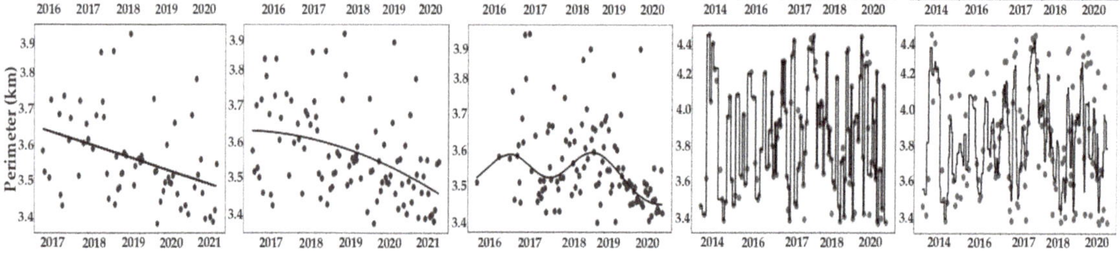

Figure 8. Regression models with greater R^2 for the area and perimeter data of Burlan Lake.

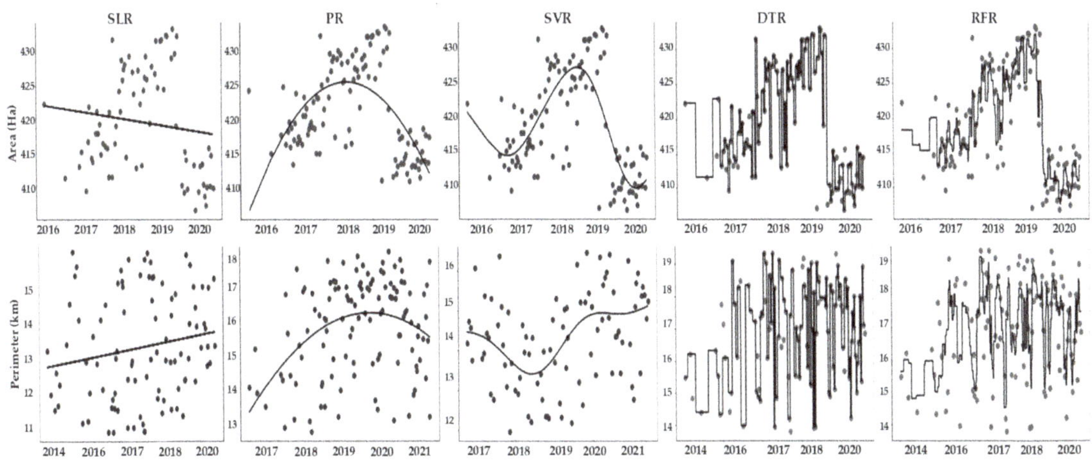

Figure 9. Regression models with higher R^2 for the area and perimeter data of Pomacochas Lake.

For Burlan Lake, with the AVV and DVV combinations classified by CART, RFR obtained the best R^2 for the area (0.46) and perimeter (0.43), respectively. In turn, for Pomacochas Lake, the combination DVH classified by SVM and AVH classified by RF obtained the best R^2 for the area (0.41) and perimeter (0.42), respectively, according to SVR.

Next, Figure 10 compares the R^2 of each regression method. It can be seen that for Burlan Lake, RFR showed higher R^2 in the area and perimeter data, thus showing an average adaptation of the model to the data, while for Pomacochas Lake the model that best fits the area and perimeter was the regression model by support vectors.

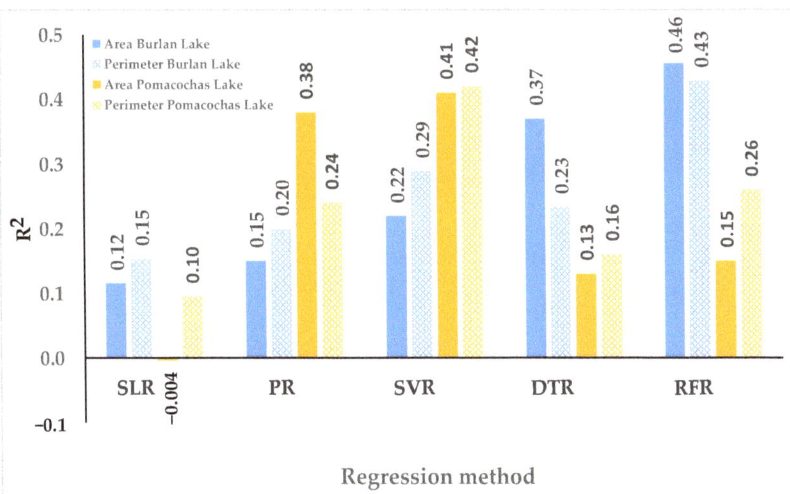

Figure 10. Comparison of the R^2 calculated for the models of SLR, PR, SVR, DTR, and RFR.

3.3.3. Validation

Figure 11 shows the polygons obtained from the classification of a SAR image of 2021-02-10, using CART (green), RF (blue) and SVM (purple) and overlaid with the orthomosaic of Burlan and Pomacochas lakes obtained by the RPAS on the same passage date.

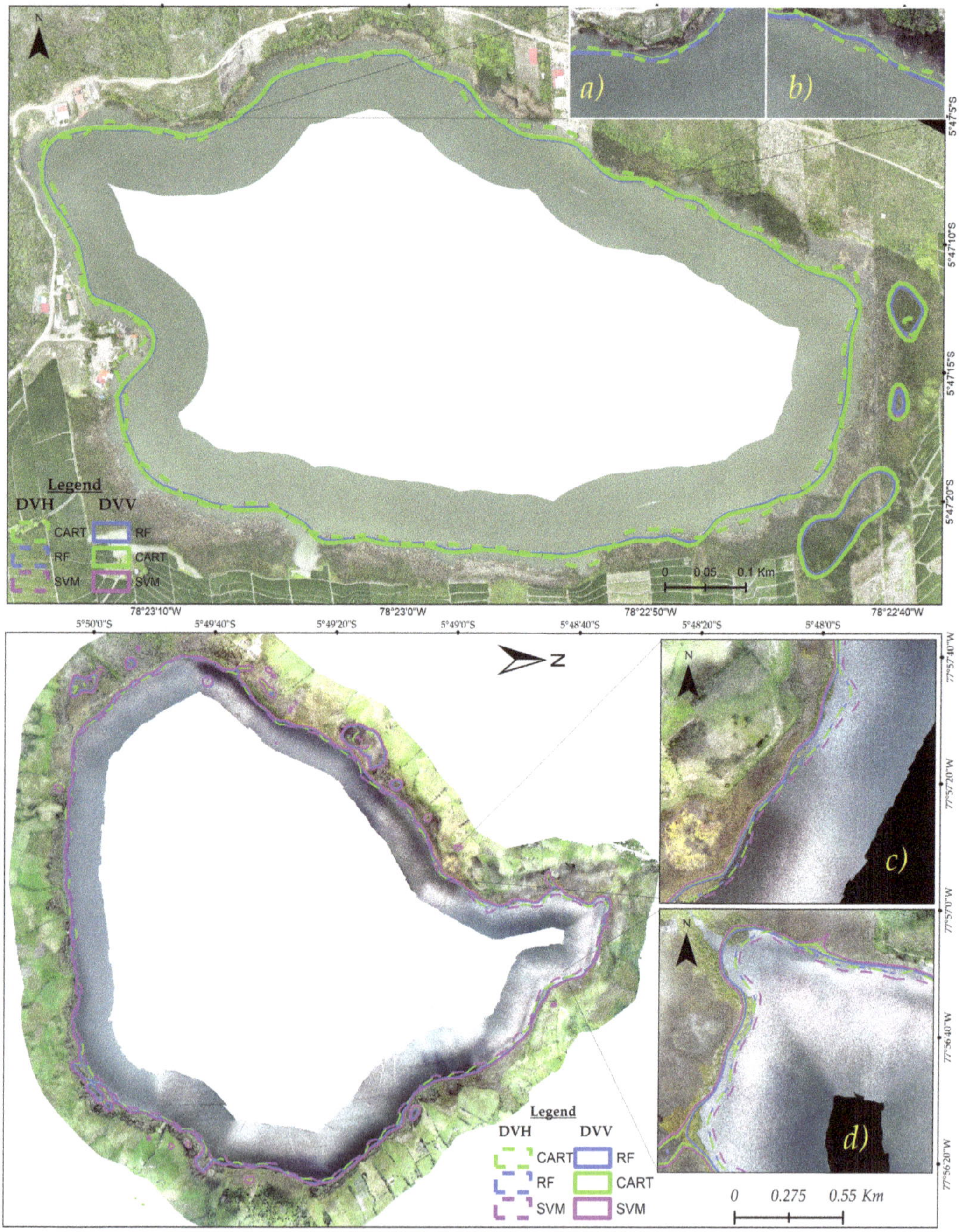

Figure 11. Overlapping of the SAR classification and orthomosaic of Burlan (top) and Pomacochas (bottom) Lakes for 2021-02-10, and subfigures (**a–d**) represent a zooming of each zone to visualize the classification result on the RPAS orthomosaic.

The continuous lines were obtained from the classification of a descending combination SAR image in VV polarization (DVV), while the discontinuous lines are the result of the descending combination in VH polarization (DVH).

The orthomosaics had a mean RMS error of 0.043 m for pomacochas and 0.008 m for Burlan lake. The area (A) and perimeter (P) in hectares and kilometres, respectively, were calculated for each polygon extracted from the SAR image of the DVH and DVV combinations. These values were compared with the estimation by the best regression method and the flight with RPAS performed on Burlan and Pomacochas Lakes. In addition, the percentage of variation of the SAR image and the regression estimation were calculated with respect to the values obtained by the RPAS, as shown in Table 4.

Table 4. Cross comparison of area and perimeter of SAR classification, the method of regression of higher R^2 with respect to photogrammetric flight.

		SAR Image												Best Regressionmethod	Δ%	RPAS
		DVV						DVH								
		CART	Δ%	RF	Δ%	SVM	Δ%	CART	Δ%	RF	Δ%	SVM	Δ%			
Burlan lake	A	43.53	−3.27	42.89	−4.69	43.42	−3.51	42.46	−5.64	42.48	−5.60	42.48	−5.60	44.47	−1.18	45.63
	P	3.4	−17.68	3.3	−20.10	3.38	−18.16	2.87	−30.51	2.87	−30.51	2.87	−30.51	3.82	−7.51	4.13
Pomacochas lake	A	434.89	1.35	430.77	0.39	437.18	1.89	420.57	−1.99	420.57	−1.99	414.23	−3.46	411.89	−4.01	429.09
	P	12.21	23.46	11.13	12.54	13.03	31.75	9.51	−3.84	9.49	−4.04	9.14	−7.58	17.46	76.54	9.89

4. Discussion

The monitoring of lakes using SAR images is very diverse, and commercial SAR products [6,52] or free access products such as those of the Sentinel-1 mission [28] can be used. In 2015, the launch of GEE [23] and the incorporation of the GRD products of Sentinel-1 facilitated the management of and access to SAR images. In this study, we used 517 Sentinel-1 A/B images for both lakes under study, having greater data availability as of 2016, we considered the period 2014–2020, as did Zijie et al. [53], but we calculated water masks by combining the polarizations and directions flight of the satellite. This approach was proposed because the retrospection in the images is different according to the direct flight or polarization considered; we based it on Table 4 and Figure 5.

To calibrate the first-level data of Sentinel-1, there are four look-up tables (LUTs). In the case of the level 1 files in Sentinel-1 GRD format, the zero sigma correction type is the most commonly used to generate the dispersion coefficient (σ°) [54]. To perform the correction of Sentinel-1 images, processes such as the application of orbit files, thermal noise removal, border noise removal, speckle filtering, and range-Doppler terrain correction are performed, all of which are performed in SNAP. In China, Zeng et al. [14] used this data processing approach for their research. For our part, we used the Sentinel-1 GRD products already available in GEE. This dataset provides images in which the pixel values are directly related to the backscatter of the radar by scene. That is, they are radiometrically calibrated, including thermal noise removal and terrain correction using Shuttle Radar Topography Mission (SRTM) and Advanced Spaceborne Thermal Emission and Reflection Radiometer (ASTER) digital elevation models (DEMs). Therefore, when using the GEE functions to homogenize the images, we opted to eliminate the noise using reducing filtering.

There are several ways to approach the extraction of water bodies from SAR images, for example, Otsu segmentation [55] and delineation through active contour models [56]. In this study, we used SAR images classified by three machine learning algorithms [32] to compare the results of the classification and to take advantage of the versatility and adaptability of GEE for the processing of SAR images, in addition to the parallel execution of the three algorithms CART, RF, and SVM.

Because similar studies have not been reported for the study lakes, we cannot compare the results of the classification, and we only lay the foundations for subsequent studies framed in the sixth sustainable development goal (target 6.6, indicator 6.6.1), which mentions that there are changes in the extent of water-related ecosystems over time [57].

Due to the geographical location of the study lakes, no marked trends were found with respect to monthly changes in area and perimeter, with the exception of the month of January, which is the month of greatest rainfall in the area. The area and perimeter values obtained by CART and Random Forest were similar, but the Support Vector Machine yields different values due to the input parameters of each algorithm, for example, the decision trees (CART and Random Forest) and the types of kernels (SMV) used in the classification. Several studies compare the performance of classifiers in different applications [58–60], and obtain different accuracies by simply modifying the number of decision trees or the type of kernel [61], so, at present, there are no defined parameters for image classification, and it is the task of each researcher to use and modify the input parameters. In our case, the accuracies were similar, but the values of the area and perimeter differed in some cases.

In China, Zijie et al. [53] found a slow upwards trend since 2014–2020 in Baiyangdian Lake and that the area of the lake was greater in spring and winter; in our case, the precipitation shows a similar behaviour with the area of Burlan lake, while for Pomacochas lake there is no defined trend regarding precipitation. Indeed, in Figure 12, we show the precipitation (mm/day) extracted from Climate Hazards Group InfraRed Precipitation with Station data (CHIRPS) for the study lakes.

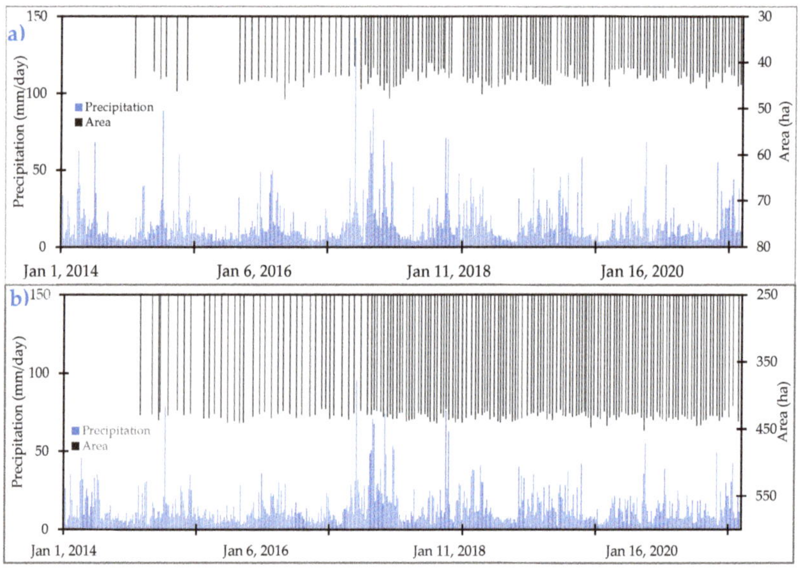

Figure 12. Daily distribution of precipitation for (**a**) Burlan and (**b**) Pomacochas lakes.

We consider that in the high-resolution orthomosaic obtained by the RPAS, the contour of the lakes is better defined than in the SAR image, these overlapping errors at the edges of the lakes are due to the different spatial resolutions. Valdez-Lazalde et al. [62] used high (Ikonos and QuickBird-2) and medium-resolution (SPOT-4 and Landsat-7 ETM^{+}) images for the estimation of the tree cover of a pine forest. In turn, Hernán et al. [63] found better results with aerial and satellite images of 1m and 2.44 m spatial resolution, respectively, for the estimation of biomass in vineyards. As shown in Figure 11 and Table 4, the area and perimeter values of the images with the VH band were lower because the waves that were transmitted vertically and those that return to the sensor horizontally are small. This means that the intensity of the VH band was lower than that of the VV band [64].

In Peru, especially for the Amazonas region, there is no geospatial information with high spatial resolution [65], which is why the regression analysis was limited regarding including other variables such as precipitation, evapotranspiration, and temperature, the same variables that influence the dynamics of a lake [66]. For the calculation of meteorological variables, established models can be used or calibrations can be performed to obtain greater precision in the estimation of these variables [67]. In this investigation, we tried to relate only the area and date of acquisition of the SAR image; for this purpose, the dates were transformed to ordinal numbers, and taking advantage of the robustness of the nonlinear regressors (RFR, SVR, and DTR), a correlation and predicted area and perimeter data with a mean R^2 fit ($\pm$0.4) were obtained.

Through the type of regression analysis used in this research, the area and perimeter values were similar to those of the validation with the RPAS, unlike the perimeter of Pomacochas lake, which was overestimated, in addition, the shape of the polygons extracted from the SAR images classified by our approach differs slightly with the shape of each lake, as shown in Figure 11. The variation in the contour shapes of each lake occurs because the spatial resolution of the RPAS used is much higher (50 cm/pixel) than that of Sentinel-1 (approximately 10 m/pixel).

The use of single polarizations can help to detect water bodies, but double polarizations have better performance [68,69]. In particular, we used dual polarizations, specifically the data obtained from VV polarization obtained better consistency according to SVR for Burlan Lake, while the VH polarization according to RFR for Pomacochas Lake. As shown in Figure 10, the maximum R^2 of the regression methods does not exceed 0.5, so data from different sensors can be used to correct this [70].

Generating geospatial information from optical data in areas of cloud cover is a challenge [71]. Additionally, analyzing the dynamics of lakes in the Amazonas using data from all the factors that influence a lake continues to be a challenge due to the temporal resolution (different acquisition dates), absence of historical climate data, and low density of meteorological stations, which are issues to be resolved in future research. It should be noted that there are various products that can be obtained from SAR images (vegetation indices, and interferograms), but our research was focused only on providing a rapid methodology for the analysis of the dynamics of two lakes using the area and perimeter and their correlation with the date of acquisition of the GRD-type SAR images.

5. Conclusions

Processing Sentinel-1 data in GEE is efficient, fast, and suitable for studies of lake dynamics located in areas with high cloud cover. In addition, the good spatial and temporal resolution of Sentinel-1 data is suitable for an analysis of changes in short periods, helping to show the multitemporal dynamics of water bodies. In particular, this research helped to show the variation in the area and perimeter of the Burlan and Pomacochas lakes, which was greater in the first months of each year.

On the other hand, GC was essential for quickly and easily executing five regression methods, showing that Random Forest Regression worked better both as a classifier and as a predictor. Variations of −1.18% and −7.51% were achieved with respect to the area and perimeter of Burlan Lake obtained through the Remote Pilot Aircraft System. On the other hand, for Pomacochas Lake, RFR underestimated the area of Pomacochas Lake by −4.01% and overestimated the value of the perimeter by 76.54%.

Finally, this research provided a general methodology for the processing of Sentinel-1 data to analyse water bodies, using Classification and Regression Trees, Random Forests and Support Vector Machines similar to a classifier. In addition, customizable scripts were provided for prediction using five regression methods in Google Colaboratory.

Supplementary Materials: The following are available online at https://www.mdpi.com/article/10.3390/ijgi11110534/s1, Table S1: Sentinel-1 imagery used for Burlan and Pomacochas Lakes, Table S2: Area, perimeter, and R^2 of each dataset for the five regression methods.

Author Contributions: Conceptualization, Darwin Gómez Fernández and Nilton Rojas Briceño; Data curation, Rolando Salas Lopez and Nilton Rojas Briceño; Formal analysis, Darwin Gómez Fernández and Jhonsy Silva López; Funding acquisition, Jhonsy Silva López and Manuel Oliva; Investigation, Rolando Salas Lopez; Methodology, Darwin Gómez Fernández, Rolando Salas Lopez, Nilton Rojas Briceño and Jhonsy Silva López; Project administration, Manuel Oliva; Resources, Rolando Salas Lopez and Manuel Oliva; Software, Darwin Gómez Fernández; Supervision, Manuel Oliva; Validation, Nilton Rojas Briceño; Visualization, Nilton Rojas Briceño, Jhonsy Silva López, and Manuel Oliva; Writing—original draft, Darwin Gómez Fernández and Jhonsy Silva López; Writing—review & editing, Darwin Gómez Fernández and Rolando Salas Lopez. All authors have read and agreed to the published version of the manuscript.

Funding: This work was carried out with the support of the Public Investment Project GEOMATICA (SNIP N°. 312235), executed by the Research Institute for Sustainable Development in Highland Forests (INDES-CES) of the National University Toribio Rodriguez de Mendoza de Amazonas (UNTRM).

Data Availability Statement: The processing codes in Google Earth Engine and Google Colaboratory are available in the following web repository: Available online: https://github.com/dargofer/SAR_image_classification (accessed on 15 October 2022).

Acknowledgments: The authors acknowledge and appreciate the support of the Research Institute for Sustainable Development in Highland Forests (INDES-CES) of the National University Toribio Rodríguez de Mendoza de Amazonas (UNTRM). To Kirill Eremenko and Hadelin de Ponteves for training in regression methods on the Google Colaboratory platform.

Conflicts of Interest: The authors declare no conflict of interest.

References

1. USGS Where Is Earth's Water? Available online: https://www.usgs.gov/special-topic/water-science-school/science/where-earths-water?qt-science_center_objects=0#qt-science_center_objects (accessed on 10 April 2021).
2. Meyer, M.F.; Labou, S.G.; Cramer, A.N.; Brousil, M.R.; Luff, B.T. The global lake area, climate, and population dataset. *Sci. Data* **2020**, *7*, 1–12. [CrossRef]
3. Messager, M.L.; Lehner, B.; Grill, G.; Nedeva, I.; Schmitt, O. Estimating the volume and age of water stored in global lakes using a geo-statistical approach. *Nat. Commun.* **2016**, *7*, 13603. [CrossRef] [PubMed]
4. Lee, Z.; Shang, S.; Qi, L.; Yan, J.; Lin, G. A semi-analytical scheme to estimate Secchi-disk depth from Landsat-8 measurements. *Remote Sens. Environ.* **2016**, *177*, 101–106. [CrossRef]
5. Liu, J.; Yang, H.; Gosling, S.N.; Kummu, M.; Flörke, M.; Pfister, S.; Hanasaki, N.; Wada, Y.; Zhang, X.; Zheng, C.; et al. Water scarcity assessments in the past, present, and future. *Earth's Future* **2017**, *5*, 545–559. [CrossRef]
6. Li, S.; Tan, H.; Liu, Z.; Zhou, Z.; Liu, Y.; Zhang, W.; Liu, K.; Qin, B. Mapping High Mountain Lakes Using Space-Borne Near-Nadir SAR Observations. *Remote Sens.* **2018**, *10*, 1418. [CrossRef]
7. Bioresita, F.; Puissant, A.; Stumpf, A.; Malet, J.P. Fusion of Sentinel-1 and Sentinel-2 image time series for permanent and temporary surface water mapping. *Int. J. Remote Sens.* **2019**, *40*, 9026–9049. [CrossRef]
8. Liao, H.-Y.; Wen, T.-H. Extracting urban water bodies from high-resolution radar images: Measuring the urban surface morphology to control for radar's double-bounce effect. *Int. J. Appl. Earth Obs. Geoinf.* **2020**, *85*, 102003. [CrossRef]
9. Barasa, B.; Wanyama, J. Freshwater lake inundation monitoring using Sentinel-1 SAR imagery in Eastern Uganda. *Ann. GIS* **2020**, *26*, 191–200. [CrossRef]
10. Musa, Z.N.; Popescu, I.; Mynett, A. A review of applications of satellite SAR, optical, altimetry and DEM data for surface water modelling, mapping and parameter estimation. *Hydrol. Earth Syst. Sci.* **2015**, *19*, 3755–3769. [CrossRef]
11. Brisco, B. Mapping and Monitoring Surface Water and Wetlands with Synthetic Aperture Radar. *Remote Sens. Wetl. Appl. Adv.* **2015**, 119–136. Available online: https://www.researchgate.net/profile/B-Brisco/publication/271765042_Remote_Sensing_of_Wetlands_Applications_and_Advances/links/59e4b1e1a6fdcc7154e140aa/Remote-Sensing-of-Wetlands-Applications-and-Advances.pdf (accessed on 15 October 2022).
12. Dewan, A.M.; Kankam-Yeboah, K.; Nishigaki, M. Using Synthetic Aperture Radar (SAR) Data for Mapping River Water Flooding in an Urban Landscape: A Case Study of Greater Dhaka, Bangladesh. *J. Jpn. Soc. Hydrol. Water Resour.* **2006**, *19*, 44–54. [CrossRef]
13. Nath, R.K.; Deb, S.K. Water-Body Area Extraction From High Resolution Satellite Images-An Introduction, Review, and Comparison. *Int. J. Image Process.* **2010**, *3*, 353–372.
14. Zeng, L.; Schmitt, M.; Li, L.; Zhu, X.X. Analysing changes of the poyang lake water area using sentinel-1 synthetic aperture radar imagery. *Int. J. Remote Sens.* **2017**, *38*, 7041–7069. [CrossRef]
15. Ding, X.W.; Li, X.F. Monitoring of the water-area variations of Lake Dongting in China with ENVISAT ASAR images. *Int. J. Appl. Earth Obs. Geoinf.* **2011**, *13*, 894–901. [CrossRef]

16. Costa, M.P.F.; Telmer, K.H. Utilizing SAR imagery and aquatic vegetation to map fresh and brackish lakes in the Brazilian Pantanal wetland. *Remote Sens. Environ.* **2006**, *105*, 204–213. [CrossRef]
17. Grunblatt, J.; Atwood, D. Mapping lakes for winter liquid water availability using SAR on the north slope of alaska. *Int. J. Appl. Earth Obs. Geoinf.* **2014**, *27*, 63–69. [CrossRef]
18. Nery, T.; Sadler, R.; Solis-Aulestia, M.; White, B.; Polyakov, M.; Chalak, M. Comparing supervised algorithms in Land Use and Land Cover classification of a Landsat time-series. In Proceedings of the International Geoscience and Remote Sensing Symposium (IGARSS), Beijing, China, 10–15 July 2016; Institute of Electrical and Electronics Engineers Inc.: Piscataway, NJ, USA, 2016; pp. 5165–5168.
19. Shetty, S. Analysis of Machine Learning Classifiers for LULC Classification on Google Earth Engine Analysis of Machine Learning Classifiers for LULC Classification on Google Earth Engine. Master's Thesis, Universidad de Twente, Enschede, The Netherlands, 2019.
20. Liaw, A.; Wiener, M. Classification and Regression by randomForest. *R News* **2002**, *2*, 18–22.
21. Schapire, R.E.; Freund, Y.; Bartlett, P.; Lee, W.S. Boosting the margin: A new explanation for the effectiveness of voting methods. *Ann. Stat.* **1998**, *26*, 1651–1686. [CrossRef]
22. Breiman, L. Bagging Predictors. *Mach. Learn.* **1996**, *2*, 123–140. [CrossRef]
23. Gorelick, N.; Hancher, M.; Dixon, M.; Ilyushchenko, S.; Thau, D.; Moore, R. Google Earth Engine: Planetary-scale geospatial analysis for everyone. *Remote Sens. Environ.* **2017**, *202*, 18–27. [CrossRef]
24. Tamiminia, H.; Salehi, B.; Mahdianpari, M.; Quackenbush, L.; Adeli, S.; Brisco, B. Google Earth Engine for geo-big data applications: A meta-analysis and systematic review. *ISPRS J. Photogramm. Remote Sens.* **2020**, *164*, 152–170. [CrossRef]
25. Brownlee, J. *Master Machine Learning Algorithms*, 2016. Available online: https://machinelearningmastery.com/master-machine-learning-algorithms/ (accessed on 15 October 2020).
26. SENAMHI Mapa Climático del Perú. Available online: https://www.senamhi.gob.pe/?p=mapa-climatico-del-peru (accessed on 22 October 2020).
27. Barboza-Castillo, E.; Maicelo-Quintana, J.L.; Vigo-Mestanza, C.; Castro-Silupú, J.; Oliva-Cruz, S.M. Análisis morfométrico y batimétrico del lago Pomacochas (Perú). *Indes* **2016**, *2*, 90–97. [CrossRef]
28. ESA Sentinel-1. Available online: https://sentinel.esa.int/web/sentinel/missions/sentinel-1 (accessed on 12 January 2021).
29. GEE Sentinel-1 Algorithms. Available online: https://developers.google.com/earth-engine/guides/sentinel1 (accessed on 15 January 2021).
30. Maitre, H. (Ed.) *Processing of Synthetic Aperture Radar Images*; ISTE Ltd Jhon Wiley & Sons, Inc.: Hoboken, NJ, USA, 2008; ISBN 978-1-84821-024-0.
31. GEE ee.Image.focal_median. Available online: https://developers.google.com/earth-engine/apidocs/ee-image-focal_median#javascript (accessed on 15 January 2021).
32. GEE Supervised Classification. Available online: https://developers.google.com/earth-engine/guides/classification (accessed on 22 January 2021).
33. Breiman, L. Random forests. *Mach. Learn.* **2001**, *45*, 5–32. [CrossRef]
34. Breiman, L.; Jerome, F.; Stone, C.J.; Olshen, R.A. *Classification and Regresion Trees*; Taylor & Francis Group: Abingdon, UK, 1984; ISBN 978-0-412-04841-8.
35. Burges, C.J.C. A tutorial on support vector machines for pattern recognition. *Data Min. Knowl. Discov.* **1998**, *2*, 121–167. [CrossRef]
36. Wei, C.-W.; Chang, C.-C.; Lin, C.-J. *A Practical Guide to Support Vector Classification 2003.* Available online: https://www.bibsonomy.org/bibtex/2c04ef97dc3c3de168e684c3e4abe061b/jil (accessed on 15 January 2021).
37. Stehman, S.V. Selecting and interpreting measures of thematic classification accuracy. *Remote Sens. Environ.* **1997**, *62*, 77–89. [CrossRef]
38. Pedregosa, F.; Varoquaux, G.; Gramfort, A.; Michael, V.; Thirion, B.; Grisel, O.; Blondel, M.; Prettenhofer, P.; Weiss, R.; Dubourg, V.; et al. Scikit-learn: Machine Learning in Python. *J. Mach. Learn. Res.* **2011**, *12*, 2825–2830.
39. Altman, N.; Krzywinski, M. Association, correlation and causation. *Nat. Methods* **2015**, *12*, 899–900. [CrossRef] [PubMed]
40. Altman, N.; Krzywinski, M. Simple linear regression. *Nat. Methods* **2015**, *12*, 999–1000. [CrossRef] [PubMed]
41. sklearn.linear_model.LinearRegression. Available online: https://scikit-learn.org/stable/modules/generated/sklearn.linear_model.LinearRegression.html (accessed on 25 January 2021).
42. sklearn.preprocessing.PolynomialFeatures. Available online: https://scikit-learn.org/stable/modules/generated/sklearn.preprocessing.PolynomialFeatures.html (accessed on 25 January 2021).
43. sklearn.preprocessing.StandardScaler. Available online: https://scikit-learn.org/stable/modules/generated/sklearn.preprocessing.StandardScaler.html (accessed on 26 January 2021).
44. Vapnik, V.N. *The Nature of Statistical Learning Theory*; Springer: New York, NY, USA, 2000.
45. sklearn.svm.SVR. Available online: https://scikit-learn.org/stable/modules/generated/sklearn.svm.SVR.html (accessed on 25 January 2021).
46. Support Vector Regression (SVR) Using Linear and Non-Linear Kernels—Scikit-Learn 1.1.2 Documentation. Available online: https://scikit-learn.org/stable/auto_examples/svm/plot_svm_regression.html#sphx-glr-auto-examples-svm-plot-svm-regression-py (accessed on 5 October 2022).

47. sklearn.ensemble.RandomForestRegressor. Available online: https://scikit-learn.org/stable/modules/generated/sklearn.ensemble.RandomForestRegressor.html#examples-using-sklearn-ensemble-randomforestregressor (accessed on 25 January 2021).
48. sklearn.tree.DecisionTreeRegressor . Available online: https://scikit-learn.org/stable/modules/generated/sklearn.tree.DecisionTreeRegressor.html (accessed on 25 January 2021).
49. Metrics and Scoring: Quantifying the Quality of Predictions. Available online: https://scikit-learn.org/stable/modules/model_evaluation.html#regression-metrics (accessed on 26 January 2021).
50. seaborn.boxenplot. Available online: https://seaborn.pydata.org/generated/seaborn.boxenplot.html#seaborn.boxenplot (accessed on 25 March 2021).
51. Hofmann, H.; Kafadar, K.; Wickham, H. Letter-value plots: Boxplots for large data. *Am. Stat.* **2011**, 22. [CrossRef]
52. Strozzi, T.; Wiesmann, A.; Kääb, A.; Joshi, S.; Mool, P. Glacial lake mapping with very high resolution satellite SAR data. *Nat. Hazards Earth Syst. Sci.* **2012**, *12*, 2487–2498. [CrossRef]
53. Jiang, Z.; Jiang, W.; Ling, Z.; Wang, X.; Peng, K.; Wang, C. Surface Water Extraction and Dynamic Analysis of Baiyangdian Lake Based on the Google Earth Engine Platform Using Sentinel-1 for Reporting SDG 6.6.1 Indicators. *Water* **2021**, *13*, 138. [CrossRef]
54. European Space Agency Radiometric Calibration of Level-1 Products. Available online: https://sentinel.esa.int/web/sentinel/radiometric-calibration-of-level-1-products (accessed on 26 September 2021).
55. Li, J.; Wang, S. An automatic method for mapping inland surface waterbodies with Radarsat-2 imagery. *Int. J. Remote Sens.* **2015**, *36*, 1367–1384. [CrossRef]
56. Horritt, M.S.; Mason, D.C.; Luckman, A.J. Flood boundary delineation from synthetic aperture radar imagery using a statistical active contour model. *Int. J. Remote Sens.* **2001**, *22*, 2489–2507. [CrossRef]
57. Indicators for the Sustainable Development Goals. Available online: https://sdg.data.gov/es/ (accessed on 24 May 2021).
58. Pande-Chhetri, R.; Abd-Elrahman, A.; Liu, T.; Morton, J.; Wilhelm, V.L. Object-based classification of wetland vegetation using very high-resolution unmanned air system imagery. *Eur. J. Remote Sens.* **2017**, *50*, 564–576. [CrossRef]
59. Statnikov, A.; Wang, L.; Aliferis, C.F. A comprehensive comparison of random forests and support vector machines for microarray-based cancer classification. *BMC Bioinform.* **2008**, *9*, 319. [CrossRef]
60. Liu, M.; Wang, M.; Wang, J.; Li, D. Comparison of random forest, support vector machine and back propagation neural network for electronic tongue data classification: Application to the recognition of orange beverage and Chinese vinegar. *Sens. Actuators B Chem.* **2013**, *177*, 970–980. [CrossRef]
61. Liu, T.; Abd-Elrahman, A.; Morton, J.; Wilhelm, V.L. Comparing fully convolutional networks, random forest, support vector machine, and patch-based deep convolutional neural networks for object-based wetland mapping using images from small unmanned aircraft system. *GISci. Remote Sens.* **2018**, *55*, 243–264. [CrossRef]
62. Valdez-Lazalde, J.R.; González-Guillén, M.d.J.; de los Santos-Posadas, H.M. Estimación de cobertura arbórea mediante imágenes satelitales multiespectrales de alta resolución. *Agrociencia* **2006**, *40*, 383–394.
63. Vila, H.; Perez Peña, J.; García, M.; Vallone, R.C.; Mastrantonio, L.; Olmedo, G.F.; Rodríguez Plaza, L.; Salcedo, C. *Congreso Internacional de la AET. "Teledetección Hacia un Mejor Entendimiento de la Dinámica Global y Regional"*; Estación Experimental Agropecuaria Mendoza INTA: Mendoza, Argentina, 2007.
64. Moreira, A.; Prats-Iraola, P.; Younis, M.; Krieger, G.; Hajnsek, I.; Papathanassiou, K. A Tutorial on Synthetic Aperture Radar. *IEEE Geosci. Remote Sens. Mag.* **2013**, *1*, 6–43. [CrossRef]
65. Yunis, C.R.C.; López, R.S.; Cruz, S.M.O.; Castillo, E.B.; López, J.O.S.; Trigoso, D.I.; Briceño, N.B.R. Land Suitability for Sustainable Aquaculture of Rainbow Trout (*Oncorhynchus mykiss*) in Molinopampa (Peru) Based on RS, GIS, and AHP. *ISPRS Int. J. Geo-Inf.* **2020**, *9*, 28. [CrossRef]
66. Jin-Ming, Y.; Li-Gang, M.; Cheng-Zhi, L.; Yang, L.; Jian-li, D.; Sheng-Tian, Y. Temporal-spatial variations and influencing factors of Lakes in inland arid areas from 2000 to 2017: A case study in Xinjiang. *Geomat. Nat. Hazards Risk* **2019**, *10*, 519–543. [CrossRef]
67. Valipour, M. Calibration of mass transfer-based models to predict reference crop evapotranspiration. *Appl. Water Sci.* **2017**, *7*, 625–635. [CrossRef]
68. Irwin, K.; Braun, A.; Fotopoulos, G.; Roth, A.; Wessel, B. Assessing Single-Polarization and Dual-Polarization TerraSAR-X Data for Surface Water Monitoring. *Remote Sens.* **2018**, *10*, 949. [CrossRef]
69. Scott, K.A.; Xu, L.; Pour, H.K. Retrieval of ice/water observations from synthetic aperture radar imagery for use in lake ice data assimilation. *J. Great Lakes Res.* **2020**, *46*, 1521–1532. [CrossRef]
70. Vickers, H.; Malnes, E.; Høgda, K.-A. Long-Term Water Surface Area Monitoring and Derived Water Level Using Synthetic Aperture Radar (SAR) at Altevatn, a Medium-Sized Arctic Lake. *Remote Sens.* **2019**, *11*, 2780. [CrossRef]
71. Zhang, H.; Zhang, Y.; Lin, H. A comparison study of impervious surfaces estimation using optical and SAR remote sensing images. *Int. J. Appl. Earth Obs. Geoinf.* **2012**, *18*, 148–156. [CrossRef]

International Journal of *Geo-Information*

MDPI

Article

Aquifer and Land Subsidence Interaction Assessment Using Sentinel-1 Data and DInSAR Technique

Fatemeh Rafiei [1], Saeid Gharechelou [2,*], Saeed Golian [1,3] and Brian Alan Johnson [4]

1 Department of Water and Environment, Faculty of Civil Engineering, Shahrood University of Technology, Shahrood 3619995161, Iran
2 Department of Surveying, Faculty of Civil Engineering, Shahrood University of Technology, Shahrood 3619995161, Iran
3 Irish Climate Analysis and Research Units (ICARUS), Maynooth University, W23 F2H6 Maynooth, Ireland
4 Natural Resources and Ecosystem Services Area, Institute for Global Environmental Strategies (IGES), Hayama 240-0115, Japan
* Correspondence: sgharachelo@shahroodut.ac.ir; Tel.: +98-2332392204

Abstract: Climate change and overpopulation have led to an increase in water demands worldwide. As a result, land subsidence due to groundwater extraction and water level decline is causing damage to communities in arid and semiarid regions. The agricultural plain of Samalghan in Iran has recently experienced wide areas of land subsidence, which is hypothesized to be caused by groundwater overexploitation. This hypothesis was assessed by estimating the amount of subsidence that occurred in the Samalghan plain using DInSAR based on an analysis of 25 Sentinel-1 descending SAR images over 6 years. To assess the influence of water level changes on this phenomenon, groundwater level maps were produced, and their relationship with land subsidence was evaluated. Results showed that one major cause of the subsidence in the Samalghan plain was groundwater overexploitation, with the highest average land subsidence occurring in 2019 (34 cm) and the lowest in 2015 and 2018 (18 cm). Twelve Sentinel-1 ascending images were used for relative validation of the DInSAR processing. The correlation value varied from 0.69 to 0.89 (an acceptable range). Finally, the aquifer behavior was studied, and changes in cultivation patterns and optimal utilization of groundwater resources were suggested as practical strategies to control the current situation.

Keywords: subsidence; DInSAR; Sentinel-1; groundwater extraction; aquifer behavior

Citation: Rafiei, F.; Gharechelou, S.; Golian, S.; Johnson, B.A. Aquifer and Land Subsidence Interaction Assessment Using Sentinel-1 Data and DInSAR Technique. *ISPRS Int. J. Geo-Inf.* **2022**, *11*, 495. https://doi.org/10.3390/ijgi11090495

Academic Editors: Wolfgang Kainz, Walter Chen and Fuan Tsai

Received: 5 July 2022
Accepted: 15 September 2022
Published: 19 September 2022

Publisher's Note: MDPI stays neutral with regard to jurisdictional claims in published maps and institutional affiliations.

1. Introduction

Groundwater aquifers are one of the most valuable water resources in many parts of the world [1]. Climate change and rapid population growth, however, are putting groundwater aquifers under pressure, particularly in densely populated areas in arid and semiarid environments [2]. Hence, frequent monitoring of groundwater level is essential to identify areas experiencing rapid depletion in groundwater resources and to move toward sustainable water management of aquifers, especially in arid areas prone to drought [3]. Land subsidence resulting from overexploitation of groundwater resources has been reported in many different parts of the world [4]. The immediate factor driving force the subsidence is considered to be the dropping of groundwater level; however, the primary factor behind this phenomenon is the existence of unconsolidated sediment deposits in the local aquifer system [5].

This phenomenon has affected many cities and regions in the world and caused serious damage to infrastructures and buildings. For instance, Beijing, the capital city of China, has suffered from land subsidence due to groundwater extraction since the 1950s [6]. The subsidence rate in this city is more than 100 mm/year [7]. Subsidence has also occurred in Mexico City in Mexico [8], California [9], Hue in Vietnam [10], Gorgan in Iran [11], Semarang in Indonesia [12], Mashhad in Iran [13], and many other cities in the world [4].

Dry climatic conditions in most inland areas of Iran have increased the dependency on groundwater resources to meet different demands [14]. More than 600 plains in Iran have confronted land subsidence due to excessive groundwater withdrawal [15]. In addition, there have been some plains identified with subsidence of 1 mm per day, indicating a critical situation [13].

A gradual or sudden settlement of the ground surface without any limitation on speed or occurrence area is called subsidence [16]. Different processes such as fluid extraction and mining result in the deformation or movement of soil materials, leading to land subsidence [17]. Excessive groundwater extraction leads to a continuous decline in groundwater level, reduces pore pressure [18], and results in compression and consolidation of the soil layer and gradual settling of the ground surface [19].

Land subsidence may result from elastic or inelastic deformation [20,21], and is in proportion to changes in the groundwater table and the compatible layer thickness [22]. For an elastic deformation in aquifers with semipermeable layers, the groundwater head must remain above the previous lowest level (stress before consolidation). If the groundwater table falls below the previous lowest level, inelastic compaction leads to permanent shifts in the solid grains, and porosity in the groundwater system will reduce [20]. Regions with massive pumping of groundwater usually experience inelastic deformation [21]. By introducing control strategies for groundwater pumping to limit its detrimental effects, a low uplift value is often observed [23–25]. However, only a small part of the initial compaction is recoverable when water is recharged into the aquifer [21].

To implement a proper analysis of the causing factor(s) of land subsidence in a particular site, a vital step is to obtain accurate measurements of the actual amount of subsidence that occurred at multiple points in time. At present, various methods can be used to measure the subsidence amplitude, including precise differential leveling [26] and permanent stations of the Global Positioning System (GPS), Light Detection and Ranging (LiDAR) [27], and Interferometric Synthetic Aperture Radar (InSAR) [28].

One of the most accurate and economic techniques based on remote sensing is radar interferometry. This technique evaluates the amount and range of subsidence over the whole area under study (rather than at a small number of sample points) and provides the possibility of repeated monitoring of the area relatively frequently (i.e., whenever new satellite images of the site are acquired) [29,30]. Differential Interferometric Synthetic Aperture Radar (DInSAR) is an efficient radar interferometry approach [31] that can be used to accurately detect ground movements and land deformation patterns at a high spatial resolution (~5 × 20 m in the case of the Sentinel-1) over large geographic areas [7]. DInSAR analysis can achieve up to millimeter accuracy, making it a very capable tool for land subsidence and uplift measurement under suitable conditions [32], such as short temporal and perpendicular baseline, and suitable atmospheric conditions [33]. This technique's ability to determine land deformations from millimeters to centimeters in size makes it a suitable method for monitoring slow-moving deformation [34].

Increasing water demands and urbanization has caused overexploitation and depletion in many aquifers. Mexico, for example, includes a number of aquifers suffering from depletion, and as a result, land subsidence combined with ground fracturing. Synthetic aperture radar proved its ability to monitor land subsidence in this area for a long-term period. Maximum settlement rates in this area were recognized as 14 cm/year in 1996 and 10 cm/year in 2010–2012, and these increased to 12 cm/year from 2015 to 2020 [35] Excessive groundwater withdrawal has led to land subsidence and sinkholes in Central Anatolia in Turkey. Interferometric Synthetic Aperture Radar was used to monitor and better understand the relation between groundwater extractions and land subsidence in this area. The investigations revealed subsidence of 70 mm/year in this area, which follows the overall groundwater level changes with over 80% cross-correlation consistency [36]. The Tuscany region has been affected by land subsidence due to water overexploitation, geothermal activity, and urbanization. This area was studied using SqueeSAR processing of Sentinel-1 between 2014 and 2018. The descending images time series showed a maximum value of 55 mm/year subsidence in the Montemurlo area [37]. Sui et al. [30] studied subsidence in the Loess Plateau of China using DInSAR due to underground extraction using ALOS PALSAR data in 2007. They monitored subsidence with maximum values of 20 cm in some areas. They proposed an approach to combine InSAR results with the support vector machine regression algorithm to predict subsidence with a high level of accuracy.

The Samalghan plain, located in the North Khorasan province of Iran, is one of the most important agricultural plains in this country. It currently suffers, however, from subsidence, causing irreversible damage to the plain. Groundwater conditions in the mentioned area have become essential to be monitored and managed, as groundwater overexploitation may be a causing factor. A prior study [38] found that subsidence in the plain could be divided into two regions: a high-risk region in the west and northwest of the plain, and a low-risk region in the eastern parts. During the past decades, surface fractures have been created in the northwest Samalghan plain that is being developed. Consequently, studying subsidence in this plain is of great importance. In the previous studies conducted on Samalghan plain, subsidence was not measured over the entire plain area, and the relationship between subsidence rates and the rate of groundwater decline was not investigated. We aimed to fill these research gaps in the present study by detecting land subsidence using the DInSAR method and quantitatively analyzing its relationship with groundwater depletion. Then, the aquifer response to water recharge and discharge in terms of changes in groundwater head and land subsidence was evaluated.

2. Materials and Methods

2.1. Study Area

The study area is an important agricultural plain located in the western part of North Khorasan Province, Iran, covering an area of over 1100 km^2 with an arid/semiarid climate. The location and extent of this area, simplified geology, and the piezometric well locations are presented as follows (Figure 1). This plain is located in the Kopeh Dagh geological zone, and its geographical location is limited within 37°21′–37°39′ N in latitude and 56°25′ to 57°06′ E in longitude. The high demand for fresh water in this area relies mostly on groundwater, which has increased due to growing agricultural consumption [39].

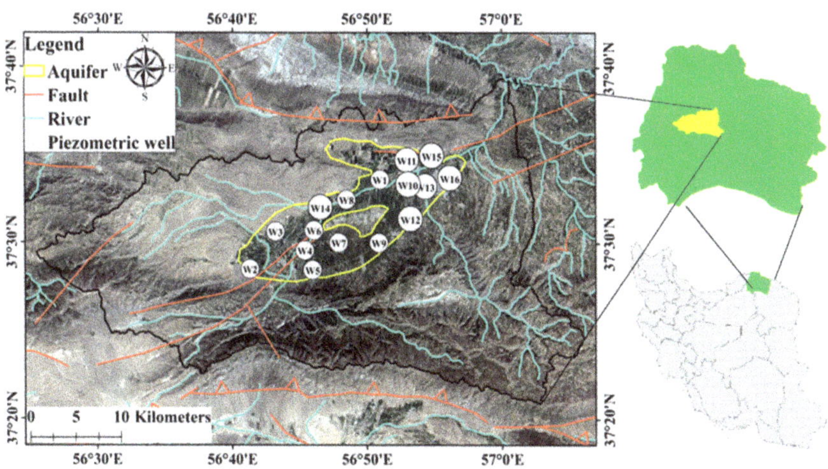

Figure 1. Location map of study area in North Khorasan Province of Iran.

The maximum elevation is in Korkhod Mountains, 2680 m above sea level, and the minimum elevation, at the outlet of Darband, is 600 m above sea level. The average annual precipitation of this basin is approximately 465 mm, with a mean annual temperature and evapotranspiration of 11.1 °C and 1132 mm/year, respectively [40].

Based on figures obtained from the regional water authority organization of North Khorasan Province (NKHRW), this plain has 333 operational wells, with a discharge rate of 37.65 million cubic meters (MCM) per year. The number of wells experiencing overextraction in this area is 86, and the total volume of this overextraction is 17.17 MCM. An additional 1.2 MCM is extracted annually from 15 aqueducts in this plain. The aquifer recharge capacity of this plain—the maximum volume of water that can be recharged after discharge each year—is about 5 MCM/year; the current amount of extraction has caused a lack of balance in the aquifer. In terms of water quality, this plain is categorized as suitable for agricultural use and acceptable for drinking.

According to data from NKHRW, 35 MCM of the annual groundwater consumption is appertained by agriculture. The allocation of groundwater resources is represented in Figure 2. The main farmed crops in this area are cotton, rice, wheat, and barley. The density of farmlands is higher in the low-lying parts of the plain [39].

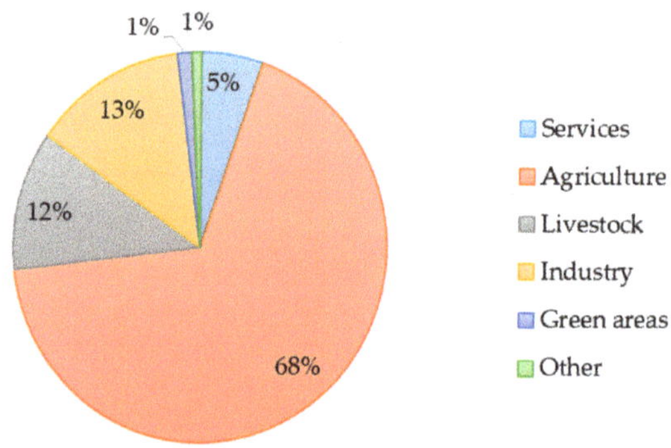

Figure 2. Percentage of different groundwater consumption sectors in Samalghan plain.

A prior study claimed that there have been fractures reported in the northwest parts of the plain in a north–south direction. These fractures continue along a line several hundred meters in length dictated by a fault in the rock. Most of these cracks were formed in the agricultural lands and caused substantial damage to farmlands due to extra water escape and leakage into them, as well as the change of slope. As a result, farmers are constantly flattening the land and trying to eliminate the effects of cracks in agricultural lands [38].

Groundwater quality deterioration is another negative impact of these cracks. The formation of cracks can be a channel for the transfer of surface pollutants, including effluent and drainage from the agricultural lands following the use of chemical fertilizers and pesticides to the aquifer [41].

The underground geological investigations of the Samalghan plain show the variable thickness of the alluvium of this plain. In the northern part of this plain, the Tirgan formation can be observed in both the surface and the depth. The groundwater level is low in this area; therefore, most of the exploitation wells are located in this part. In the eastern part of the plain, Neogene sediments are located on the Tirgan formation, and the main lithology of these sediments is conglomerate, marl, and evaporite rocks.

The underground geological layers investigated using the log information of the existing wells in the region in three directions is shown in Figure 3a. Examining the well logs in the direction of AA′ shows that the sediments in the northeast of the aquifer are fine-grained, and the sediments change to coarse-grained towards the southwest of the area. In this regard, the thickness of the alluvium is so high that it did not meet the bedrock in the SW15 well, with a depth of 140 m. The SW2 well has encountered limestone due to its proximity to the Tirgan limestone formation. This means that along the AA′ direction, the bedrock has a great depth, and in some places it is formed of limestone.

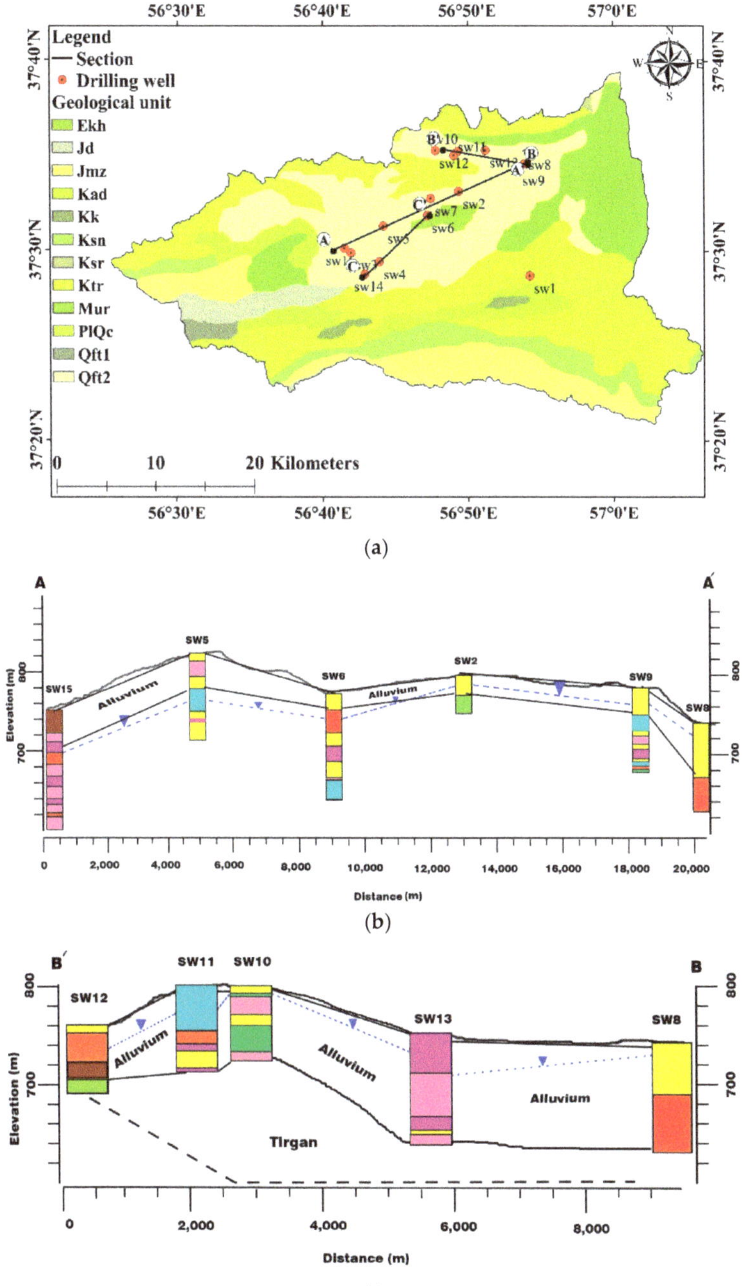

Figure 3. *Cont.*

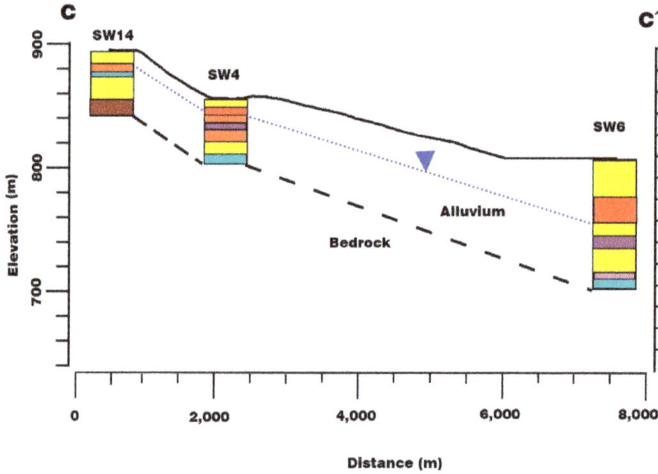

(d)

Figure 3. The location of drilling wells for the investigation of well logs (**a**); (**b**) the log of wells in AA′; (**c**) the log of wells in BB′ direction; and (**d**) the log of wells in CC′ section.

Excavation of the drilling wells in the direction of BB′ indicates that the sediments become mostly coarse, and as the thickness of the fine-grained sediments decreases, the thickness of the coarse-grained sediments increases from east to west. In the western part of the Tirgan plain, limestone can be seen on the ground; it can also be seen in the deep parts of the northwest of the plain.

Surveys in the direction of CC′ reveal that the topographical changes decrease from the south to the north of the plain. There is a height difference of 94 m between the two wells; the water depth increases from the south to the north of the plain. Investigations in this direction indicate that the thickness of coarse-grained sediments increase from the south to the north of the plain [42].

2.2. Dataset

To study land subsidence in this plain, two series of datasets have been used: groundwater time-series data and SAR remote sensing imagery. In this section, these two datasets will be introduced.

2.2.1. Groundwater Data

To prepare and study the groundwater level change maps, piezometric data of the studying area are needed. In this study, the piezometric data consist of monthly groundwater level time series at 16 well sites covering the period of 2008–2018, provided by the Iran Water Resources Management Company (http://wrs.wrm.ir/ (accessed on 27 December 2019)). First, the historical groundwater level data were collected and sorted by year, then the trend changes in water level over each well, and also the average changes in water level in the area were studied. The groundwater level for 2019 and 2020 was predicted with the use of historical data and the best-fitted trend line in Excel. Therefore, various trendlines were fitted in this process, and the most accurate estimation was implemented according to the available data. To carry this out, exponential, linear, logarithmic, polymonal with power 2, and power trendlines were evaluated using cross-validation. The well locations can be seen as presented in Figure 1. In this study, the IDW (inverse distance weighted) method with a power value of 1, 2, 3, and 4, and ordinary Kriging and Spline methods were conducted for interpolation of groundwater levels, and their performances were evaluated by cross-validation. Then, the most accurate method was chosen as the interpolation method to develop wall-to-wall groundwater level maps of the study site.

2.2.2. SAR Data

Utilizing the radar interferometry method, the rate and trend of subsidence in Samalghan plain were investigated using the Single Look Complex (SLC) radar data stacks of the Sentinel-1 Satellite. In this study, Sentinel-1A data provided in the Copernicus database Sentinel-1 SAR, which is freely available through https://asf.alaska.edu/ (accessed on 22 January 2021), were used. The 12-day repeat orbit cycle is expected to increase the coherence value of interferometric pairs for land deformation monitoring. The Sentinel-1 data were acquired in interferometric wide (IW) swath mode, which is made up of three subswathes (IW1, IW2, and IW3) with a total swath of 250 km [8]. This mode enables coverage of a large area with a good spatial resolution [43] of 5 × 20 m for SLC-type data [44].

A total of 24 pairs of images in VV polarization were processed with the use of the conventional DInSAR technique. Appropriate selection of the data pair is a vital step because the results are highly dependent on coherence values [34]. In order to increase the coherence value and reduce the speckle noise in the interferograms and enhance the results, images with the least perpendicular baseline and the temporal baseline difference between acquisition dates were selected. The data are used in a way that every 4 interferograms contain 4 seasons for each year, with the optimal perpendicular baseline value. The interferometric pairs were chosen as given in Table 1.

Table 1. Analyzed descending datasets information, including interferogram ID (Int ID), master date, slave date, temporal baseline (days), and perpendicular baseline (meters).

Int ID	Master	Slave	TB (Days)	PB (m)	Int ID	Master	Slave	TB (Days)	PB (m)
D1	2020.11.07	2020.08.27	73	19	D13	2017.11.11	2017.08.19	84	20
D2	2020.08.27	2020.05.23	96	7	D14	2017.08.19	2017.05.27	74	17
D3	2020.05.23	2020.02.29	84	55	D15	2017.05.27	2017.02.08	108	33
D4	2020.02.29	2019.12.07	84	1	D16	2017.02.08	2016.11.04	96	36
D5	2019.12.07	2019.08.21	108	61	D17	2016.11.04	2016.08.24	72	66
D6	2019.08.21	2019.05.29	84	11	D18	2016.08.24	2016.05.20	96	0
D7	2019.05.29	2019.02.22	96	55	D19	2016.05.20	2016.02.14	96	26
D8	2019.02.22	2018.11.06	108	30	D20	2016.02.14	2015.11.10	96	6
D9	2018.11.06	2018.08.14	84	52	D21	2015.11.10	2015.08.06	96	19
D10	2018.08.14	2018.05.10	96	10	D22	2015.08.06	2015.05.26	168	38
D11	2018.05.10	2018.03.11	60	64	D23	2015.05.26	2015.02.07	108	13
D12	2018.03.11	2017.11.11	120	63	D24	2015.02.07	2014.11.03	180	66

The 30 m Shuttle Radar Topography Mission (SRTM) digital elevation model was used for co-registration of the data stack. Due to the nonexistence of either previous land subsidence measurement data in the study area, GPS measurement data, or precise leveling data, the displacement results could not be validated; however, the results could be deemed to be acceptable, since DInSAR is known as an acceptable methodology giving accurate results for deformation in most areas [5]. Therefore, 11 pairs of ascending-direction images from 2018 to 2020, which were available with the optimum baseline value in the same path, were used for the validation of DInSAR the process.

2.2.3. Additional Data

In order to investigate the hypothesis that groundwater has an impact on land subsidence in the Samalghan plain, the impact of other factors such as railways and earthquakes was also examined. In order to investigate the effect of tectonic conditions and continuous earthquakes on subsidence in the region, earthquakes with a magnitude of 4 [45] or above on the Richter scale were prepared from the Seismological Center of the Institute of Geophysics, University of Tehran. In the period under study, there is no earthquake above magnitude of 4 with its center in Samalghan plain or a short distance from the region. These earthquakes do not have much effect (in centimeters) on the displacement of the study area. In addition, there is no rail line in the study area, so the impact of these two cases was ignored.

Soil type data were not available in the area, so geological information from the area was used. The information obtained from a previous study [42]. Finally, according to the study of geological maps in the study area, the relationship between subsidence loss and soil granulation was studied.

2.3. DInSAR Method

As mentioned earlier, the method employed to monitor subsidence in the Samalghan plain was DInSAR, due to its high accuracy and speed of monitoring. With the usage of this technique, accuracy in millimeters or centimeters is achievable for the velocity of land deformation [46].

The DInSAR process is conducted using the SNAP software developed by the European Space Agency (ESA) (https://step.esa.int/main/toolboxes/snap/ (accessed on 31 March 2020)). The conventional DInSAR process is shown in Figure 4 for a single-pair DInSAR. To reduce the decorrelation in interferograms and its relevant errors, the master acquisition was chosen based on the measured temporal and spatial baselines to achieve an appropriate coherence in interferograms [47]. For the descending data, the image taken later (after the event, e.g., 2020) was used as the master. In order to perform interferometric processing, two or more images were coregistered as a stack [48,49]. This step causes each ground target in both images to be identified in the same position [50]. At this point, to correct azimuth and amplitude estimations, enhanced spectral diversity (ESD) [51] was used.

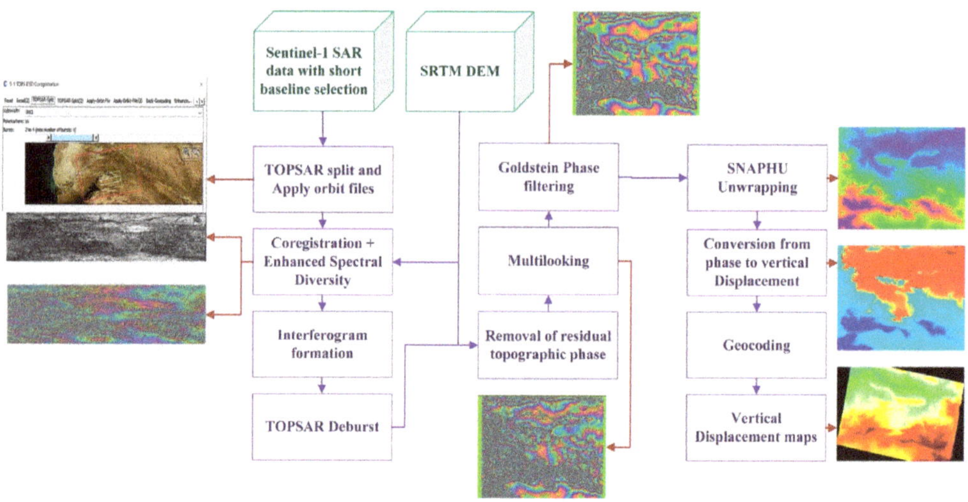

Figure 4. Conventional DInSAR workflow in SNAP.

An interferogram contains information about both topographic and surface movement. In other words, the interferometric phase is a measure of the difference in the path length between the target and the two sensor positions [28,52]. The DInSAR technique aims to separate the contribution of the topographic phase of the earth's surface and the portion of the displacement phase to show the extent of the displacement. In order to remove the topographic phase effect, an additional interferogram or a digital elevation model is required [53]. The 30 m SRTM Dem was used to remove the flat earth effect [54,55] and phase pixels due to the topography.

The interferograms were then filtered by Goldstein phase filtering [56] in order to remove the noise from radar instruments and temporal decorrelation [5]. Subsequently, they were unwrapped using the minimum cost flow (MCF) algorithm with SNAPHU [57,58]. Then, the unwrapped interferograms were converted into line-of-sight (LOS) displacement maps.

The vertical displacement map of the earth was prepared and geocoded. After developing Samalghan land subsidence maps, a cumulative displacement map, which is the sum of all displacements of all segments, was produced for each year. Next, the maps were examined, and the relationship between land subsidence and groundwater changes in the study area was investigated.

3. Results and Discussion

To investigate the interaction behavior of the aquifer and to evaluate the land subsidence in the Samalghan plain, radar images were processed. Secondly, the piezometric well data in the area were used to investigate groundwater level changes, then the relation between groundwater level changes and land subsidence was studied. This section will discuss the findings of the research.

3.1. Groundwater Level Variation

The trend in average annual groundwater level in the study period was plotted, and the best trend line was fitted, as shown in Figure 5. From this, it can be seen that the average depth of groundwater increased from 2008 to 2018 (i.e., groundwater level decreased). Thus, the recharge rate is significantly less than the rate of groundwater extraction. Exponential, linear, logarithmic, polymonal with power 2, and power trendlines were evaluated. The results showed that the best fitted trend line is the polymonal trendline, with a root-mean-square error (RMSE) of 0.4 (m) and correlation coefficient (CC) of 0.95 (Table 2). Using the fitted diagram trendline, the average groundwater level in 2019 and 2020 was estimated to be 24.1 and 24.9 m, respectively.

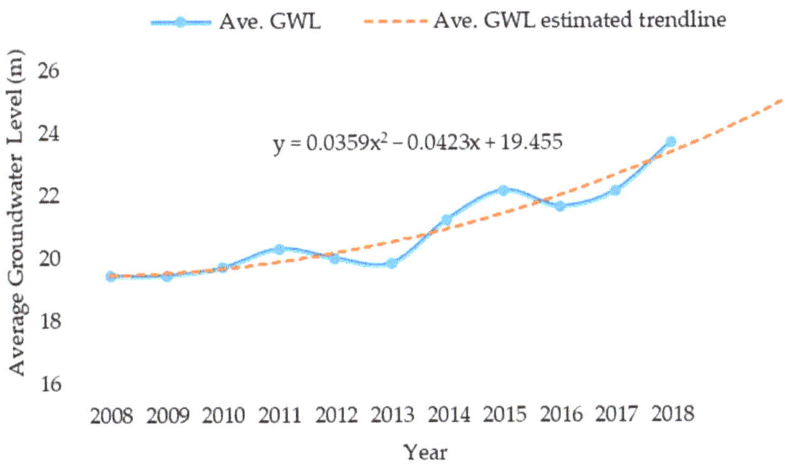

Figure 5. Variations in average groundwater level in Samalghan plain.

Table 2. The CC and RMSE value of the selected trendlines to estimate groundwater level.

Trendline Method	CC	RMSE (m)
Exponential	0.93	0.58
Linear	0.92	0.51
Logarithmic	0.81	1.00
Polymonal—power 2	**0.95**	**0.40**
Power	0.82	0.76

Of the different interpolation methods and weighting parameters tested (Table 3), the ordinary Kriging method showed the lowest RMSE (1.6 m). This result is similar to that of other studies' findings that Kriging tends to outperform IDW methods [59].

Table 3. The RMSE value of the interpolation methods to generate the groundwater level maps.

Trendline Method	RMSE (m)
IDW—power1	2.3
IDW—power2	2.1
IDW—power3	2.1
IDW—power4	2.2
Spline	2.7
Ordinary Kriging	**1.6**

Thus, using the observed well data, wall-to-wall groundwater level maps of the plain were generated using the ordinary Kriging interpolation method in low-height parts of the area (less than 1100 m). By overlaying the 2008 and 2018 groundwater level maps and calculating the difference (Figure 6), it is clear that the northern area saw the most significant groundwater level drop, being up to 20 m in some areas. In general, the whole plain was found to have been affected by groundwater fall. It is worthwhile to mention the fact that there are uncertainties in some parts of the region due to the lack of data and the absence of observational wells (e.g., north-west and south-east of the plain). Nevertheless, some external data (i.e., piezometric wells in a northern adjacent plain) were used to increase the accuracy in these areas.

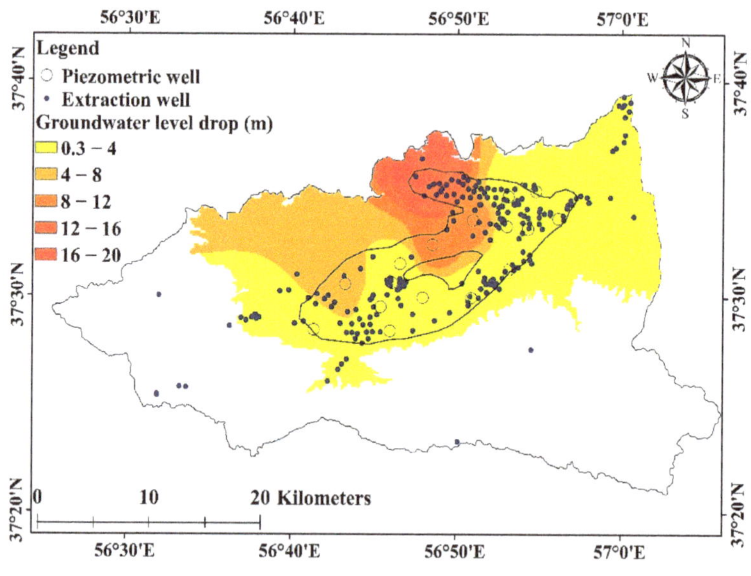

Figure 6. Groundwater level change map between 2008 and 2018.

3.2. Displacement Maps

The results of DInSAR processing and investigation of the annual displacement maps show the pattern of subsidence over time in the study area (Figure 7). In a previous study [38], the authors pointed out that cracks have been created in the northwest of the plain, moving towards the east of the region. It can be seen that land subsidence in the Samalghan plain mainly occurred in the northern and northwestern parts of the region.

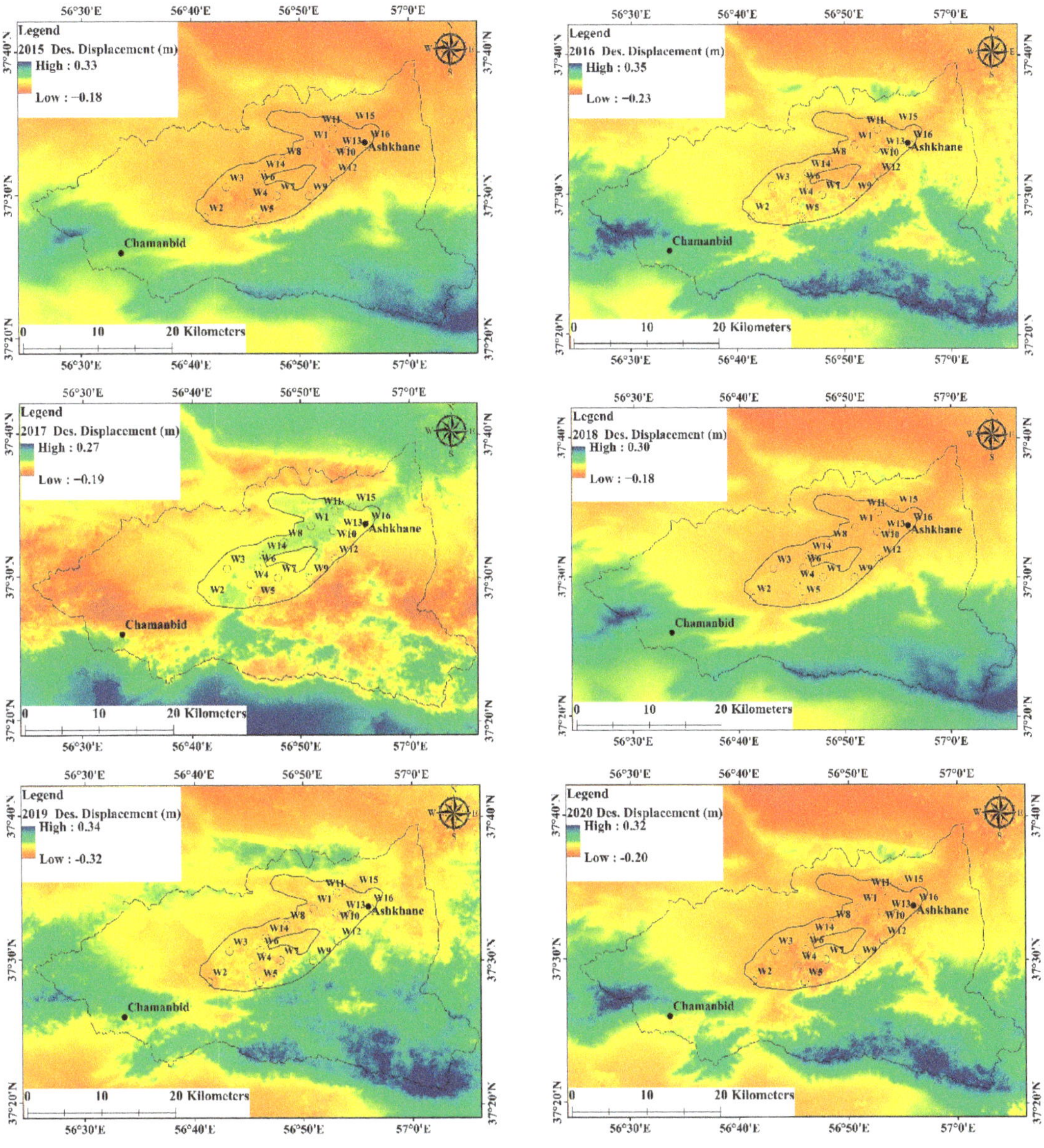

Figure 7. Cumulative annual vertical displacement maps for 2015–2020 using descending data.

Based on deformation maps in the region, the land subsidence occurred in 2015 in all northern parts of the Samalghan plain. There was no subsidence in the south and southwest of the area in 2015. This pattern also took place in 2016. Moreover, by 2016, the maximum subsidence had reached a new peak compared to the previous year. In 2017, the land subsidence pattern changed slightly and spread to almost the entire plain. In the north of the city of Ashkhane, a slight uplift spread lightly towards the northeast–southwest of the plain. In 2018, a subsidence pattern similar to the pattern in 2015 and 2016 was detected in the Samalghan plain, which happened in all northern parts of Samalghan plain, and the pattern continues towards the central parts of the plain. Examination of the displacement maps shows the recurrence of subsidence in the north and northwest of Samalghan plain, the central part containing an aquifer, and the western part near Chamanbid in recent years. The subsidence in western parts is developing towards the Chamanbid, and if not properly managed will cause damage to residential areas. The results of field research in the city of Ashkhane and the surrounding areas showed a lot of damage to these areas, including cracks in the walls, well casing protruding, and holes in the ground.

The uplift observed in mountains is sought to have various reasons. According to information from organizations, major activities had been performed in highly elevated areas in this region. Firstly, afforestation is being conducted on an annual scale of around 70 hectares. A part of the vegetation growth can be misrecognized by uplift, as an error. Moreover, there are many villages located in mountainous areas, expanding their farmlands and plowing the bare lands, and transforming them into rainfed cultivation areas. The agricultural and urban activities in these villages are detected as uplift as well.

Following the trend of land surface changes during the study period, it is possible to observe the uplift of the ground surface after its subsidence in some places. For instance, around Chamanbid, Ashkhane, some parts of the southern part of the plain, and even the northern part of the plain in 2016, the ground surface recovered after its previous subsidence. In addition, in some areas that had been involved in subsidence in 2017, including south of the Samalghan aquifer, west and north of the Samalghan plain, and south-east of the plain, an uplift was observed in the following years. As a result, the possibility of the elastic ground behavior is high, which needs further investigation. In Figure 8, the land subsidence in the two major cities in the study area has been shown. As can be seen from the provided figure, in the city of Ashkhane, after two consecutive years of subsidence, in 2017 and 2018, an uplift was observed, and after that, two years of subsidence occurred again; however, the effect of subsidence was greater, and overall, a cumulative subsidence of 39 cm occurred at this point. In Chamanbid, the situation was different: from 2015 to 2018, an uplift was detected, while in 2019 and 2020, the trend changed and subsidence occurred.

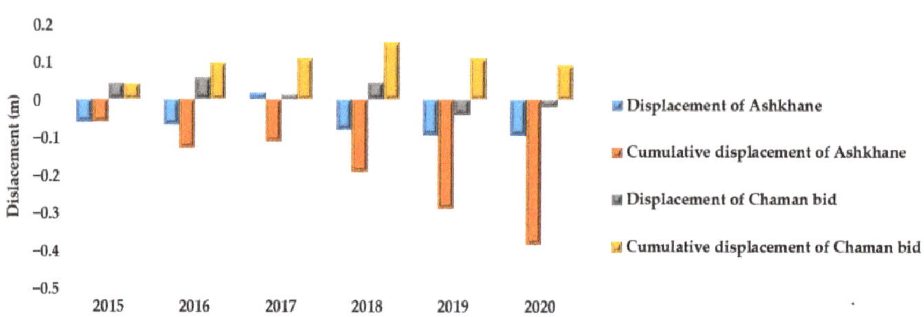

Figure 8. The trend of annual and cumulative land deformation in Ashkhane and Chamanbid cities.

3.3. Validation

Since neither precise leveling nor GPS station data were available during the study period over the area, it is not possible to evaluate the results of InSAR processing using Earth observations. Moreover, there were no other sufficient SAR data available for the plain. Hence, a comparative validation method was used to validate the processing results in this region. For this purpose, the InSAR processing was performed using data with a different path and direction. The results were deemed acceptable if the interferometric processing results of ascending and descending satellite images of one sensor were similar [60]. The subsidence values at the piezometric well locations were also extracted in the ascending and descending displacement maps for comparison. The results showed a satisfactory result quantitatively, with a correlation of 0.69–0.89 between the subsidence estimates given by the ascending/descending data (Table 4), and a similar visual pattern in the Figure 9 maps. The slight differences between the datasets may be partially explained by the time differences between the image acquisitions.

Table 4. Correlation of monitored subsidence for ascending and descending data at the position of piezometric wells in Samalghan plain.

Year	CC
2018	0.72
2019	0.69
2020	0.89

Furthermore, to determine the validity of the research findings and also to collect field evidence of subsidence in the Samalghan plain, a field operation was conducted. Subsidence has had various effects on plain areas (Figure 10). From this, we found that the effects of subsidence can be seen in the form of cracks in the ground, cracks in the walls, well casing protruding, and holes in the ground. The subsidence in some parts of the plain is accompanied by surface cracks, but in other places, it is observed as a uniform and homogeneous subsidence.

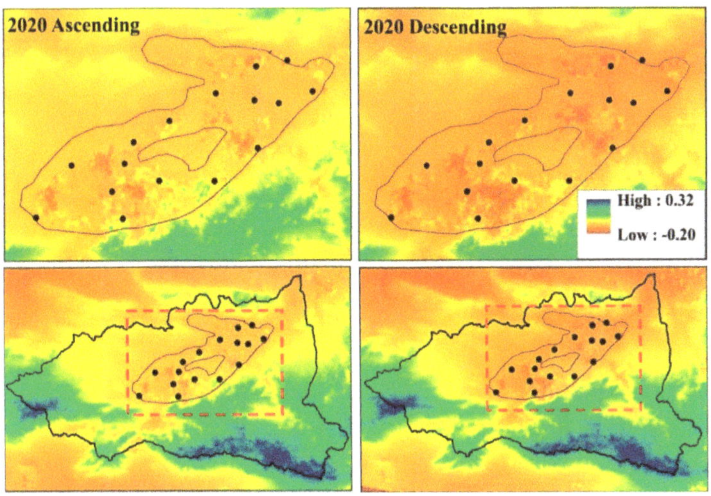

(a)

Figure 9. *Cont.*

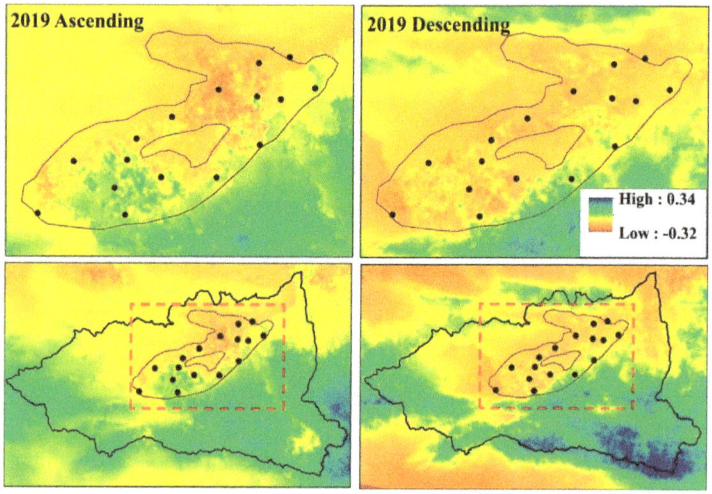

(b)

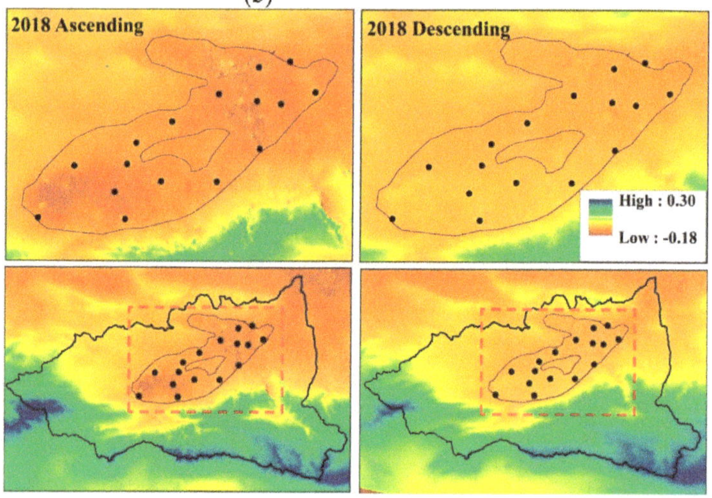

(c)

Figure 9. Annual cumulative displacement maps for the ascending and descending data in (**a**) 2020, (**b**) 2019, and (**c**) 2018.

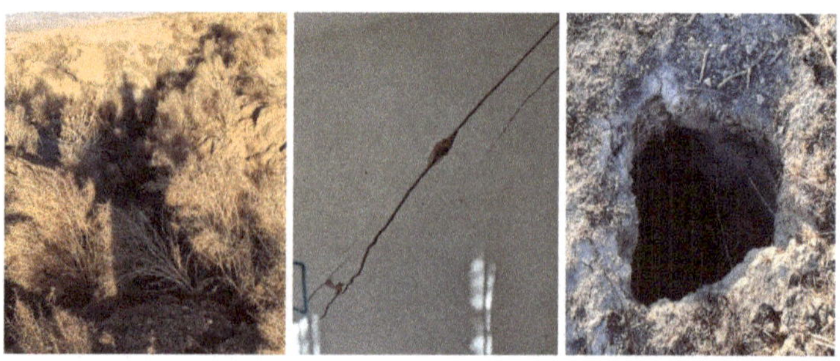

Figure 10. Evidence of land subsidence in the study area as cracks and holes (Photographs were taken at Ashkhane on 25 February, 2021 by Rafiei, F.).

3.4. Groundwater Related Subsidence

Since this paper aimed to study the whole plain, we decided to investigate the groundwater level drop over the whole area, and although there are piezometric wells in the aquifer, there are some extraction wells outside of the aquifer (Figure 6). In addition, as can be seen, there is land subsidence in the north of the plain, even in the areas without piezometric data, but the groundwater level maps indicate that a groundwater level drop is happening there. However, to increase the accuracy, we investigated the relationship between the groundwater drop and subsidence in well locations in more detail.

To investigate the relation between water level decline and land subsidence, water level variations in the piezometric wells were overlaid with the cumulative displacement map. As shown in Figure 11, subsidence occurred in all points experiencing water level falls. In addition, in the areas where the water decline was lower, the subsidence radius value was smaller around the well. Conversely, the subsidence around the wells with a higher water level drop was spread with a greater radius. The subsidence has entirely affected the low-height parts of Samalghan, which contain a high concentration of agricultural lands and experience noticeable water extraction.

To better understand the relationship between water level changes and land subsidence, two sections (profiles) were considered in the Samalghan aquifer. Their locations are indicated by the two lines shown in Figure 11a. The first line was drawn to cover a longer length of the aquifer in the plain and passes through the position of more piezometers. The graphs of land surface change and water level decline were then plotted in Figure 11b. It can be seen that the higher the drop in water level, the more subsidence occurs in the area. A few examples of extents with a stronger connection between water level drop and land subsidence are shown in Figure 11, indicated with the red boxes. With the lower water level fall, the rate of subsidence also decreased.

In the next step, the temporal evolution of groundwater changes and land deformation at piezometric wells are drawn on a dual-axis chart. According to Figure 12, at some wells, both changes follow almost a similar trend (Figure 12a,b), and with subsidence of groundwater, subsidence is observed. Similarly, with the recovery of the water level in the wells, an uplift is detected. However, in some wells, the ground surface with a one-year lag showed similar behavior to the water surface, as shown in Figure 12b,c.

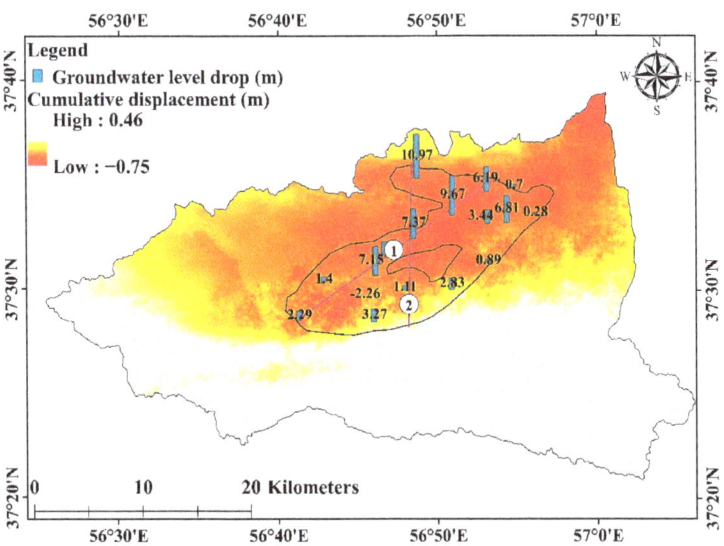

(a)

Figure 11. *Cont.*

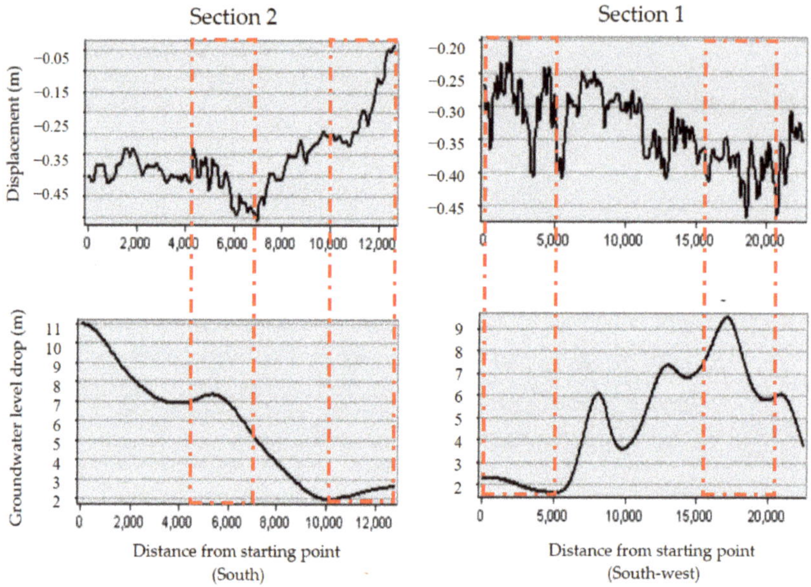

(b)

Figure 11. (**a**) Change in water level and position of the piezometric wells overlaid with subsidence in the region from 2015 to 2020, with purple lines as section. (**b**) Sections in the aquifer showing the groundwater level drop and displacement plots.

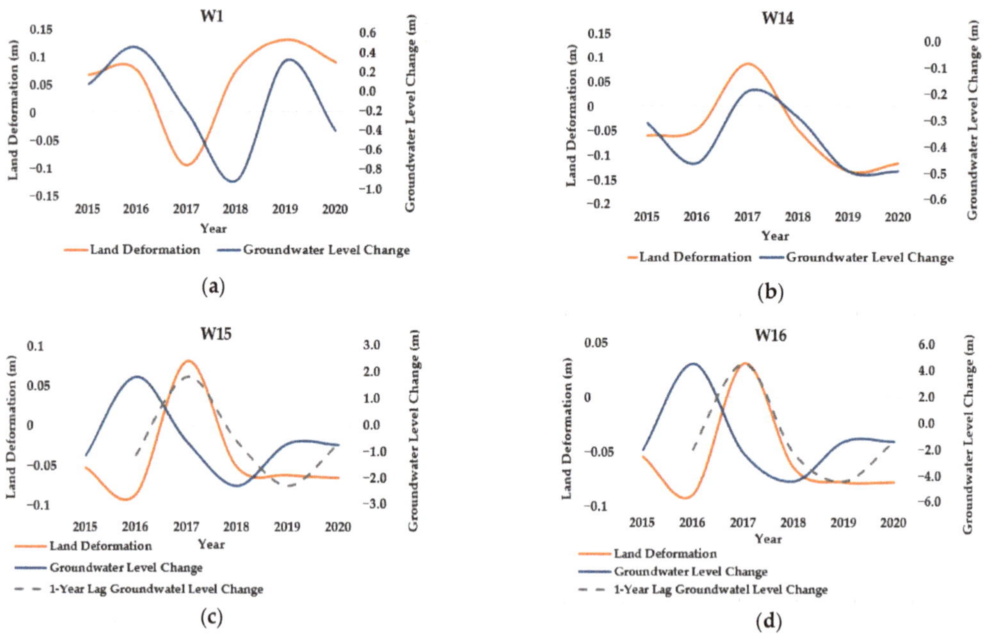

Figure 12. Temporal evolution of deformation (InSAR estimations) (orange line), groundwater level change (blue line) at (**a**) well 1, (**b**) well 14, (**c**) well 15, and (**d**) well 16, and the groundwater level change with a one-year lag in (**b**,**c**) (gray-dashed line).

3.5. Aquifer Behavior

The following section discusses subsidence in piezometric wells in the plain, considering changes in subsidence over time.

One procedure for monitoring the relationship between the groundwater level variations and land surface changes is to plot these two parameters (groundwater level and land deformation). The authors of [61] claimed that a method can be used to estimate the approximate value of the storage coefficient of an aquifer. Based on this technique, the y-axis indicates the water level change, while the ground level changes (derived from the Interferometric processing) are plotted on the x-axis. In the next step, a linear trendline is fitted on these datasets, and the inverse slope of this line represents the aquifer storage coefficient. The trendline can estimate the approximate value of subsidence for a certain amount of water drop [61,62]. Based on this method, the storage coefficient value around the piezometric wells was calculated with the minimum value of 0.04, and the maximum value of 0.36. The higher the computational storage coefficient, the more sensitivity of the ground surface to respond to the water level changes. After that, using the Kriging interpolation method, a map of changes in the approximate storage coefficient of the aquifer was calculated (Figure 13).

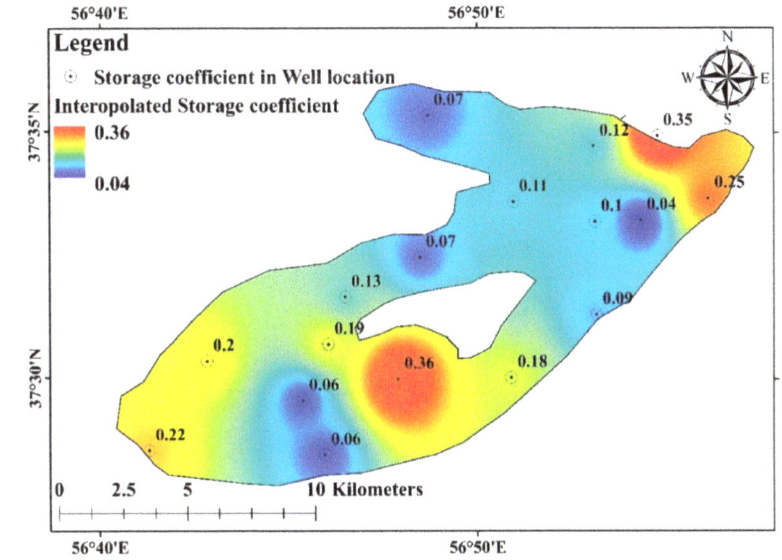

Figure 13. The value of the interpolated storage coefficient of the Samalghan aquifer, using the ordinary Kriging method.

To study the aquifer behavior, the relationship between the DInSAR displacements for every extraction–recovery period concerning the distance to the wells was measured. For this purpose, in every extraction–recovery cycle at a radial distance varying from 250 to 4000 m from the observation wells (Figure 14), the average displacement was measured in the desired buffer area. Then, the maximum and minimum subsidence rates for the generated maps and their average were plotted on a graph. It was noted that the results were not significant and effective at a distance of 4000 m and they were rather homogeneous and intense from 1000 to 4000 m. On the other hand, at a radius of 1000 m, the effect of adjacent wells could be seen to some extent. Therefore, the greatest influence area of the aquifer exploitation is limited to the 1000 m-radius circle around the wells, and the values extracted from a radial distance of 4000 m were excluded from the analysis. The averages of the displacement data collected for each year were then plotted at a distance of 250 m, which was closest to each well, as shown in Figure 15.

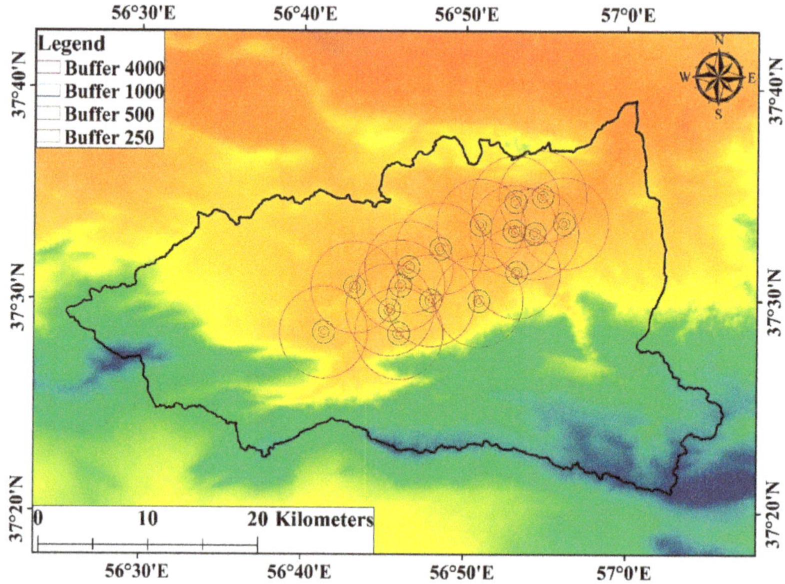

Figure 14. Buffer areas with different radius values and locations around well fields in the study area on the displacement maps.

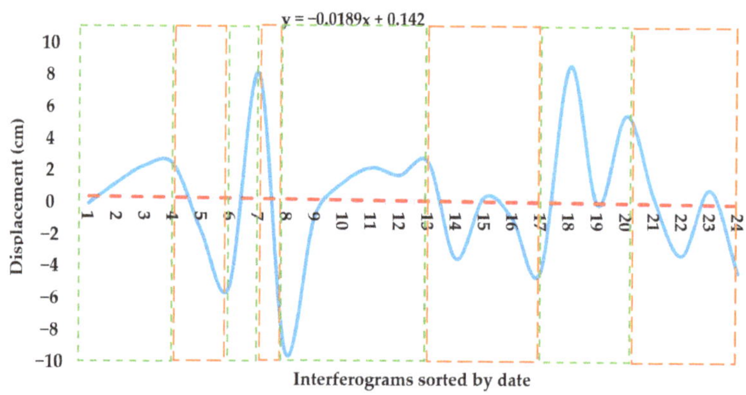

Figure 15. Graph of average surface changes at a distance of 250 m from observation wells (green and orange-dashed rectangles representing recovery and extraction periods, respectively).

According to the trend of land surface changes in the study period (Figure 15), the whole time range was divided into four periods, including both uplift and subsidence. The data order in Figure 15 is ascending, with 1 representing the first interferogram produced in 2014, and 24 representing the last interferogram produced in 2020 (Table 5). A trend line was also fitted on the chart, representing the inelastic behavior in the region.

Table 5. Interferogram ID and its corresponding number in Figure 15 *X*-axis.

Interferogram ID	Number	Interferogram ID	Number	Interferogram ID	Number
D24	1	D23	2	D22	3
D21	4	D20	5	D19	6
D18	7	D17	8	D16	9
D15	10	D14	11	D13	12
D12	13	D11	14	D10	15
D9	16	D8	17	D7	18
D6	19	D5	20	D4	21
D3	22	D2	23	D1	24

The expected inelastic behavior in the region should be in accordance with the fitting line as represented in Figure 15. Due to the noncompliance of the aquifer system with this trend or behaving at the closest distance to this trend, it can be said that the behavior of the Samalghan aquifer might be elastic.

After the investigation of the selected periods, the relation between differential ground surface displacements and the duration of the extraction–recovery phases for every cycle was also analyzed. To reach this goal, the ratio of uplift–subsidence ratio (SR) and the cycle temporal ratio (TR) were calculated according to Table 6. The SR ratio represents the dependence between ground surface uplift during recovery and the subsidence that occurred in the extraction phase [29]. The results not only showed the highest average of subsidence in cycle 2, but also included the highest amount of uplift in the same period. The highest ratio of uplift to subsidence is in the fourth period, with a value of 1800% with a radius of 250 m around the wells. As is advised by [29], to achieve optimal aquifer management, the best time ratio is between 2 and 4. The closer the value obtained to this interval, the better the time frame for managing the area. Therefore, the first period has the best time interval, with a recovery period of 276 days and a subsidence period of 192 days. The third cycle also had a good value of TR, in which the recovery period was 444 days and the extraction period was 360 days. According to SR values, the smaller the time ratio in each period, the higher the monitored uplift to subsidence ratio. In agreement with the optimal number mentioned earlier, to reduce the damages caused by subsidence with proper management of water resources, for each year of uncontrolled extraction of plain resources, two years of plain could be considered as the recovery phase.

According to Table 6, for cycle 1, ground surface subsidence related to the extraction phase represented the minimum value compared to the other cycles' SR values of 32%, 20%, and 33% for 250 m, 500 m, and 1000 m from well fields, respectively. Meanwhile, the maximum TR value was represented in cycle 1, with the optimum amount of 1.43. In cycle 2, although the amount of uplift was on average about 3 times bigger than the phase 1 recovery, the value of subsidence compensated for both values of uplift in cycles 1 and 2. However, it was noticeable that the SR increased significantly. In the third cycle, the SR value decreased, while the TR value in this cycle rose. This procedure was reversed in the fourth cycle. While the time ratio decreased, the value of uplift increased significantly compared to subsidence in the extraction phase. The regional water company of North Khorasan proposed to rehabilitate water resources in 2016 and took action in 2018. Strict enforcement of groundwater extraction seems to have been effective in the study area.

Table 6. Average changes of the land surface in the four selected periods studied at different distances from observation wells, the ratio of uplift to subsidence values, and the ratio of uplift to subsidence phase.

Cycle	Phase	250 m from Well Field	500 m from Well Field	1000 m from Well Field
Cycle 1 Recovery: 3 November 2014–6 August 2015 Extraction: 6 August 2015–14 February 2016	Recovery (277 days) Extraction (193 days) SR	1.19 −3.65 32%	0.93 −4.45 20%	1.34 −3.95 33%
	TR	1.43	1.43	1.43
Cycle 2 Recovery: 14 February 2016–20 May 2016 Extraction: 20 May 2016–24 August 2016	Recovery (97 days) Extraction (97 days) SR	7.65 −9.35 81%	8.15 −9.55 85%	9.15 −9.35 97%
	TR	1.00	1.00	1.00
Cycle 3 Recovery: 24 August 2016–11 November 2017 Extraction: 11 November 2017–6 November 2018	Recovery (445 days) Extraction (371 days) SR	0.83 −2.26 36%	0.88 −2.45 35.9%	1.23 −2.99 41%
	TR	1.19	1.19	1.19
Cycle 4 Recovery: 6 November 2018–21 August 2019 Extraction: 21 August 2019–7 November 2020	Recovery (289 days) Extraction (445 days) SR	4.27 −1.83 233%	3.78 −2.05 184%	3.51 −2.42 145%
	TR	0.65	0.65	0.65

Lastly, the stress–displacement relationship, which delivers a relationship between the measured displacement (ΔD) caused by a groundwater level fall (Δh), was computed for the water level drop–subsidence temporal series shown in Table 7 using Equation (1) [63,64]:

$$S_{ke} = \frac{\Delta D}{\Delta h} \tag{1}$$

Table 7. Computed elastic storage coefficient for wells shown in Figure 1.

Well ID	S_{ke}	Well ID	S_{ke}	Well ID	S_{ke}
W1	2.1×10^{-5}	W7	5.3×10^{-5}	W13	1.6×10^{-4}
W2	1.6×10^{-5}	W8	1.7×10^{-4}	W14	4.3×10^{-5}
W3	1.6×10^{-5}	W9	5.4×10^{-5}	W15	1×10^{-4}
W4	2.1×10^{-4}	W10	3×10^{-6}	W16	1.9×10^{-4}
W5	1.5×10^{-5}	W11	8×10^{-6}	W17	4.7×10^{-5}
W6	1.3×10^{-5}	W12	1.6×10^{-4}		

The calculated value represents the deformability, and adopts different values with the stress rate according to the piezometric level [64]. When the stress induced by the hydraulic head variation overreaches the maximum pre-existing stress, as with preconsolidation stress, deformation rates could be largely high and irrecoverable in most cases due to the changes in arrangement and compaction of the soils. However, if induced stress does not exceed the preconsolidation stress, the deformations are much smaller and mostly elastic. This different soil behavior can be introduced to Equation 1 by assigning two different skeletal specific storages, elastic (S_{ske}) and inelastic (S_{skv}), according to the state of stress concerning the preconsolidation stress.

In this work, the elastic storage coefficients S_{ke} of the Samalghan aquifer were calculated for the period 2015–2020 using piezometric data for the 17 available wells where DInSAR retrieved deformations were also observed (Figure 16). These data applied for plotting the stress–strain curves that represent the relationship between piezometric level changes and aquifer system deformations, from which elastic storage coefficients were determined consisting of the slope of the branch of the stress–strain curve [29,64].

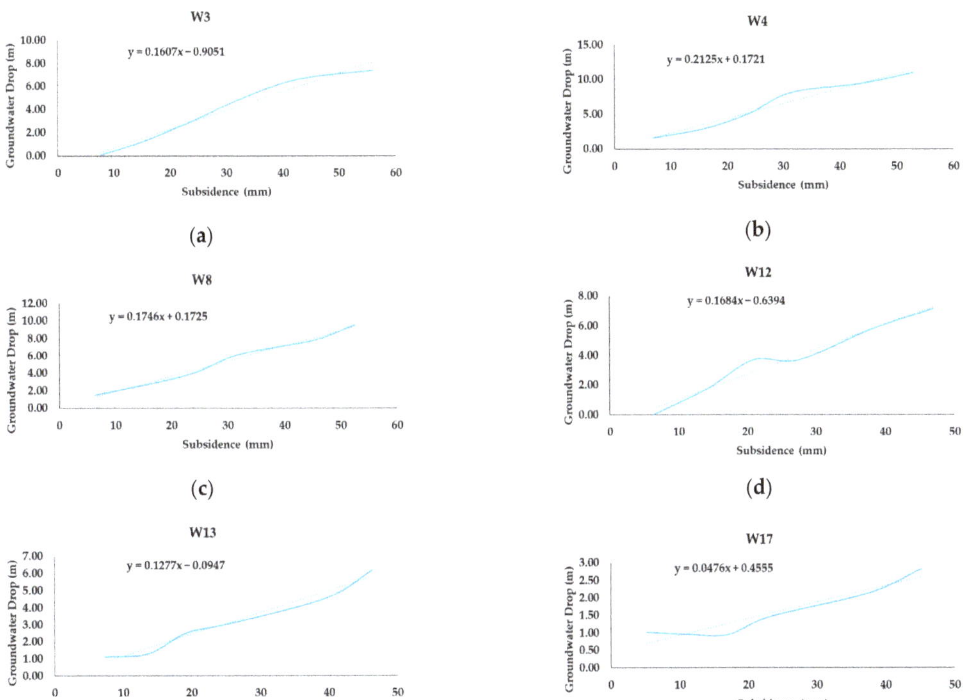

Figure 16. Stress–strain analysis for wells W3 (**a**), W4 (**b**), W8 (**c**), W 12 (**d**), W13 (**e**), and W17 (**f**).

Figure 16 shows the stress–strain relationship at six well locations in the study area. Groundwater level drops define the stresses, and the ground displacements represent the vertical deformation of the aquifer system. The stress–strain trajectories during the period are similar in most of the wells. The S_{ke} values of each well site obtained using the slope of the trendline for each well are shown in Table 7. S_{ke} values vary from 3×10^{-6} at well 10 to 2.1×10^{-4} at well 4.

If the aquifer system is not stressed exceeding its preconsolidation stress, the S_{ke} values are quite stable and independent of the considered time interval. The S_{ke} computed values can be applied to estimate groundwater levels at well locations during the desired SAR acquisition time interval and compared with observed groundwater level values [3].

3.6. Suggestions

One of the complex issues that will arise with population growth in the future is the management of water resources in order to meet the demands and reduce the damage caused by the excessive extraction of groundwater.

Various factors have an impact on groundwater level drop [65], which can be controlled by proper management. First and foremost, one crucial step is increasing public awareness and attracting public participation [66] and implementing operational strategies. Water resources can be used optimally and these valuable resources can be protected. In order to manage the groundwater drop, several steps can be taken, some of which are addressed in this section.

3.6.1. Aquifer Conservation

The first action is the protection of groundwater aquifers. This is an important step to reduce the rate of groundwater depletion. There are several steps that can be taken to achieve this aim:

- Decentralization of exploitation wells [67];
- Increasing infiltration by restoring vegetation [68] in pastures;
- Construction control in aquifer recharge areas and preventing reduction of irrigation level;
- Increasing aquifer recharge through the use of injection wells [69], increasing infiltration of rivers and canals [70], flood infiltration in dried aqueducts, and recharging through infiltration of the surface water from natural pits.

3.6.2. Reduce Water Consumption

Reducing water consumption helps to reduce water loss in the region in several ways: first, by reducing groundwater extraction, and second, by reducing surface water consumption and increasing aquifer infiltration. There are several ways to do this:

- Explaining the leading problems and also increasing the level of awareness of consumers [71];
- Treatment of wastewater and effluents and their reuse [72];
- Improving soil conditions by using modern irrigation methods [73] and reducing evaporation;
- Reducing water transmission losses;
- Promoting greenhouse cultivation [74] in high-consumption areas;
- Promoting and developing hydroponics [75] and providing budgets for these facilities.

3.6.3. Soil Amendation

By improving collapsible soils, the development process of cracks in the Samalghan plain can be reduced. Due to the fact that the fractures in the plain have reached some roads and caused damage to them, it is possible to reduce further damages during the construction or repair of roads by using some techniques. Soil collection, replacement, and compaction; collection of moisture-sensitive soils; chemical stabilization by injection; and the use of piles or foundations are some of the steps that can be taken in this regard.

3.7. Geological Investigation

By comparing the results of annual cumulative displacement with the geological map (Figure 17), it is possible to identify the areas in which more subsidence has occurred. According to the given information in the previous study [42], the soil texture of the plain can be identified. Land deformation has occurred in different parts of the Samalghan plain, which is discussed as follows:

1. East of the plain, on the Marl Formation, including fine sediments;
2. North of the plain, on the Tirgan formation (Ktr), and composed of orbitoline limestone;
3. Sarcheshmeh formation (Ksr) in the eastern Samalghan plain, containing marl (consisting of a high percentage of clay);

4. The center of the plain where the aquifer is located, on new deposits (Qft2) composed of coarse- to fine-grained sediments including clay, sand, and silt, and Khangiran formation (Ekh), formed of sandstone;
5. Some parts of eastern Chamanbid at a radius of 14 kms, and the south of Chamanbid at a distance of 2 km from the city on the Pliocene conglomerate (P1QC), which consists of sandstone and conglomerate, and some parts of Kalat Formation (Kk), including fine sand and erodible limestone;
6. West of the plain at a distance of 4 km to the north of Chamanbid on a small part of the Jmz formation, composed of lime and dolomite;
7. At 5 km to the northeast of Chamanbid, and southwest of Samalghan aquifer, which is located on a part of the Chamanbid formation (Jd).

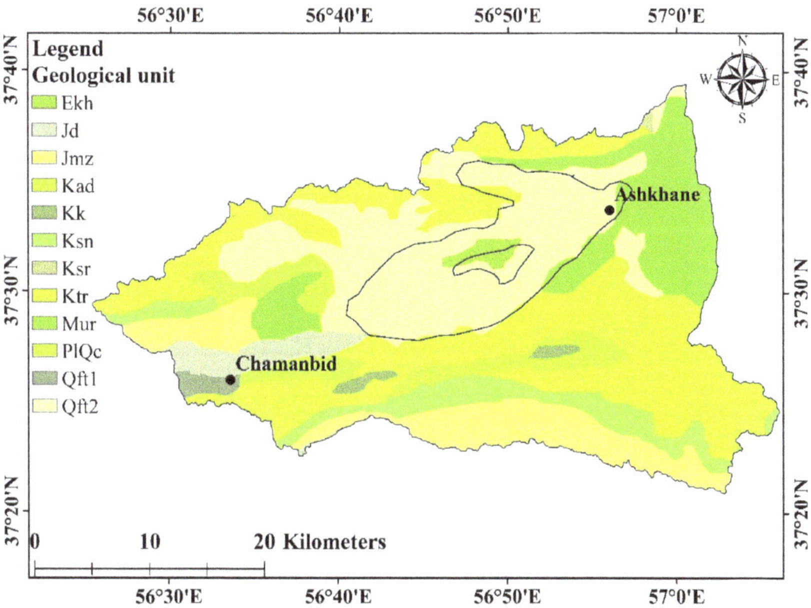

Figure 17. Geological map of the Samalghan plain [42].

The textures of the mentioned formations are mostly fine-grained and soluble rocks. Most of the constituent particles of these sections—lime, clay, and silty sediments—are fine-grained, which confirms that subsidence occurs mostly in fine-grained soils. Due to the clayey texture of the soil in most of these areas, the elastic behavior of the ground cannot be expected. Therefore, in these areas (Sanganeh, Marl, Qft2), it is not possible to propose new suggestions for the land recovery. In addition, the groundwater flow direction [39] is towards the center of the plain, and these areas outside the aquifer are not recharged at a high velocity; so in these areas, groundwater should be extracted under management.

According to the given information, from east to west and from south to north of the Samalghan aquifer, the soil texture varies from fine-grained to coarse-grained. For this reason, decision making in these sections should be different. In the western and northern aquifer, elastic behavior can be expected, so the subsidence can be recovered to an acceptable range by recharging this area or transferring surface water from the highlands to these areas. In contrast, the eastern and southern Samalghan aquifer, subsidence should be managed by overseeing and controlling the withdrawal from wells and managing the consumption of water resources, for example, reducing farmers' water needs by modifying the cultivation pattern and increasing the irrigation system efficiency. To control subsidence in the Samalghan aquifer, the aquifer layers can be examined more closely using well log information, which will be available in extended studies.

3.8. Soft Soil Thickness

It is well-known that when the piezometric level falls, the land subsidence is greater if the accumulated compressible layer is thicker [76]. In this section, the distribution of soft soil thickness derived from boreholes located in the aquifer is compared to the land subsidence measured by DInSAR. The land subsidence in the Samalghan plain is developed in the sand, gravel, and clayey layer. Figure 18 represents the distribution of the mentioned soft soil layer existing in the Samalghan aquifer overlayed with the slope map. The clayey layer is one of the biggest contributors to land subsidence. As can be seen from Figure 18, the layer thickness from the surface ranges between ~30 to 140 m, and as shown in the deformation maps, all of these areas are affected by land subsidence.

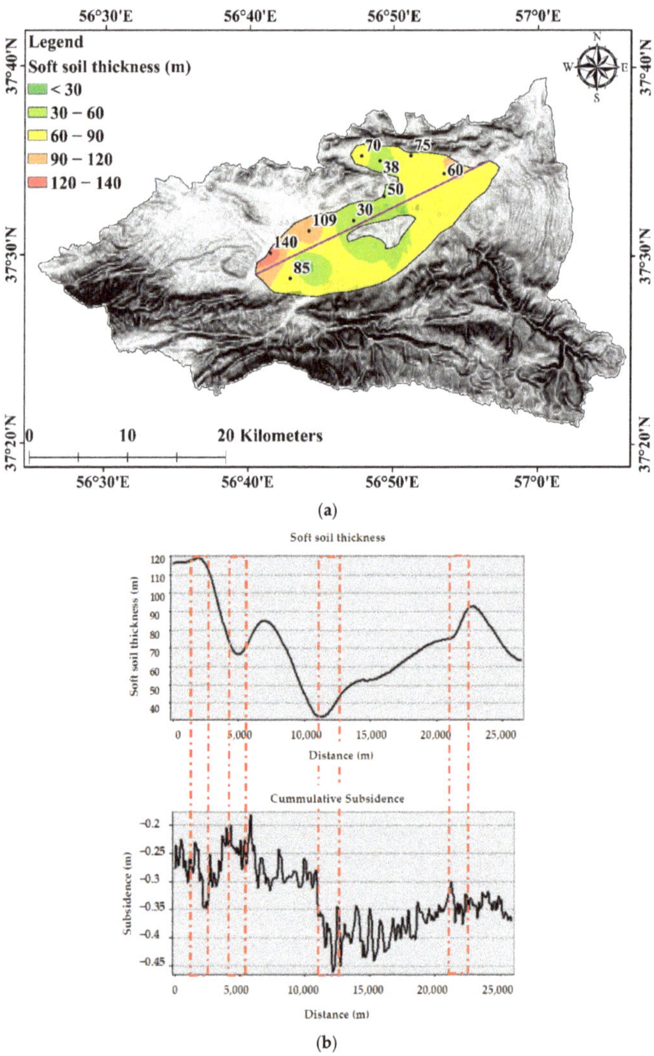

Figure 18. Soft soil thickness map of the Samalghan plain (**a**); the line shows the section used for investigating the soft soil section along the southwest–northeast direction (**b**).

For better investigation, a cross section was employed to better interpret the relation between the soft soil and the deformation. Although there is some correlation between these two (as shown in red boxes in some places), there are some areas that are not completely affected by soft soil. For example, in the first box, the figure for accumulated subsidence reached 40 cm, where the soil thickness increased to its highest value, and in the second box, the accumulated subsidence decreased to approximately 25 cm as the thickness decreased to just below 70 m. However, this trend is not always the same along the area, e.g., in the distance of 15000–2000 m, which shows that the subsidence is not completely affected by the soil thickness. Although subsidence affected some areas according to their thickness, the relation between the subsidence and the groundwater drop was stronger in the deformed area.

4. Conclusions

Most of the previous studies have illustrated plastic or elastoplastic deformation in aquifers [4,6,9,11]. Overexploitation and continuous piezometric level decline [7] act together with seasonal variations [11] as the main driving factors of land subsidence. These variations are usually the reason for continuous plastic deformation and cyclic elastic deformations in aquifers [29].

The present study aimed to monitor land subsidence using the Sentinel-1 radar images in Samalghan plain and investigate its relationship with changes in groundwater level and aquifer behavior. The spatial and temporal evolution of ground surface displacement was evaluated using radar interferometry by processing datasets of Sentinel-1 SLC from November 2014 to November 2020, with results showing that subsidence occurred over most of the Samalghan plain, while in a few areas uplift was also observed in one or more years. It shows that the aquifer behavior showed an elastic deformation in some areas, but in other parts of the region, plastic deformation was detected. With the examination of groundwater level changes in Samalghan plain and land deformation in this area, it can be concluded that the subsidence in this plain is affected by groundwater depletion and uncontrolled extraction from the aquifer. The study of the aquifer interaction in the area concerning the variation of land deformation caused by changes in the groundwater level represents an elastic behavior of the land in the Samalghan aquifer and around Chamanbid city, and the inelastic behavior of the land surface in other parts of the plain. It is undeniable that understanding the land surface response to water level decline is essential in arid and semiarid areas to reduce damage and achieve sustainable water resource management. These results showed that the quasi-elastic aquifer deformational behavior is influenced by groundwater withdrawal in Samalghan plain. Applying groundwater management exploitation seems reasonable because the piezometric level in wells is mostly recovered after the extraction periods. To illustrate, it can be advised that after 1–2 years of overexploitation, the aquifer should enter 2–4 years of recovery. In addition, the identified displacements show a modest subsidence phenomenon influencing a broad area. With reference to the continuous groundwater depletion in the Samalghan plain, both aquifer conservation and reducing water consumption would alleviate the current situation. However, some actions could be taken specifically in areas pursuant to their subsidence situation. We advise that according to the subsidence pattern and the geological structures (Figures 7 and 17), proposed methods of reducing water consumption should be applied to the Ktr, Ekh, and Mur sections to manage water consumption. Furthermore, aquifer conservation would be practical in some parts of the PlQC, Kad, Jmz, Qft2, and northern parts of Ktr geological sections. The investigation of the relationship between land subsidences showed that although there is a small correlation between the soft soil thickness and the land deformation, the subsidence in this area is mostly affected by the groundwater level variations.

Author Contributions: For research contributions, Fatemeh Rafiei performed data processing and the first drafting of the manuscript; Saeid Gharechelou methodologically advised, investigated, and validated the result and edited the manuscript; Saeed Golian supervised; and Brian Alan Johnson revised the final manuscript grammatically and technically. All authors have read and agreed to the published version of the manuscript.

Funding: This research received no external funding.

Data Availability Statement: Data are available on request.

Acknowledgments: The authors acknowledge the support of the Regional Water Authority of North Khorasan province of Iran for groundwater data and some water resources reports sharing. We would also like to thank the experts for their contribution in progressing this study.

Conflicts of Interest: The authors declare no conflict of interest.

References

1. Custodio, E. Aquifer overexploitation: What does it mean? *Hydrogeol. J.* **2002**, *10*, 254–277. [CrossRef]
2. Famiglietti, J.S. The global groundwater crisis. *Nat. Clim. Chang.* **2014**, *4*, 945–948. [CrossRef]
3. Béjar-Pizarro, M.; Ezquerro, P.; Herrera, G.; Tomás, R.; Guardiola-Albert, C.; Hernández, J.M.R.; Fernández, M.; Miguel, M.; Martínez, R. Mapping groundwater level and aquifer storage variations from InSAR measurements in the Madrid aquifer, Central Spain. *J. Hydrol.* **2017**, *547*, 678–689. [CrossRef]
4. Chitsazan, M.; Rahmani, G.; Ghafoury, H. Investigation of subsidence phenomenon and impact of groundwater level drop on alluvial aquifer, case study: Damaneh-Daran plain in west of Isfahan province, Iran. *Model. Earth Syst. Environ.* **2020**, *6*, 1145–1161. [CrossRef]
5. Bhattarai, R.; Alifu, H.; Maitiniyazi, A.; Kondoh, A. Detection of land subsidence in Kathmandu Valley, Nepal, using DInSAR technique. *Land* **2017**, *6*, 39. [CrossRef]
6. Bai, Z.; Wang, Y.; Balz, T. Beijing Land Subsidence Revealed Using PS-InSAR with Long Time Series TerraSAR-X SAR Data. *Remote Sens.* **2022**, *14*, 2529. [CrossRef]
7. Gao, M.; Gong, H.; Chen, B.; Li, X.; Zhou, C.; Shi, M.; Yuan, S.; Zheng, C.; Duan, G. Regional land subsidence analysis in eastern Beijing plain by insar time series and wavelet transforms. *Remote Sens.* **2018**, *10*, 365. [CrossRef]
8. Cigna, F.; Tapete, D. Present-day land subsidence rates, surface faulting hazard and risk in Mexico City with 2014–2020 Sentinel-1 IW InSAR. *Remote Sens. Environ.* **2021**, *253*, 112161. [CrossRef]
9. Miller, M.M.; Jones, C.E.; Sangha, S.S.; Bekaert, D. Rapid drought-induced land subsidence and its impact on the California aqueduct. *Remote Sens. Environ.* **2020**, *251*, 112063. [CrossRef]
10. Braun, A.; Hochschild, V.; Pham, G.T.; Nguyen, L.H.K.; Bachofer, F. Linking land subsidence to soil types within Hue city in Central Vietnam. *J. Vietnam. Environ.* **2020**, *12*, 1–6. [CrossRef]
11. Rezaei, A.; Mousavi, Z. Characterization of land deformation, hydraulic head, and aquifer properties of the Gorgan confined aquifer, Iran, from InSAR observations. *J. Hydrol.* **2019**, *579*, 124196. [CrossRef]
12. Yastika, E.; Shimizu, N.; Abidin, H.Z. Monitoring of long-term land subsidence from 2003 to 2017 in coastal area of Semarang, Indonesia by SBAS DInSAR analyses using Envisat-ASAR, ALOS-PALSAR, and Sentinel-1A SAR data. *Adv. Space Res.* **2019**, *63*, 1719–1736. [CrossRef]
13. Gharechelou, S.; Akbari Ghoochani, H.; Golian, S.; Ganji, K. Evaluation of land subsidence relationship with groundwater depletion using Sentinel-1 and ALOS-1 radar data (Case study: Mashhad plain). *J. GIS RS for Nat. Res.* **2021**, *12*, 11–14. Available online: http://dorl.net/dor/20.1001.1.26767082.1400.12.3.3.8 (accessed on 1 October 2021). (In Persian).
14. Mesgaran, M.B.; Madani, K.; Hashemi, H.; Azadi, P. Iran's land suitability for agriculture. *Sci. Rep.* **2017**, *7*, 7670. [CrossRef]
15. Mohammady, M.; Pourghasemi, H.R.; Amiri, M.; Tiefenbacher, J. Spatial modeling of susceptibility to subsidence using machine learning techniques. *Stoch. Environ. Res. Risk Assess.* **2021**, *35*, 1689–1700. [CrossRef]
16. Sayyaf, M.; Mahdavi, M.; Barani, O.R.; Feiznia, S.; Motamedvaziri, B. Simulation of land subsidence using finite element method: Rafsanjan plain case study. *Nat. Hazards* **2014**, *72*, 309–322. [CrossRef]
17. Dyskin, A.V.; Pasternak, E.; Shapiro, S.A. Fracture mechanics approach to the problem of subsidence induced by resource extraction. *Eng. Fract. Mech.* **2020**, *236*, 107173. [CrossRef]
18. Abou Zaki, N.; Torabi Haghighi, A.M.; Rossi, P.J.; Tourian, M.; Kløve, B. Monitoring groundwater storage depletion using gravity recovery and climate experiment (GRACE) data in Bakhtegan Catchment, Iran. *Water* **2019**, *11*, 1456. [CrossRef]
19. Wang, D.; Li, M.; Chen, J.; Xia, X.; Zhang, Y. Numerical study on groundwater drawdown and deformation responses of multi-layer strata to pumping in a confined aquifer. *J. Shanghai Jiaotong Univ.* **2019**, *24*, 287–293. [CrossRef]
20. Poland, J.F.; Davis, G.H. Land Subsidence Due to Withdrawal of Fluids. In *Reviews in Engineering Geology*; Varnes, D.J., Kiersch, G., Eds.; Geological Society of America: Onslow Beach, NC, USA, 1969; Volume 2, pp. 187–269. [CrossRef]
21. Guzy, A.; Malinowska, A.A. State of the art and recent advancements in the modelling of land subsidence induced by groundwater withdrawal. *Water* **2020**, *12*, 2051. [CrossRef]

22. Wilson, A.M.; Gorelick, S. The effects of pulsed pumping on land subsidence in the Santa Clara Valley, California. *J. Hydrol.* **1996**, *174*, 375–396. [CrossRef]
23. Zhang, Y.; Wu, J.; Xue, Y.; Wang, Z.; Yao, Y.; Yan, X.; Wang, H. Land subsidence and uplift due to long-term groundwater extraction and artificial recharge in Shanghai, China. *Hydrogeol. J.* **2015**, *23*, 1851–1866. [CrossRef]
24. Hammond, W.C.; Burgette, R.J.; Johnson, K.M.; Blewitt, G. Uplift of the western transverse ranges and Ventura area of Southern California: A four-technique geodetic study combining GPS, InSAR, leveling, and tide gauges. *J. Geophys. Res. Solid Earth* **2018**, *123*, 836–858. [CrossRef]
25. Hu, X.; Lu, Z.; Wang, T. Characterization of hydrogeological properties in salt lake valley, Utah, using InSAR. *J. Geophys. Res. Earth Surf.* **2018**, *123*, 1257–1271. [CrossRef]
26. Galloway, D.L.; Burbey, T.J. Regional land subsidence accompanying groundwater extraction. *Hydrogeol. J.* **2011**, *19*, 1459–1486. [CrossRef]
27. Khan, S.D.; Huang, Z.; Karacay, A. Study of ground subsidence in northwest Harris county using GPS, LiDAR, and InSAR techniques. *Nat. Hazards* **2014**, *73*, 1143–1173. [CrossRef]
28. Bürgmann, R.; Rosen, A.; Fielding, E.J. Synthetic aperture radar interferometry to measure Earth's surface topography and its deformation. *Annu. Rev. Earth Planet Sci.* **2000**, *28*, 169–209. [CrossRef]
29. Ezquerro, P.; Herrera, G.; Marchamalo, M.; Tomás, R.; Béjar-Pizarro, M.; Martínez, R. A quasi-elastic aquifer deformational behavior: Madrid aquifer case study. *J. Hydrol.* **2014**, *519*, 1192–1204. [CrossRef]
30. Sui, L.; Ma, F.; Chen, N. Mining subsidence prediction by combining support vector machine regression and interferometric synthetic aperture radar data. *ISPRS Int. J. Geo-Inf.* **2020**, *9*, 390. [CrossRef]
31. Martins, B.H.; Suzuki, M.; Yastika, E.; Shimizu, N. Ground surface deformation detection in complex landslide area—Bobonaro, Timor-Leste—Using SBAS DinSAR, UAV photogrammetry, and field observations. *Geosciences* **2020**, *10*, 245. [CrossRef]
32. Braun, A. Retrieval of digital elevation models from Sentinel-1 radar data–open applications, techniques, and limitations. *Open Geosci.* **2021**, *13*, 532–569. [CrossRef]
33. Raspini, F.; Bardi, F.; Bianchini, S.; Ciampalini, A.; Del Ventisette, C.; Farina, P.; Ferrigno, F.; Solari, L.; Casagli, N. The contribution of satellite SAR-derived displacement measurements in landslide risk management practices. *Nat. Hazards* **2017**, *86*, 327–351. [CrossRef]
34. Fárová, K.; Jelének, J.; Kopačková-Strnadová, V.; Kycl, P. Comparing DInSAR and PSI techniques employed to Sentinel-1 data to monitor highway stability: A case study of a massive Dobkovičky landslide, Czech Republic. *Remote Sens.* **2019**, *11*, 2670. [CrossRef]
35. Cigna, F.; Tapete, D. Satellite InSAR survey of structurally-controlled land subsidence due to groundwater exploitation in the Aguascalientes Valley, Mexico. *Remote Sens. Environ.* **2021**, *254*, 112254. [CrossRef]
36. Orhan, O.; Oliver-Cabrera, T.; Wdowinski, S.; Yalvac, S.; Yakar, M. Land subsidence and its relations with sinkhole activity in Karapınar region, Turkey: A multi-sensor InSAR time series study. *Sensors* **2021**, *21*, 774. [CrossRef]
37. Del Soldato, M.; Solari, L.; Raspini, F.; Bianchini, S.; Ciampalini, A.; Montalti, R.; Ferretti, A.; Pellegrineschi, V.; Casagli, N. Monitoring ground instabilities using SAR satellite data: A practical approach. *ISPRS Int. J. Geo-Inf.* **2019**, *8*, 307. [CrossRef]
38. Khosropanah, E.; Karami, G.; Jeyhooni, S. Effects of the excessive withdrawals of groundwater and subsidence in the Semalghan plain. In Proceedings of the 7th Iranian Conference of Engineering Geology and the Environment, Shahrud, Iran, 5 September 2011. (In Persian).
39. Rafiee, F.; Gharechelou, S.; Golian, S.; Nozarpour, N. Recharge and Discharge Zones Identification using GIS (Case study: Semalghan plain). In Proceedings of the 12th International Congress on Civil Engineering, Mashhad, Iran, 12–14 July 2021.
40. Nasiri, S.; Ansari, H.; Ziaei, A.N. Simulation of water balance equation components using SWAT model in Samalqan Watershed (Iran). *Arab. J. Geosci.* **2020**, *13*, 421. [CrossRef]
41. Singh, N.S.; Sharma, R.; Parween, T.; Patanjali, K. Modern Age Environmental Problems and their Remediation. In *Pesticide Contamination and Human Health Risk Factor*; Oves, M., Zain Khan, M., Ismail, I.M.I., Eds.; Springer: Cham, Switzerland, 2018; pp. 49–68. [CrossRef]
42. Ajam, H.; Mohammadzadeh, H.; Karami, Q.; Kazemi Gelian, R. Investigating of Samalqan aquifer groundwater quality base on underground variations of alluvial and rock facies. *Sci Semiannu. J. Sediment. Facies* **2018**, *10*, 291–310. (In Persian) [CrossRef]
43. Liu, Z.; Liu, W.; Massoud, E.; Farr, T.G.; Lundgren, P.; Famiglietti, J.S. Monitoring groundwater change in California's central valley using sentinel-1 and grace observations. *Geosciences* **2019**, *9*, 436. [CrossRef]
44. Bui, L.K.; Le, V.; Dao, D.; Long, N.Q.; Pham, H.V.; Tran, H.H.; Xie, L. Recent land deformation detected by Sentinel-1A InSAR data (2016–2020) over Hanoi, Vietnam, and the relationship with groundwater level change. *Gisci. Remote Sens.* **2021**, *58*, 161–179. [CrossRef]
45. Koukouvelas, I.; Mpresiakas, A.; Sokos, E.; Doutsos, T. The tectonic setting and earthquake ground hazards of the 1993 Pyrgos earthquake, Peloponnese, Greece. *J. Geol. Soc.* **1996**, *153*, 39–49. [CrossRef]
46. Chen, B.; Gong, H.; Chen, Y.; Lei, K.; Zhou, C.; Si, Y.; Li, X.; Pan, Y.; Gao, M. Investigating land subsidence and its causes along Beijing high-speed railway using multi-platform InSAR and a maximum entropy model. *Int. J. Appl. Earth Obs. Geoinf.* **2021**, *96*, 102284. [CrossRef]
47. Khorrami, M.; Hatami, M.; Alizadeh, B.; Khorrami, H.; Rahgozar, P.; Flood, I. Impact of ground subsidence on groundwater quality: A case study in Los Angeles, California. In Proceedings of the Computing in Civil Engineering 2019: Smart Cities,

Sustainability, and Resilience, Atlenta, GA, USA, 17–19 June 2019; American Society of Civil Engineers: Reston, VA, USA, 2019; pp. 162–170.
48. Veci, L. TOPS Interferometry Tutorial. Sentinel-1 Toolbox; Array Systems Computing Inc. and ESA, 2015. Available online: http://step.esa.int/docs/tutorials/S1TBX%20Stripmap%20Interferometry%20with%20Sentinel-1%20Tutorial.pdf (accessed on 27 September 2018).
49. Sansosti, E.; Berardino, P.; Manunta, M.; Serafino, F.; Fornaro, G. Geometrical SAR image registration. *IEEE Trans. Geosci. Remote Sens.* **2006**, *44*, 2861–2870. [CrossRef]
50. Kelany, K.A.H.; Baniasadi, A.; Dimopoulos, N.; Gara, M. Improving InSAR Image Quality and Co-Registration through CNN-Based Super-Resolution. In Proceedings of the IEEE International Symposium on Circuits and Systems (ISCAS), Seville, Spain, 28 September 2020; pp. 1–5.
51. Fattahi, H.; Agram, P.; Simons, M. A network-based enhanced spectral diversity approach for TOPS time-series analysis. *IEEE Trans. Geosci. Remote Sens.* **2016**, *55*, 777–786. [CrossRef]
52. Wegmttller, U.; Strozzi, T.; Werner, C.; Wiesmann, A.; AG, G.R.S. Sar interferometry for topographic mapping and surface deformation monitoring. *FJP* **2002**, *18*, 24–32.
53. Tolomei, C.; Caputo, R.; Polcari, M.; Famiglietti, N.A.; Maggini, M.; Stramondo, S. The use of interferometric synthetic aperture radar for isolating the contribution of major shocks: The case of the thessaly, Greece, seismic sequence. *Geosciences* **2021**, *11*, 191. [CrossRef]
54. Ai, B.; Liu, K.; Li, X.; Li, D.H. Flat-earth phase removal algorithm improved with frequency information of interferogram. In Proceedings of the Geoinformatics and Joint Conference on GIS and Built Environment: Classification of Remote Sensing Images, Guangzhou, China, 7 November 2008; Volume 7147, pp. 418–427. [CrossRef]
55. Xu, B.; Li, Z.; Zhu, Y.; Shi, J.; Feng, G. SAR interferometric baseline refinement based on flat-earth phase without a ground control point. *Remote Sens.* **2020**, *12*, 233. [CrossRef]
56. Goldstein, R.M.; Werner, C.L. Radar interferogram filtering for geophysical applications. *Geophys. Res. Lett.* **1998**, *25*, 4035–4038. [CrossRef]
57. Chen, C.W.; Zebker, H.A. Two-dimensional phase unwrapping with statistical models for nonlinear optimization. In Proceedings of the IGARSS 2000—IEEE 2000 International Geoscience and Remote Sensing Symposium. Taking the Pulse of the Planet: The Role of Remote Sensing in Managing the Environment. Proceedings (Cat. No. 00CH37120), Honolulu, HI, USA, 6 August 2000; Volume 7, pp. 3213–3215. [CrossRef]
58. Huang, Q.; Zhou, H.; Dong, S.; Xu, S. Parallel branch-cut algorithm based on simulated annealing for large-scale phase unwrapping. *IEEE Trans. Geosci. Remote Sens.* **2015**, *53*, 3833–3846. [CrossRef]
59. Johnson, B.; Tateishi, R.; Kobayashi, T. Remote sensing of fractional green vegetation cover using spatially-interpolated endmembers. *Remote Sens.* **2012**, *4*, 2619–2634. [CrossRef]
60. Maghsoudi, Y.; Amani, R.; Ahmadi, H. A study of land subsidence in west of Tehran using Sentinel-1 data and permanent scatterer interferometric technique. *Arab. J. Geosci.* **2021**, *14*, 30. [CrossRef]
61. Sadeghi, Z.; Veldan Zoj, M.J.; Dehghani, M. Introduction and comparison of two radar interferometry methods based on permanent scatterer to measure land subsidence (Case study: Southwest of Tehran plain). *Iran. J. Remote Sens. GIS* **2012**, *4*, 97–110. (In Persian)
62. Bonì, R.; Cigna, F.; Bricker, S.; Meisina, C.; McCormack, H. Characterisation of hydraulic head changes and aquifer properties in the London Basin using Persistent Scatterer Interferometry ground motion data. *J. Hydrol.* **2016**, *540*, 835–849. [CrossRef]
63. Navarro-Hernández, M.I.; Tomás, M.; Lopez-Sanchez, J.M.; Cárdenas-Tristán, A.; Mallorquí, J.J. Spatial analysis of land subsidence in the San Luis Potosi valley induced by aquifer overexploitation using the coherent pixels technique (CPT) and sentinel-1 InSAR observation. *Remote Sens.* **2020**, *12*, 3822. [CrossRef]
64. Tomás, R.; Herrera, G.; Delgado, J.; Lopez-Sanchez, J.M.; Mallorquí, J.J.; Mulas, J. A ground subsidence study based on DInSAR data: Calibration of soil parameters and subsidence prediction in Murcia City (Spain). *Eng. Geol.* **2010**, *111*, 19–30. [CrossRef]
65. Dinh, B.H.; Go, G.H.; Kim, Y.S. Performance of a horizontal heat exchanger for ground heat pump system: Effects of groundwater level drop with soil–water thermal characteristics. *Appl. Therm. Eng.* **2021**, *195*, 117203. [CrossRef]
66. Samani, S. Analyzing the groundwater resources sustainability management plan in Iran through comparative studies. *Groundw. Sustain. Dev.* **2021**, *12*, 100521. [CrossRef]
67. Joshi, D.; Kulkarni, H.; Aslekar, U. Bringing aquifers and communities together: Decentralised groundwater governance in rural India. In Proceedings of the Water Governance: Challenges and Prospects, Singapore, 1 February 2019. [CrossRef]
68. Lozano-Baez, S.E.; Cooper, M.; Meli, P.; Ferraz, S.F.; Rodrigues, R.R.; Sauer, T.J. Land restoration by tree planting in the tropics and subtropics improves soil infiltration, but some critical gaps still hinder conclusive results. *For. Ecol. Manag.* **2019**, *444*, 89–95. [CrossRef]
69. Parker, A.L.; Pigois, J.P.; Filmer, M.S.; Featherstone, W.E.; Timms, N.E.; Penna, N.T. Land uplift linked to managed aquifer recharge in the Perth Basin, Australia. *Int. J. Appl. Earth Obs. Geoinf.* **2021**, *105*, 102637. [CrossRef]
70. Villeneuve, S.; Cook, G.; Shanafield, M.; Wood, C.; White, N. Groundwater recharge via infiltration through an ephemeral riverbed, central Australia. *J. Arid Environ.* **2015**, *117*, 47–58. [CrossRef]
71. Stavenhagen, M.; Buurman, J.; Tortajada, C. Saving water in cities: Assessing policies for residential water demand management in four cities in Europe. *Cities* **2018**, *79*, 187–195. [CrossRef]

72. Zhang, H.; Xu, Y.; Kanyerere, T. A review of the managed aquifer recharge: Historical development, current situation and perspectives. *Phys. Chem. Earth* **2020**, *118*, 102887. [CrossRef]
73. Porhemmat, J.; Nakhaei, M.; Dadgar, M.A.; Biswas, A. Investigating the effects of irrigation methods on potential groundwater recharge: A case study of semiarid regions in Iran. *J. Hydrol.* **2018**, *565*, 455–466. [CrossRef]
74. Raeisi, L.G.; Morid, S.; Delavar, M.; Srinivasan, R. Effect and side-effect assessment of different agricultural water saving measures in an integrated framework. *Agric. Water Manag.* **2019**, *223*, 105685. [CrossRef]
75. Edgerton, C.; Estrada, A.; Fairchok, K.; Parker, M.T.; Jezak, A.; Pavelka, C.; Lee, H.; Doyle, L.; Feldmeth, A. Addressing water insecurity with a greywater hydroponics system in South Africa. In Proceedings of the IEEE Global Humanitarian Technology Conference (GHTC), Seattle, WA, USA, 29 October–1 November 2020; pp. 1–4. [CrossRef]
76. Chen, M.; Tomás, R.; Li, Z.; Motagh, M.; Li, T.; Hu, L.; Gong, H.; Li, X.; Yu, J.; Gong, X. Imaging land subsidence induced by groundwater extraction in Beijing (China) using satellite radar interferometry. *Remote Sens.* **2016**, *8*, 468. [CrossRef]

International Journal of *Geo-Information*

MDPI

Article

Assessment of Morphometric Parameters as the Basis for Hydrological Inferences in Water Resource Management: A Case Study from the Sinú River Basin in Colombia

Alvaro López-Ramos [1], Juan Pablo Medrano-Barboza [1], Luisa Martínez-Acosta [1], Guillermo J. Acuña [2], John Freddy Remolina López [3] and Alvaro Alberto López-Lambraño [4,5,6,*]

[1] GICA Group, Department of Civil Engineering, Universidad Pontificia Bolivariana Campus Montería, Carrera 6 # 97A-99, Montería 230002, Córdoba, Colombia
[2] CAMHA Group, Department of Sanitary and Environmental Engineering, Universidad Pontificia Bolivariana Campus Montería, Carrera 6 # 97A-99, Montería 230002, Córdoba, Colombia
[3] ITEM Group, Department of Electronic Engineering, Universidad Pontificia Bolivariana Campus Montería, Carrera 6 # 97A-99, Montería 230002, Córdoba, Colombia
[4] Faculty of Engineering, Architecture and Design, Universidad Autónoma de Baja California, Mexicali 22860, Mexico
[5] Hidrus S.A. de C.V., Ensenada 22760, Mexico
[6] Grupo Hidrus S.A.S., Montería 230002, Córdoba, Colombia
* Correspondence: alopezl@uabc.edu.mx; Tel.: +521-442-194-6654 or +521-646-134-5766

Abstract: The geomorphology of a basin makes it possible for us to understand its hydrological pattern. Accordingly, satellite-based remote sensing and geo-information technologies have proven to be effective tools in the morphology analysis at the basin level. Consequently, this present study carried out a morphological analysis of the Sinú river basin, analyzing its geometric characteristics, drainage networks, and relief to develop integrated water resource management. The analyzed zone comprises an area of 13,971.7 km^2 with three sub-basins, the upper, the middle, and the lower Sinú sub-basins, where seventeen morphometric parameters were evaluated using remote sensing (RS) and geographical information system (GIS) tools to identify the rainwater harvesting potential index. The Sinú basin has a dendritic drainage pattern, and the results of the drainage network parameters make it possible for us to infer that the middle and lower Sinú areas are the ones mainly affected by floods. The basin geometry parameters indicate an elongated shape, implying a lesser probability of uniform and homogeneous rainfall. Additionally, the hypsometric curve shape indicates that active fluvial and alluvial sedimentary processes are present, allowing us to conclude that much of the material has been eroded and deposited in the basin's lower zones as it could be confirmed with the geological information available. The obtained results and GIS tools confirm the basin's geological heterogeneity. Furthermore, they were used to delimit the potential water harvesting zones following the rainwater harvesting potential index (RWHPI) methodology. The research demonstrates that drainage morphometry has a substantial impact on understanding landform processes, soil characteristics, and erosional characteristics. Additionally, the results help us understand the relationship between hydrological variables and geomorphological parameters as guidance and/or decision-making instruments for the competent authorities to establish actions for the sustainable development of the basin, flood control, water supply planning, water budgeting, and disaster mitigation within the Sinú river basin.

Keywords: watershed management; watershed land surface; geo-information technology; morphology; relief

Citation: López-Ramos, A.; Medrano-Barboza, J.P.; Martínez-Acosta, L.; Acuña, G.J.; Remolina López, J.F.; López-Lambraño, A.A. Assessment of Morphometric Parameters as the Basis for Hydrological Inferences in Water Resource Management: A Case Study from the Sinú River Basin in Colombia. *ISPRS Int. J. Geo-Inf.* **2022**, *11*, 459. https://doi.org/10.3390/ijgi11090459

Academic Editors: Walter Chen, Fuan Tsai and Wolfgang Kainz

Received: 9 July 2022
Accepted: 18 August 2022
Published: 24 August 2022

1. Introduction

A hydrographic basin is an area partly or entirely drained by several watercourses and delimited by an imaginary line formed by points of the highest topographic elevation called watershed, which separates it from neighboring basins [1]. To this extent, just as the hydrological cycle is the fundamental concept in hydrology, the hydrographic basin is a naturally defined hydrological unit that becomes the basis for any study of water resource management [2]. In a basin, morphometry makes it possible to quantitatively study, from a mathematical point of view, the surface configuration as well as the shape and relief [3]. Horton [4] proposed the bases for the quantitative description of a hydrographic basin's shape and drainage network in addition to the interrelationships between morphometry, climate, vegetation, and soil properties. Similarly, it has been established that given the unidirectional water flow, processes in the upper parts of the basin invariably have repercussions in the lower part and the basin morphometric characteristics have a decisive influence on its hydrological response [5]. It is also paramount to note that through morphometry study, it is possible to describe and even predict flow behavior corresponding to the water courses that drain it, and thus, quantify the surface and underground water potential, making it possible to analyze alternatives to the use of water resources in areas where it is required [6,7].

Contrastingly, the hydrological response of a basin and its geological history can be described from morphometric parameters such as area, altitude, slope, shape, drainage density, and length of streams by correlating them with hydrological phenomena such as runoff [8]. In other words, the basin response to a series of precipitation events depends, on one hand, on the rain properties (intensity, duration, frequency, and so on) and, on the other, on the basin's morphological and geological features.

Therefore, morphometric characterization is significant in hydrological research and in studies regarding the management and conservation of natural resources. Such is the case of the Adnan et al. [9] research, which estimated the morphometric parameters of the Karnaphuli and Sangu basins in Bangladesh to assess the region's susceptibility to flash flood events and obtain a flood risk map for the area.

In the case of Gajbhiye et al., Malik et al., Nitheshnirmal et al., and Rahmati [10–13], they determined the morphometric parameters in different drainage basins to assess the susceptibility to erosion and concluded that it was necessary to establish effective practices for land use and water resource management in these basins. In a similar manner, using geographic information systems and the morphometric characteristics of a basin, potential rainwater harvesting (RWH) zones can be determined through multi-criteria decision analysis, which involves establishing suitability criteria and Boolean logic [14,15]. Regarding the study of the rain-runoff phenomenon in hydrographic basins, Jena and Tiwari [16] performed a correlation analysis between morphometric parameters and the features of the unit hydrographs of the Tarafeni and Bhairabbanki basins in India, thus obtaining nonlinear regression models to generate synthetic unit hydrographs.

Meanwhile, Viramontes-Olivas et al. [17] analyzed the morphometric parameters of the San Pedro Conchos river basin in Chihuahua, Mexico, noticing that drainage density is influenced by vegetation cover and lithology, since it regulates infiltration rates and feeding of the subsurface flow, thus reducing the effects and impact of possible floods in the basin.

The geomorphological features of a basin can also be used to determine the homogeneous regions in which hydrological response to a precipitation event is similar. In that aspect, from the correlation of these features with the hydrological variables, it becomes possible to transfer hydrological information in sites with missing or incomplete data [18]. Hence, based on the methods proposed by Horton [4,19], Schumm [20], Strahler [21], and Shreve [22], the characterization of multiple basins has been successfully achieved to obtain reliable information on the hydrological response and physical characteristics of the soil, such as permeability and even from the present parent rock.

Nonetheless, these methods are usually complex and time-consuming (weeks, even months) when used to analyze large areas, especially when there are basins that can exceed 1000 km^2 [23]. As a result, in recent years, hydrological research has been supported through geographic information systems (GIS) since it is a technique specialized in handling large data sets, allowing time optimization for its analysis and understanding of the spatial distribution of the variables to analyze. This represents an improvement in the systematization of the description, comparison, and classification of hydrographic basins regardless of their extension, this being of great use and applicability in hydrology [24–29].

Consequently, this study aims to; (1) determine the morphometric parameters of the Sinú river basin in Colombia through GIS and existing cartography to characterize its drainage network and morphology; (2) analyze the morphometric parameters relationship with erosion and floods in the Sinú river basin; and (3) based on the obtained results, determine the areas with rainwater harvesting potential within the basin as an alternative to mitigate floods and drought situations in the study area. Moreover, considering the lack of hydrological information in the basin, these analyses provide valuable information to improve hydrological models and to integrally manage the basin's water resources, since these results can be the foundations for decision-making standards and guidelines focused on sustainable development, natural resource conservation, land use planning and management, flood risk mitigation, and water supply for the communities of the study area.

2. Area of Study

The Sinú river basin is located in the northwestern part of Colombia, between 9°30′ N to 7°05′ N and 76°35′ W to 75°15′ W. It has an area of 13,972 km^2, in the jurisdiction of the departments of Córdoba, Sucre, and Antioquia, and according to the provisions of the Regional Autonomous Corporation of the Sinú and San Jorge Valleys, CVS (for its acronym in Spanish), it is divided in accordance with its geographical and biotic characteristics in three zones: the upper, middle and lower Sinú zones or regions, CVS [30].

The geographical location, municipalities, and departments in which the Sinú river basin is located can be seen in Figure 1. Mainly, the basin consists of elevations below 300 masl in the lower and middle Sinú region, while in the upper Sinú region there are heights higher than 1000 masl. The Sinú river has its source in this area, specifically in the Nudo del Paramillo at an altitude of 3400 masl, crossing the basin from south to north until reaching its mouth at the Caribbean Sea in Boca de Tinajones through three mouths called Mireya, Medio, and Corea, in the municipality of San Bernardo del Viento, CVS [30].

Conversely, the main population centers of the basin are: Tierralta and Valencia in the upper basin, Montería (capital of the Córdoba department) in the middle, and Lorica in the lower basin. Furthermore, in the upper basin there is also the URRÁ hydroelectric plant and the Paramillo National Natural Park [31].

According to the Agustín Codazzi Geographical Institute, IGAC (for its acronym in Spanish) [32], there are different types of landscapes in the basin distributed within the two large geo-structures, where the region's great geomorphological diversity is evident. In the mountain range (Cordillera), landscapes of high hills and ridges represent 19.72% of the area, while in the sedimentation mega basin, the greatest diversity of landscapes are found, with a predominance of lowland landscape, occupying 70.02% of the basin, and the hillside, 7.99% of the territory.

Figure 2 presents the map of the chronostratigraphic units (CSU) and geological faults of the study area, identified by bearing in mind the International Chronostratigraphic Chart [33], corresponding to a code formed by the geochronological age notation separated with a hyphen of an acronym that indicates the rock type and its formation environment (V: volcanic, H: hypabyssal, P: plutonic, VC: volcanoclastic, S: sedimentary, and M: metamorphic) followed by a lower case letter representing its composition, metamorphism grade, or accumulation environment depending on whether igneous, metamorphic, or sedimentary rocks are involved, respectively, e.g., u: ultramafic, lg: low grade of metamorphism, ct: continental-transitional [34]. In the case of the study area, it is evident that UC Q-al, e3e4-Sm, and n6n7-Sm prevail, corresponding, respectively to alluvial and alluvial plain deposits, grainy-decreasing conglomeratic lithic arenites and intercalations of mudstones, and calcareous arenites and coarse-grained to conglomeratic quartz sandstones.

It is estimated that the 2010–2011 flood event affected 178,124 people in the Sinú River basin, representing 18% of the population of the affected municipalities, and 30,257 houses, representing 15.9% of the households. In addition, it is estimated that 50–60% of this population was in a critical condition, with percentage of unmet basic needs (UBN) values of 40%, suggesting that the greatest impacts were received by municipalities with highly deficient structural conditions [35].

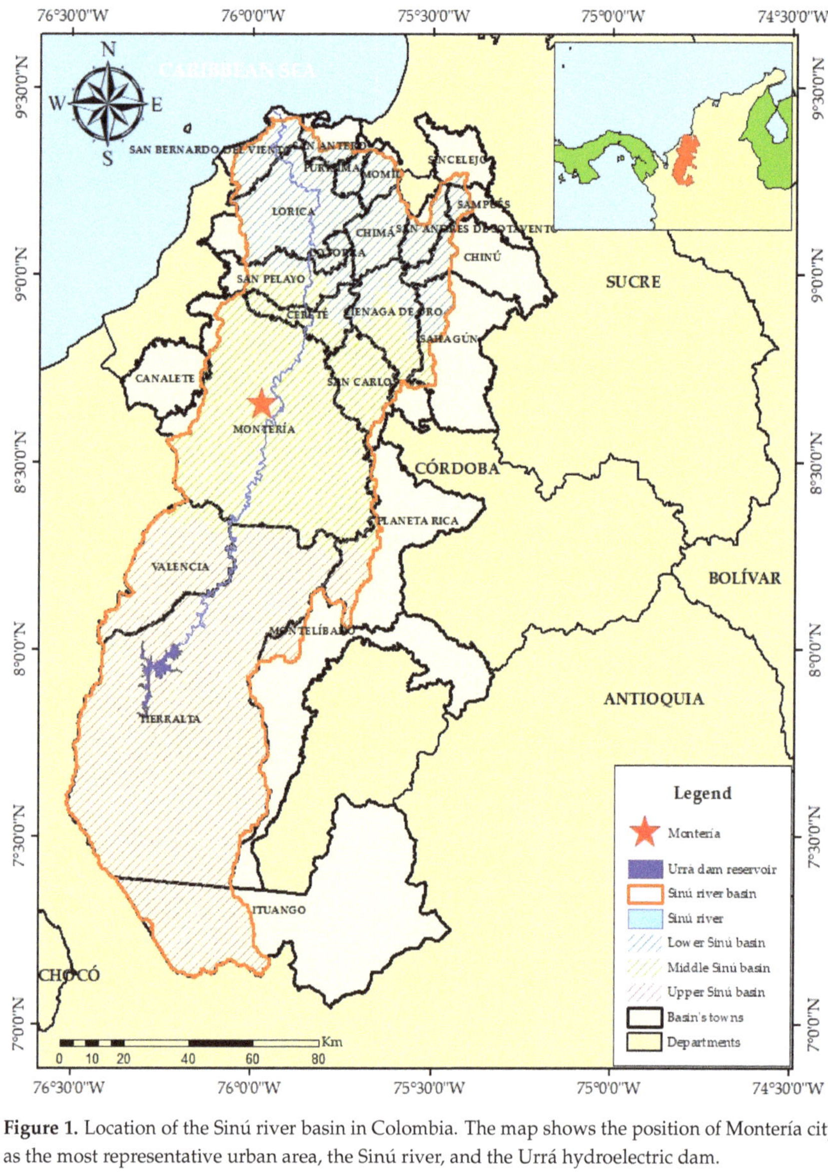

Figure 1. Location of the Sinú river basin in Colombia. The map shows the position of Montería city as the most representative urban area, the Sinú river, and the Urrá hydroelectric dam.

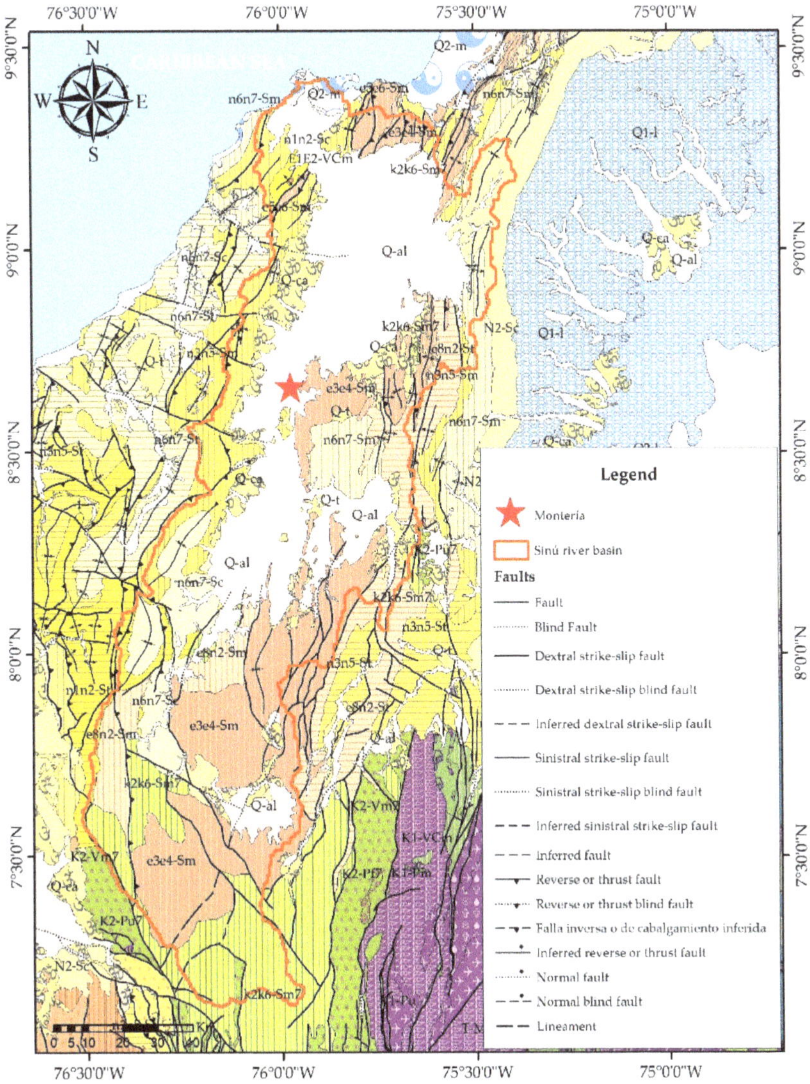

Figure 2. Chronostratigraphic units (CSU) and geological faults of the study area.

This impact caused by floods in the Sinú basin is closely linked to the development of activities in potentially floodable zones (PFZ). The National Water Study prepared by the Institute of Hydrology, Meteorology, and Environmental Studies, IDEAM (for its acronym in Spanish) [36] estimated that for the Sinú basin, about 76% of the PFZ has been transformed into agricultural territories or artificial zones. According to information from CVS [37], the points identified by drainage erosion problems and those present threats due to flooding on the Sinú river are located in the municipalities of Tierralta and Valencia in the upper Sinú sub-basin; Montería, Cereté, and San Pelayo in the middle Sinú, and Cotorra, Lorica, and San Bernardo del Viento in the lower Sinú sub-basin, occurring mostly in Lorica (24%), Tierralta (22%), and Montería (21%). The largest number of people who have been affected due to flood events are in the San Pelayo Municipality, followed by the Lorica and Cotorra municipalities. Regarding economic affectations, they are related to the productive and agro-industrial system and the most affected municipalities are: San

Pelayo, San Bernardo del Viento, and Tierralta. Finally, in terms of territorial effects, the municipalities in the basin with the largest number of flooded hectares are usually Cotorra and Lorica.

3. Materials and Methods

For the morphometric analysis of the basin through GIS, the use of a digital elevation model (DEM) was required, which, according to Felicísimo [38], is a numerical data structure that represents the spatial distribution of altitude on the earth's surface, i.e., it is a set of matrices resulting from superimposing a grid on the terrain and extracting the average altitude of each cell, so they are a regular square mesh network with equally spaced rows and columns. To delimit the watershed, different elevation models were used: SRTM from the Shuttle Radar Topographic Mission with 30-m resolution, the ALOS PALSAR RTC with 12.5 m resolution, and the Hydroshed Digital Elevation Model developed by the United States Geological Survey—USGS, which was validated by the Inter-American Development Bank (IDB) in the framework of the Integrated Model of Climate Change and Water Resources with a 460 m × 460 m resolution [39,40].

To evaluate the accuracy of the DEM, the generation of stream definition was performed using several thresholds, and the result was compared with the official cartographic plates, determining that, since most of the study area is characterized by being flat and with little slope, there was a considerable difference between the real drainage network and the one obtained using the GIS tools, especially when generating the main stream, the Sinú river in this case. Additionally, control points were taken in order to estimate the mean squared error (MSE) of the elevations, concluding that the DEMs had MSEs in the order of 5 to 20 m in some sectors, especially in the middle and lower basin of the Sinú river. Finally, it was found that none of the digital elevation models (DEMs) satisfactorily represented the drainage network; however, although the HydroSHED Digital Elevation Model was the one that best represented the basin polygon, no significant differences were found regarding the other analyzed models, the reason for which it was decided to use the ALOS PALSAR RTC DEM to delineate the basin and use the digitized drainage network of the cartographic plates to estimate the basin morphometric parameters. This DEM made it possible to know the spatial distribution of the terrain elevations in the study area and, from this, estimate slopes and other relief aspects. In this manner, with the GIS, it was possible to determine the directions of flow, the delimitation of sub-basins, and the schematization of streams that drain it from a threshold.

The methodology used to obtain the drainage network of the Sinú river basin is shown in Figure 3, where the input parameters are shown in blue, the procedures carried out on the information in yellow, and the files names generated at each step in green. Thus, through GIS, a filling of the sinks was carried out to correct errors that could occur due to the data resolution or the rounding of elevations to the nearest integer value, and, once a DEM without sinks was obtained, the flow direction from the differences in elevation and slope was determined [41]. Later, to outline the water courses that drain the Sinú river basin, flow accumulation was estimated, which is just the number of cells on the slope that flow towards each cell, and thus the streams were defined.

Afterwards, the order of the streams was established with the Horton classification system [4] to subsequently generate the polygons of the sub-basins and make manual corrections to calibrate the model, obtaining a total of 65 sub-basins: 6 corresponding to the lower Sinú, 18 to the middle Sinú, and the remaining 41 to the upper Sinú region (Figure 4).

With the information generated from the terrain modeling, the morphometric parameters were estimated from the methods and formulas proposed by Horton [4,19], Schumm [20], Strahler [21], and Mueller [42] (Table 1). Results were also used to determine the potential rainwater harvesting areas using GIS-based multi-criteria decision analysis [14,43]. Finally, the existing cartography for the area was consulted to compare and validate the results of the morphometric parameters of the drainage network, such as drainage density and channel frequency, with those obtained from the modeling data, given that these parameters are sensitive to scale and, thus, results of the parameters estimated from the analysis with a threshold of 25 km^2 would not be the same as those estimated when using the 1:25,000 scale cartographic plates from the IGAC [44].

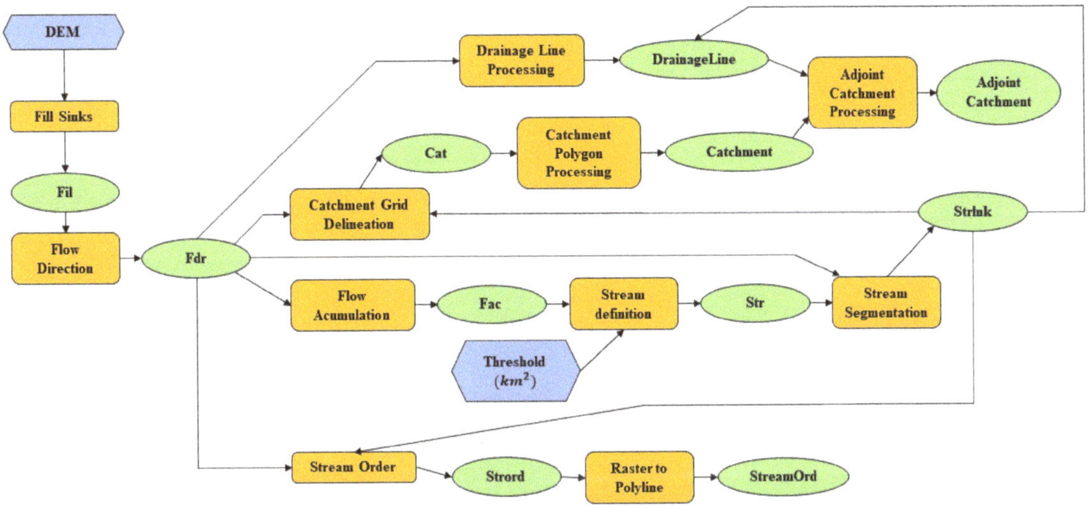

Figure 3. Flowchart of the methodology for watershed characterization through GIS.

The morphometric parameters calculated can be defined as follows:

Table 1. Morphometric Parameters used in the Study.

Morphometric Parameter	Formula/Definition	Reference
Drainage network		
Stream order	Hierarchical rank.	[4,21]
Stream length (L)	Main channel length of the stream (km)	[4]
Stream length (L_u)	Total stream length (km)/GIS software analysis	[4]
Number of streams (N_u)	Total stream number of a given order/GIS software analysis	[4]
Stream axial length (L_a)	Shortest distance between the beginning and the outlet of a stream (km)/GIS software analysis	[42]
Drainage density (D_d)	$D_d = \frac{L}{A_d}$, km/km^2	[4]
Length of overland flow (L_{OF})	$L_{ft} = \frac{1}{2D_d}$, km	[4]
Constant of channel maintenance (C_m)	$C_m = \frac{1}{D_d}, \frac{km^2}{km}$	[20]
Stream frequency (F)	$F = \frac{N}{A_d}$, $streams$/km^2	[4]
Average stream length (L_m)	$L_m = \frac{L_u}{N_u}$	[4]
Sinuosity (S)	$S = \frac{L}{L_a}$	[42]
Bifurcation ratio (R_b)	$R_b = \frac{N_u}{N_{u+1}}$	[4]
Mean bifurcation ratio (R_{bm})	$R_{bm} = \frac{\sum(R_{bu}/R_{bu+1})(N_u+N_{u+1})}{\sum(N_u+N_{u+1})}$	[20]
Stream length ratio (R_l)	$R_l = \frac{L_u}{L_{u-1}}$	[4]
Basin geometry		
Basin area (A_d)	Plan area of the watershed (km^2)	[4]
Basin perimeter (P)	GIS software analysis (km)	[4]
Basin length (L_c)	Maximum basin length (km)/GIS software analysis.	[19]
Form factor (F_f)	$F_f = \frac{A_d}{L_c^2}$	[19]
Circularity ratio (R_c)	$R_c = \frac{A_d}{\frac{P^2}{4\pi}}$	[45]

Table 1. *Cont.*

Morphometric Parameter	Formula/Definition	Reference
Elongation ratio (R_e)	$R_e = \frac{D_c}{L_c}$	[20]
Compactness coefficient (K_c)	$K_c = \frac{P}{2\sqrt{\pi A_d}}$	[19]
Basin relief		
Minimum basin height (H_{min})	GIS software analysis (masl).	[20]
Maximum basin height (H_{max})	GIS software analysis (masl).	[20]
Mean basin slope (S_c)	GIS software analysis (%).	[20]
Basin relief (H)	$H = H_{max} - H_{min}$	[20]
Relief ratio (F_H)	$F_H = \frac{H}{L_c}$	[20]

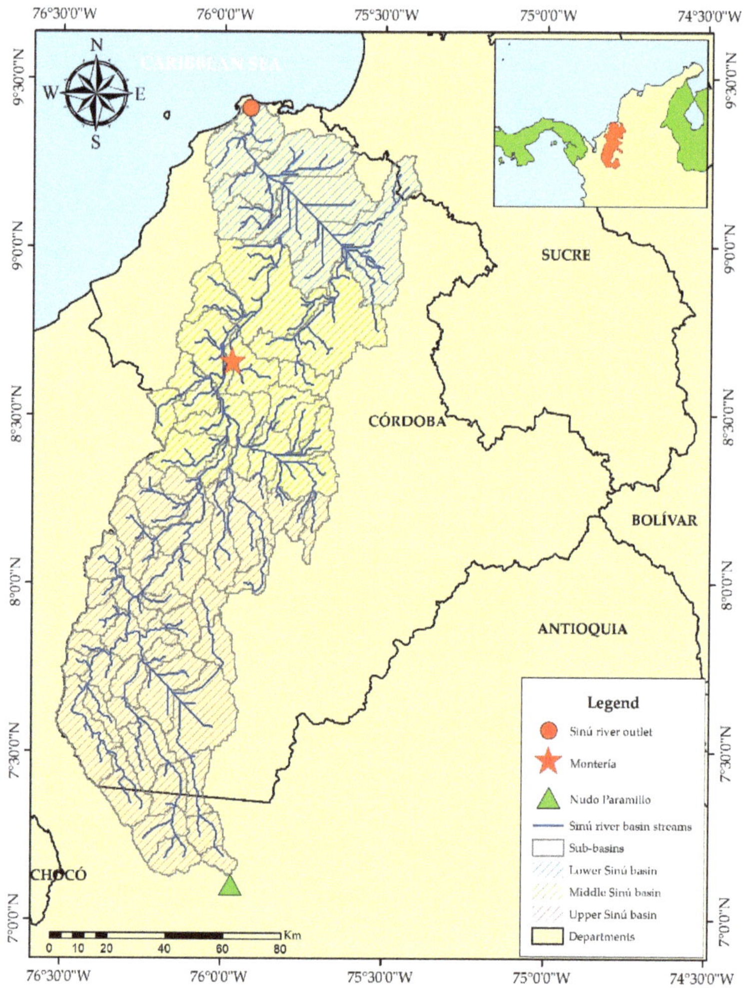

Figure 4. Characterization of the Sinú river basin in Colombia through GIS. The drainage network of the basin and the 65 sub-basins obtained are shown.

3.1. Stream Order

The stream order makes it possible to explain the hydrological behavior of a basin, since it is directly proportional to the area, the cross section of the course, and the flow that it transports. This way, it is expected that the channels of higher order drain larger areas and, thus, transport a much higher flow [6,46].

3.2. Number of Channels of Order u (N_u)

According to the stream number law established by Horton [4], "the number of streams of different orders in any basin tends to be estimated as inverse geometric series, of which the first term is unity and the ratio is the bifurcation ratio".

3.3. Length of Channels of Order u (L_u)

Length of channels of order u is calculated by measuring the length of all streams of a given order within the catchment [4].

3.4. Drainage Density (D_d)

Drainage density corresponds to the total length of streams per unit area and makes quantitative determination of whether a basin is well or poorly drained possible [4]. This parameter is related (together with the number of channels of order u) to aspects such as soil erosion and runoff, since the flow is directly proportional to the drainage density, which translates into rapid runoff that implies an increase in the peak flow of the hydrograph. Additionally, drainage density has an inverse relationship with infiltration, since high infiltration tends to inhibit the development of longer drainages, i.e., the lowest drainage densities correspond to regions with permeable soil types, dense vegetation, and low relief, while the high drainage density prevails in regions with impermeable soils, sparse vegetation, and high relief [10,21,47].

3.5. Overland Flow Length (L_{OF})

Horton [4] defined the length of the overland flow as the distance that the water must travel on the ground surface before reaching the channels of the drainage network, and also estimated that it is approximately equal to half the reciprocal of the density drainage.

3.6. Constant of Channel Maintenance (C_m)

This parameter corresponds to the inverse of the drainage density and makes it possible to estimate the amount of area in km^2 necessary for the maintenance of 1 km of channel [20,48].

3.7. Stream Frequency (F)

Stream frequency was defined by Horton [4] as the number of stream channels per unit area.

3.8. Sinuosity of Currents (S)

Considering that all watercourses must adjust to the terrain irregularities along their route, Mueller [42] proposed sinuosity as an index that allows these variations in the course of the channel to be measured from the ratio between the stream total length (km) and the shortest distance between its beginning and mouth (km).

3.9. Bifurcation Ratio (R_b)

The bifurcation ratio is the ratio of the number of channels of a specific order and the number of streams of the next order [4]. It is a parameter that reflects both the complexity of the ramifications in the basin and its geometric shape, and it is also related to factors such as slope and area physiography.

3.10. Length Ratio (R_l)

According to Horton's research [4], length ratio is the result of dividing average length of the flow of any order by average length of the next lower order.

3.11. Form Factor (F_f)

The shape factor of the basin is a parameter that makes it possible to know the geometry of the basin and is also related to the flows of the drainage network [19].

3.12. Circular Ratio (R_c)

Miller [45] defines this parameter as the ratio between area of the basin and area of a circle that has the same circumference as the basin perimeter and in this manner, if it takes values close to the unit, it indicates that the shape of the basin resembles a circle.

3.13. Elongation Ratio (R_e)

The elongation ratio corresponds to the ratio between the diameter of a circle with the same basin area (D_c) and the maximum basin length (L_c), and like R_c, it has a maximum value of one for perfectly round basins [20]. The R_e generally varies from 0.6 to 1.0 and depends on climate and the geology of the study area. Furthermore, these values can be grouped into: circular for $R_e > 0.9$, oval for $0.9 > R_e > 0.8$, and less elongated when $R_e < 0.7$ [49].

3.14. Compactness Coefficient (K_c)

According to Horton's research [19], compactness coefficient is the relationship between the basin's perimeter and a circle with the same area, thus, the closer the result is to the unit, the more circular the basin will be.

3.15. Mean Slope of the Basin (S_c)

The study of slope distribution is important because it provides data for activities such as planning of engineering works, reforestation, mechanization of agriculture, and others [50]. Additionally, it makes the evaluation of the volumes and direction of surface runoff possible [51].

3.16. Relief Factor (F_H)

The basin relief features play an important role in the development of the drainage network, superficial flow, permeability, and susceptibility to soil erosion. The basin relief (H) is then defined as the difference between maximum and minimum height.

3.17. Hypsometric Curve of the Basin

The hypsometric curve is the graphic representation of the basin relief, where its ordinate represents elevation in meters above sea level, and its abscissa, the area in km^2 that is between two levels.

3.18. Rainwater Harvesting Potential Index (RWHPI)

According to the conducted literature review, rainwater harvesting (RWH) is one of the most usual practices as an alternative to mitigate water scarcity and other environmental problems; moreover, as stated by Singh et al. [14], the planning and implementation of water harvesting projects is a multi-criteria and multi-objective problem because it depends on several factors.

In order to identify sites where the implementation of water harvesting techniques is viable, GIS tools were used to delimit the potential water harvesting zones following the methodology of rainwater harvesting potential index (RWHPI) proposed by Singh et al. [14]. Thus, the runoff coefficient, slope, and drainage density layers were used.

4. Results

4.1. Drainage Network

The DEM was used to determine the basin relief analysis including the hypsometric curve and slope calculation. From the information mentioned, it was found that the maximum order of the basin is 5, corresponding in this case to the courses of 4 and 5 order for the Sinú river (Figure 5).

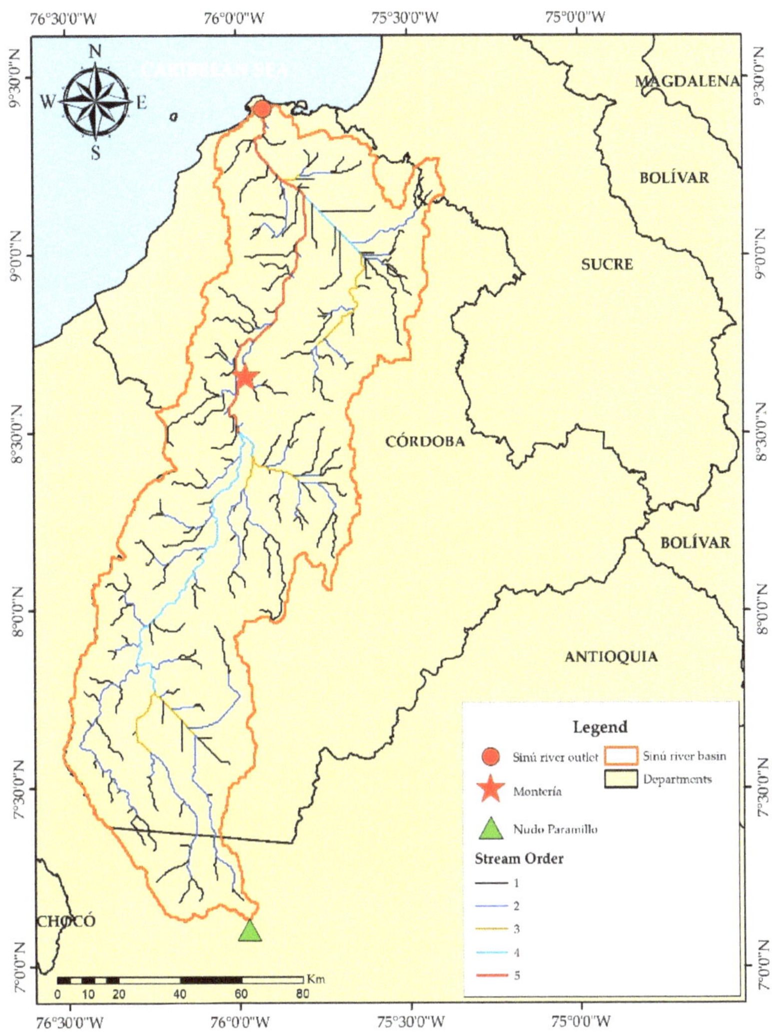

Figure 5. Stream order of the drainage network of the Sinú river basin (Colombia), schematized by GIS.

In the analysis, the Sinú river basin shows an order of 5 value with a dendritic drainage pattern, characterized by having a tree-like distribution with tributary branches in many directions and with variable angles [52]. It was also found that 66.2% of the sub-basins are of orders 1 and 2 (Table 2). Overall, the number of streams tends to decrease as the order increases, as does the number of sub-basins corresponding to each order. On the other hand, it was shown that the area of the sub-basins of order 5 was greater than those of order 1, despite the fact that the amount is much smaller. Consequently, it can be said that the stream order is directly proportional to the drained area under normal conditions, and thus to the flow rate, as previously described.

Table 2. Stream order of the Sinú river basin.

Order	Stream Number	Number of Sub-Basins	Sub Sub-Basins (km^2)	Total Streams Length (km)	Average Stream Length (km)	Stream Length Ratio (R_l)	Bifurcation Ratio (R_b)
1	166	22	4290.49	680.47	4.10		
						8.30	7.22
2	23	21	4952.43	782.76	34.03		
						1.05	3.83
3	6	8	1183.88	214.21	35.7		
						1.19	2.00
4	3	6	832.1	127.39	42.46		
						11.41	3
5	1	8	2712.76	484.66	484.66		
	199	65	13,971.7	2289.50		5.50	7.60

Additionally, during the analysis it was found that the number of streams progressively decreased as the order increased (Table 2). Out of the total of 199 streams found in the basin, 83% (166) are of order 1, 11.5% (23) of order 2, 3% (6) are of order 3, 1.5% (3) of order 4, and the main current is order of 5.

The values of the total and average length of the channels for each order can be seen in Table 2, where it is evident that 54.4% (1245.5 km) of the streams in the basin are of orders 1 and 2. These results corroborate the law of Horton stream lengths [4], which states that: "the average stream lengths of each order in a drainage basin tend to approximate a direct geometric series in which the first term is the average length of streams of order 1". This is because, when making the graph of the logarithm of the length means against the stream order (Figure 6), it is evident that the points tend to a straight line. In this manner, it can be stated that the drainage network of the Sinú river complies with what was established by Horton [4] in terms of the behavior of the quantity and average length of the streams of a certain order. This implies that in the study area, the streams of order 1, having a lower mean length than the others, prevail in the areas where the steepest slopes are found.

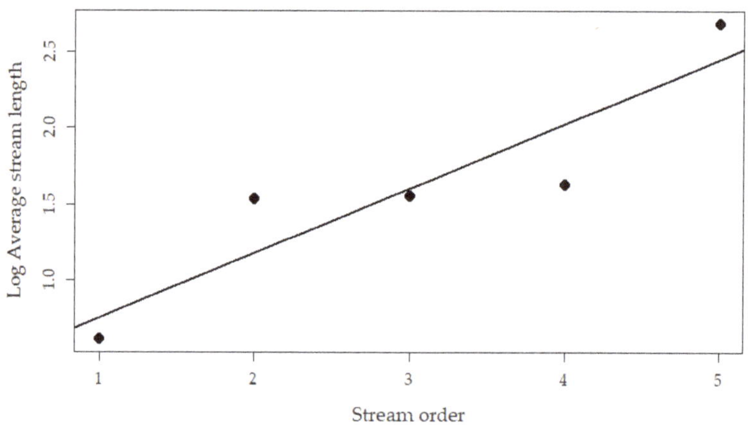

Figure 6. Horton's law of stream length for Sinú river basin, Colombia.

The obtained values of R_l for the different stream orders in the study area can be seen in Table 2, where it is evident that they range from 1.05 to 11.41.

In addition, the highest R_b was 7.22 and was calculated between the first and second order streams (Table 2), indicating that the longest overland flow lengths and the highest flows occur in these channels. The weighted average of the bifurcation ratio R_{bm} allows a representative value to be determined for the entire study area, estimated by multiplying the bifurcation ratio for each successive pair of orders times the total number of flows involved in the ratio and taking the mean of the sum of these values [20,21]. For the study area, the R_{bm} was 7.6, which is higher than the arithmetic average of the estimated R_b (4.01).

When estimating drainage density from the results of the terrain modeling, it was found that it ranges between 0.06 km/km^2 and 0.24 km/km^2 for the 65 sub-basins obtained, 0.16 km/km^2 being the value obtained when analyzing the basin as a single area.

That said, it is necessary to clarify that this parameter is sensitive to the used threshold, so its interpretation must be carried out considering the other basin morphometric features. As a result, it was decided to corroborate the values obtained for these parameters using the cartographic plates at a scale of 1: 25,000 from the IGAC [44]. In this sense, when estimating the drainage density of the Sinú river basin from the available cartography, it was found that Dd is 1.59 km/km^2, i.e., the drainage density obtained through modeling is not consistent with the one from the cartography. This is due, as explained above, to the resolution of the DEM and the threshold used in the analysis, since these parameters are sensitive to scale, as mentioned by Londoño [53] in his research. Even though these values are higher than those obtained from the terrain modeling, according to what is mentioned by the CVS [30], it can be affirmed that the Sinú river basin shows a low drainage density for the magnitudes of flows found.

On the other hand, the L_{OF} provides information on the hydrological response and the basin topography. In the case of the study area, it was found that the L_{OF} has a value of 0.31 km (Table 3), while in the sub-basins it ranges between 0.14 km and 29 km, reflecting that there is a significant variation in the basin relief features such as elevations and slope (See Figures 7 and 8). Correspondingly, Figure 7 shows that the upper Sinú is characterized by elevations higher than 150 masl, classified as hillside, mountain range, and ridges, whereas in the lower and middle Sinú there is a predominance of elevations below 350 masl (lowland and hillside).

Table 3. Morphometric Parameters: Drainage network of Sinú River Watershed.

Morphometric Parameter	Results
Drainage network	
Stream order	5.00
Stream length (L) in km	243.80
Stream length (L_u)	-
Number of streams (N_u)	-
Stream axial length (L_a) in km	184.90
Drainage density (D_d) km/km^2	1.59
Length of overland flow (L_{OF}) in km	0.31
Constant of channel maintenance (C_m) in km/km^2	0.63
Stream frequency (F) in channels/km^2	1.82
Average stream length (L_m) in km	-
Sinuosity (S)	1.32
Bifurcation ratio (R_b)	-
Mean bifurcation ratio (R_{bm})	2.88
Stream length ratio (R_l)	-

The C_m depends on factors such as relief and lithology, which in the case of the study area ranges between 0.28 km/km^2 and 59 km/km^2 for the sub-basins, with prevailing values of less than 10 km/km^2. The value obtained for the entire basin (total area) was 0.63 km/km^2, which is below the average of the values obtained in the sub-basins. This confirms what was previously described, where it was established that there are areas in the basin more susceptible to erosion due to the high slopes.

When analyzing the stream frequency (F) for the case under study, it was found that in the sub-basins of the Sinú river, values lower than 4 channels/km^2 prevail, with the arithmetic mean being 1.68 channels/km^2 and the value for the entire basin being 1.82 channels/km^2 (Table 3).

In the study area, it was found that the stream sinuosity of the Sinú river is 1.32, indicating that the stream is sinuous according to what was found by Ahmed [52]. Additionally, it was observed that 87.7% (57) of the main streams of the sub-basins obtained in the modeling have a sinuosity lower than 1.25, indicating that their alignments tend to be straight and that the runoff speed is relatively high compared to the rest of the sub-basins.

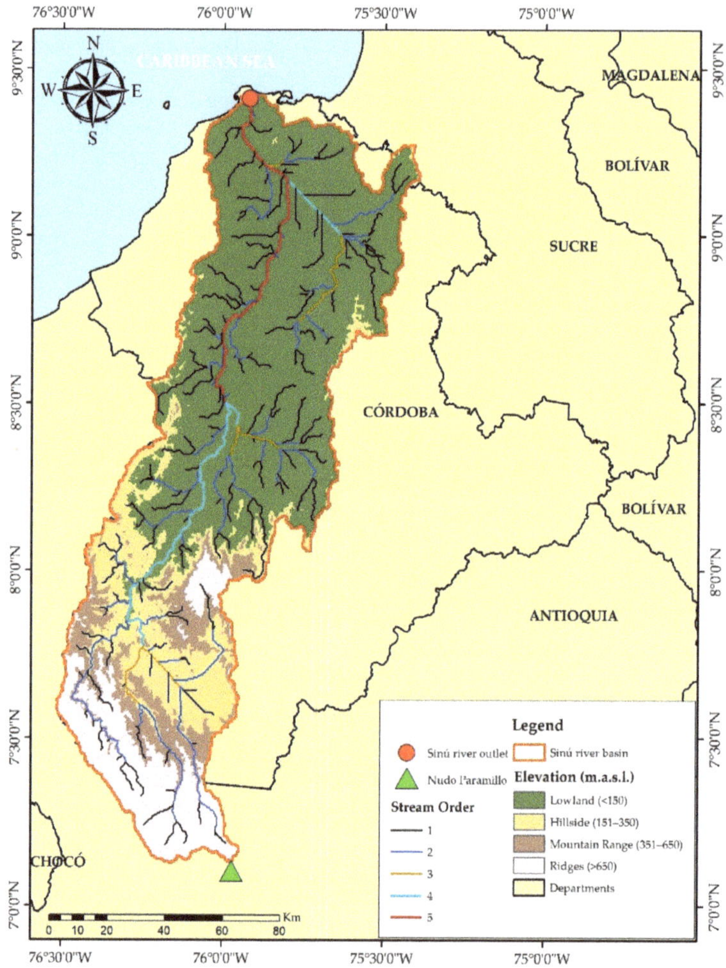

Figure 7. Elevation map of the Sinú river basin, Colombia.

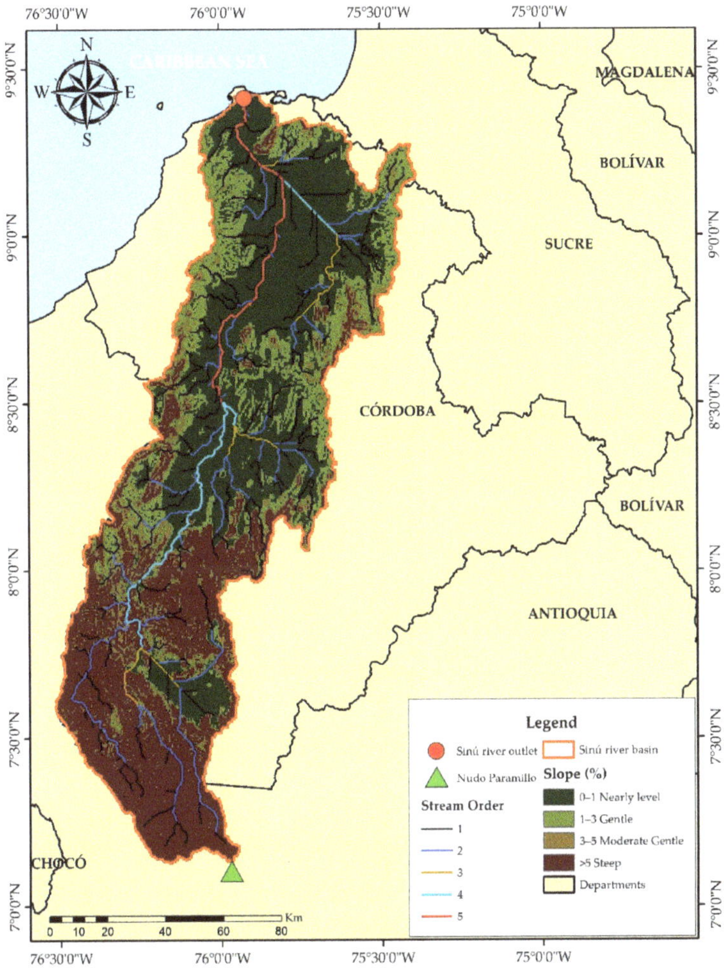

Figure 8. Slope distribution map of the Sinú river basin, Colombia.

4.2. Basin Geometry

The calculated geometry parameters can be found in Table 4. In the case of this research, the F_f for the sub-basins ranged between 0.186 and 0.721, with 90.8% (59 sub-basins) being less than 0.57, while the entire basin has an F_f of 0.22, indicating that it has an elongated shape. Similarly, the R_c values obtained for the 65 sub-basins vary between 0.17 and 0.49, 0.17 being for the complete basin. Additionally, the study area has an R_e of 0.53, making it an elongated basin, while the sub-basins have values that oscillate between 0.49 and 0.96, where 81.5% (53 sub- basins) show an R_e of less than 0.79. The estimated value of K_c was 2.43, while in the sub-basins values were between 1.42 and 2.40. This way, it becomes evident that all the sub-basins have elongated shapes according to this parameter.

Table 4. Morphometric Parameters: Geometry of Sinú River Watershed.

Morphometric Parameter	Results
Basin geometry	
Basin area (A_d) in km^2	13,971.70
Basin perimeter (P) in km	1025.90
Basin length (L_c) in km	252.40
Form factor $\left(F_f\right)$	0.22
Circularity ratio (R_c)	0.17
Elongation ratio (R_e)	0.53
Compactness coefficient (K_c)	2.43

4.3. Basin Relief

The relief of the basin was analyzed through GIS, the basin average slope was estimated as the average of the values of the slope grid generated from the elevation map (Figure 7), obtaining a value of 5.5%. Similarly, the map of the slopes of the study area was obtained (Figure 8) in which it can be observed that the upper Sinú basin is where the highest slopes occur, which may be responsible for the highest flows, while in the lower and middle Sinú basins, they decrease, forming extensive plains.

Additionally, H is 3493 m above sea level and 0 m above sea level, respectively, so the relief of the basin is 3493 m (Table 5).

Table 5. Morphometric Parameters: Relief of Sinú River Watershed.

Morphometric Parameter	Results
Basin relief	
Minimum basin height (H_{min}) in masl	3493
Maximum basin height (H_{max}) in masl	0.00
Mean basin slope (S_c) in percentage	5.50
Basin relief (H)	3493
Relief ratio (F_H)	13.83

Similarly, Schumm [20] defined the relief factor (F_H) as the ratio between the basin relief (H) and its maximum length (L_c). For the Sinú river basin, the estimated F_H was 13.83.

On the other hand, Figure 9 shows the hypsometric curve of the Sinú river basin, where the highest slopes occur at the maximum elevations, which translates into areas with steep and mountainous terrain. Similarly, it is evident that most of the terrain has gentle slopes and uniform heights, which, according to Richardson et al. [54], suggests the existence of plains within the basin.

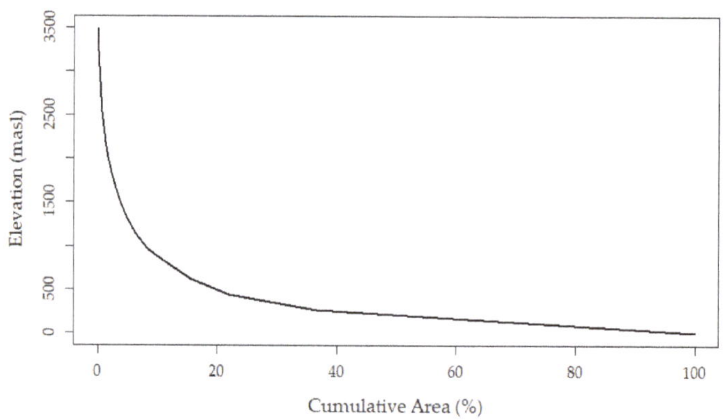

Figure 9. Hypsometric curve of the Sinú river basin.

4.4. Rainwater Harvesting Potential Index (RWHPI)

The obtained results for RWHPI were classified using the quantile technique. Thus, for the Sinú river basin, the potential rainwater harvesting zones are classified as (a) 'Very high' (RWHPI = 0.266–0.334), (b) 'High' (RWHPI = 0.265–0.259), (c) 'Moderate' (RWHPI = 0.250–0.258), and (d) 'Poor' (RWHPI = 0.189–0.249). Figure 10 shows the rainwater harvesting potential zones for the Sinú river basin.

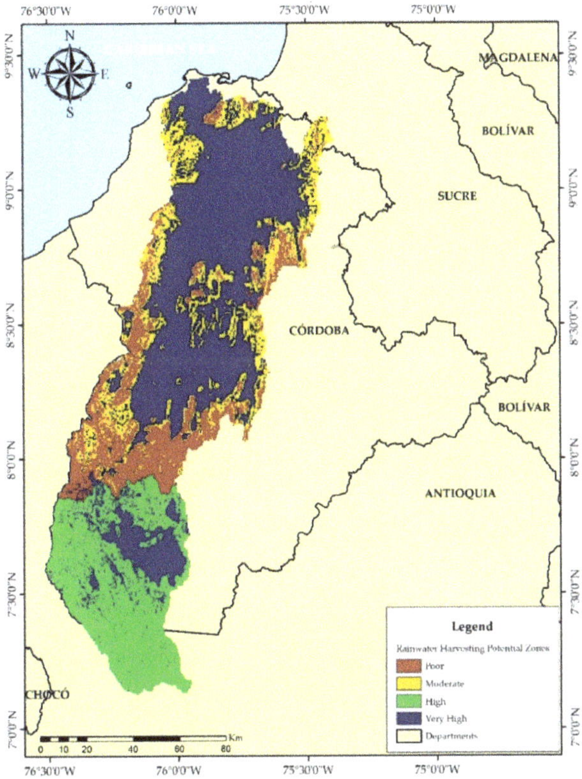

Figure 10. Map of the Sinú River Basin depicting Rainwater Harvesting Potential Zones.

5. Discussion

5.1. Drainage Network

The results of the morphometric parameters of the drainage network allow us to understand the basin hydrological behavior. First, the number of channels of order u is relevant, given that, if two basins are compared, the number of channels is an important factor, because if there is a greater number of streams, it can be inferred that there is better drainage and thus less permeability and infiltration, as mentioned by Rai et al. [6].

In accordance with the research of Ahmed et al. [55] and Rai et al. [6], the obtained results from the analysis of drainage density indicate that it is a basin with poor drainage where surface runoff cannot be quickly evacuated and, therefore, it is highly susceptible to floods. This parameter is also related to aspects such as soil erosion and runoff, since flow is directly proportional to drainage density, which translates into rapid runoff that implies, in turn, a flow of greater magnitude. Furthermore, it has an inverse relationship with infiltration, since high infiltration tends to inhibit the development of longer drains, i.e., as mentioned by Strahler [21], Bhagwat et al. [47], and Gajbhiye et al. [10,13], the lowest drainage densities correspond to regions with permeable soil types, dense vegetation, and low relief, while high drainage density prevails in regions with impermeable soils, sparse vegetation, and high relief.

Figure 11 shows the histogram built to analyze the drainage density behavior in the 65 sub-basins, finding that values less than 1.6 km/km^2 predominate in the basin and that there are some sub-basins where the value of Dd is less than unity. These results indicate that there are areas within the basin, such as the middle and lower Sinú areas, characterized by permeable soils and floodplains, while the upper Sinú is a region where impermeable soils with scarce vegetation that favor erosive processes and rapid flows prevail. It should be noted that for the variation analysis of the morphometric parameters in the sub-basins of the study area, histograms were constructed for each.

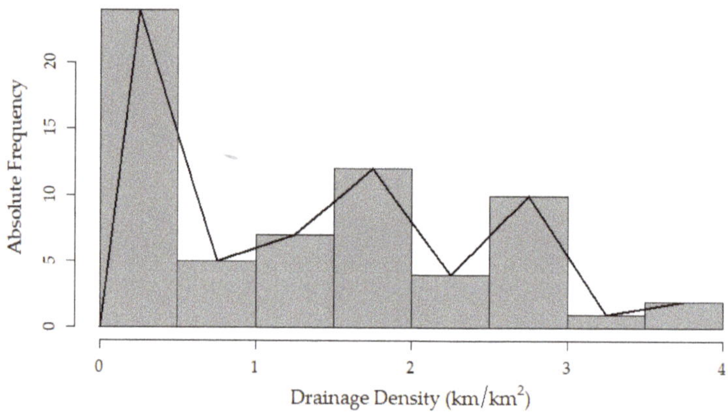

Figure 11. Histogram for drainage density analysis in the Sinú river basin.

Similarly, according to the research of Gayen et al. [56], Abboud and Nofal [57], Rai et al. [23], and Ameri et al. [7], it can be stated that the areas where the length of the overland flow is greater, are more susceptible to erosion and also reflect the presence of steep slopes. While the regions where this parameter is lower, are areas where less rain is required to generate significant flows. With these results, the heterogeneity of the geomorphological features of the basin is confirmed and it is inferred that the areas with the greatest susceptibility to erosion correspond to the upper Sinú, while the middle and lower Sinú are susceptible to flooding due to the fact that less precipitation is required to generate surface runoff.

Likewise, the variation in the results of the constant of channel maintenance corroborates the previous statement, due to the fact that values that exceed the mean value allow us to infer that a larger area is required to produce surface runoff towards the drainage network, which, as reported by Bhagwat et al. [47], favors losses by evaporation and infiltration. Otherwise, it happens with values below the average, since these favor rapid runoff and minimize the likelihood of these losses to occur. From the above, it follows that the high values correspond to the lower and middle Sinú where the terrain is relatively flat, while the low values are linked to the upper Sinú where slopes are higher.

According to what was found by Ozdemir and Bird [58] and Ameri et al. [7], the stream frequency is related to relief, vegetation cover, soil infiltration capacity, and susceptibility to erosion of a basin. Thus, it can be inferred that the areas where the highest drainage frequencies occur imply the presence of rocky surfaces, with low infiltration capacity and susceptibility to erosion. In the case of the Sinú river basin, values of F less than 1.68 channels/km^2 predominate in the middle and lower Sinú regions, where the low slope and the uniformity of the elevations inhibit the generation of water currents, while the upper Sinú shows higher values of F due to the high slopes, coinciding with what was found when analyzing D_d.

Furthermore, the bifurcation ratio between the first and second order streams is much higher than those obtained for the rest of the orders, so it is indicative that in the areas where streams of these orders predominate, runoff is faster than in the rest of the basin, and thus, said areas are more susceptible to flooding during precipitation events, as mentioned by Hajam et al. [59]. Additionally, according to the research by Magesh et al. [60], R_{bm} values range between 3 and 5 for basins where the influence of the geological structure on the drainage network is negligible, so it is inferred that the characteristics of the drainage network of the Sinú river basin are strictly linked to geology and relief. Similarly, as reported by Arulbalaji and Gurugnanam [61], R_b variation, which in this case ranges from 2 to 7.22, indicates that there is geological heterogeneity in the basin, so that the lowest R_b values indicate high permeability while higher R_b imply low permeability.

Moreover, there was a significant variation in the values obtained for length ratio, which, according to Singh et al. [59], is indicative of the fact that there is an important change in the hydrological characteristics of the underlying rock surfaces over the areas of consecutive flow orders. In other words, the behavior of the Rl indicates that there is geological heterogeneity in the basin and also, according to Bali et al. [62], that there are considerable differences in the topography and slope of the different sub-basins in the study area.

It can be stated that the obtained results for the parameters of the drainage network are consistent with the available information for the study area. The estimated parameters for the middle and lower Sinu indicate that these are areas characterized by low relief, slope, and susceptibility to floods such as the one occurred in 2010 [37] in accordance with the topography, given that, as mentioned in the study area description, lowland landscapes prevail in the region. Moreover, the prevalence of alluvial deposits (Q-al) confirms the erosion and sedimentation processes inferred from the estimated parameters, since these deposits are the result of erosion and deposition of materials associated with the dynamics of rivers, in times of both high flow and dry periods, and which are typical of meandering rivers such as the Sinú river [63].

5.2. Basin Geometry

The geometric parameters of the basin show that it has an elongated shape. In the investigations of Javed et al. [64], Iqbal et al. [1], Patel et al. [65], and Nanda et al. [66], it is established that the basins with a high F_f (circular) have a maximum flow of greater magnitude and shorter duration, while the basins with a low F_f (elongated) have a maximum flow of lesser magnitude with a longer duration. This means that there is an inversely proportional relationship between concentration time and F_f. The obtained results for this parameter indicate that the sub-basins are characterized by having elongated shapes, and thus, high concentration times compared to circular basins that have the same area. Nonetheless, the variation in the obtained values implies that there are areas within the basin that are more susceptible to flooding, such as the middle and lower Sinú basins.

As for the obtained results for the circular ratio, it is established that the basin, in addition to having an elongated shape, shows variation in the R_c values, which, considering that it is influenced by multiple characteristics such as length and frequency of the channels, geology, climate, and slope, reflects the heterogeneity of the physical features of the area of study, as mentioned by Ameri et al. [7].

The elongation ratio allows to study a basin hydrological response, since for a given precipitation, the less elongated basins will have a higher peak discharge and higher flow velocities [67]. Consequently, according to the results of Magesh [60], the results obtained indicate the presence of steep slopes in the study area, as observed in the upper Sinú area (Figure 8).

Additionally, considering that the regions with circular shapes require less time to produce a maximum flow, and that low values of compactness coefficient imply greater susceptibility to erosion, the variation in the results confirms that there is an important variability in the relief of the study area [7,64].

5.3. Basin Relief

According to Reddy et al. [68] and Sreedevi et al. [50], high values of H, such as the one obtained, indicate that there are conditions of low infiltration and high surface runoff in the Sinú river basin. Additionally, the obtained the results for relief factor reflects the presence of steep slopes and consequently, the high intensity of erosion processes that occur on the slopes and the sediment load downstream according to the work of Thomas et al. [69].

The results shown in the slope maps allows us to infer that the areas where the slope is low is where the mirrors or surface water bodies of the basin are located (lower and middle Sinú), while the highest percentage of the flow of the main currents comes from the southern area where the elevations and slopes are higher, favoring high drainage density and channel frequency [65,68].

Similarly, the hypsometric curve reflects the geomorphology of the area, with the upper Sinú region being the area characterized by the presence of steep mountains and the middle and lower Sinú, the areas of plains. According to Strahler [21,70], through features such as the area under the hypsometric curve, slope, inflection point, and sinuosity, information about the geology of the basin can be inferred, given that we generally have the same family of curves for a specific geological and climatic combination. In the case of the Sinú river basin, it is inferred that having an upwardly concave hypsometric curve, active fluvial and alluvial sedimentary processes prevail, the reason for which the material has been eroded and deposited in the lower parts of the basin. Moreover, these types of curves indicate that the basin area is concentrated in the lower parts, which implies the presence of deep boxed valleys characteristic of foothills and savannahs.

5.4. Rainwater Harvesting Potential Zones

According to the analysis conducted, more than 70% of the basin area has characteristics suitable for rainwater harvesting. Particularly, it was identified that the areas with a 'very high' RWHPI correspond mainly to the middle and lower Sinú regions. Most of the upper Sinú is characterized by a 'High' RWHPI; however, it is also evident that most of the zones with 'Poor' RWHPI are concentrated in the upper Sinú, specifically in the region close to the middle Sinú. On the other hand, the zones with 'moderate' RWHPI are scattered in small patches throughout the middle and lower Sinú. Finally, it is important to note that the areas with a 'very high' RWHPI coincide with the presence of water bodies such as reservoirs and swamps.

6. Conclusions

The morphometric analysis carried out from the terrain modeling in the Sinú river basin through GIS, the existing cartography review, and the estimation of physiographic parameters made it possible to establish the following conclusions:

The obtained results for morphometric parameters, such as hypsometric curve of the basin, D_d, F, stream order, and L_{OF} corroborated that the lower and middle Sinu sub-basins are susceptible to floods as evidenced by the flood events that have historically occurred in the study area. Concomitantly, the results indicate that the upper Sinú sub-basin is susceptible to erosive processes, which was confirmed by the review of the available information on the geology and geomorphology of the study area, establishing that the basin is framed within two large geo-structures: the mountain range or Cordillera (upper Sinú) and the sedimentation mega-basin (middle and lower Sinú). Thus, it is established that the prevalence of alluvial deposits in the basin is due to the material resulting from erosive processes in the upper Sinú that ends up being transported to the rest of the basin. Therefore, the results validate the basin geological heterogeneity.

The results allow us to affirm that the study area shows favorable characteristics for implementing water catchment techniques that contribute to the integrated management of the basin's water resources. However, so far, adequate management has not been carried out, leaving all this water to be lost when it reaches the ocean, even though there are problems in most of the lower and middle Sinú basins, especially during the dry season, where dead animals and the lack of water availability for human consumption are evident in some sectors. Considering the characteristics of the study area, techniques such as furrows and ridges built along contour lines and dams would be strategies that would make it possible to take advantage of and store water resources during the winter season to be used during the dry season. From the review of the results and the literature, it can be inferred that, under normal conditions, there is a directly proportional relationship between drainage order and flow, since stream order is directly proportional to watershed size, channel dimensions, and stream flow. Nonetheless, sometimes smaller areas have greater flow due to rainfall behavior or other factors. Runoff is a complex variable that depends on multiple conditions. In the specific case of the Sinú river basin, it was possible to establish that, due to flow regulation by the Urrá hydroelectric plant (upper Sinú), flows remain stable throughout the basin.

Finally, these findings will help for the further modeling of an integrated watershed for sustainable hydrological and hydrograph models that, besides, will help understand the relationship between hydrological variables and geomorphological parameters as guidance and decision-making instruments for the competent authorities to establish actions for the sustainable development of the watershed, flood control, water supply planning, water budgeting, and disaster mitigation within the Sinú river basin.

Author Contributions: Conceptualization, Juan Pablo Medrano-Barboza, Alvaro Alberto López-Lambraño, Alvaro López-Ramos and Luisa Martínez-Acosta; methodology, Juan Pablo Medrano-Barboza, Alvaro Alberto López-Lambraño, John Freddy Remolina López and Luisa Martínez-Acosta; software, Juan Pablo Medrano-Barboza and Guillermo J. Acuña; validation, Juan Pablo Medrano-Barboza, Alvaro Alberto López-Lambraño, Alvaro López-Ramos and Luisa Martínez-Acosta; formal analysis, Juan Pablo Medrano-Barboza, Alvaro Alberto López-Lambraño, Alvaro López-Ramos, John Freddy Remolina López, Guillermo J. Acuña and Luisa Martínez-Acosta; investigation, Juan Pablo Medrano-Barboza, Alvaro Alberto López-Lambraño, Alvaro López-Ramos, John Freddy Remolina López, Guillermo J. Acuña and Luisa Martínez-Acosta; writing—original draft preparation, Juan Pablo Medrano-Barboza, Alvaro Alberto López-Lambraño, Alvaro López-Ramos, John Freddy Remolina López, Guillermo J. Acuña and Luisa Martínez-Acosta; writing—review and editing, Juan Pablo Medrano-Barboza, Alvaro Alberto López-Lambraño and Luisa Martínez-Acosta. All authors have read and agreed to the published version of the manuscript.

Funding: This research was funded by Hidrus S.A. de C.V., Grupo Hidrus S.A.S., Universidad Pontificia Bolivariana Seccional Montería, grant number 267-07/22-G012, and Universidad Autónoma de Baja California (https://ror.org/05xwcq167) and the APC was funded by Universidad Pontificia Bolivariana (https://ror.org/02dxm8k93).

Institutional Review Board Statement: Not applicable.

Informed Consent Statement: Not applicable.

Data Availability Statement: The data presented in this study are openly available in the following sources: (a) The digital elevation model used is freely available on the U.S. Geological Survey website [39]; (b) The 1:25,000 scale cartographic plates of the Sinú river basin are freely available on the Agustín Codazzi Geographical Institute website [44].

Conflicts of Interest: The authors declare no conflict of interest.

References

1. Iqbal, M.; Sajjad, H.; Bhat, F.A. Morphometric Analysis of Shaliganga Sub Catchment, Kashmir Valley, India Using Geographical Information System. *Int. J. Eng. Trends Technol.* **2013**, *4*, 10–21.
2. Sujatha, E.R.; Selvakumar, R.; Rajasimman, B. Watershed Prioritization of Palar Sub-Watershed Based on the Morphometric and Land Use Analysis. *J. Mt. Sci.* **2014**, *11*, 906–916. [CrossRef]
3. Chitra, C.; Alaguraja, P.; Ganeshkumari, K.; Yuvaraj, D.; Manivel, M. Watershed Characteristics of Kundah Sub Basin Using Remote Sensing and GIS Techniques. *Int. J. Geomat. Geosci.* **2011**, *2*, 311–335.
4. Horton, R.E. Erosional Development of Streams and Their Drainage Basins, Hydrophyrical Approach to Quantitative Morphology. *Bull. Geol. Soc. Am.* **1945**, *40*, 275–370. [CrossRef]

5. Nneka, C.; Philip, A.; Nnadozie, O.P.; Ayogu, O. Morphometric Analysis and the Validity of Hortonian Postulations in Anambra Drainage Basin, Nigeria. *Spat. Inf. Res.* **2019**, *27*, 505–520. [CrossRef]
6. Rai, P.K.; Mishra, V.N.; Mohan, K. A Study of Morphometric Evaluation of the Son Basin, India Using Geospatial Approach. *Remote Sens. Appl. Soc. Environ.* **2017**, *7*, 9–20. [CrossRef]
7. Ameri, A.A.; Pourghasemi, H.R.; Cerda, A. Erodibility Prioritization of Sub-Watersheds Using Morphometric Parameters Analysis and Its Mapping: A Comparison among TOPSIS, VIKOR, SAW, and CF Multi-Criteria Decision Making Models. *Sci. Total Environ.* **2018**, *613–614*, 1385–1400. [CrossRef]
8. Rai, P.K.; Singh, P.; Mishra, V.N.; Singh, A.; Sajan, B.; Shahi, A.P. Geospatial Approach for Quantitative Drainage Morphometric Analysis of Varuna River Basin, India. *J. Landsc. Ecol.* **2019**, *12*, 1–25. [CrossRef]
9. Adnan, M.S.G.; Dewan, A.; Zannat, K.E.; Abdullah, A.Y.M. The Use of Watershed Geomorphic Data in Flash Flood Susceptibility Zoning: A Case Study of the Karnaphuli and Sangu River Basins of Bangladesh. *Nat. Hazards* **2019**, *99*, 425–428. [CrossRef]
10. Gajbhiye, S.; Mishra, S.K.; Pandey, A. Prioritizing Erosion-Prone Area through Morphometric Analysis: An RS and GIS Perspective. *Appl. Water Sci.* **2014**, *4*, 51–61. [CrossRef]
11. Malik, A.; Kumar, A.; Kandpal, H. Morphometric Analysis and Prioritization of Sub-Watersheds in a Hilly Watershed Using Weighted Sum Approach. *Arab. J. Geosci.* **2019**, *12*, 118. [CrossRef]
12. Nitheshnirmal, S.; Thilagaraj, P.; Rahaman, S.A.; Jegankumar, R. Erosion Risk Assessment through Morphometric Indices for Prioritisation of Arjuna Watershed Using ALOS-PALSAR DEM. *Model. Earth Syst. Environ.* **2019**, *5*, 907–924. [CrossRef]
13. Rahmati, O.; Samadi, M.; Shahabi, H.; Azareh, A.; Rafiei-Sardooi, E.; Alilou, H.; Melesse, A.M.; Pradhan, B.; Chapi, K.; Shirzadi, A. Geoscience Frontiers SWPT: An Automated GIS-Based Tool for Prioritization of Sub-Watersheds Based on Morphometric and Topo-Hydrological Factors. *Geosci. Front.* **2019**, *10*, 2167–2175. [CrossRef]
14. Singh, L.K.; Jha, M.K.; Chowdary, V.M. Multi-Criteria Analysis and GIS Modeling for Identifying Prospective Water Harvesting and Artificial Recharge Sites for Sustainable Water Supply. *J. Clean. Prod.* **2017**, *142*, 1436–1456. [CrossRef]
15. Ezzeldin, M.; Konstantinovich, S.E.; Igorevich, G.I. Determining the Suitability of Rainwater Harvesting for the Achievement of Sustainable Development Goals in Wadi Watir, Egypt Using GIS Techniques. *J. Environ. Manag.* **2022**, *313*, 114990. [CrossRef]
16. Jena, S.K.; Tiwari, K.N. Modeling Synthetic Unit Hydrograph Parameters with Geomorphologic Parameters of Watersheds. *J. Hydrol.* **2006**, *319*, 1–14. [CrossRef]
17. Viramontes-Olivas, O.A.; Escoboza-García, L.F.; Pinedo-Álvarez, C.; Pinedo-Álvarez, A.; Reyes-Gómez, V.M.; Román-Calleros, J.A.; Pérez-Márquez, A. Morfometria de La Cuenca Del Río San Pedro, Conchos, Chihuahua. *Tecnociencia* **2007**, *I*, 21–31.
18. Escalante, C.A.; Reyes, L. *Tecnicas Estadisticas En Hidrología*; Facultad de Ingeniería, Universidad Autónoma de México: Mexico City, Mexico, 2002.
19. Horton, R.E. Drainage-Basin Characteristics. *Eos Trans. Am. Geophys. Union* **1932**, *13*, 350–361. [CrossRef]
20. Schumm, S. Evolution of Drainage Systems and Slopes in Badland at Peth Amboy, New Jersey. *Geol. Soc. Am. Bull.* **1956**, *67*, 597–646. [CrossRef]
21. Strahler, A.N. Quantitative Classification of Watershed Geomorphology. *Trans. Am. Geophys. Union* **1957**, *38*, 913–920. [CrossRef]
22. Shreve, R.L. Statistical Law of Stream Numbers. *J. Geol.* **1966**, *74*, 17–37. [CrossRef]
23. Rai, P.K.; Chaubey, P.K.; Mohan, K.; Singh, P. Geoinformatics for Assessing the Inferences of Quantitative Drainage Morphometry of the Narmada Basin in India. *Appl. Geomat.* **2017**, *9*, 167–189. [CrossRef]
24. Alqahtani, F.; Qaddah, A.A. GIS Digital Mapping of Flood Hazard in Jeddah-Makkah Region from Morphometric Analysis. *Arab. J. Geosci.* **2019**, *12*, 199. [CrossRef]
25. Mangan, P.; Haq, M.A.; Baral, P. Morphometric Analysis of Watershed Using Remote Sensing and GIS—A Case Study of Nanganji River Basin in Tamil Nadu, India. *Arab. J. Geosci.* **2019**, *12*, 202. [CrossRef]
26. Panda, B.; Venkatesh, M.; Kumar, B. A GIS-Based Approach in Drainage and Morphometric Analysis of Ken River Basin and Sub-Basins, Central India. *J. Geol. Soc. India* **2019**, *93*, 75–84. [CrossRef]
27. Prakash, K.; Rawat, D.; Singh, S.; Chaubey, K.; Kanhaiya, S.; Mohanty, T. Morphometric Analysis Using SRTM and GIS in Synergy with Depiction: A Case Study of the Karmanasa River Basin, North Central India. *Appl. Water Sci.* **2019**, *9*, 1–10. [CrossRef]
28. Shivhare, V.; Gupta, C.; Mallick, J.; Singh, C.K. Geospatial Modelling for Sub-Watershed Prioritization in Western Himalayan Basin Using Morphometric Parameters. *Nat. Hazards* **2022**, *110*, 545–561. [CrossRef]
29. Shrivatra, J.R.; Manjare, B.S.; Paunikar, S.K. A GIS-Based Assessment in Drainage Morphometry of WRJ-1 Watershed in Hard Rock Terrain of Narkhed Taluka, Maharashtra, Central India. *Remote Sens. Appl. Soc. Environ.* **2021**, *22*, 100467. [CrossRef]
30. Corporación Autónoma Regional de los Valles del Sinú y San Jorge (CVS). *Fases de Prospección y Formulación Del Plan de Ordenamiento y Manejo Integral de La Cuenca Hidrográfica Del Río Sinú (POMCA-RS)*; CVS: Montería, Colombia, 2006.
31. Valbuena, D.L. *Geomorfología y Condiciones Hidráulicas Del Sistema Fluvial Del Río Sinú. Integración Multiescalar. 1945–1999–2016*; Universidad Nacional de Colombia: Bogotá, Colombia, 2017.
32. Instituto Geográfico Agustín Codazzi (Ed.) *Estudio General de Suelos y Zonificación de Tierras, Departamento de Córdoba*, 1st ed.; Imprenta Nacional de Colombia: Bogotá, Colombia, 2009.
33. Remane, J. *Explanatory Note to the International Stratigraphic Chart*; International Union of Geological Sciences: Trondheim, Norway, 2000.

34. Gómez, J.; Nivia, Á.; Montes, N.E.; Almanza, M.F.; Alcárcel, F.A.; Madrid, C.A. Notas Explicativas: Mapa Geológico de Colombia. In *Compilando la Geología de Colombia: Una Visión a 2015*; Gómez, J., Almanza, M.F., Eds.; Servicio Geológico Colombiano: Bogotá, Colombia, 2015; Volume 33, pp. 9–33.
35. BID; CEPAL; DNP. *Valoración de Daños y Pérdidas. Ola Invernal En Colombia 2010–2011*; Misión BID—Cepal: Bogotá, Colombia, 2012; 240p.
36. IDEAM. *Estudio Nacional Del Agua 2018*; Instituto de Hidrología, Meteorología y Estudios Ambientales: Bogotá, Colombia, 2019; ISBN 9789585489127.
37. CVS. *Puntos Críticos Cuenca Sinú*; CVS: Montería, Colombia, 2019.
38. Felicísimo, A.M. *Modelos Digitales Del Terreno. Introducción y Aplicaciones En Las Ciencias Ambientales*; Pentalfa Ediciones: Oviedo, Spain, 1994.
39. U.S. Geological Survey (USGS). HydroSHEDS 15 Arc-Second DEMs for South America. Available online: http://hydrosheds.cr.usgs.gov (accessed on 28 May 2018).
40. Rineer, J.; Bruhn, M.; Miralles-Wilhelm, F. *Nota Técnica Base de Datos de Hidrología Analítica Para América Latina y El Caribe*; Banco Interamericano de Desarrollo (BID): Washington, DC, USA, 2014.
41. Jenson, K.; Domingue, O. Extracting Topographic Structure from Digital Elevation Data for Geographic Information System Analysis. *Photogramm. Eng. Remote Sens.* **1988**, *54*, 1593–1600.
42. Mueller, J.E. An Introduction to the Hydraulic and Topographic Sinuosity Indexes. *Ann. Assoc. Am. Geogr.* **1968**, *58*, 371–385. [CrossRef]
43. Martínez-Acosta, L.; López-Lambraño, A.A.; López-Ramos, A. Design Criteria for Planning the Agricultural Rainwater Harvesting Systems: A Review. *Appl. Sci.* **2019**, *9*, 5298. [CrossRef]
44. Instituto Geográfico Agustín Codazzi Datos Abiertos Cartografía y Geografía-Cartografía Base Escala 1:25,000. Available online: https://geoportal.igac.gov.co/contenido/datos-abiertos-cartografia-y-geografia (accessed on 27 May 2018).
45. Miller, V.C. *A Quantitative Geomorphologic Study of Drainage Basin Characteristics in the Clinch Mountain Area, Virginia and Tennessee*; Technical report; Columbia University: New York, NY, USA, 1953.
46. Magesh, N.S.; Jitheshlal, K.V.; Chandrasekar, N.; Jini, K.V. GIS Based Morphometric Evaluation of Chimmini and Mupily Watersheds, Parts of Western Ghats, Thrissur District, Kerala, India. *Earth Sci. Inform.* **2012**, *5*, 111–121. [CrossRef]
47. Bhagwat, T.N.; Shetty, A.; Hegde, V.S. Spatial Variation in Drainage Characteristics and Geomorphic Instantaneous Unit Hydrograph (GIUH); Implications for Watershed Management-A Case Study of the Varada River Basin, Northern Karnataka. *Catena* **2011**, *87*, 52–59. [CrossRef]
48. Pareta, K.; Pareta, U. Quantitative Morphometric Analysis of a Watershed of Yamuna Basin, India Using ASTER (DEM) Data and GIS. *Int. J. Geomat. Geosci.* **2011**, *2*, 248–269.
49. Chopra, R.; Dhiman, R.D.; Sharma, P.K. Morphometric Analysis of Sub-Watersheds in Gurdaspur District, Punjab Using Remote Sensing and GIS Techniques. *J. Indian Soc. Remote Sens.* **2005**, *33*, 531–539. [CrossRef]
50. Sreedevi, P.D.; Subrahmanyam, K.; Ahmed, S. The Significance of Morphometric Analysis for Obtaining Groundwater Potential Zones in a Structurally Controlled Terrain. *Environ. Geol.* **2005**, *47*, 412–420. [CrossRef]
51. Hajam, R.A.; Hamid, A.; Bhat, S. Application of Morphometric Analysis for Geo-Hydrological Studies Using Geo-Spatial Technology—A Case Study of Vishav Drainage Basin. *J. Waste Water Treat. Anal.* **2013**, *4*, 12. [CrossRef]
52. Ahmed, A.A.; Fawzi, A. Meandering and Bank Erosion of the River Nile and Its Environmental Impact on the Area between Sohag and El-Minia, Egypt. *Arab. J. Geosci.* **2011**, *4*, 1–11. [CrossRef]
53. Londoño Arango, C. *Cuencas Hidrográficas: Bases Conceptuales, Caracterización, Administración*; Universidad del Tolima: Ibagué, Colombia, 2001.
54. Richardson, J.C.; Hodgson, D.M.; Wilson, A.; Carrivick, J.L.; Lang, A. Testing the Applicability of Morphometric Characterisation in Discordant Catchments to Ancient Landscapes: A Case Study from Southern Africa. *Geomorphology* **2016**, *261*, 162–176. [CrossRef]
55. Ahmed, S.A.; Chandrashekarappa, K.N.; Raj, S.K.; Nischitha, V.; Kavitha, G. Evaluation of Morphometric Parameters Derived from ASTER and SRTM DEM—A Study on Bandihole Sub-Watershed Basin in Karnataka. *J. Indian Soc. Remote Sens.* **2010**, *38*, 227–238. [CrossRef]
56. Gayen, S.; Bhunia, G.S.; Shit, P.K. Morphometric Analysis of Kangshabati-Darkeswar Interfluves Area in West Bengal, India Using ASTER DEM and GIS Techniques. *J. Geol. Geosci.* **2013**, *2*, 1–10. [CrossRef]
57. Abboud, I.A.; Nofal, R.A. Morphometric Analysis of Wadi Khumal Basin, Western Coast of Saudi Arabia, Using Remote Sensing and GIS Techniques. *J. Afr. Earth Sci.* **2017**, *126*, 58–74. [CrossRef]
58. Ozdemir, H.; Bird, D. Evaluation of Morphometric Parameters of Drainage Networks Derived from Topographic Maps and DEM in Point of Floods. *Environ. Geol.* **2009**, *56*, 1405–1415. [CrossRef]
59. Singh, P.; Thakur, J.K.; Singh, U.C. Morphometric Analysis of Morar River Basin, Madhya Pradesh, India, Using Remote Sensing and GIS Techniques. *Environ. Earth Sci.* **2013**, *68*, 1967–1977. [CrossRef]
60. Magesh, N.S.; Jitheshlal, K.V.; Chandrasekar, N.; Jini, K.V. Geographical Information System-Based Morphometric Analysis of Bharathapuzha River Basin, Kerala, India. *Appl. Water Sci.* **2013**, *3*, 467–477. [CrossRef]
61. Arulbalaji, P.; Gurugnanam, B. Geospatial Tool-Based Morphometric Analysis Using SRTM Data in Sarabanga Watershed, Cauvery River, Salem District, Tamil Nadu, India. *Appl. Water Sci.* **2017**, *7*, 3875–3883. [CrossRef]

62. Bali, R.; Agarwal, K.K.; Ali, S.N.; Rastogi, S.K.; Krishna, K. Drainage Morphometry of Himalayan Glacio-Fluvial Basin, India: Hydrologic and Neotectonic Implications. *Environ. Earth Sci.* **2012**, *66*, 1163–1174. [CrossRef]
63. Servicio Geológico Colombiano; Universidad Nacional de Colombia. *Memoria Técnica Explicativa Del Mapa Geomorfológico Analítico Aplicado a La Zonificación de Amenaza Por Movimientos En Masa Escala 1:100.000 Plancha 89bis–Río Salaquí, Departamento de Chocó*; Universidad Nacional de Colombia: Bogotá, Colombia, 2015.
64. Javed, A.; Yousuf, M.; Rizwan, K. Prioritization of Sub-Watersheds Based on Morphometric and Land Use Analysis Using Remote Sensing and GIS Techniques. *J. Indian Soc. Remote Sens.* **2009**, *37*, 261–274. [CrossRef]
65. Patel, D.P.; Gajjar, C.A.; Srivastava, P.K. Prioritization of Malesari Mini-Watersheds through Morphometric Analysis: A Remote Sensing and GIS Perspective. *Environ. Earth Sci.* **2013**, *69*, 2643–2656. [CrossRef]
66. Nanda, A.M.; Ahmed, P.; Kanth, T.A.; Hajam, R.A. Morphometric Analysis of Sandran Drainage Basin (J & K) Using Geo—Spatial Technology. *Earth Sci. India* **2014**, *7*, 55–66.
67. Bajabaa, S.; Masoud, M.; Al-Amri, N. Flash Flood Hazard Mapping Based on Quantitative Hydrology, Geomorphology and GIS Techniques (Case Study of Wadi Al Lith, Saudi Arabia). *Arab. J. Geosci.* **2014**, *7*, 2469–2481. [CrossRef]
68. Reddy, G.P.O.; Maji, A.K.; Gajbhiye, K.S. Drainage Morphometry and Its Influence on Landform Characteristics in a Basaltic Terrain, Central India—A Remote Sensing and GIS Approach. *Int. J. Appl. Earth Obs. Geoinf.* **2004**, *6*, 1–16. [CrossRef]
69. Thomas, J.; Joseph, S.; Thrivikramji, K.P.; Abe, G.; Kannan, N. Morphometrical Analysis of Two Tropical Mountain River Basins of Contrasting Environmental Settings, the Southern Western Ghats, India. *Environ. Earth Sci.* **2012**, *66*, 2353–2366. [CrossRef]
70. Strahler, A.N. Hypsometric (Area-Altitude) Analysis of Erosional Topography. *Geol. Soc. Am. Bull.* **1952**, *63*, 1117–1142. [CrossRef]

International Journal of
Geo-Information

MDPI

Article

Vegetation Greenness Trend in Dry Seasons and Its Responses to Temperature and Precipitation in Mara River Basin, Africa

Wanyi Zhu [1,2], Zhenke Zhang [1,2,*], Shuhe Zhao [1,3], Xinya Guo [1,2], Priyanko Das [1,2], Shouming Feng [1,2] and Binglin Liu [1,2,4]

1 School of Geography and Ocean Science, Nanjing University, Nanjing 210023, China; mg20270036@smail.nju.edu.cn (W.Z.); zhaosh@nju.edu.cn (S.Z.); dg21270014@smail.nju.edu.cn (X.G.); dg1927501@smail.nju.edu.cn (P.D.); fengsm@smail.nju.edu.cn (S.F.); DG1827018@smail.nju.edu.cn (B.L.)
2 Institute of African Studies, Nanjing University, Nanjing 210023, China
3 Jiangsu Center for Collaborative Innovation in Geographical Information Resource Development and Application, Nanjing 210023, China
4 School of Geography and Planning, Nanning Normal University, Nanning 530001, China
* Correspondence: zhangzk@nju.edu.cn

Abstract: The Mara River Basin of Africa has a world-famous ecosystem with vast vegetation, which is home to many wild animals. However, the basin is experiencing vegetation degradation and bad climate change, which has caused conflicts between people and wild animals, especially in dry seasons. This paper studied the vegetation greenness (VG), vegetation greenness trends (VGT), and their responses to climate change in dry seasons in the Mara River Basin, Africa. Firstly, based on Google Earth Engine (GEE) platform and Sentinel-2 images, the vegetation distribution map of the Mara River Basin was drawn. Then dry seasons MODIS NDVI data (January to February and June to September) were used to analyze the VGT. Finally, a random forest regression algorithm was used to evaluate the response of VG and VGT to temperature and precipitation derived from ERA5 from 2000 to 2019 at a resolution of 250 m. The results showed that the VGT was fluctuating in dry seasons, and the spatial differentiation was obvious. The greenness increasing trends both upstream and downstream were significantly larger than that of in the midstream. The responses of VG to precipitation were almost twice larger than temperature, and the responses of VGT to temperature were about 1.5 times larger than precipitation. The climate change trend of rising temperature and falling precipitation will lead to the degradation of vegetation and the reduction of crop production. There will be a vegetation degradation crisis in dry seasons in the Mara River Basin in the future. Identifying the spatiotemporal changes of VGT in dry seasons will be helpful to understand the response of VG and VGT to climate change and could also provide technical support to cope with climate-change-related issues for the basin.

Keywords: Mara River Basin; dry seasons; vegetation greenness; random forest regression; spatiotemporal differentiation

Citation: Zhu, W.; Zhang, Z.; Zhao, S.; Guo, X.; Das, P.; Feng, S.; Liu, B. Vegetation Greenness Trend in Dry Seasons and Its Responses to Temperature and Precipitation in Mara River Basin, Africa. *ISPRS Int. J. Geo-Inf.* **2022**, *11*, 426. https://doi.org/10.3390/ijgi11080426

Academic Editors: Wolfgang Kainz and Walter Chen

Received: 12 June 2022
Accepted: 24 July 2022
Published: 28 July 2022

1. Introduction

The dynamic of vegetation greenness (VG) is important for understanding the effect of climate change and human encroachment on land surfaces [1]. In general, vegetation greenness can be explained as the growth of surface green vegetation. The East African region is classified as semi-arid land that is sensitive to human intrusion on vegetation and climate variation [2]. This region is one of the most important land ecosystems specified by the codominance of grasses, forests, and shrubs, and it covers 20% of the world's land [3]. However, vegetation is almost brown in the late dry seasons [4]. It was found that the deforestation rate increased from 0.22% (1900) to 0.39% (2000) in the East African region [5]. In recent years, frequent drought disasters have caused a continuous decline in vegetation greenness from Central Kenya to Central Tanzania during the El-Nino period [6]. Therefore, it is important to monitor vegetation greenness change to mitigate the natural disaster in the East African region.

Remote sensing is the most widely used tool for monitoring vegetation change, desertification, and agriculture activity on a global scale [7]. In recent decades, various satellite products at different temporal, spatial, and spectral resolutions have provided accurate datasets for the earth's surface [8]. The high-spatial-resolution satellite imagery, such as LANDSAT (30 m) and Sentinel-2 (10 m), provides timely and accurate monitoring of vegetation classes [9]. Sentinel-2 optical image especially gives an additional advantage in monitoring vegetation changes due to their red edge bands [10]. Moreover, the shortwave infrared and optical bands in the satellite sensor have the capability to construct a band-ratio such as the normalized difference vegetation index (NDVI) [11]. However, due to their short observation period, the Landsat and Sentinel-2A sensors cannot be used for near-real-time monitoring and long-term changes in VG [12]. The Moderate Resolution Imaging Spectroradiometer (MODIS) sensor has 36 spectral bands, including two vegetation indices (NDVI and EVI) for comparison of global vegetation change [13]. In addition, the MODIS NDVI product allows for the monitoring of vegetation dynamics for a long period of time, at a high spatial (250 m) and temporal (16-day composite) resolution [14].

The normalized difference vegetation index (NDVI) has been proved to be the best indicator of vegetation greenness [15,16]. Most of the previous studies used MODIS NDVI products for long-term VG monitoring at a both the regional and global scale. Hmimina et al. [17] investigated the usefulness of MODIS NDVI data for monitoring the seasonal changes of VG in the African savanna, including terrestrial biomes and deciduous and evergreen forests. This study found that the 16-day composite of MODIS NDVI allows for accurate estimation of greenness in the spring season. Potter [18] analyzed the recovery rates of VG by using MODIS NDVI in Alaska's severely burned wetland ecosystem. Fang et al. [19] reported vegetation dynamics by using the Breaks for Additive Seasonal and Trend (BFAST) method and MODIS NDVI product in Quebec, Canada, from 2000 to 2011. Wang et al. [20] monitored VG change by using MODIS NDVI data at three river source regions in China. Similarly, Gillespie et al. [21] investigated the spatial and temporal pattern of vegetation changes in Southern California by using MODIS NDVI data from 2000 to 2016. In recent years, Touhami et al. [22] evaluated the MODIS NDVI time series to monitor the vegetation dynamic over the Mediterranean forest region in Northeast Tunisia in response to the climatic variable. Therefore, this study considers MODIS NDVI products to monitor VG over the Mara River Basin in East Africa.

The greenness of grass and shrub was more sensitive to climate change than forests [5]. Generally, the NDVI had a linear relationship with average annual precipitation. However, the response of NDVI to precipitation was more significant than temperature [1]. The increase in precipitation can promote a vegetation greenness trend (VGT), while the increase in temperature will inhibit the VGT of a region [6]. Nicholson et al. [2] showed that NDVI variability is closely related to climate factors. Therefore, it is important to monitor climate factors on VG. For African Savannas, most of the previous studies proved that the VG presented a decreasing trend in dry seasons, using NDVI, which is retrieved from various satellite sensors, such as Advanced Very High Resolution Radiometer (AVHRR), the Global Inventory Monitoring Modeling System (GIMMS), and MODIS [1,5,23]. The Maasai Nara National Park in East Africa especially showed decreasing rainfall and a rising temperature in the dry seasons, thus causing the forests and shrubs to show a preceding greening trend, and the grass showed a browning trend [24,25]. In general, climate change significantly impacted VG in the Mara River Basin [26], and vegetation degradation seriously threatened the survival of livestock and wildlife in the Mara River Basin [27]. However, there is still little known about the VGT in dry seasons and its responses to climate change in the Mara River Basin. Currently, most studies consider only the temporal differentiation and responses of VG to environmental factors and ignore the spatial differences. In addition, most studies focus only on the Maasai/Serengeti ecosystem, whereas few studies consider the whole basin. In addition, most studies choose a large area to establish the relationship between VG and climatic variables (precipitation and temperature). However, no studies have shown the VG on a basin scale. Therefore, this paper considers the Mara River Basin for analyzing the responses of VG to climate change.

Based on the spatial vegetation distribution maps of the Mara River Basin drawn on the GEE platform, this paper uses the MODIS NDVI product and climate data for dry seasons in the Mara River Basin from 2000 to 2019. Our study analyzed the relationship between VG and climate change. The main objectives of this study were (i) to estimate the spatial and temporal distribution of VG by using the random forest (RF) regression algorithm in the Mara River Basin; (ii) to investigate the trend of VG and climatic variables (precipitation and temperature), using the Sen+Mann–Kendall test of the studied region; and (iii) to provide a clear view of the spatial distribution of VG, VGT, and their responses to climate change on a basin scale. In addition, this paper provides a theoretical basis for scientific assessment and reference to formulate ecological protection policies in the Mara River Basin.

2. The Study Area and Data

2.1. The Study Area

The Mara River (Figure 1) is the transboundary between Tanzania and Kenya in East Africa (Location: 33°88′ E to 35°90′ E and 0°28′ S to 1°97′ S). The basin contributed 65% of its area in Kenya and 35% in Tanzania. The river originates at the Mau Forest Escarpment and merges at the rural Musuma in Tanzania to Lake Victoria, passing group ranches, Maasai-Mara National Reserve, and Serengeti National Park. The Mara River is the only perennial river in the region that plays an important role in the ecohydrology of the basin [28]. The upstream of the Mau Forest Escarpment is a protected complex forest, and the rest of the areas in the upstream are almost farmland. The midstream mainly consists of grass and ranches, including two international wildlife reserves, which are the main wildlife tourist attractions in the basin. The downstream includes the Mara wetland and Mara mine. Moreover, it is the main source of production and living materials for residents and wild animals [29].

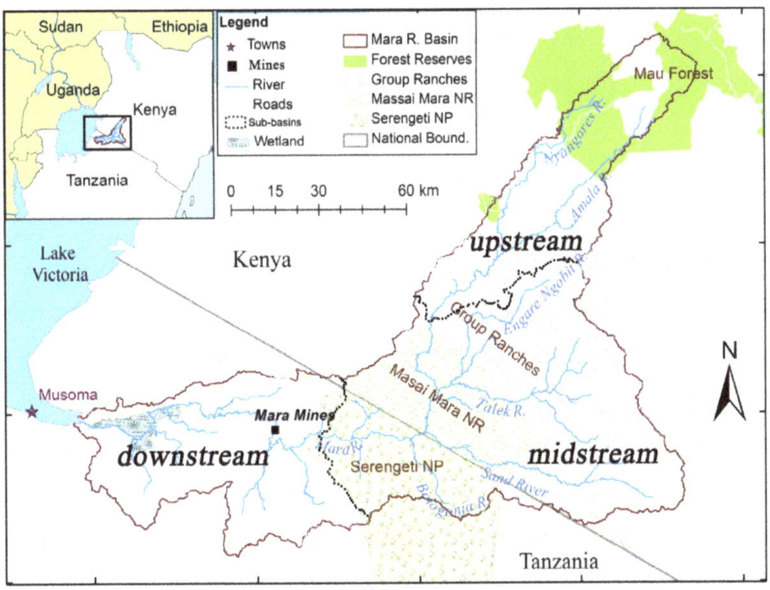

Figure 1. Location Map of the Mara River Basin.

The precipitation in the basin is bimodal with two rainy and dry seasons. The long dry season is from June to September, and the short dry season is from January to February. The average temperature is 18~23 °C, and the annual total precipitation is 200~500 mm in the dry season [30]. The dry season is the time for the great migration of wild animals [31], and also the time for the largest water requirement for agricultural irrigation upstream [32]. The more frequent and severe drought disasters in the dry seasons [24] have caused rapidly increasing water demand for agricultural irrigation [32]. The ecological environment in this region is fragile, which has led to a terrible impact on the basin's ecosystem.

2.2. Data

The data used in this paper include Sentinel-2 images and MODIS NDVI, temperature, and precipitation data. The time series of Sentinel-2 images were from June 2015 to June 2020 and were obtained from the Google Earth Engine (GEE) platform (https://earthengine.google.com/, accessed on 30 July 2021). This paper used Sentinel-2 images to draw vegetation distribution map referring to some relevant references [33,34]. The VG (NDVI) data were derived from MODIS sensor (MOD13Q1) from Aeronautics and Space Administration (NASA) (https://modis.gsfc.nasa.gov/, accessed on 10 August 2021). The time series of the MODIS NDVI data were from January to February and June to September during 2000–2019. The temperature and precipitation data were obtained from ERA5-Land monthly averaged data from 1950 to present from the European Centre for Medium-Range Weather Forecasts (https://www.ecmwf.int/en/forecasts/datasets/reanalysis-datasets/era5, accessed on 20 August 2021), during January/February, and June–September from 2000 to 2019. The temperature and precipitation data were resampled to a 250 m resolution to match with the spatial resolution of MODIS NDVI products.

3. Methods

The overall workflow of our study was structured with three sections (Figure 2). First, this study used the random forest (RF) algorithm to classify Sentinel-2 images into four main vegetation types (forest, crop, shrub, and grass) and obtained the vegetation distribution map in GEE platform. Second, the RF regression algorithm was used for MODIS NDVI, temperature, and precipitation data to analyze the response of VG to climate change. Then Thiel–Sen/Mann–Kendall trend-testing was used for the VG, temperature and precipitation. Finally, the RF regression algorithm was used again to investigate the response of VGT to temperature and precipitation.

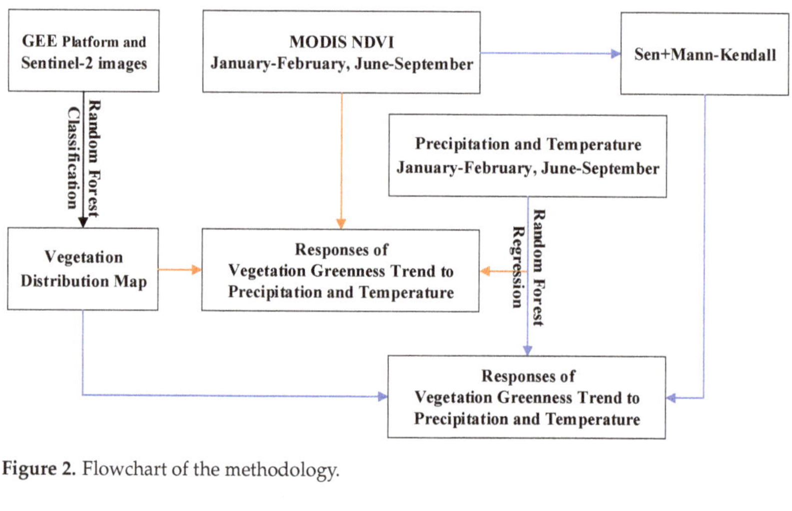

Figure 2. Flowchart of the methodology.

3.1. Random Forest Algorithm

The random forest algorithm is an algorithm based on the classification tree, combining bagging, random subspace, and decision tree methods [35,36]. The algorithm integrates the bootstrap aggregation method to generate subsets; that is, M (M = 1, 2, 3, ... , n) training sample sets with the same size as the original sample sets are randomly selected from the original sample set through bootstrap aggregation, and multiple decision trees are constructed accordingly. When splitting each node of the decision trees, the random subspace method is introduced to evenly and randomly extract a feature subset from all K features, and then an optimal splitting feature from the subset is selected. Finally, the mean value of multiple decision trees is taken as the final result. The RF algorithm has been widely used in classification and regression [37,38]. When the dependent variable is a classified variable, the algorithm is a classification algorithm. When the dependent variable is a continuous variable, the algorithm is a regression algorithm. The random forest algorithm can be briefly expressed as follows:

$$\hat{f} = \frac{1}{B}\sum_{b=1}^{B} f_b(x) \quad (1)$$

where $\hat{f}$ represents the final results, B is the number of the trees, f_b is the classification or regression function, and x represents the training sample values.

The RF algorithm improves the accuracy of results using bootstrap to alleviate high variance and weak the correlation between decision trees. It is easily operated by only adjusting the number of trees in the forest and debugging the number of features of each node is needed to generate a reasonable model quickly and efficiently. Compared with other machine learning algorithms, the RF algorithm can incorporate nonlinear relationships and explain complex relationships between variables [39]. Therefore, this study used RF algorithms to draw a vegetation distribution map and to analyze the responses of VG and VGT to climate change in dry seasons of the Mara River Basin.

3.2. Vegetation Distribution Mapping Based on GEE Platform

The RF algorithm was used to create a vegetation distribution map from Sentinel-2 images on the GEE platform. The classification process was as follows.

Savanna is dominated by grass, forests, and shrubs, so the vegetation was planned to be classified into four types (forest, crop, shrub, and grass), according to the references [33,34] and the consideration for saving classification time and observability. Then we extracted ROIs for each vegetation type. Considering the above ROIs should be evenly distributed in the whole basin and proportional to the area of each vegetation, 20,000 ROIs were finally extracted: forest (2240), crop (3620), grass (10,500), and shrub (3640), respectively, on the GEE platform as the training samples for the supervised classification. After exporting Sentinel-2 images into the GEE platform, the QA60 band was used to remove the clouds from the study area, and we calculated the 5-day cycle of NDVI. The Max Value Compound (MVC) method [40] was used to synthesize the maximum monthly NDVI to generate 12 months of monthly NDVI for every year. To further eliminate clouds and smooth the filtering, we used the time series harmonic analysis method (HANTS) to reconstruct the monthly NDVI [41]. Before the supervised classification, the importance of monthly NDVI was evaluated by RF regression, and only the months with the importance greater than 50% (January, February, June, July, August, and November) were selected as the characteristic months [42]. The 20,000 samples were randomly split into two groups during classification by the RF algorithm: 70% for training and 30% for validation. Then the yearly vegetation distribution maps from 2015 to 2019 were drawn. Based on the vegetation distribution maps from 2015 to 2019, the final vegetation distribution map of the Mara River Basin was created from the dominant vegetation type of each grid.

3.3. Thiel–Sen/Mann–Kendall Trend-Testing Approach

Thiel–Sen/Mann–Kendall trend-testing [43] was used to investigate the trends of VG, temperature, and precipitation. This approach uses the Sen trend degree S-value to reduce the interference of images' noise and to judge whether the images show increased or decreased trends. The S-value is calculated by Formula (2):

$$S = \sum_{i=1}^{n-1} \sum_{j=i+1}^{n} sgn(x_j - x_i) \tag{2}$$

$$sgn(x_j - x_i) = \begin{cases} +1, \ x_j - x_i > 0 \\ 0, \ x_j - x_i = 0 \\ -1, \ x_j - x_i < 0 \end{cases} \tag{3}$$

For vegetation, a positive value of S-value indicated a greening trend, and a negative S-value indicated a browning trend. For temperature and precipitation, a positive S-value presented an increasing trend for climate factors, while a negative S-value presented a decreasing trend.

Then this approach uses Mann–Kendall Z-value to test the significance of the long-term sequence trend, which can better detect areas with minor changes and judge the change trends more accurately. The length of time series in this paper is 20 ($n \geq 10$); the S-value approximately obeys the standard normal distribution, and then the Z-value can be used for trend testing.

$$Z = \begin{cases} \frac{S-1}{\sqrt{VAR(S)}}, \ S > 0 \\ 0, \ S = 0 \\ \frac{S+1}{\sqrt{VAR(S)}}, \ S < 0 \end{cases} \tag{4}$$

$$VAR(S) = \frac{n(n-1)(2n+5) - \sum_{i=1}^{m} t_i(t_i-1)(2t_i+5)}{18} \tag{5}$$

where n is the number of time series, m is the number of ties (repeated data groups) in the time series, t_i is the extent of any given tie (numbers of repeated data in group i), and the bilateral trend test is performed for Z-value.

When taking significance level α as 0.05, the significant change trend |Z-value| is 1.96. When the |Z-value| is more than 1.96, this means that the VG or climate factors changed rapidly, and when the |Z-value| is less than 1.96, this means that there was no rapid change.

3.4. The Relationship between VG and Climate Factors

To analyze the responses of VG and VGT to temperature and precipitation, this study first converted all the rasters to ASCIIs in ArcGIS, including dry seasons' MODIS NDVI, temperature and precipitation, and NDVI trend during 2000–2019. Then a RF regression model was built for MODIS NDVI and climate factors to analyze the response of VG to temperature and precipitation. The *importance* function in the *Random Forest* R package was used to investigate the contribution of temperature and precipitation to the VG. The importance of the responses was quantified by how much the model accuracy decreases (%IncMSE) when the variable was excluded. Finally, after all the ASCIIs were reconverted into rasters in ArcGIS, the importance of the four vegetation types (forest, crop, grass, and shrub) was extracted, respectively, and the mean importance was used as the response. The same work was performed to analyze of VGT to temperature and precipitation by using NDVI trend, temperature, and precipitation data. We used the coefficient of determination (R^2) and root means square error (RMSE) to evaluate the model-fitting results.

4. Results

4.1. Vegetation Distribution Mapping of Mara River Basin

The random forest algorithm was used to classify the vegetation into four types (forest, shrub, grass, and crop), and then we used producer's accuracies, user's accuracies, Kappa coefficients, and overall accuracies to evaluate the classification results of each year. The producer's accuracies from 2015 to 2019 for forest were 0.98%, 0.95%, 0.96%, 0.98%, and 0.90%; for crop, they were 0.95%, 0.88%, 0.90%, 0.96%, and 0.84%; for grass, they were 0.88%, 0.85%, 0.87%, 0.90%, and 0.83%; and for shrub, they were 0.82%, 0.80%, 0.80%, 0.85%, and 0.87%. The user's accuracies from 2015 to 2019 for forest were 0.90%, 0.86%, 0.89%, 0.95%, and 0.83%; for crop, they were 0.86%, 0.81%, 0.84%, 0.82%, and 0.88%; for grass, they were 0.83%, 0.84%, 0.86%, 0.85%, and 0.85%; and for shrub, they were 0.80%, 0.82%, 0.85%, 0.87%, and 0.80%. The overall accuracies from 2015 to 2019 were 90.00%, 87.25%, 88.75%, 91.25%, and 84.37%, respectively, and the Kappa coefficients were all above 0.8. The vegetation distribution map of the Mara River Basin was drawn by the dominant vegetation type of each grid (Figure 3). The basin's forest, shrub, grass, and crop areas were 1541.68, 2488.49, 7214.55, and 2505.44 km^2, respectively. Forests were distributed in upstream and downstream areas, crops were distributed in upstream area, and grass and shrubs were mainly distributed in the midstream area.

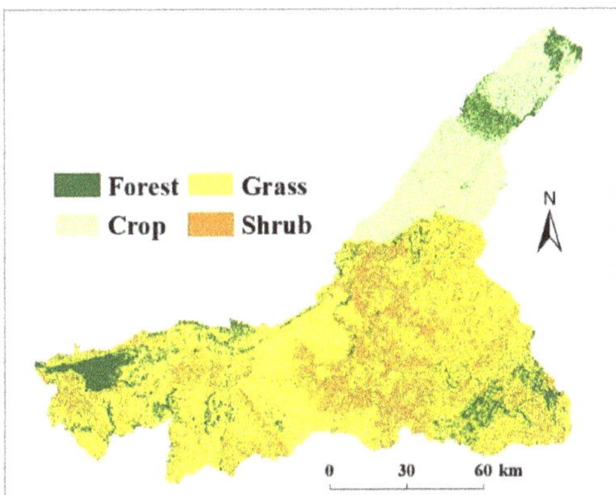

Figure 3. Vegetation distribution map of Mara River Basin.

4.2. Spatiotemporal Trend of VG in Dry Seasons

The average NDVI in the dry seasons (January/February and June–September) from 2000 to 2019 was 0.53. The maximum NDVI (0.63) was observed in the year 2007. The minimum NDVI (0.39) was observed in the year 2000. The mean NDVI in the dry seasons showed a fluctuating trend (Figure 4). The greenness in the upstream and downstream vegetation was relatively high, and the greenness of the midstream was low (Figure 5b). The greenness of each vegetation type from high to low was forest (0.665), crop (0.629), shrub (0.532), and grass (0.474), respectively. The VG in the reclamation area was higher than 0.6 and lower than 0.8 in Mau Forest Escarpment due to irrigation, sufficient heat, and relatively sufficient rainfall in the lower latitude. The VG was mostly 0.7~0.8 in the Mara wetland, the highest in the basin due to Lake Victoria's plenty of water supply. However, the VG was lower than 0.5 due to insufficient water supply in the reserves in the midstream area. Generally, the spatial distribution of greenness in the Mara River Basin is closely related to climate conditions. The VG was high in the regions with more precipitation and sufficient heat and low in the regions with less rainfall.

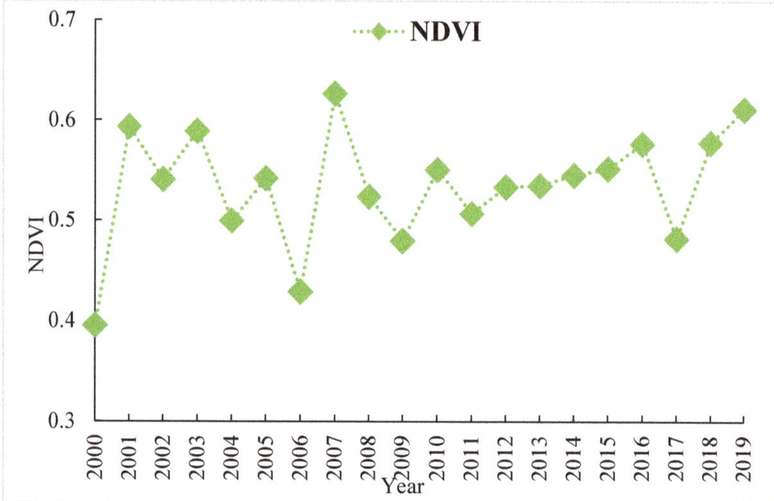

Figure 4. Temporal trend of VG in dry seasons of Mara River Basin.

Figure 5 shows the spatial distribution of VGT in dry seasons in Mara River basin. The VG in the dry seasons showed a greening trend in the Mara River Basin, especially in areas where forests and crops are widely distributed (Figure 5c). Meanwhile, in the midstream, where dominant vegetation types were shrub and grass, the VG often presented a browning trend (Figure 5c). In terms of the significant changes, most of the areas showed a non-significant increase trend. The significant change trend presented concentrated in the upstream and downstream areas, where forests and crops were the dominant vegetation types (Figure 5d). The significant decrease change trend and non-significant decrease change trend were often staggered in the midstream and downstream areas (Figure 5d).

Table 1 shows the proportions of area for vegetation in significantly different change trends. Forest had the most proportion in significant increase change trend of all vegetation types, and grass had the most proportion in significant decline trend. In general, the most vegetation was in non-significant increase change trend (62.17%), and the least vegetation was in significant decline change trend (2.26%).

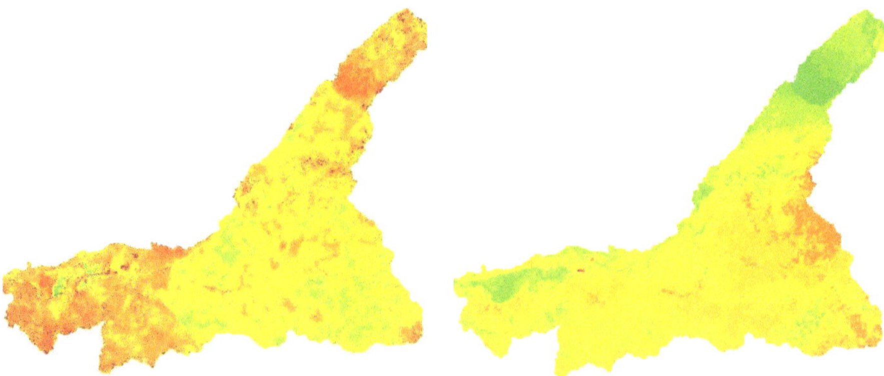

(a) (b)

Figure 5. *Cont.*

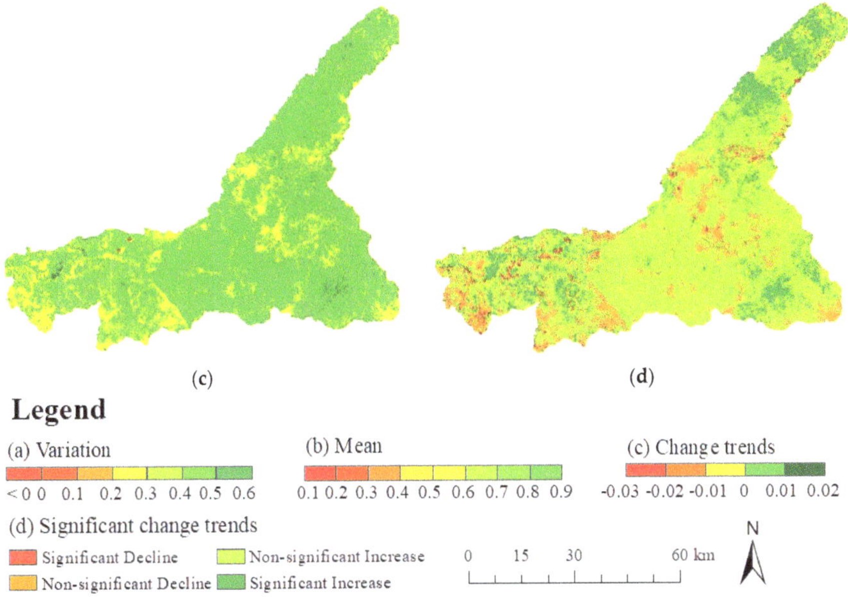

Figure 5. Spatial distribution of VGT in dry seasons in Mara River Basin: (**a**) NDVI variation in dry seasons; (**b**) average NDVI in dry seasons; (**c**) NDVI change trends in dry seasons (S-value); (**d**) NDVI significant change trends in dry seasons (Z-value).

Table 1. Proportion of area with different VG trends (%).

Change Trend	Forest	Crop	Grass	Shrub	Total Area (km^2)
Significant Decline	2.02	1.20	2.62	2.47	312.10
Non-significant Decline	6.58	7.72	16.12	19.73	1950.59
Non-significant Increase	37.54	46.34	68.62	74.44	8547.90
Significant Increase	53.87	44.74	12.64	3.36	2939.57
Total Area (km^2)	1541.68	2488.49	7214.55	2505.44	13,750.16

4.3. Responses of VG and VGT to Climate Change in Dry Seasons

The average temperature in the dry seasons (January to February and June to September) in the Mara River Basin was 19.4 °C from 2000 to 2019, whereas the highest temperature was recorded in the year 2017 (20.5 °C), and the lowest temperature was recorded in the year 2001 (18.4 °C). The average total precipitation in the dry seasons was 445.8 mm, the maximum precipitation was observed in the year 2007 (660.8 mm), and the minimum was observed in the year 2015 (318.2 mm). Moreover, the temperature and precipitation fluctuated during 2000–2019 (Figure 6). Compared with the VG, the precipitation was highest at the same time in the year 2007. The fluctuation trend of precipitation and VG was more similar compared to the temperature. The VG was smaller when there was a high temperature and less precipitation. On the contrary, the VG had a greater probability to show a high value with low temperature and much precipitation.

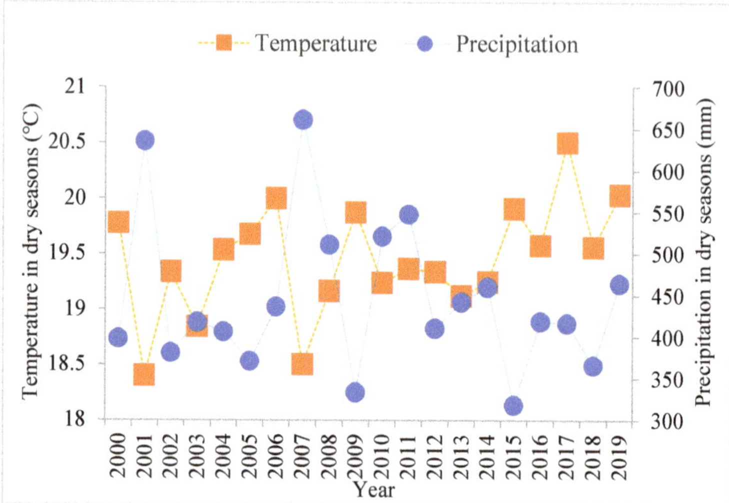

Figure 6. Annual average temperature and total precipitation of dry seasons in Mara River Basin.

Figure 7 shows the spatial trend of precipitation and temperature in the dry seasons from 2000 to 2019. A total of 84.16% of the precipitation in the basin showed a declining trend, especially in the downstream area, while only 15.84% of the precipitation showed an increasing trend. The eastern region of the basin had the largest increasing trend. For the significance of the precipitation change trend, 0.29% of the precipitation in the basin showed a significant decline trend, and 83.87% of the precipitation showed a non-significant decrease trend. Only the precipitation in the eastern region showed increase trends, whereas 15.42% of the precipitation in the basin showed a non-significant increase trend, and 0.42% in the basin showed a significant increasing trend. The temperature showed an increasing trend in the whole basin. For the significance of the temperature change trends, 82.88% of the temperature in the basin showed a non-significant increase trend, and 17.12% of the of the temperature in the basin showed a significant increasing trend, which was concentrated in the upstream and midstream areas.

Figure 8 showed the density scatter plots of RF regression for VG and VGT, the R^2 values were 0.95 and 0.91, respectively; the RMSE values were both 0.023. The fitting degree of the models were high, which verified the feasibility of the RF algorithm in analyzing the responses of VG and VGT to temperature and precipitation in the Mara River Basin.

To clearly understand the spatial responses of VG and VGT to precipitation and temperature, we used the natural breakpoint method to divide the importance into four segments and named the importance from low to high as I, II, III, and IV, respectively (Figure 9). The high importance indicated great responses of VG or VGT to precipitation and temperature.

The VG in the east of the basin had the most responses to precipitation, while the greenness concentrated in the midstream had the least responses to precipitation. The greenness that had the most responses to precipitation scattered around the greenness had the least responses in the downstream. Specifically, crops in the upstream area and some grass in the mid-low stream area had more responses to precipitation. In contrast, the forest in and around the Mara wetland and the shrub in Masai Mara National Park and Serengeti National Park had few responses to precipitation (Figure 9a). The VG in the upstream area had the most responses to temperature, and the greenness in the midstream had the least responses to temperature. Specific to vegetation types, shrub in the midstream had the least responses to temperature changes, while crop had the most responses to temperature changes. The response to temperature in the downstream area was low, expect in the Mara wetland (Figure 9b). The spatial responses of greenness to temperature were similar to precipitation in the upstream and midstream areas, with both showing a great response. The spatial responses of VG to temperature were opposite to the responses to precipitation in the downstream.

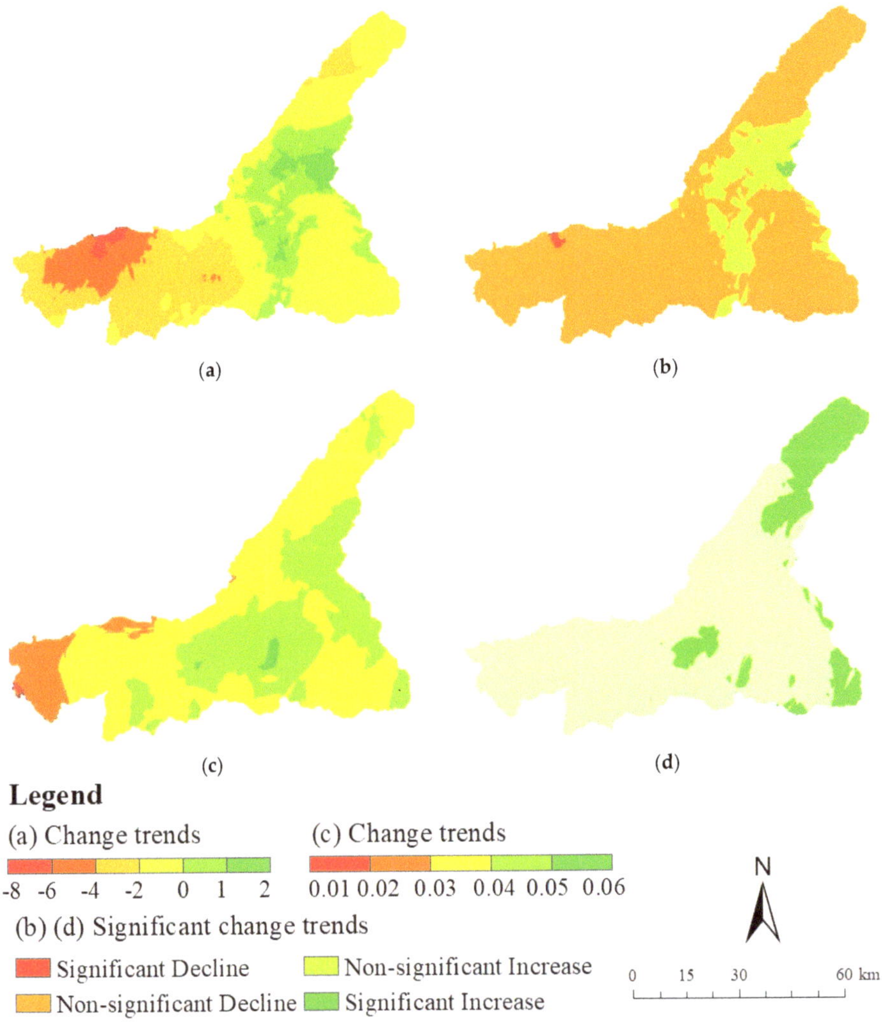

Figure 7. Spatial distribution of temperature and precipitation change trends in dry seasons in Mara River Basin: (**a**) Precipitation change trend in dry seasons (*S*-value); (**b**) precipitation significant change trend in dry seasons (*Z*-value); (**c**) temperature change trend in dry seasons (*S*-value); (**d**) temperature significant change trend in dry seasons (*Z*-value).

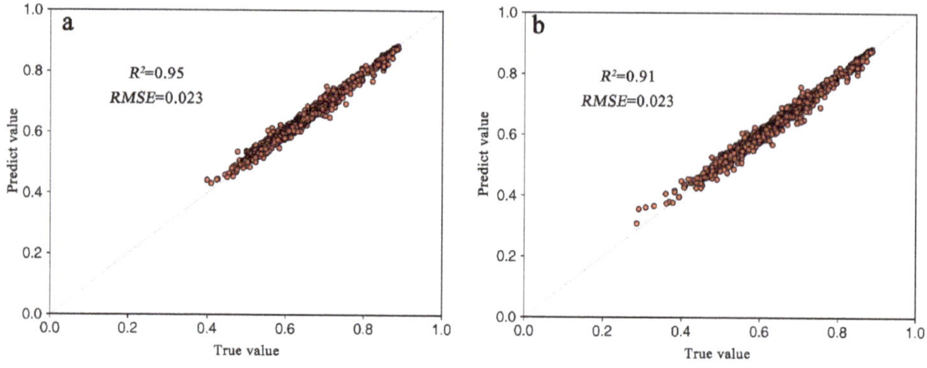

(a) (b)

Figure 8. Performance of random forest regression on the responses of VG and VGT to precipitation and temperature in dry seasons: (**a**) VG and (**b**) VGT.

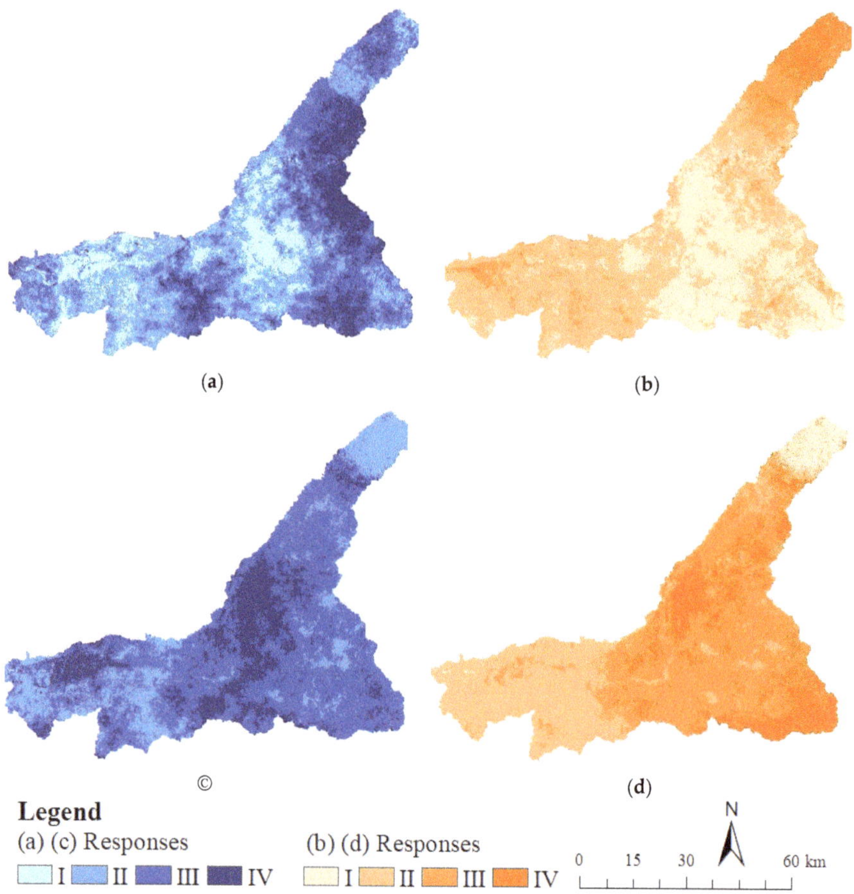

Figure 9. Responses of VG and VGT to climate factors in Mara River Basin in dry seasons: (**a**) responses of greenness to precipitation in dry seasons; (**b**) responses of greenness to temperature in dry seasons; (**c**) responses of greenness trend to precipitation in dry seasons; and (**d**) responses of greenness trend to temperature in dry seasons.

The spatial responses of VGT to climate were not consistent. The VGT in national parks had great responses to precipitation, while VGT in the upstream area and around the Mara wetland had less responses to precipitation. Specific to vegetation types, crops north of the forest and grass downstream had the least responses to precipitation, while forest and grass in the midstream area had the most responses to precipitation (Figure 9c). The response of VGT to temperature was great from the Mau Forest down to all of the midstream, and crops north of the Mau Forest had the least response to temperature (Figure 9d). The responses of VGT to precipitation were consistent with temperature in the upstream and midstream areas. However, there was a slight difference in the downstream.

Tables 2 and 3 showed the responses of VG and VGT to climate change, respectively. We considered the importance of temperature and precipitation. The greenness of all vegetation types had more responses to precipitation than to temperature. Crop greenness had the most responses to precipitation, and forest greenness had the least responses to precipitation. Forest greenness had the most responses to temperature, and grass greenness had the least responses to temperature. The Pearson correlation showed that precipitation had a positive impact on VG, while temperature had a negative impact on the greenness of forests and crops and had a positive impact on the greenness of grass and shrubs (Table 2). The crop is rain-fed and more affected by human activities, such as irrigation, so it had less correlation with climate. As a result, the crop greenness will be increased even under the climate change condition, where the temperature is rising and precipitation is falling. The greenness of grass and shrub can decrease if there is no interference from human activities. The VG in the Mara River Basin would decline rapidly because the VG had many more responses to decreasing precipitation than rising temperature.

Table 2. Response of VG to climate change in dry seasons in Mara River Basin.

Vegetation	r	Temperature	Precipitation
Forest	0.42	25.68(−)	35.36(+)
Crop	0.28	30.65(−)	70.23(+)
Grass	0.62	18.78(+)	60.04(+)
Shrub	0.45	22.43(+)	47.55(+)

r is the Pearson's coefficient between the predicted and observed cover based on the independent validation samples. The signs in the brackets represent the sign of Pearson's coefficient between the given vegetation and climate factors (n = non-significant).

Table 3. Responses of VGT to climate change in dry seasons in Mara River Basin.

Vegetation	r	Temperature	Precipitation
Forest	0.36	40.10(−)	27.46(−)
Crop	0.42	26.64(−)	70.23(+)
Grass	0.57	44.81(−)	34.64(+)
Shrub	0.48	42.27(−)	50.48(n)

r is the Pearson's coefficient between the predicted and observed cover based on the independent validation samples. The signs in the brackets represent the sign of Pearson's coefficient between the given vegetation and climate factors (n = non-significant).

The responses of VGT to temperature were greater than to precipitation, except for crops. The grass greenness trend had the most responses to temperature, and the crop greenness trend had the least responses to temperature. The shrub greenness trend had the most responses to precipitation, and the crop greenness trend had the least responses to precipitation. The responses were twice as different. The Pearson correlation showed that temperature had negative impacts on all VGTs. Precipitation had a negative impact on forests, it had no significant impact on shrubs, and it had a positive impact on grass and crops (Table 3). The forests in the Mara River Basin are dominated by sparse forests with strong drought resistance; this means that, no matter how much worse the climate gets, forests can maintain green in dry seasons. However, in general, the increasing temperature and decreasing precipitation were a disadvantage for VGTs. Climate change will lead to vegetation degradation, which will seriously reduce the food sources of wild animals and residents, threaten the production and living of residents, hinder economic development, and ultimately lead to further aggravation of poverty.

5. Discussion

High temperature accelerates the transpiration of vegetation and inhibits the growth of vegetation. As a result, when the temperature was high, the VG was low. Precipitation in the Mara River Basin is controlled by the Indian Ocean El Nino. The precipitation increases in El Nino, and it decreases in La Nina. Under the strong control of La Nina in year 2000, the basin experienced extreme arid conditions all year long; it was the severest drought in nearly a century [44]. The temperature in the year 2000 was also high. Therefore, the VG was lowest in the year 2000. Controlled by the extremely strong El Nino, the precipitation in the dry season increased significantly in year 2007, which was an extremely humid year [45]. Moreover, the temperature was very low, so the VG was highest in year 2007.

The spatial differentiation of VG in the dry seasons of the Mara River Basin was significantly different. The agricultural reclamation in the upstream area in Kenya accounted for the largest proportion of the significant increasing trend, and the grass in the midstream accounted for the largest decline trend. There was an obvious correlation between VG and climate factors [23]. The greenness of grass and shrubs was more sensitive to climate change than forests, as was consistent with the research results of Ghebrezgabher et al. [5]. The responses of VG to precipitation were greater than those to temperature [1]. Increased precipitation will promote greening, and rising temperatures will lead to the greening of grass and shrubs, but the browning of forests and crops.

The climate change trend of decreasing precipitation and increasing temperature in the dry seasons in the Mara River Basin will lead to significant vegetation browning, a finding that is consistent with the findings of Ogutu et al. [44]. The VG in the upstream area should show a significant browning due to the reduction of precipitation, but the crop greenness increased significantly due to a large amount of irrigation water added to ensure the growth of crops in dry seasons. The shrub and grass were staggered around the forest in Serengeti National Park, due to the manmade drainage channel in the southeast which supplied sufficient water sources to promote vegetation greening [46]. Affected by the reclamation in the upstream area and overgrazing in the midstream area, the runoff and suspended sediment increased greatly in the Mara River [33,47]. As a result, the downstream diverged, and the runoff increased greatly [48], leading to the expansion of the Mara wetland. The rich swamps in the wetlands provided sufficient water and nutrients for vegetation growth, promoting the VG near the Mara wetlands to increase rapidly. In order to support the mining of the Mara Mine, the Tanzanian government destroyed the vegetation and built corresponding infrastructure [49], which led to a significant decrease in VG near the Mara Mine.

The responses of VGT to temperature were more significant than those to precipitation. The increasing temperature will inhibit VGT, and decreasing precipitation will promote the grass greening trend and lead to the forest and crop browning trend, as is consistent with the findings of Li et al. [25]. The VGT had little response to climate due to sufficient irrigation water. The rising temperature in the dry seasons strongly impacted the greenness trend of shrubs and forests in the midstream. The reduction of precipitation led to obvious browning in the Mara wetland. This VGT will threaten the ecological environment and economy [33]. The rain-fed crop relies on the rainfall in the rainy seasons to irrigate, and the growth in the dry seasons requires large-scale irrigation. Due to the rapid population growth in Kenya subbasin, the irrigated area has grown rapidly to ensure food production [50]. For instance, the irrigated area had reached 1000 ha, with a total annual irrigation water volume of 12.25 million m^3 in year 2015, and the irrigated area maintains an annual growth rate of 100 ha [29]. The decrease in runoff and the rapid increase in water demand in irrigation caused a water shortage in the midstream and downstream [51]. Degraded grass in the midstream reduces food sources for wildlife and forces animals in areas where the grass degrades most to travel farther to obtain enough food.

The climate change trend of the Mara River Basin is consistent with that of Africa. It is expected that, by the end of this century, the average temperature in the dry seasons will increase by up to 7 °C, and precipitation will decline [52]. Climate change has already had an enormous impact on the habitats of Africa. Reduced rainfall in the dry season can lead to vegetation degradation and threaten livestock and wildlife survival, even in national parks [53]. Climate change has led to a significant reduction of species [54]. The rising temperature and more severe drought will destroy suitable habitats and increase the risk of species extinction [55,56]. For the Mara River Basin, climate change will increase the probability of droughts and floods [57], which will profoundly negatively impact biodiversity, agriculture, animals, and even human society. Accelerated drying of wildlife habitats has resulted in degraded vegetation and changes in savannah phenology [58,59], which will change the migration routes of wildlife [39] and increase animal diseases such as anthrax and Rift Valley fever [51]. Therefore, corresponding water resources and environmental management policies should be formulated to adapt to climate change and achieve sustainable development of the Mara River Basin.

6. Conclusions

The vegetation greenness of the Mara River Basin in the dry season was fluctuating, especially in the agricultural area in the upstream, from 2000 to 2019. The response of VG to precipitation was much greater than temperature. The increase in precipitation can promote vegetation greening, and the increase in temperature can inhibit greening. The vegetation browning was related to the increasing temperature and the decreasing precipitation, while vegetation greening was related to the increasing precipitation. There were quite differences in the greening trend for different vegetations. The grass browning trend in the midstream was the most obvious due to the decreasing precipitation and the increasing temperature. The decreasing precipitation and increasing temperature in the Mara River Basin will accelerate vegetation degradation and threaten the ecological security of the Masai Mara National Park. Vegetation degradation will make it difficult for animals to obtain food, seriously destroying wild animals' habitat and safety and causing more frequent ecological disasters. Considering that the decreasing trend of precipitation is much greater than the increasing trend of temperature, VG will experience serious degradation in the dry seasons. Therefore, it is necessary to actively formulate related environmental management policies to address climate change.

Author Contributions: Wanyi Zhu, Zhenke Zhang and Shuhe Zhao designed the paper. Wanyi Zhu conducted the data processing and statistical analysis and wrote the paper. Xinya Guo, Priyanko Das, Shouming Feng and Binglin Liu made great efforts in the writing and figures optimizationr. All the authors contributed to paper revision. All authors have read and agreed to the published version of the manuscript.

Funding: Authors acknowledge the support from the National key R&D projects (No. 2018YFE0105900).

Institutional Review Board Statement: Not applicable.

Informed Consent Statement: Not applicable.

Data Availability Statement: The Sentinel-2 images are supported by the Google Earth Engine (https://earthengine.google.com/, accessed on 30 July 2021). The ERA5-Land monthly averaged data from 1950 to present date are from the European Centre for Medium-Range Weather Forecasts (https://www.ecmwf.int/en/forecasts/datasets/reanalysis-datasets/era5, accessed on 10 August 2021). The MODIS MOD13Q1 data are from NASA (https://modis.gsfc.nasa.gov/, accessed on 20 August 2021).

Acknowledgments: The authors are thankful for the generous support of the funding from National key R&D projects (No. 2018YFE0105900) and would like to thank for Google Earth Engine, the European Centre for Medium-Range Weather Forecasts, and NASA for their valuable data.

Conflicts of Interest: The authors declare no conflict of interest.

References

1. Lamchin, M.; Wang, S.W.; Lim, C.H.; Ochir, A.; Pavel, U.; Gebru, B.M.; Choi, Y.; Jeon, S.W.; Lee, W.K. Understanding global spatio-temporal trends and the relationship between vegetation greenness and climate factors by land cover during 1982–2014. *Glob. Ecol. Conserv.* **2020**, *24*, e01299. [CrossRef]
2. Nicholson, S.E.; Davenport, M.L.; Malo, A.R. A comparison of the vegetation response to rainfall in the Sahel and East Africa, using normalized difference vegetation index from NOAA AVHRR. *Clim. Chang.* **1990**, *17*, 209–241. [CrossRef]
3. Li, W.; Buitenwerf, R.; Munk, M.; Amoke, I.; Bøcher, P.K.; Svenning, J.C. Accelerating savanna degradation threatens the Maasai Mara socio-ecological system. *Glob. Environ. Chang.* **2020**, *60*, 102030. [CrossRef]
4. Sankaran, M.; Hanan, N.P.; Scholes, R.J.; Ratnam, J.; Augustine, D.J.; Cade, B.S.; Gignoux, J.; Higgins, S.I.; Le Roux, X.; Ludwig, F.; et al. Determinants of woody cover in African savannas. *Nature* **2005**, *438*, 846–849. [CrossRef] [PubMed]
5. Ghebrezgabher, M.G.; Yang, T.; Yang, X.; Eyassu Sereke, T. Assessment of NDVI variations in responses to climate change in the Horn of Africa. *Egypt. J. Remote Sens. Space Sci.* **2020**, *23*, 249–261. [CrossRef]
6. Musau, J.; Patil, S.; Sheffield, J.; Marshall, M. Spatio-temporal vegetation dynamics and relationship with climate over East Africa. *Hydrol. Earth Syst. Sci. Discuss.* **2016**, *502*, 1–30.
7. Barbosa, H.A.; Lakshmi Kumar, T.V.; Silva, L.R.M. Recent trends in vegetation dynamics in the South America and their relationship to rainfall. *Nat. Hazards* **2015**, *77*, 883–899. [CrossRef]
8. Dhillon, M.S.; Dahms, T.; Kübert-Flock, C.; Steffan-Dewenter, I.; Zhang, J.; Ullmann, T. Spatiotemporal Fusion Modelling Using STARFM: Examples of Landsat 8 and Sentinel-2 NDVI in Bavaria. *Remote Sens.* **2022**, *14*, 677. [CrossRef]
9. Chakhar, A.; Hernández-López, D.; Ballesteros, R.; Moreno, M.A. Improving the accuracy of multiple algorithms for crop classification by integrating sentinel-1 observations with sentinel-2 data. *Remote Sens.* **2021**, *13*, 243. [CrossRef]
10. Khoirunnisa, F.; Wibowo, A. Using NDVI algorithm in Sentinel-2A imagery for rice productivity estimation (Case study: Compreng sub-district, Subang Regency, West Java). *IOP Conf. Ser. Earth Environ. Sci.* **2020**, *481*, 012064. [CrossRef]
11. Ma, C.; Johansen, K.; McCabe, M.F. Monitoring Irrigation Events and Crop Dynamics Using Sentinel-1 and Sentinel-2 Time Series. *Remote Sens.* **2022**, *14*, 1205. [CrossRef]
12. Mao, D.; Wang, Z.; Luo, L.; Ren, C. Integrating AVHRR and MODIS data to monitor NDVI changes and their: Relationships with climatic parameters in Northeast China. *Int. J. Appl. Earth Obs. Geoinf.* **2012**, *18*, 528–536. [CrossRef]
13. Kawamura, K.; Akiyama, T.; Yokota, H.; Tsutsumi, M.; Yasuda, T.; Watanabe, O.; Wang, G.; Wang, S. Monitoring of forage conditions with MODIS imagery in the Xilingol steppe, Inner Mongolia. *Int. J. Remote Sens.* **2005**, *26*, 1423–1436. [CrossRef]
14. Petus, C.; Lewis, M.; White, D. Monitoring temporal dynamics of Great Artesian Basin wetland vegetation, Australia, using MODIS NDVI. *Ecol. Indic.* **2013**, *34*, 41–52. [CrossRef]
15. Walker, D.A.; Epstein, H.E.; Raynolds, M.K.; Kuss, P.; Kopecky, M.A.; Frost, G.V.; Danils, F.J.A.; Leibman, M.O.; Moskalenko, N.G.; Matyshak, G.V.; et al. Environment, vegetation and greenness (NDVI) along the North America and Eurasia Arctic transects. *Environ. Res. Lett.* **2012**, *7*, 015504. [CrossRef]
16. Ju, J.; Masek, J.G. The vegetation greenness trend in Canada and US Alaska from 1984–2012 Landsat data. *Remote Sens. Environ.* **2016**, *176*, 1–16. [CrossRef]
17. Hmimina, G.; Dufrêne, E.; Pontailler, J.Y.; Delpierre, N.; Aubinet, M.; Caquet, B.; de Grandcourt, A.; Burban, B.; Flechard, C.; Granier, A.; et al. Evaluation of the potential of MODIS satellite data to predict vegetation phenology in different biomes: An investigation using ground-based NDVI measurements. *Remote Sens. Environ.* **2013**, *132*, 145–158. [CrossRef]
18. Potter, C. Recovery rates of Wetland Vegetation Greenness in severely burned ecosystems of Alaska derived from satellite image analysis. *Remote Sens.* **2018**, *10*, 1456. [CrossRef]
19. Fang, X.; Zhu, Q.; Ren, L.; Chen, H.; Wang, K.; Peng, C. Large-scale detection of vegetation dynamics and their potential drivers using MODIS images and BFAST: A case study in Quebec, Canada. *Remote Sens. Environ.* **2018**, *206*, 391–402. [CrossRef]
20. Wang, Z.; Liu, X.; Wang, H.; Zheng, K.; Li, H.; Wang, G.; An, Z. Monitoring vegetation greenness in response to climate variation along the elevation gradient in the three-river source region of China. *ISPRS Int. J. Geo-Inf.* **2021**, *10*, 193. [CrossRef]
21. Gillespie, T.W.; Ostermann-Kelm, S.; Dong, C.; Willis, K.S.; Okin, G.S.; MacDonald, G.M. Monitoring changes of NDVI in protected areas of southern California. *Ecol. Indic.* **2018**, *88*, 485–494. [CrossRef]
22. Touhami, I.; Moutahir, H.; Assoul, D.; Bergaoui, K.; Aouinti, H.; Bellot, J.; Andreu, J.M. Multi-year monitoring land surface phenology in relation to climatic variables using MODIS-NDVI time-series in Mediterranean forest, Northeast Tunisia. *Acta Oecologica* **2022**, *114*, 103804. [CrossRef]
23. Couteron, P.; Kokou, K. Woody vegetation spatial patterns in a semi-arid savanna of Burkina Faso, West Africa. *Plant Ecol.* **1997**, *132*, 211–227. [CrossRef]
24. Mutiti, C.M.; Medley, K.E.; Mutiti, S. Using GIS and remote sensing to explore the influence of physical environmental factors and historical land use on bushland structure. *Afr. J. Ecol.* **2017**, *55*, 477–486. [CrossRef]
25. Ogutu, J.O.; Piepho, H.P.; Dublin, H.T.; Bhola, N.; Reid, R.S. El Niño-Southern Oscillation, rainfall, temperature and Normalized Difference Vegetation Index fluctuations in the Mara-Serengeti ecosystem. *Afr. J. Ecol.* **2008**, *46*, 132–143. [CrossRef]
26. Li, W.; Buitenwerf, R.; Munk, M.; Bøcher, P.K.; Svenning, J.C. Deep-learning based high-resolution mapping shows woody vegetation densification in greater Maasai Mara ecosystem. *Remote Sens. Environ.* **2020**, *247*, 111953. [CrossRef]
27. Ogutu, J.O.; Owen-Smith, N. ENSO, rainfall and temperature influences on extreme population declines among African savanna ungulates. *Ecol. Lett.* **2003**, *6*, 412–419. [CrossRef]

28. McClain, M.E.; Subalusky, A.L.; Anderson, E.P.; Dessu, S.B.; Melesse, A.M.; Ndomba, P.M.; Mtamba, J.O.D.; Tamatamah, R.A.; Mligo, C. Comparaison du régime d'écoulement, de l'hydraulique en rivière et des communautés biologiques en vue de déduire les relations débit-écologie de la rivière Mara au Kenya et en Tanzanie. *Hydrol. Sci. J.* **2014**, *59*, 801–819. [CrossRef]
29. George Marcellus Metobwa, O. Water Demand Simulation Using WEAP 21: A Case Study of the Mara River Basin, Kenya. *Int. J. Nat. Resour. Ecol. Manag.* **2018**, *3*, 9. [CrossRef]
30. WREM International Inc. *Mara River Basin Monograph: Final Report*; WREM International Inc.: Atlanta, GA, USA, 2008; p. 446.
31. Mnaya, B.; Mtahiko, M.G.G.; Wolanski, E. The Serengeti will die if Kenya dams the Mara River. *Oryx* **2017**, *51*, 581–583. [CrossRef]
32. Dessu, S.B.; Melesse, A.M.; Bhat, M.G.; McClain, M.E. Assessment of water resources availability and demand in the Mara River Basin. *Catena* **2014**, *115*, 104–114. [CrossRef]
33. Mango, L.M.; Melesse, A.M.; McClain, M.E.; Gann, D.; Setegn, S.G. Land use and climate change impacts on the hydrology of the upper Mara River Basin, Kenya: Results of a modeling study to support better resource management. *Hydrol. Earth Syst. Sci.* **2011**, *15*, 2245–2258. [CrossRef]
34. Zermoglio, F.; Scott, O.; Said, M.; Incs, I.C. Vulnerability and Adaptation Assessment in the Mara river basin. *Who* **2019**, *100*, 102–105.
35. Breiman, L. Random forests. *Mach. Learn.* **2001**, *45*, 5–32. [CrossRef]
36. Ho, T.K. The random subspace method for constructing decision forests. *IEEE Trans. Pattern Anal. Mach. Intell.* **1998**, *20*, 832–844. [CrossRef]
37. Eisavi, V.; Homayouni, S.; Yazdi, A.M.; Alimohammadi, A. Land cover mapping based on random forest classification of multitemporal spectral and thermal images. *Environ. Monit. Assess.* **2015**, *187*, 291. [CrossRef] [PubMed]
38. Jhonnerie, R.; Siregar, V.P.; Nababan, B.; Prasetyo, L.B.; Wouthuyzen, S. Random Forest Classification for Mangrove Land Cover Mapping Using Landsat 5 TM and Alos Palsar Imageries. *Procedia Environ. Sci.* **2015**, *24*, 215–221. [CrossRef]
39. Davies, A.B.; Asner, G.P. Elephants limit aboveground carbon gains in African savannas. *Glob. Chang. Biol.* **2019**, *25*, 14585. [CrossRef]
40. Chen, J.; Jönsson, P.; Tamura, M.; Gu, Z.; Matsushita, B.; Eklundh, L. A simple method for reconstructing a high-quality NDVI time-series data set based on the Savitzky-Golay filter. *Remote Sens. Environ.* **2004**, *91*, 332–344. [CrossRef]
41. Verhoef, W. Application of harmonic analysis of NDVI time series (HANTS). *Fourier Anal. Temporal NDVI S. Afr. Am. Continents.* **1996**, *108*, 19–24.
42. Hao, P.; Zhan, Y.; Wang, L.; Niu, Z.; Shakir, M. Feature selection of time series MODIS data for early crop classification using random forest: A case study in Kansas, USA. *Remote Sens.* **2015**, *7*, 5347–5369. [CrossRef]
43. Mann, H.B. Nonparametric Tests Against Trend. *Econometrica* **1945**, *13*, 245. [CrossRef]
44. Ogutu, J.O.; Owen-Smith, N.; Piepho, H.P.; Said, M.Y.; Kifugo, S.C.; Reid, R.S.; Gichohi, H.; Kahumbu, P.; Andanje, S. Changing wildlife populations in nairobi national park and adjoining athi-kaputiei plains: Collapse of the migratory wildebeest. *Open Conserv. Biol. J.* **2013**, *7*, 11–26. [CrossRef]
45. Das, P.; Zhang, Z.; Ren, H. Evaluation of four bias correction methods and random forest model for climate change projection in the Mara River Basin, East Africa. *J. Water Clim. Chang.* **2022**, *13*, 1900–1919. [CrossRef]
46. Reed, D.N.; Anderson, T.M.; Dempewolf, J.; Metzger, K.; Serneels, S. The spatial distribution of vegetation types in the Serengeti ecosystem: The influence of rainfall and topographic relief on vegetation patch characteristics. *J. Biogeogr.* **2009**, *36*, 770–782. [CrossRef]
47. Dutton, C.L.; Subalusky, A.L.; Anisfeld, S.C.; Njoroge, L.; Rosi, E.J.; Post, D.M. The influence of a semi-Arid sub-catchment on suspended sediments in the Mara River, Kenya. *PLoS ONE* **2018**, *13*, e0192828. [CrossRef]
48. Bregoli, F.; Crosato, A.; Paron, P.; McClain, M.E. Humans reshape wetlands: Unveiling the last 100 years of morphological changes of the Mara Wetland, Tanzania. *Sci. Total Environ.* **2019**, *691*, 896–907. [CrossRef]
49. Mwemezi, B.R.; Luvara, V.G.M. Reliability of the Environmental Feasibility Studies to the Mining and Construction Projects: A Case of Mara River Basin in Tanzania. *Am. J. Environ. Eng.* **2017**, *7*, 65–72. [CrossRef]
50. Pruijssen, M.J. *FLEX-Topo Modelling of Water Use and Demand in the Mara River Basin, Kenya*; Delft University of Technology: Delft, The Netherlands, 2015.
51. Bartzke, G.S.; Ogutu, J.O.; Mukhopadhyay, S.; Mtui, D.; Dublin, H.T.; Piepho, H.P. Rainfall trends and variation in the Maasai Mara ecosystem and their implications for animal population and biodiversity dynamics. *PLoS ONE* **2018**, *13*, e0202814. [CrossRef]
52. Wainwright, C.M.; Black, E.; Allan, R.P. Future Changes in Wet and Dry Season Characteristics in CMIP5 and CMIP6 simulations. *J. Hydrometeorol.* **2021**, *9*, 2339–2356. [CrossRef]
53. Ogutu, J.O.; Owen-Smith, N. Oscillations in large mammal populations: Are they related to predation or rainfall? *Proc. Afr. J. Ecol.* **2005**, *43*, 332–339. [CrossRef]
54. Sintayehu, D.W. Impact of climate change on biodiversity and associated key ecosystem services in Africa: A systematic review. *Ecosyst. Health Sustain.* **2018**, *4*, 225–239. [CrossRef]
55. Brown, J.D. *Biogeography, Sinauer Associates*; TTESOL International Association: Alexandria, VA, USA, 2013; Volume 17, pp. 44–50.
56. Thuiller, W.; Broennimann, O.; Hughes, G.; Alkemade, J.R.M.; Midgley, G.F.; Corsi, F. Vulnerability of African mammals to anthropogenic climate change under conservative land transformation assumptions. *Glob. Chang. Biol.* **2006**, *12*, 424–440. [CrossRef]

57. Barros, V.R.; Field, C.B.; Dokken, D.J.; Mastrandrea, M.D.; Mach, K.J.; Bilir, T.E.; Chatterjee, M.; Ebi, K.L.; Estrada, Y.O.; Genova, R.C.; et al. *Climate Change 2014 Impacts, Adaptation, And Vulnerability Part B: Regional Aspects: Working Group II Contribution to The Fifth Assessment Report of The Intergovernmental Panel On Climate Change*; Cambridge University Press: Cambridge, UK, 2014; ISBN 9781107415386.
58. Walther, G.R.; Post, E.; Convey, P.; Menzel, A.; Parmesan, C.; Beebee, T.J.C.; Fromentin, J.M.; Hoegh-Guldberg, O.; Bairlein, F. Ecological responses to recent climate change. *Nature* **2002**, *416*, 389–395. [CrossRef] [PubMed]
59. Root, T.L.; Price, J.T.; Hall, K.R.; Schneider, S.H.; Rosenzweig, C.; Pounds, J.A. Fingerprints of global warming on wild animals and plants. *Nature* **2003**, *421*, 57–60. [CrossRef] [PubMed]

International Journal of
Geo-Information

MDPI

Article

Identifying Suitable Watersheds across Nigeria Using Biophysical Parameters and Machine Learning Algorithms for Agri–Planning

Pranay Panjala [1], Murali Krishna Gumma [1,2,*], Hakeem Ayinde Ajeigbe [3], Murtala Muhammad Badamasi [4], Kumara Charyulu Deevi [1] and Ramadjita Tabo [3]

1 Geospatial Data and Big Data Sciences, ICRISAT, Hyderabad 502324, India; p.pranay@cgiar.org (P.P.); d.kumaracharyulu@cgiar.org (K.C.D.)
2 Geospatial Data and Big Data Sciences, ICRISAT, Niamey BP 12404, Niger
3 Geospatial Data and Big Data Sciences, ICRISAT, Kano PMB 3491, Nigeria; h.ajeigbe@cgiar.org (H.A.A.); r.tabo@cgiar.org (R.T.)
4 Centre for Dryland Agriculture (CDA), Bayero University, Kano PMB 3491, Nigeria; mmbadamasi.geog@buk.edu.ng
* Correspondence: m.gumma@cgiar.org

Abstract: Identifying suitable watersheds is a prerequisite to operationalizing planning interventions for agricultural development. With the help of geospatial tools, this paper identified suitable watersheds across Nigeria using biophysical parameters to aid agricultural planning. Our study included various critical thematic layers such as precipitation, temperature, slope, land-use/land-cover (LULC), soil texture, soil depth, and length of growing period, prepared and modeled on the Google Earth Engine (GEE) platform. Using expert knowledge, scores were assigned to these thematic layers, and a priority map was prepared based on the combined weighted average score. We also validated priority watersheds. For this, the study area was classified into three priority zones ranging from 'high' to 'low'. Of the 277 watersheds identified, 57 fell in the high priority category, implying that they are highly favorable for interventions. This would be useful for regional-scale water resource planning for agricultural landscape development.

Keywords: water; watershed proritization; agriculture; dryland; Google Earth Engine

Citation: Panjala, P.; Gumma, M.K.; Ajeigbe, H.A.; Badamasi, M.M.; Deevi, K.C.; Tabo, R. Identifying Suitable Watersheds across Nigeria Using Biophysical Parameters and Machine Learning Algorithms for Agri–Planning. *ISPRS Int. J. Geo-Inf.* **2022**, *11*, 416. https://doi.org/10.3390/ijgi11080416

Academic Editors: Walter Chen, Fuan Tsai and Wolfgang Kainz

Received: 31 May 2022
Accepted: 8 July 2022
Published: 22 July 2022

1. Introduction

The population of the world is projected to reach 10 billion by 2050, which means that we will require a higher rate of food production than we have now (World Population Data Sheet 2020). In Nigeria, the rapidly expanding and urbanizing population—which is expected to more than double in the next 35 years—has long exceeded domestic food production capability [1,2]. This makes it imperative that activities that help in attaining a high rate of food production and food self-sufficiency are more sharply focused. As part of the efforts needed to regain food self-sufficiency, Natural Resource Management (NRM) development programs must be conducted at the watershed level [3]. Moreover, there should be a focus on the fundamental principles of land and water resources management, such as watershed development and development of catchments and sub-catchments, which are critical to securing Nigeria's environmental and agricultural resilience [4]. Presently, irrigation covers only 7% of the irrigable land in Nigeria [1]. While rapid expansion of agricultural capacity, including through private investment [2], is indeed making more lands productive as an objective toward bridging the food deficit, there are warning signals like drought, gully formation, overgrazing, and erosion that need to be taken into account in agricultural initiatives across Africa (World Bank 2012). Identification of hotspots integrating various parameters like population, land-use/land-cover (LULC), and drainage networks can lead us to better solutions in agricultural development [5,6]. This approach takes into account the possible social aspects of the challenge too. Further, running decision tools can give satisfying results by aiding decision-making in relation to the implementation and development of natural resources. However, NRM has thus far been poorly implemented for agriculture development as well as for water supply. While Africa has rich natural resources and Nigeria has abundant water resources, there is an absence of efficient use of such resources. Preparation of watershed prioritization maps can help us enhance efficient utilization of natural resources, which currently are largely untapped in Nigeria [7].

Characterization of natural resources needs multidisciplinary investigations carried out by experts from different areas of expertise. In the present study, we prioritized watershed areas based on different biophysical parameters, such as population, soil, precipitation, landscape, LULC, and social parameters. Climate parameters, such as temperature and precipitation, highly influence the performance of watersheds: Low and very high rainfall negatively affects agriculture, as do extremes of temperature. Moderate climatic conditions are better for rainfed agriculture. In general, land resource management acknowledges the association between social and biophysical factors in attaining satisfying results [8–11].

Several studies have used the approach of integrating various thematic maps using geospatial tools for locating potential groundwater zones [12–16]. Similarly, studies have also been carried out on aspects of natural resources and development planning using remote sensing and GIS technologies [14,17–22]. Using various biophysical, socioeconomic, and technical parameters with a multi-criteria approach, geospatial techniques have been widely used in the assessment of land suitability for prioritization [23–29]. Specifically, several studies have shown that the weighted sum method is the most efficient method for prioritizing watersheds in developing countries [5,30].

The purpose of prioritizing watersheds is to identify focus watersheds for restoration activities that can address their critical needs and for intervention planning. It is a useful tool for decision-makers as it combines all the necessary information and allows a comparison of watersheds within the same cluster. This approach allows researchers to develop a summary of the watersheds of interest by spatially locating them and obtaining relevant information about their vulnerability. This process can also help in locating multiple watersheds with regard to prioritizing watershed protection and restoration.

In this study, we conducted a prioritization of watersheds across Nigeria to support natural resource management and agricultural planning. We identified, on the basis of biophysical parameters, an optimum number of watersheds ranging from low to high priority so that specific watersheds could be targeted for interventions. Further, with the help of geospatial inputs, thematic spatial data layers were used to construct a spatial model. We identified priority watersheds by allotting different weights based on the opinion of subject matter specialists (SMS). This scientific approach allowed us to prioritize watersheds strategically using multiple biophysical parameters at a time. This high-precision technique helps in delineating watersheds with utmost care and confidence.

2. Materials and Methods

2.1. Study Area

Nigeria lies between latitudes 4° N and 14° N and longitudes 4° E and 15° E. It is bordered on the north, east, west, and south by the Republic of Niger, the Republic of Benin, Cameroon, and the Gulf of Guinea, respectively (Figure 1). This location in West Africa gives the country a very wide range of climatic patterns. According to Odekunle (2004), Nigeria's climate is dominated by the influence of three major atmospheric phenomena: Maritime tropical (mT) air mass, continental tropical air mass, and equatorial easterlies. Rainfall varies within the country with a mean annual rainfall in the range of 1000–2000 mm in humid areas and 300–1100 mm in semi-arid areas. There is a slight variability of climate from south to north. In the north, the mean maximum temperatures are higher (32 °C) than in the south, while the mean minimum temperatures are lower (24 °C). As per the FAO's soil taxonomy, the major soil types in Nigeria are Fluvisols, Regosols, Gleysols,

Acrisols, Ferrasols, Alisols, Lixisols, Cambisols, Luvisols, Nitosols, Arenosols, and Vertisols with varied potential for agricultural use. The Niger and Benue rivers are the major rivers in Nigeria. The Niger River has an irrigation potential of 1.68 million hectares (Mha) in Nigeria, but its use is limited to only 0.68 Mha. The country has six distinct agroecological zones varying from the Atlantic coast to the arid savanna of the Sahel. The major staple crops in the humid parts of Nigeria are cassava, yam, cocoyam, and maize, whereas in the subhumid and semi-arid parts, maize, sorghum, millet, cowpea, and groundnut are grown. The major commercial crops include cocoa, oil palm, cotton, ginger, and sesame.

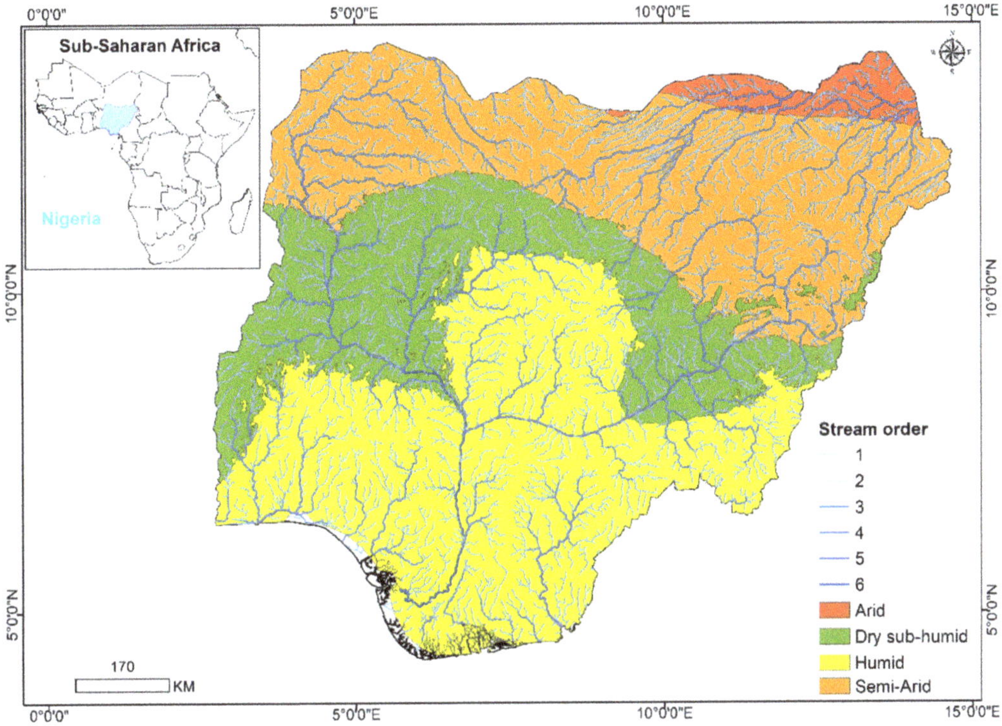

Figure 1. Location map of Nigeria with stream networks and agroecological zones (FAO).

2.2. Methodology

For identifying priority watersheds, we applied the methodology of weighted integration of multiple thematic layers using the geographic information system (GIS) (Figure 2). We used thematic spatial layers of both biophysical and social parameters that are important for agriculture. The priority order, i.e., ranking, of every spatial layer was obtained from subject matter experts, including NARS scientists in Nigeria. The priority classes were decided on the basis of the multi-criteria decision rule.

For thematic layers, such as LULC, a map of the year 2014 was prepared from MODIS 250 m satellite imagery using Normalized Difference Vegetation Index (NDVI) time-series data. The slope map was prepared from SRTM 30 m data. Similarly, other thematic spatial layers were acquired from the public domain using Google Earth Engine. The weightage and scores for the values in the thematic layers were given in relation to their positive effect on watershed and agricultural development. Thematic layers with a high positive value were given the highest weightage and vice versa. Upon integration of multiple spatial layers, the sum of all weights was calculated. High priority was given to the thematic layer that obtained the highest score and vice versa.

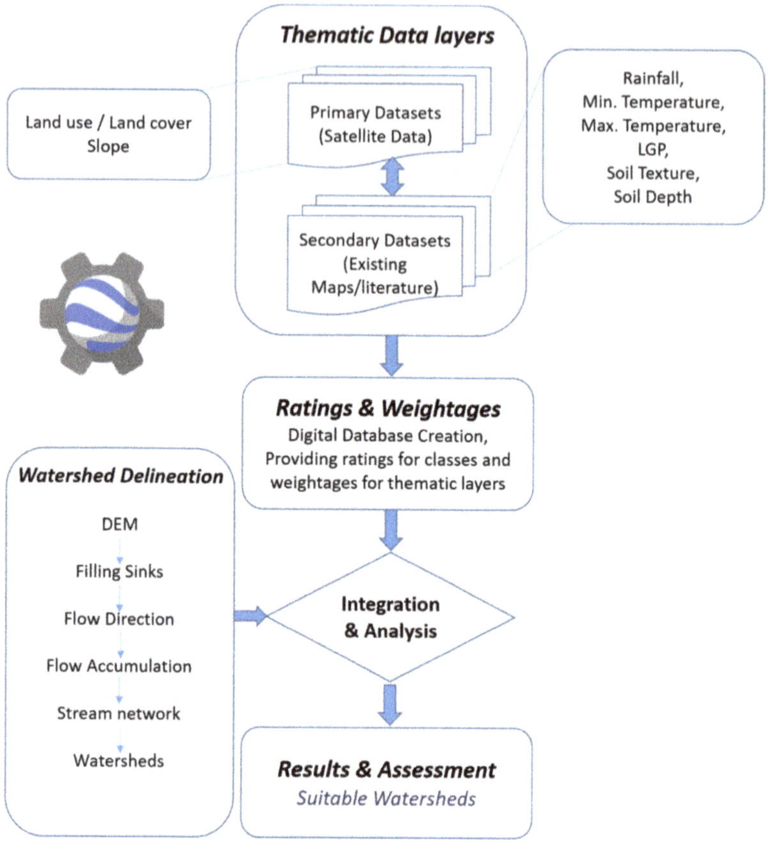

Figure 2. Methodology of watershed prioritization using geo-spatial layers.

2.3. Criteria and Determining Factors

Various thematic layers, such as soil, slope, LULC, rainfall, maximum and minimum temperature, length of growing period (LGP) (see Appendix A) were considered for the prioritization analysis based on their importance and relationship with other thematic layers. Based on the rating given by subject matter experts, the criteria to define prioritization was the sum of weights for all thematic layers (Table 1).

Table 1. Priority levels for thematic layers.

Suitability Criteria	Priority Level		
	Low (1)	Moderate (2)	High (3)
Average min. temp (°C)	0–15	15–20	20–25
Average max. temp (°C)	Up to 20 and >40	20–30	30–40
Average precipitation (mm)	Up to 250	250–1000	>1000
Slope (% rise)	>20	5–20	<5
Soil texture (class values)	5, 6, 8	2, 3, 4, 7	1
Soil depth (mm)	<5	5–20	>20
LGP (No. of Days)	>240	150–240	60–150

2.3.1. Land-Use/Land-Cover

Land-use/land-cover (LULC) patterns were mapped for the year 2014 using MODIS 250 m resolution satellite imagery, targeting major land-use classes like croplands (Figure 3), shrub lands, water bodies, and built-up/open lands [31,32]. Among these LULC classes, the dominant class with the highest score was cropland. Rainfed croplands were chosen rather than irrigated cropland because of their higher priority in watershed development. Classes like built-up land and water bodies were given less priority, whereas shrub lands and grassland were given medium priority because of their vegetation status. The LULC layer was assigned the weightage of 3.

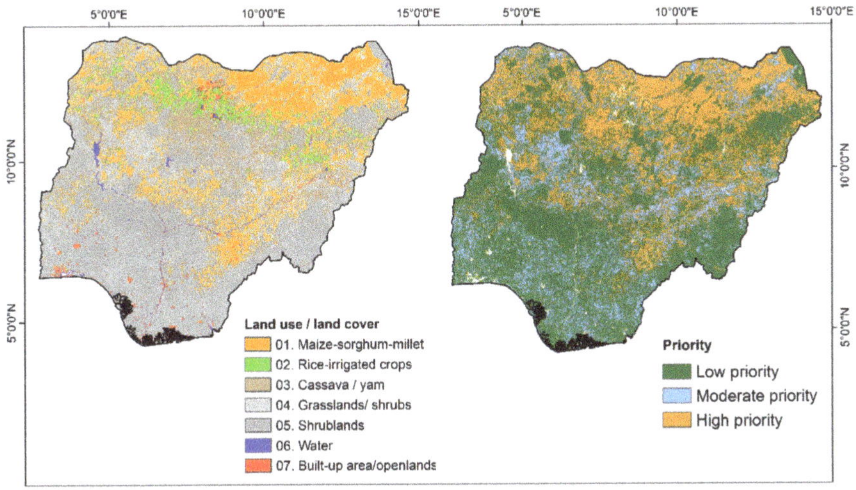

Figure 3. LULC classes in Nigeria.

2.3.2. Slope

The slope map was derived from SRTM 30 m DEM data (Figure 4). The map was stratified in terms of percentage change showing the rise or fall of land surface, which is a crucial factor in determining water flow. Lower percent change of elevation, i.e., slope, was given a high priority because of ease during cultivation and high groundwater potential. High percent change was given low priority in the estimation. This layer was given a low weightage of 1.

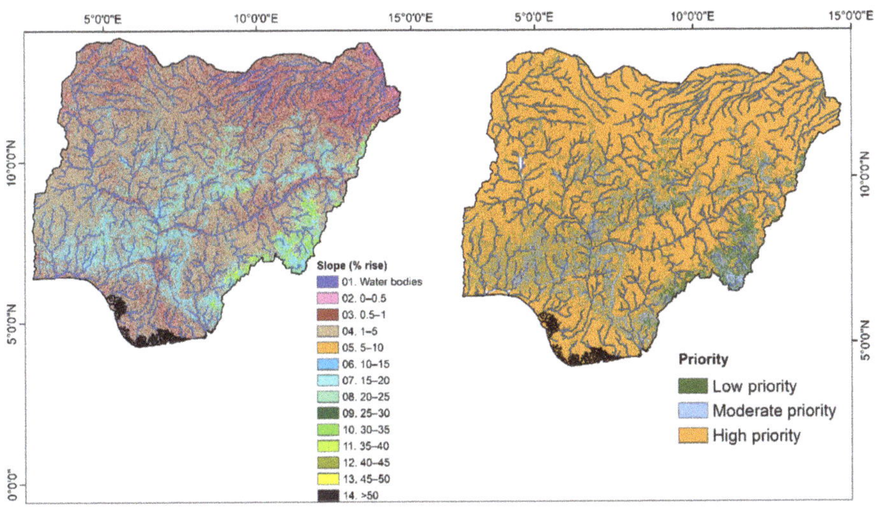

Figure 4. Slope map of Nigeria (SRTM DEM: http://srtm.sci.cgiar.org/) (accessed on 11 January 2022).

2.3.3. Soils

Soil parameters [33] (soil texture and soil depth) play a vital role in watershed prioritization because of their critical role in runoff. The water withstanding capacity of a location depends upon the soil type/texture and permeability at that location. The experts' scores were assigned for both layers, i.e., soil texture and soil depth, based on priority. Soil texture was classified into eight types (clay, clay loam, loamy sand, loam, sand, sandy clay loam, sandy clay, and sandy loam). Clay soils were given high priority, and sandy soils were given low priority (Figure 5a). Soil depth was classified into six classes (Figure 5b). Deeper soils were given a higher priority than lower-depth soils. These layers were assigned a weightage of 3.

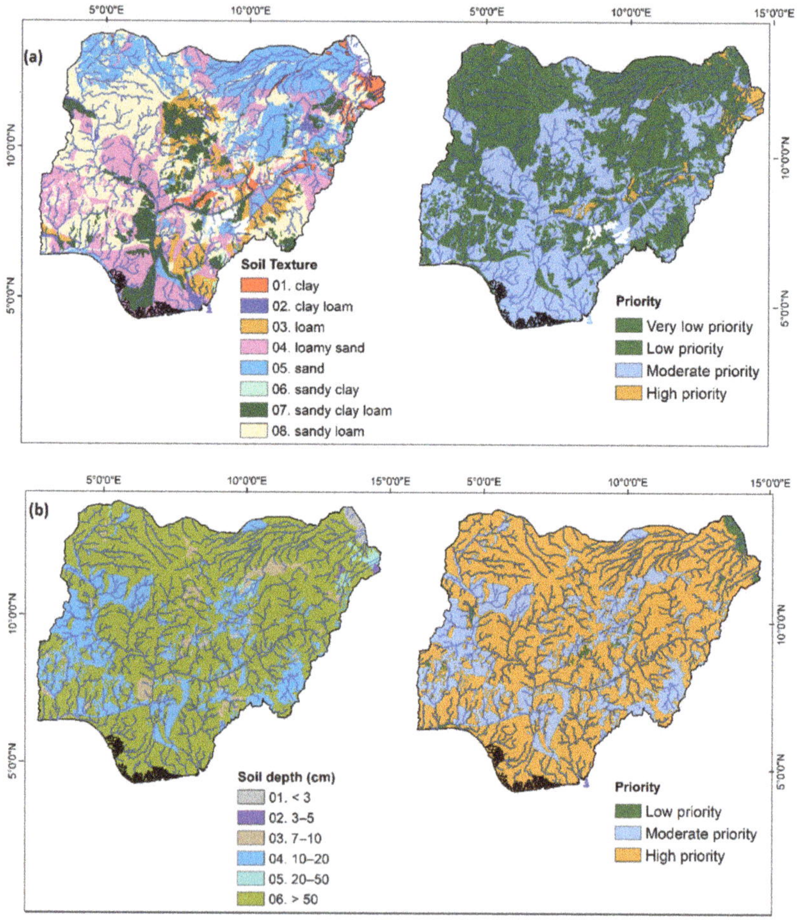

Figure 5. (**a**) Soil texture and its priority map and (**b**) soil depth and its priority map.

2.3.4. Rainfall

The annual rainfall data (2010–2018) were downloaded from Terra Climate [34] (Figure 6). Average rainfall was classified into 10 classes. The areas receiving less than 250 mm of rainfall were given a low priority and areas with rainfall greater than 1000 mm were given high priority, and medium range of rainfall was allotted moderate priority. A weightage of 3 was given to this layer.

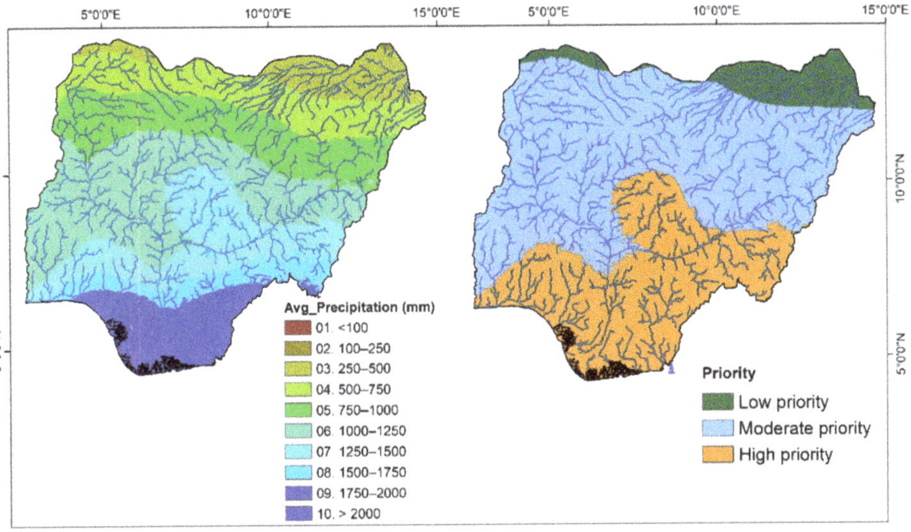

Figure 6. Mean annual rainfall in Nigeria (TerraClimate).

2.3.5. Length of Growing Period (LGP)

The length of the growing period (LGP) is one of the factors that determine the vegetation in an area in a year [35]. LGP was classified into seven classes in which two classes, <60 days and >240 days, were given low priority, while the LGP class 60–150 days was given high priority and 150–240 days moderate priority (Figure 7). A weightage of 2 was given to this layer. The LGP product was prepared by FAO as a part of the World Bank's review of its rural development strategy. It was prepared using vegetation indices as well as annual rainfall.

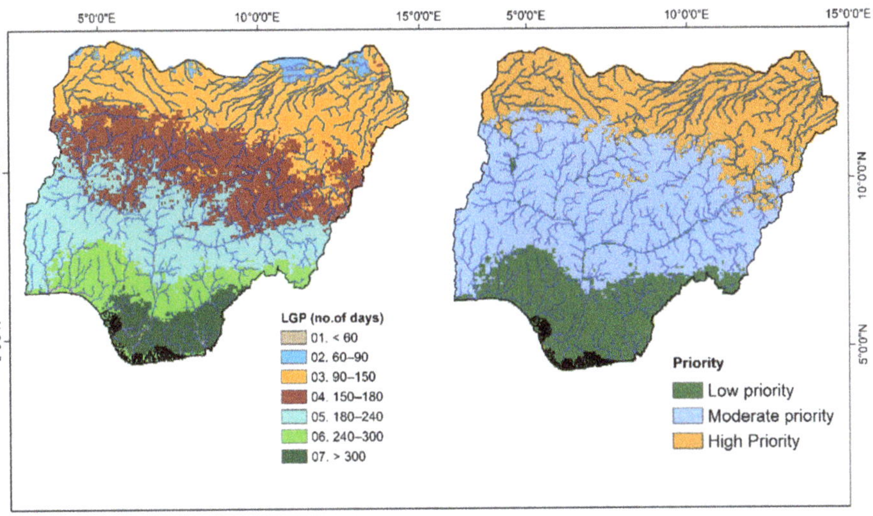

Figure 7. Length of the growing period (LGP) in Nigeria.

2.3.6. Temperature

Minimum temperature: Average minimum temperature data were downloaded from WorldClim and classified into four classes with 5 °C intervals (Figure 8a). The areas with an average minimum temperature <5 °C were allotted a very low priority, and those between 5 and 15 °C were given low priority. Areas with average minimum temperatures between 20 and 25 °C were given a high priority, whereas those with 15–20 °C were assigned moderate priority. This layer was given a weightage of 2.

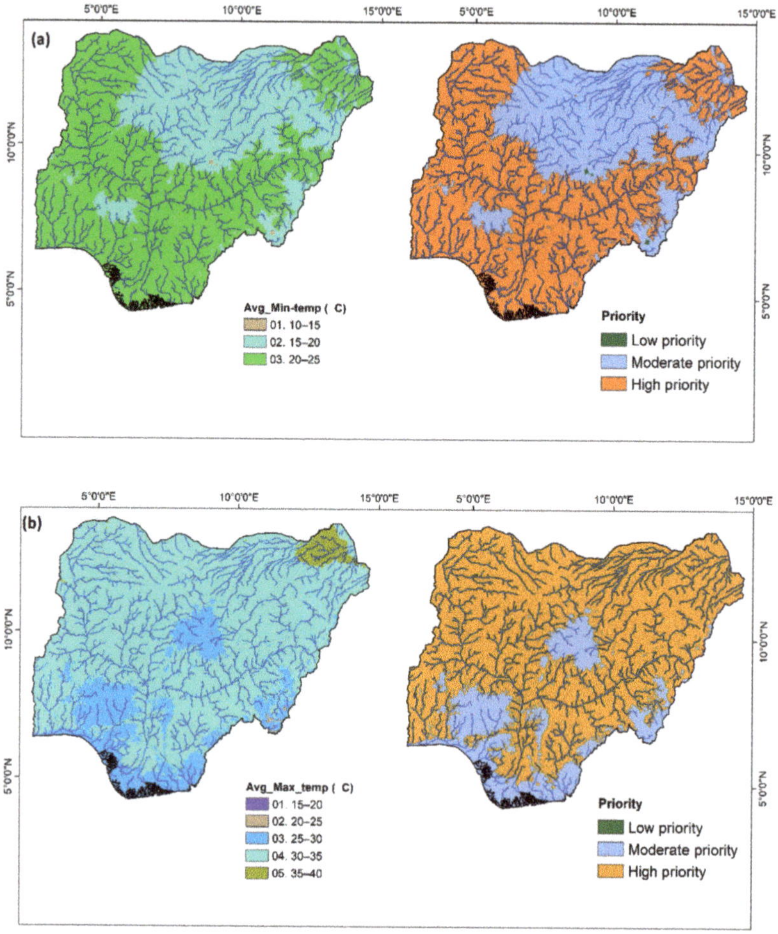

Figure 8. (**a**) Average annual minimum temperature and it priority map. (**b**) Average annual maximum temperature and its priority map.

Maximum temperature: Average maximum temperature data were downloaded from WorldClim and classified into six classes (Figure 8b). Areas having a mean maximum temperature of <20 °C or >40 °C were given low priority. Those areas with a mean maximum temperature of 20–30 °C were given moderate priority, whereas areas with maximum temperature varying in the 30–40 °C range were given a high priority. This layer was given a weightage of 3.

2.4. Determining Thematic Layer Weights

On the basis of expert/scientists' knowledge and a review of published papers [8,14,16,36,37], weights were allotted to different layers. The layers most favorable to watershed interventions were those that received a high weightage of 3. The layers least favorable to interventions were those that had a weightage of 1, while a weightage of 2 indicated moderately favorable layers. Layers like average annual maximum temperature, annual average precipitation, LULC, soil texture, and soil depth were given a high weightage of 3.

Annual average minimum temperature and LGP were given a weightage of 2. The slope map was given a low weightage of 1.

2.5. Integration of Thematic Layers Using Spatial Models

The integration of these thematic layers was carried out by developing a spatial model on GEE. The classes within each layer were reclassified on the basis of their scores given by experts (Equation (1)). Then, using the raster calculator, the weightages given by experts were multiplied by the respective layers (Equation (2)).

$$T_{sw} = T_r \times W \tag{1}$$

T_{sw} = Thematic layer with weighted score
T_r = Reclassified thematic layer
W = Weights

Then, all the weighted thematic layers were summed up and integrated to get the priority map

$$P_m = \Sigma T_{sw} \tag{2}$$

P_m = Priority map

2.6. Spatial Modeling Using Machine Learning Algorithms on Google Earth Engine Platform

Layers such as rainfall and temperature from WorldClim and slope maps from SRTM DEM were available on the GEE platform. Other layers, such as LULC, LGP, and soil maps, were ingested into GEE assets.

The layers were reclassified using decision tree algorithms incorporating the expert-given values using code as in the example below.

Example for rainfall reclassification:

```
"
var DTstring_prep =
['1) root 9999 9999 9999', '2) prec<=250 9999 9999 1 *',//Allocated value 1
'3) prec>250 9999 9999 9999', '6) prec<=1000 9999 9999 2 *',//Allocated value 2
'7) prec>1000 9999 9999 3 *'].join("\n"); Allocated value 3
var classifier_prep = ee.Classifier.decisionTree(DTstring_prep);
var reclassifiedImage_prep = prep.select('prec').classify(classifier_prep);
"
```

In the above example of a decision tree algorithm, it reclassified pixels with a value <250 m as 1, whereas values between 250 and 1000 mm were reclassified as 2 and those >1000 mm were 3.

A similar procedure was used for all the layers by giving scores to the respective pixels that are favorable to watershed interventions. The weightages are then multiplied with the scores of respective layers as per expert opinion and were summed up as in the example below.

For example:

```
"var weighted=
reclassifiedImage_minTem.add(reclassifiedImage_maxTem).add(reclassifiedImage_slop).add
(reclassifiedImage_prep)
"
```

The above example shows the addition of the reclassified layers of minimum temperature, maximum temperature, and precipitation. Then, the summed-up layer is reclassified as per priority, low, medium, or high, based on the values attained by each pixel.

2.7. Watershed Delineation

The major input data for delineating the watersheds were drawn from SRTM 30 m horizontal resolution DEM obtained from the web portal of the Consortium for Spatial Information [38] (http://srtm.csi.cgiar.org/) (accessed on 11 January 2022). These data were utilized to delineate the stream network and the slope map using ArcGIS tools. The sequence of steps followed to delineate the stream network, as well as watersheds, is illustrated in Figure 2.

The process starts with filling the sinks by comparing the values of neighboring cells. The filled sinks help in the generation of flow direction by finding the steepest descent of every cell. Then, flow accumulation is calculated using flow direction by counting the number of cells that are flowing to a particular cell. A set of thresholds for flow accumulation and flow direction generates the stream network.

The generation of pour points at the sixth stream order for the entire study area helps in the generation of watersheds (Figure 9a,b).

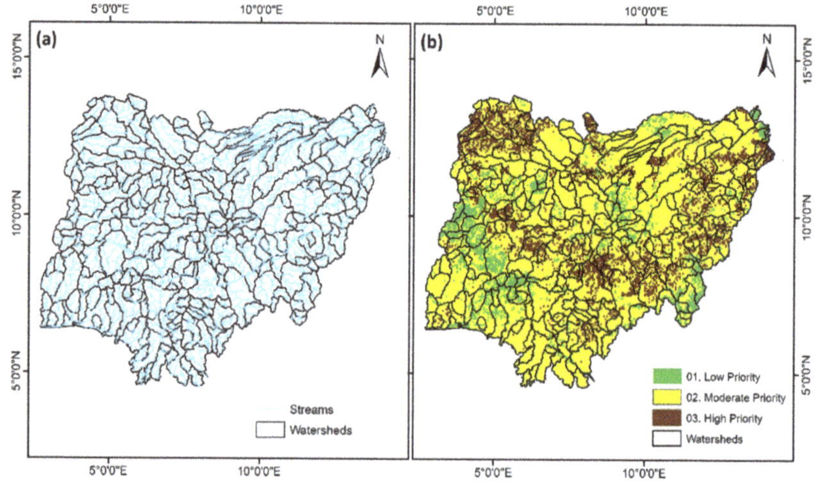

Figure 9. (**a**) Watershed and stream network delineation in Nigeria. (**b**) Spatial distribution of watersheds and their priority in Nigeria.

3. Results and Discussion

3.1. Watershed Analysis and Prioritization of Watersheds

Among all the watersheds identified throughout Nigeria, 277 were identified as having an area greater than 100 ha. Out of these, 144 watersheds were found to have an area less than 0.2 Mha, 71 were in the range of 0.2–0.4 Mha, and 26 in the range of 0.4–0.6 Mha. Only about 30 watersheds have an area greater than 0.6 Mha (Table 2).

Table 2. Area-wise classification of watersheds in Nigeria.

Area (Mha)	No. of Watersheds
<0.2	144
0.2 to 0.4	71
0.4 to 0.6	26
0.6 to 0.8	15
0.8 to 1.0	8
1.0 to 1.2	6
1.2 to 1.4	3
1.4 to 1.6	2
>1.6	2

The watershed prioritization map of Nigeria was derived after integration of the allocated priority values for different thematic layers. The priority map was categorized into three classes: High, moderate/medium, and low priority. The areas identified as high-priority are very favorable to watershed development, and the low-priority zones are the least favorable. Most of the watersheds in Nigeria fell in the moderate-priority class. The defined watershed map of Nigeria was overlaid on the priority map to identify strategic watersheds for agricultural development (Figure 9a,b).

3.2. Integration of Watershed Map with Thematic Layers

For a more detailed understanding of the watersheds, priority maps were prepared as per each thematic layer (Figure 10a–g). Table 3 shows the number of watersheds in every thematic category.

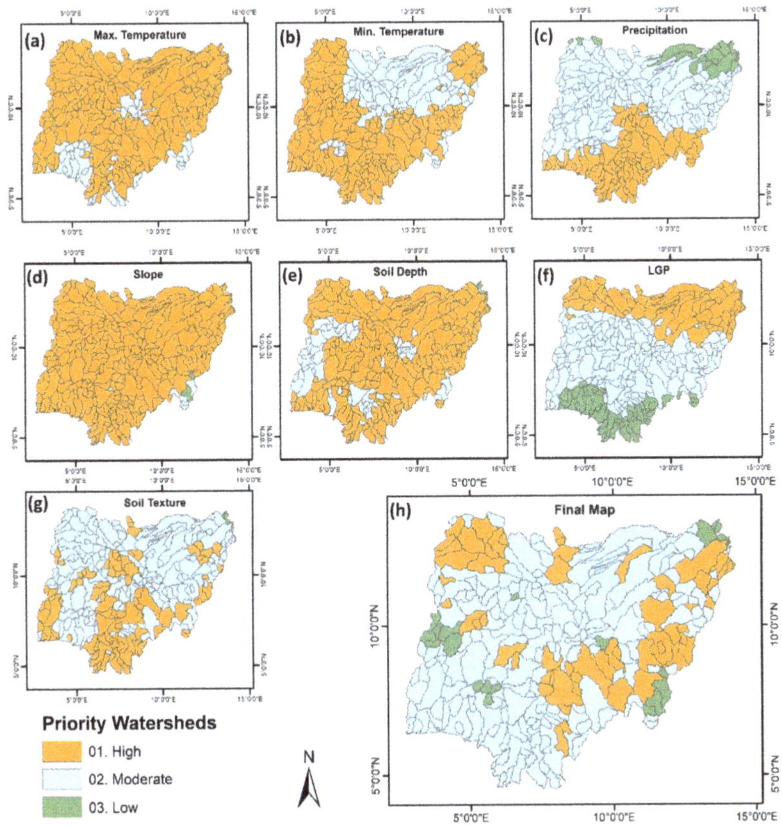

Figure 10. (**a**–**g**) Prioritized watershed maps as per thematic layer. (**h**) Prioritized watersheds after integration of all thematic layers in the study area.

The watershed map with the integration of all thematic layers is shown in Figure 10h. Considering only the precipitation layer, we found that only 98 watersheds had highly suitable rainfall conditions, which is a crucial layer for agriculture planning. About 159 watersheds fell in the moderate-priority class. For maximum and minimum temperature, almost all watersheds had moderately suitable or highly suitable conditions. Soil conditions too showed a favorable tendency. These findings show the importance of watersheds in this country.

Table 3. Number of proritized watersheds by thematic layer.

Thematic Layers	Number of Watersheds		
	Low Priority	Moderate Priority	High Priority
Maximum temperature	0	47	230
Minimum temperature	0	98	179
Precipitation	20	159	98
Slope	2	6	269
Soil depth	6	59	212
Soil texture	5	153	119
LGP	51	148	78
Final map	21	199	57

3.3. Validation of Priority Watersheds

On the basis of the available data, we validated the priority watersheds in relation to dams constructed in Nigeria (Figure 11). The details of dams and their purpose are illustrated in Table 4.

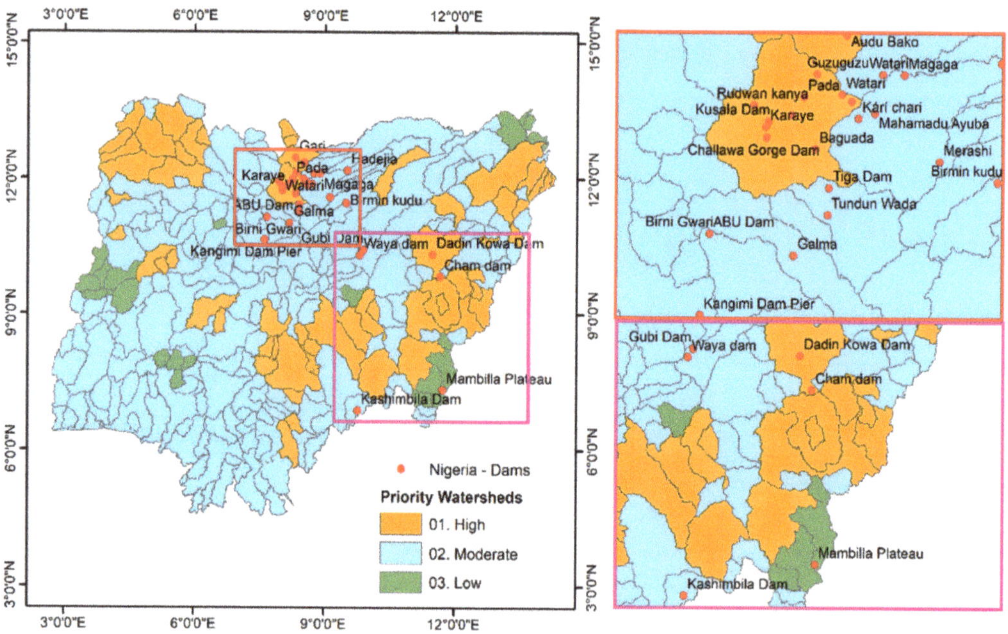

Figure 11. Map showing the location of dams in Nigeria.

We found that most of the dams constructed for the purpose of irrigation lie within moderate and high-priority watersheds. Dams constructed for multiple purposes, such as irrigation, as well as hydroelectric power generation, were mostly in moderate-priority watersheds, whereas dams located within high-priority watersheds were those built only for irrigation. The Mambila Plateau dam constructed for hydroelectric power generation lies in a low-priority zone. This validation indicated that the study correctly prioritized watersheds for agricultural planning and development.

Table 4. Locations of dams and their purpose.

FID	Long.	Lat.	Name	Objective	State	Priority
0	8.52168	12.340262	Audu Bako	Irr/ws	Kano	High
1	8.319344	12.449709	Gari	Irr/ws	Kano	High
2	8.750295	12.101573	Watari	Irr/ws	Kano	Moderate
3	8.026064	11.816966	Kusala Dam	Irr/ws	Kano	High
4	8.013794	11.719821	Challawa Gorge Dam	Irr/ws	Kano	High
5	8.55439	11.939492	Adu Bayero	Irr/ws	Kano	High
6	8.32209	11.650109	Baguada	Irr/ws	Kano	High
7	8.010376	11.785614	Karaye	Irr/ws	Kano	High
8	8.491756	11.982099	Watari	Irr/ws	Kano	High
9	8.170203	11.856063	Tomas dam	Irr/ws	Kano	High
10	7.933704	11.915973	Rudwan kanya	Irr/ws	Kano	High
11	8.2476	11.967202	Pada	Irr/ws	Kano	High
12	8.329988	12.10452	Guzuguzu	Irr/ws	Kano	High
13	9.50311	12.173312	Hadejia	Irr	Jigawa	Moderate
14	8.701458	11.866207	Mahamadu Ayuba	Irr/ws	Kano	Moderate
15	9.108804	11.575692	Merashi	Irr/ws	Kano	Moderate
16	8.885244	12.100326	Magaga	Irr/ws	Kano	Moderate
17	9.481497	11.450196	Birmin kudu	Irr/ws	Jigawa	Moderate
18	8.409686	11.413934	Tiga Dam	Irr/ws	Kano	Moderate
19	8.402408	11.25013	Tundun Wada	Irr/ws	Kano	Moderate
20	8.184803	11.004238	Galma	Irr/ws/HP	Kaduna	Moderate
21	7.6548271	11.135078	Birni Gwari	Irr/ws	Kaduna	Moderate
22	7.6548271	11.135078	ABU Dam	Irr/ws	Kaduna	Moderate
23	7.596162	10.639892	Kangimi Dam Pier	Irr/ws	Kaduna	Moderate
24	9.881125	10.418291	Gubi Dam	Irr/ws	Bauchi	Moderate
25	9.80136	10.30249	Waya dam	Irr/ws/HP	Bauchi	Moderate
26	11.481694	10.322154	Dadin Kowa Dam	Irr/ws/HP	Gombe	High
27	11.6613	9.8329	Cham dam	Irr/ws	Gombe	High
28	11.72	7.33	Mambilla Plateau	HP	Taraba	Low
29	9.761765	6.873387	Kashimbila Dam	Irr/ws	Taraba	Moderate

Irr = Irrigation; ws = Water Supply; HP = Hydroelectric power.

4. Discussion

Natural resource management plays a crucial role in the sustainable utilization of the available natural resources. In the context of watershed management, prioritization of watersheds helps in the effective use of natural resources for agricultural development in a shorter period of time. Watershed prioritization using remote sensing and GIS techniques is an easy and convenient approach based on weighted scores provided by SMS/scientists. In past studies, watershed prioritization was carried out using quantitative analysis, statistical methods, fuzzy and AHP techniques [39–41], morphometric analysis [42], delineation of groundwater potential zones [43], prioritization of sub-watersheds [44,45], prioritization of semi-arid agricultural watersheds [46], spatial assessment of soil erosion risk [47,48], and many other parameters. Our study considered biophysical parameters and major LULC classes to carry out watershed prioritization in Nigeria as a tool for agricultural development and planning. These parameters included average minimum temperature, average maximum temperature, average precipitation, slope, soil depth, and length of the growing period, which have a major role in watershed development and management. Analyzing these biophysical parameters and rating them with the help of subject experts, we carried out prioritization of watersheds in Nigeria using SRTM DEM-delineated data. Various studies have employed different methods of watershed prioritization for expansion of agriculture [5,49], critical sub-basins in mountainous watersheds [50], natural resource management [40,51], sediment yield index [52], LULC change impacts [53], assessment of flash flood risk with the help of weighted-sum models [54], etc. However, in all these methods, prioritization of watersheds was analyzed based on individual biophysical parameters such as topographical information, LULC, weather data, soil texture, soil depth and slope, etc. Nevertheless, the multi-criteria decision-making approach depends on the total score obtained after applying each thematic layer, and the accuracy of analysis of each input parameter.

It is very important to identify high-priority watersheds in Africa as land resource development programs are generally designed on a watershed basis. Therefore, appropriate prioritization is required for proper intervention and management. In our study, based on priority classification for every parameter, priority-wise watersheds were delineated and mapped. This helps various stakeholders in making decisions appropriate to their requirements. Various stakeholders in Nigeria will significantly benefit from the findings of this study. Integration of slope, soil depth, and soil texture maps and prioritization on the basis of those parameters should help in planning for soil conservation measures and watershed interventions. The maximum and minimum temperature layers in our study indicate the direct or indirect effects on soil moisture as well as evapotranspiration [55]. Prioritization of watersheds as per the precipitation layer clearly indicates the water-sufficient and water-deficient areas. Flood-prone and drought-prone watersheds can also be identified by considering the relevant parameters. Prioritization of watersheds in terms of the LGP indicated the vegetation levels throughout the year. Every parameter has a favorable and non-favorable relation with the watershed. Some parameters positively impact the watershed and others negatively. The integration of all such parameters can provide insights to mitigate risks. Integration of all parameters in a systematic and scientific manner can help in precise targeting of watershed interventions and agricultural development plans.

High-priority and moderate-priority watersheds are the best-suited sites for NRM interventions, such as construction of water structures, whereas low-priority areas have less a suitable environment potential for agricultural development. High-priority watersheds are highly suitable for constructing structures for irrigation, whereas moderate-priority watersheds can be utilized for multipurpose projects. Low-priority watersheds can be used for other purposes. The identification and delineation of such watershed areas help in better agricultural development planning, as well as implementation of appropriate interventions.

5. Conclusions

Identifying watersheds suitable for interventions is important for efficient utilization of natural resources. Prioritization is an important step for efficient natural resource management and increasing crop-water productivity. Using data generated from satellite imagery and information adapted from available open-source global data sets and national sources, we prepared spatial maps of watersheds in Nigeria. From this, we identified and prioritized suitable watersheds across the country for better agricultural, as well as livelihood, development. We integrated thematic layers prevailing in these watersheds and gave weighted scores to them with the help of experts and published papers. By the integration of these weighted layers, we generated a priority map of watersheds in Nigeria. The analysis showed that most of the areas in Nigeria fall in the class of moderate priority. Higher-resolution datasets can further improve these maps, and the method can be applicable to small areas to implement watershed interventions.

Author Contributions: Conceptualization, Murali Krishna Gumma; methodology, Murali Krishna Gumma and Pranay Panjala; validation, Pranay Panjala, Kumara Charyulu Deevi; writing—original draft preparation, Murali Krishna Gumma, Pranay Panjala and other authors; writing—review and editing, Murali Krishna Gumma, Pranay Panjala, Ramadjita Tabo and other authors; project administration, Hakeem Ayinde Ajeigbe; resources, Murtala Muhammad Badamasi. All authors have read and agreed to the published version of the manuscript.

Funding: This research received no external funding.

Institutional Review Board Statement: Not applicable.

Informed Consent Statement: Not applicable.

Data Availability Statement: The data that support the findings of this study are available from the corresponding author upon reasonable request.

Acknowledgments: The authors would like to thank the International Crops Research Institute for The Semi-Arid Tropic (ICRISAT), Centre for Dryland Agriculture Bayero University Kano, Nigeria for providing institutional funding to conduct the experimental work. We are grateful for the help of the field and the staff of the Agronomy Unit at ICRISAT Kano for assisting with field operations.

Conflicts of Interest: The authors declare no conflict of interest.

Appendix A

Table A1. Parameters related information and their sources.

Variable	Year	Type	Resolution	Source
Maximum Temperature	2010–2018	Raster	~4 km	TerraClimate
Minimum Temperature	2010–2018	Raster	~4 km	TerraClimate
Precipitation	2010–2018	Raster	~4 km	TerraClimate
Slope	2014	Raster	90 m	SRTM
Soil	1994	Vector	1:5,000,000	FAO [33]
LULC	2014	Raster	250 m	LULC [31]
LGP	2011	Raster	8 km	LGP [35]

References

1. Groninger, J.W. Building watershed management capacity in Nigeria: Expanding the role of agriculture colleges. *J. Contemp. Water Res. Educ.* **2016**, *158*, 78–84.
2. Adesina, A. Investing in Nigeria's Agricultural Value Chain. Presented at the Bank of Industry's Nigerian Investment Forum, London, UK, 30 July 2012; Available online: http://www.newworldnigeria.com/pdf/AdesinaLondonBOIForumfinal.pdf (accessed on 4 May 2020).
3. Khan, M.A.; Gupta, V.P.; Moharana, P.C. Watershed prioritization using remote sensing and geographical information system: A case study from Guhiya, India. *J. Arid. Environ.* **2001**, *49*, 465–475. [CrossRef]
4. Jewitt, G. Can integrated water resources management sustain the provision of ecosystem goods and services? *Phys. Chem. Earth Parts A/B/C* **2002**, *27*, 887–895. [CrossRef]
5. Gumma, M.K.; Birhanu, B.Z.; Mohammed, I.A.; Tabo, R.; Whitbread, A.M. Prioritization of watersheds across Mali using remote sensing data and GIS techniques for agricultural development planning. *Water* **2016**, *8*, 260. [CrossRef]
6. Gumma, M.; Thenkabail, P.S.; Fujii, H.; Namara, R. Spatial models for selecting the most suitable areas of rice cultivation in the Inland Valley Wetlands of Ghana using remote sensing and geographic information systems. *J. Appl. Remote Sens.* **2009**, *3*, 033537.
7. Bazzi, H.; Baghdadi, N.; Ienco, D.; El Hajj, M.; Zribi, M.; Belhouchette, H.; Escorihuela, M.J.; Demarez, V. Mapping irrigated areas using Sentinel-1 Time series in Catalonia, Spain. *Remote Sens.* **2019**, *11*, 1836. [CrossRef]
8. Moore, I.D.; Grayson, R.; Ladson, A. Digital terrain modelling: A review of hydrological, geomorphological, and biological applications. *Hydrol. Process.* **1991**, *5*, 3–30. [CrossRef]
9. Panhalkar, S.; Pawar, C. Watershed Development Prioritization by applying WERM model and GIS techniques in Vedganga Basin (INDIA). *ARPN J. Agric. Biol. Sci.* **2011**, *6*, 38–44.
10. Vittala, S.S.; Govindaiah, S.; Gowda, H.H. Prioritization of sub-watersheds for sustainable development and management of natural resources: An integrated approach using remote sensing, GIS and socio-economic data. *Curr. Sci.* **2008**, *95*, 345–354.
11. Iqbal, M.; Sajjad, H. Prioritization based on Morphometric Analysis of Dudhganga Catchment, Kashmir Valley, India using Remote Sensing and Geographical Information System. *Afr. J. Geo-Sci. Res.* **2014**, *2*, 1–6.
12. Kamaraju, M.V.V.; Bhattacharya, A.; Sreenivasa, R.; Chandrasekhar, R.; Murthy, G.S.; Rao, T.C.H.M. Ground-water potential evaluation of West Godavari district, Andhra Pradesh State, India-a GIS Approach. *Ground Water* **1996**, *34*, 318–325. [CrossRef]
13. Mattikalli, H.M.; Devereux, B.J.; Richards, K.S. Integration of remote sensedsatellite images with a Geographical Information System. *Comput. Geosci.* **1995**, *21*, 947–956. [CrossRef]
14. Murthy, K.S.R. Groundwater potential in a semi-arid region of Andhra Pradesh—A geographical information system approach. *Int. J. Remote Sens.* **2000**, *21*, 1867–1884. [CrossRef]

15. Sidhu, R.S.; Mehta, R.S. Delineation of groundwater potential zones in Kushawati river watershed a tributary of Zauri river in Goa, using remotely sensed data. In Proceedings of the National Symposium on Engineering Applications of Remote Sensing and Recent Advantages, Indore, India, 21–23 December 1989; pp. 41–46.
16. Gumma, M.K.; Pavelic, P. Mapping of groundwater potential zones across Ghana using remote sensing, geographic information systems, and spatial modeling. *Environ. Monit. Assess.* **2013**, *185*, 3561–3579. [CrossRef]
17. Hellden, U.; Olsson, L.; Stern, M. Approaches to desertification monitoring in Sudan. In *Satellite Remote Sensing in Developing Counties*; Lery, L.G., Ed.; European Space Agency: Paris, France, 1982; pp. 131–144.
18. Kushwaha, S.P.S. *Application of Remote Sensing in Shifting Cultivation Areas*; Technical Report; Abteilung Luftbildmessung and Fernerkundung, Universitat Freiburg: Freiburg, Germany, 1993; pp. 23–28.
19. Smith, A.Y.; Blackwell, R.J. Development of an information data base for watershed monitoring. *Photogramm. Eng. Remote Sens.* **1980**, *46*, 1027–1038.
20. Trotter, C.M. Remotely sensed data as information source for geographical information system in natural resources management: A review. *Int. J. Remote Sens.* **1991**, *5*, 225–239. [CrossRef]
21. Said, S.; Siddique, R.; Shakeel, M. Morphometric analysis and sub-watersheds prioritization of Nagmati River watershed, Kutch District, Gujarat using GIS based approach. *J. Water Land Dev.* **2018**, *39*, 131–139. [CrossRef]
22. Zheng, J.; Fu, H.; Li, W.; Wu, W.; Zhao, Y.; Dong, R.; Yu, L. Cross-regional oil palm tree counting and detection via a multi-level attention domain adaptation network. *ISPRS J. Photogramm. Remote Sens.* **2020**, *167*, 154–177. [CrossRef]
23. Singh, L.; Saravanan, S.; Jennifer, J.J.; Abijith, D. Application of multi-influence factor (MIF) technique for the identification of suitable sites for urban settlement in Tiruchirappalli City, Tamil Nadu, India. *Asia-Pac. J. Reg. Sci.* **2021**, *5*, 797–823. [CrossRef]
24. Roy, S.; Hazra, S.; Chanda, A.; Das, S. Assessment of groundwater potential zones using multi-criteria decision-making technique: A micro-level case study from red and lateritic zone (RLZ) of West Bengal, India. *Sustain. Water Resour. Manag.* **2020**, *6*, 4. [CrossRef]
25. Serele, C.; Pérez-Hoyos, A.; Kayitakire, F. Mapping of groundwater potential zones in the drought-prone areas of south Madagascar using geospatial techniques. *Geosci. Front.* **2020**, *11*, 1403–1413. [CrossRef]
26. Abd Manap, M.; Nampak, H.; Pradhan, B.; Lee, S.; Sulaiman, W.N.A.; Ramli, M.F. Application of probabilistic-based frequency ratio model in groundwater potential mapping using remote sensing data and GIS. *Arab. J. Geosci.* **2014**, *7*, 711–724. [CrossRef]
27. Rahmati, O.; Samadi, M.; Shahabi, H.; Azareh, A.; Rafiei-Sardooi, E.; Alilou, H.; Melesse, A.M.; Pradhan, B.; Chapi, K.; Shirzadi, A. SWPT: An automated GIS-based tool for prioritization of sub-watersheds based on morphometric and topo-hydrological factors. *Geosci. Front.* **2019**, *10*, 2167–2175. [CrossRef]
28. Kumar, K.C.A.; Reddy, G.P.O.; Masilamani, P.; Sandeep, P. Spatial modelling for identification of groundwater potential zones in semi-arid ecosystem of southern India using Sentinel-2 data, GIS and bivariate statistical models. *Arab. J. Geosci.* **2021**, *14*, 1362. [CrossRef]
29. Pandey, M.; Sharma, P.K. Remote sensing and GIS based watershed prioritization. In Proceedings of the 2017 IEEE International Geoscience and Remote Sensing Symposium (IGARSS), Fort Worth, TX, USA, 23–28 July 2017; pp. 6182–6185.
30. Aher, P.; Adinarayana, J.; Gorantiwar, S. Quantification of morphometric characterization and prioritization for management planning in semi-arid tropics of India: A remote sensing and GIS approach. *J. Hydrol.* **2014**, *511*, 850–860. [CrossRef]
31. Xiong, J.; Thenkabail, P.S.; Gumma, M.K.; Teluguntla, P.; Poehnelt, J.; Congalton, R.G.; Yadav, K.; Thau, D. Automated cropland mapping of continental Africa using Google Earth Engine cloud computing. *ISPRS J. Photogramm. Remote Sens.* **2017**, *126*, 225–244. [CrossRef]
32. Gumma, M.K.; Thenkabail, P.S.; Hideto, F.; Nelson, A.; Dheeravath, V.; Busia, D.; Rala, A. Mapping irrigated areas of Ghana using fusion of 30 m and 250 m resolution remote-sensing data. *Remote Sens.* **2011**, *3*, 816–835. [CrossRef]
33. FAO. AQUASTAT—FAO Water Report 29. 2005. Available online: http://www.fao.org/nr/water/aquastat/countries/ghana/index.stm (accessed on 11 February 2022).
34. Abatzoglou, J.T.; Dobrowski, S.Z.; Parks, S.A.; Hegewisch, K.C. TerraClimate, a high-resolution global dataset of monthly climate and climatic water balance from 1958–2015. *Sci. Data* **2018**, *5*, 170191. [CrossRef]
35. Vrieling, A.; De Leeuw, J.; Said, M.Y. Length of growing period over Africa: Variability and trends from 30 years of NDVI time series. *Remote Sens.* **2013**, *5*, 982–1000. [CrossRef]
36. Murthy, K.S.R.; Mamo, A.G. Multi-criteria decision evaluation in groundwater zones identification in Moyale-Teltele subbasin, South Ethiopia. *Int. J. Remote Sens.* **2009**, *30*, 2729–2740. [CrossRef]
37. Moran, M.S.; Peters-Lidard, C.D.; Watts, J.M.; McElroy, S. Estimating soil moisture at the watershed scale with satellite-based radar and land surface models. *Can. J. Remote Sens.* **2004**, *30*, 805–826. [CrossRef]
38. Belgiu, M.; Csillik, O. Sentinel-2 cropland mapping using pixel-based and object-based time-weighted dynamic time warping analysis. *Remote Sens. Environ.* **2018**, *204*, 509–523. [CrossRef]
39. Baja, S.; Chapman, D.M.; Dragovich, D. A conceptual model for defining and assessing land management units using a fuzzy modeling approach in GIS environment. *Environ. Manag.* **2002**, *29*, 647–661. [CrossRef]
40. Chowdary, V.; Chakraborthy, D.; Jeyaram, A.; Murthy, Y.K.; Sharma, J.; Dadhwal, V. Multi-criteria decision making approach for watershed prioritization using analytic hierarchy process technique and GIS. *Water Resour. Manag.* **2013**, *27*, 3555–3571. [CrossRef]
41. Jaiswal, R.; Thomas, T.; Galkate, R.; Ghosh, N.; Singh, S. Watershed prioritization using Saaty's AHP based decision support for soil conservation measures. *Water Resour. Manag.* **2014**, *28*, 475–494. [CrossRef]

42. Gopinath, G.; Nair, A.G.; Ambili, G.; Swetha, T. Watershed prioritization based on morphometric analysis coupled with multi criteria decision making. *Arab. J. Geosci.* **2016**, *9*, 129. [CrossRef]
43. Fashae, O.A.; Tijani, M.N.; Talabi, A.O.; Adedeji, O.I. Delineation of groundwater potential zones in the crystalline basement terrain of SW-Nigeria: An integrated GIS and remote sensing approach. *Appl. Water Sci.* **2014**, *4*, 19–38. [CrossRef]
44. Welde, K. Identification and prioritization of subwatersheds for land and water management in Tekeze dam watershed, Northern Ethiopia. *Int. Soil Water Conserv. Res.* **2016**, *4*, 30–38. [CrossRef]
45. Biswas, S.; Sudhakar, S.; Desai, V.R. Prioritisation of subwatersheds based on morphometric analysis of drainage basin: A remote sensing and gis approach. *J. Indian Soc. Remote Sens.* **1999**, *27*, 155. [CrossRef]
46. Farhan, Y.; Anbar, A.; Al-Shaikh, N.; Mousa, R. Prioritization of semi-arid agricultural watershed using morphometric and principal component analysis, remote sensing, and GIS techniques, the Zerqa River Watershed, Northern Jordan. *Agric. Sci.* **2016**, *8*, 113–148. [CrossRef]
47. Farhan, Y.; Nawaiseh, S. Spatial assessment of soil erosion risk using RUSLE and GIS techniques. *Environ. Earth Sci.* **2015**, *74*, 4649–4669.
48. Gajbhiye, S.; Mishra, S.; Pandey, A. Prioritizing erosion-prone area through morphometric analysis: An RS and GIS perspective. *Appl. Water Sci.* **2014**, *4*, 51–61. [CrossRef]
49. Kalin, L.; Hantush, M.M. An auxiliary method to reduce potential adverse impacts of projected land developments: Subwatershed prioritization. *Environ. Manag.* **2009**, *43*, 311. [CrossRef] [PubMed]
50. Besalatpour, A.; Hajabbasi, M.A.; Ayoubi, S.; Jalalian, A. Identification and prioritization of critical sub-basins in a highly mountainous watershed using SWAT model. *Eurasian J. Soil Sci.* **2012**, *1*, 58–64.
51. Aouragh, M.H.; Essahlaoui, A. A TOPSIS approach-based morphometric analysis for sub-watersheds prioritization of high Oum Er-Rbia basin, Morocco. *Spat. Inf. Res.* **2018**, *26*, 187–202. [CrossRef]
52. Rawat, K.S.; Tripathi, V.K.; Mishra, A.K. Sediment yield index mapping and prioritization of Madia subwatershed, Sagar District of Madhya Pradesh (India). *Arab. J. Geosci.* **2014**, *7*, 3131–3145. [CrossRef]
53. Kundu, S.; Khare, D.; Mondal, A. Landuse change impact on sub-watersheds prioritization by analytical hierarchy process (AHP). *Ecol. Inform.* **2017**, *42*, 100–113. [CrossRef]
54. Prasad, R.N.; Pani, P. Geo-hydrological analysis and sub watershed prioritization for flash flood risk using weighted sum model and Snyder's synthetic unit hydrograph. *Model. Earth Syst. Environ.* **2017**, *3*, 1491–1502. [CrossRef]
55. Ekness, P.; Randhir, T.O. Effect of climate and land cover changes on watershed runoff: A multivariate assessment for storm water management. *J. Geophys. Res. Biogeosci.* **2015**, *120*, 1785–1796. [CrossRef]

International Journal of
Geo-Information

MDPI

Article

Landslide Susceptibility Prediction Based on High-Trust Non-Landslide Point Selection

Yizhun Zhang [1] and Qisheng Yan [2,*]

[1] School of Earth Sciences, East China University of Technology, Nanchang 330013, China; 2020120017@ecut.edu.cn
[2] School of Science, East China University of Technology, Nanchang 330013, China
* Correspondence: 199760023@ecut.edu.cn

Abstract: Landslide susceptibility prediction has the disadvantages of being challenging to apply to expanding landslide samples and the low accuracy of a subjective random selection of non-landslide samples. Taking Fu'an City, Fujian Province, as an example, a model based on a semi-supervised framework using particle swarm optimization to optimize extreme learning machines (SS-PSO-ELM) is proposed. Based on the landslide samples, a semi-supervised learning framework is constructed through Density Peak Clustering (DPC), Frequency Ratio (FR), and Random Forest (RF) models to expand and divide the landslide sample data. The landslide susceptibility was predicted using high-trust sample data as the input variables of the data-driven model. The results show that the area under the curve (AUC) valued at the SS-PSO-ELM model for landslide susceptibility prediction is 0.893 and the root means square error (RMSE) is 0.370, which is better than ELM and PSO-ELM models without the semi-supervised framework. It shows that the SS-PSO-ELM model is more effective in landslide susceptibility. Thus, it provides a new research idea for predicting landslide susceptibility.

Keywords: landslide susceptibility prediction; semi-supervised learning; clustering by fast search and finding density peaks; random forest; extreme learning machine

Citation: Zhang, Y.; Yan, Q. Landslide Susceptibility Prediction Based on High-Trust Non-Landslide Point Selection. *ISPRS Int. J. Geo-Inf.* **2022**, *11*, 398. https://doi.org/10.3390/ijgi11070398

Academic Editors: Walter Chen, Fuan Tsai and Wolfgang Kainz

Received: 2 June 2022
Accepted: 12 July 2022
Published: 13 July 2022

1. Introduction

Landslide is a complex geological phenomenon, determined by how the rock mass on the slope is affected by rainwater soaking and artificial factors and how it slides down due to gravity. It is the most common geological disaster in the world [1]. Landslides cause severe casualties and economic losses every year, seriously restricting the economic development of some regions. In many areas, disasters have hindered the development of cities and become a barrier to poverty alleviation in various countries [2,3]. Therefore, how to effectively predict the susceptibility to landslides has become a hotspot in current landslide research [4]. Drawing accurate landslide susceptibility maps can provide essential guidance for early warning and prevention and provide a basis and suggestions for disaster prevention and mitigation work.

Many scholars have researched landslide disasters, including susceptibility prediction, disaster risk assessment, landslide mechanism analysis, and detection [5–9]. Landslide susceptibility prediction comprehensively analyzes various geological and environmental factors, historical landslide data, and physical laws of landslides in the study area to identify the probability of future landslides in the study area [10]. The principal methods of landslide susceptibility prediction are empirical models, statistical models, and machine learning models. Lyu et al. [11] used the analytic hierarchy process to predict the susceptibility to geological disasters and assess the disaster risk in Lanzhou. They provided suggestions and a basis for disaster prevention work in Lanzhou. In a statistical model, Khan et al. [12] used frequency ratio techniques to map landslide susceptibility in the northern region of Pakistan. They drew a landslide susceptibility map based on the relationship between landslide inventories and landslide causative factors compiled from visual interpretations of SPOT-5 images, providing a basis for relevant agencies to formulate and implement landslide mitigation measures.

The development of machine learning, compared with previous empirical and statistical models, has a better nonlinear predictive ability in landslide susceptibility prediction [13]. Nevertheless, from the current research results, almost all machine learning methods for analyzing the potential risk of landslides rely heavily on inventory datasets of the known spatial extent of landslides or the characteristic GPS location of each known landslide in the target study area [14]. Therefore, landslide susceptibility prediction requires more detailed and accurate maps and inventories [15]. Thus, evaluating the application of different machine learning methods and deep learning convolutional neural networks in landslide detection and susceptibility prediction has become an essential task for landslide applications. For example, Ghorbanzadeh et al. used deep learning models to study landslide detection and the development and validation of methods for systematically updating landslide lists [16]. Balogun et al. [17] used the gray wolf optimization algorithm, the bat algorithm, and the cuckoo algorithm to jointly optimize the support vector machine regression model's parameters, which improved the landslide susceptibility prediction accuracy in western Serbia. Ivan et al. [18] employed a statistically calibrated Bayesian framework and introduced an approximate likelihood formulation, leading to the improved prediction accuracy of landslide susceptibility. Guo et al. [19] proposed a prediction model of back propagation neural network based on wavelet analysis and a gray wolf optimization algorithm. Taking China's Three Gorges Reservoir area as an example, landslide displacement was predicted, providing the basis for landslide warning. Zhang et al. [20] proposed a BP neural network model optimized by a new water cycle algorithm. The model was used to predict landslides in the Three Gorges Reservoir area. It has a faster convergence speed and higher prediction accuracy than the traditional BPNN model. Benbouras used particle swarm optimization (PSO), genetic algorithm (GA), and nine other hybrid meta-heuristic algorithms to spatially predict landslide susceptibility in the Sahel region of Algeria. Moreover, it draws an accurate map to help land-use managers and policymakers mitigate landslide hazards [21].

Although machine learning models have achieved a series of results in predicting landslide susceptibility, there are still some deficiencies in the use of machine learning models in landslide susceptibility prediction. For example, when using machine learning models to predict landslide susceptibility. It is difficult to obtain landslide sample data in the wild [22]. Moreover, the existing research is challenging to select valuable non-landslide raster data [23]. In a previous study, non-landslide points were randomly selected in the study area or based on expert experience. This decision can lead to bias and overfitting, leading to immeasurable errors in data processing, resulting in reduced model prediction accuracy [24]. This paper proposes an extreme learning machine model based on a semi-supervised framework and uses the particle swarm optimization algorithm to optimize the parameters of the extreme learning machine (SS-PSO-ELM). The model is used to expand the landslide sample data and divide the high-trust non-landslide sample data, which solves the shortcomings of the existing model and further improves the accuracy of landslide sensitivity mapping.

The semi-supervised learning method divides the unlabeled sample data according to the labeled sample data. The method's core assumes that the unlabeled samples can provide helpful feature space distribution information [25]. Using the clustering algorithm to realize the pre-training and classification of data pseudo-labels can alleviate the difficulty in obtaining accurate sample data to the greatest extent. Semi-supervised learning has been widely used in sample data analysis and evaluation [26–28]. In landslide susceptibility prediction and landslide detection, supervised learning frameworks, semi-supervised learning frameworks, and unsupervised learning frameworks have also demonstrated their superiority [29–31]. This paper selects Fu'an City, Fujian Province, China, as the research area. The SS-PSO-ELM and the ELM and PSO-ELM models without a semi-supervised framework are used for comparative analysis to explore the semi-supervised learning framework's modeling effect.

2. Overview and Data of the Study Area

2.1. Study Area

The study area is Fu'an City, Fujian Province, China, located in the northeastern part of Fujian province. As shown in Figure 1, between 119°23′~119°52′ E and 26°41′~27°24′ N, the total area is 1880 km^2. The study area is located near the ocean, the climate is warm and humid, and the climate is a subtropical marine monsoon climate. The study area contains three major mountain ranges: the southeast slope of the Jiufeng mountains, the southwest Taimu mountains, and the Donggong mountains. The mountain trend is roughly northeast-southwest, and the terrain is inclined to the south. The east and west are high, and the middle is low-lying. The study area forms a north-south valley. The stratigraphic Mesozoic in the study area has an extensive distribution range, and the Cenozoic and Sinian sub-world are only exposed in small spaces. Landforms are divided into five types: mountains, hills, valleys, plains, and beaches [32].

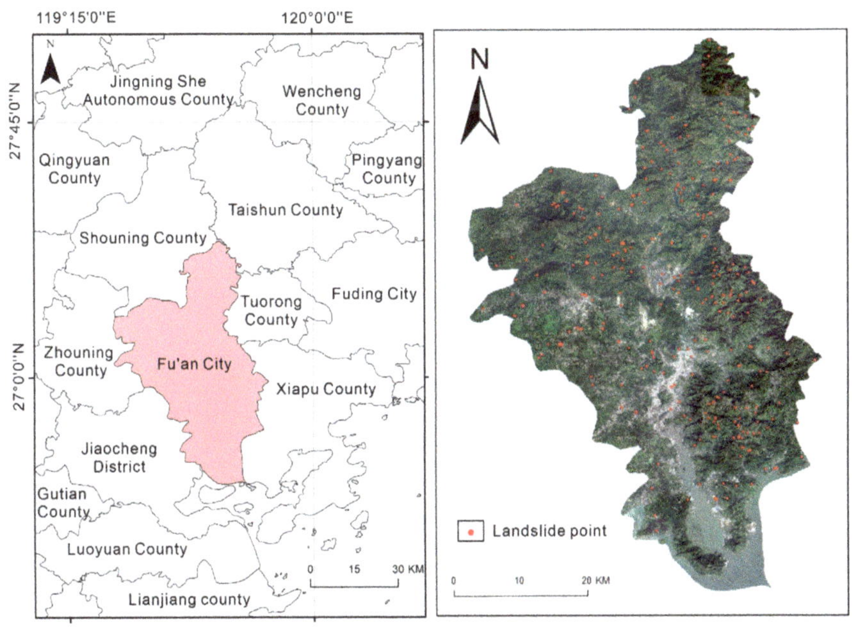

Figure 1. Geographical location and landslide location map of Fu'an.

2.2. Data Sources

The primary data sources are: (1) Field investigation and relevant landslide data obtained by Fu'an Natural Resources Bureau; (2) From Geospatial Data Cloud (https://www.gscloud.cn/) (accessed date: 19 November 2021), 30 m resolution DEM data and Landsat 8 remote sensing images to extract elevation, slope, NDVI, plane curvature, profile curvature, river system distance, slope aspect, and other information, as shown in Figure 2; (3) A 1:200,000 geological map to obtain the lithological data of the study area, as shown in Figure 2h.

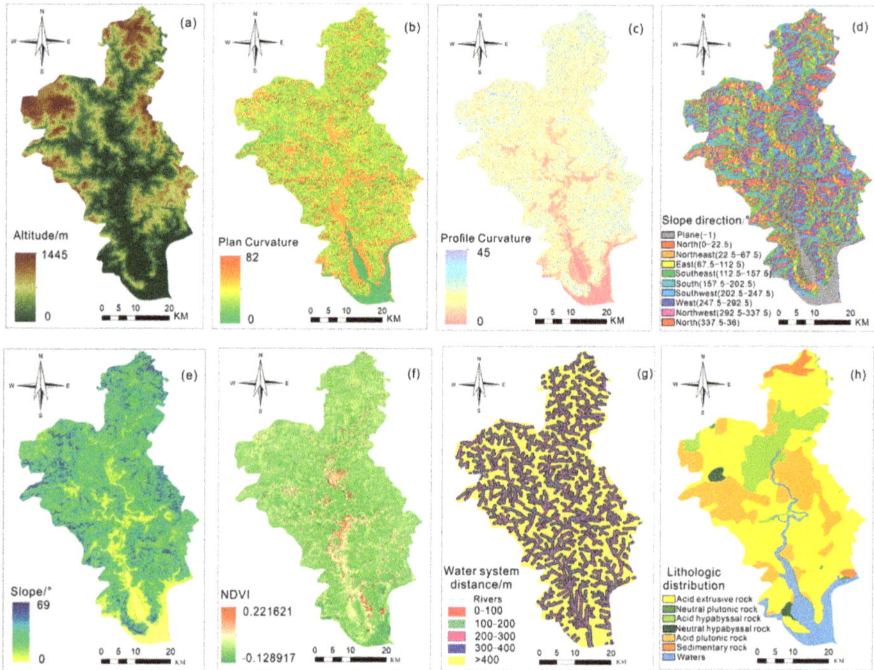

Figure 2. Environmental factors of landslides in Fu'an. (**a**) Elevation map. (**b**) Plane curvature map. (**c**) Profile curvature map. (**d**) Slope direction map. (**e**) Slope map. (**f**) NDVI map. (**g**) Water system distance map. (**h**) Lithology map.

2.3. Environmental Factors

According to the geographical situation of Fu'an City, the existing landslide research, and the introduction of relevant references, most of the landslides in Fu'an City are located in relatively high terrain. Landslides are mainly distributed over the eastern and surrounding areas and more minor in the central and western regions. This paper extracts eight landslide environmental factors: elevation, slope, NDVI, plane curvature, section curvature, water system distance, slope aspect, and lithology.

3. Methods

The flow of the SS-PSO-ELM model proposed in this paper is shown in Figure 3: (1) Landslide location information and environmental factor data in the study area are obtained based on field surveys; (2) A semi-supervised learning framework is constructed based on a density peak clustering algorithm, random forest model, and frequency ratio method, and using a semi-supervised learning framework to convert landslide information and environmental factor data from field surveys into high-trust non-landslide data and landslide data; (3) High-trust data are weighted using a max-correlation min-redundancy algorithm; (4) The weighted data are substituted onto the PSO-ELM model to predict landslide susceptibility and draw a landslide susceptibility map; (5) Using ROC curve, landslide susceptibility index, and root mean square error, the prediction accuracy of the landslide is evaluated, to provide new research ideas and theoretical guidance for landslide susceptibility prediction.

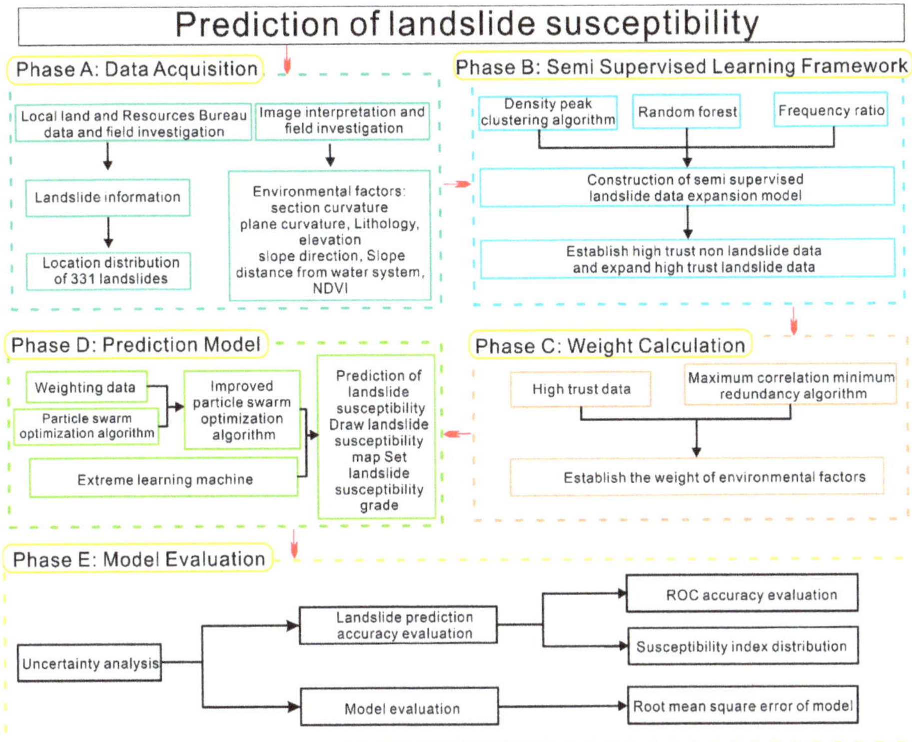

Figure 3. Landslide susceptibility prediction modeling flowchart.

3.1. Density Peak Clustering Algorithm

The density peak clustering algorithm is a new clustering algorithm proposed by Rodriguez in 2014 [33]. The algorithm has the advantages of the mean clustering method, hierarchical clustering method, grid clustering algorithm, and density clustering algorithm, which are fast and straightforward, and insensitive to noise, and overcomes the shortcomings of the high computational complexity of the existing traditional clustering algorithms. The density peak clustering algorithm defines new clustering metrics: Minimum Density Distance and Local Density. The algorithm uses low-density areas to distinguish high-density areas and can quickly and effectively identify cluster centers in many data. This is suitable for data of any distribution type [34].

Let the local density be ρ, the minimum density distance be δ, the local density of the data point x_a be ρ_a, and the reach of data point x_a to the nearest data point x_b whose local density is more significant than itself be δ_a.

The local density formula of the density peak clustering algorithm is:

$$\rho_a = \sum_{a \neq b} \chi(d_{ab} - d_c) \tag{1}$$

The minimum density distance formula is:

$$\delta_a = \min_{b:\rho_b > \rho_a} (d_{ab}) \tag{2}$$

In the procedure, $\chi(\bullet)$ is the logical judgment function, $(\bullet) < 0$, $\chi(\bullet) = 1$, otherwise $\chi(\bullet) = 0$, d_{ab} is the distance between x_a and x_b, and d_c is the cut-off distance. Take δ as the abscissa and ρ as the ordinate to get the clustering decision diagram of the density peak clustering algorithm. Select several points of the relatively large distance between local density and minimum density as cluster center points and remove noise points with relatively low local density but rather large minimum density distance. Finally, the data close to the cluster center are grouped into a cluster to complete the clustering.

3.2. Max-Correlation Min-Redundancy Algorithm

The maximum correlation minimum redundancy algorithm was first proposed by Peng [35] to extract optimal eigenvalues. The algorithm's core is to find the feature of the most significant correlation between the dependent variable and the little correlation between the independent variables in a part set to delete and simplify the feature set and eliminate redundant variables.

The maximum correlation minimum redundancy algorithm calculates the correlation and redundancy between features based on mutual information. Let the two variables be X and Y. The mutual information formula is:

$$I(X;Y) = \int\int P(x,y)\log\frac{P(X,Y)}{P(X)P(Y)}\mathrm{d}X\mathrm{d}Y \quad (3)$$

where $P(X,Y)$ is the joint probability function of X and Y and $P(X)$ and $P(Y)$ are the probability density functions of X and Y, respectively. Mutual information can be understood as the amount of data that contains the Y variable in the X variable.

The maximum correlation is defined as:

$$\begin{cases} \max D(S,\mathrm{c}) \\ D = \frac{1}{|\mathrm{s}|}\sum\limits_{i=1}^{s} I(x_i;c) \end{cases} \quad (4)$$

The minimum redundancy is defined as:

$$\begin{cases} \min R(S) \\ R = \frac{1}{|S|^2}\sum\limits_{i=1}^{s-1}\sum\limits_{j=i+1}^{s} I(x_i,x_j) \end{cases} \quad (5)$$

where S is the feature set composed of factors, c is the target value, and $I(X_i;c)$ is the mutual information between the factor features and the target.

The feature selection criteria of the maximum correlation minimum redundancy algorithm are:

Information subtraction:

$$\begin{cases} \max\varphi(D,R) \\ \varphi(D,R) = D - R \end{cases} \quad (6)$$

Information entropy:

$$\begin{cases} \max\varphi(D,R) \\ \varphi(D,R) = D/R \end{cases} \quad (7)$$

According to information entropy or information subtraction, the total score of correlation and redundancy between factors is obtained. Then, factors are selected to be removed based on the score or the weight of each factor is calculated.

3.3. Extreme Learning Machine

Huang [36] proposed the extreme learning machine, which improved the traditional feedforward neural network's slow learning speed, making it easy to fall into a local minimum, making it easy to overtrain, and causing the generalization performance to decline. Extreme Learning Machine is a machine learning based on a feedforward neural network. The innovations include: (1) The connection weights and thresholds of the input layer and the hidden layer can be set randomly; there is no need to adjust after setting, reducing the amount of calculation. (2) The weight between the hidden layer and the output layer does not need to be iteratively adjusted and is converted into a method for solving the equation system.

The calculation process of extreme learning machine can be expressed as:

$$f_L(x) = \sum_{i=1}^{L} \beta_i g(w_i * x_j + b_j), j = 1, \ldots, N \tag{8}$$

where L is the number of hidden units, β_i is the weight vector between the ith hidden layer and the output layer, g is the activation function, b is the bias vector, w_i is the weight vector between the input layer and the hidden layer, and N is the number of training samples.

$$T = H \cdot \beta \tag{9}$$

Formula (8) can be transformed into Formula (9), where H is the output matrix of the hidden layer, β is the output weight, and T is the output result. Once the input weight w_i and the paranoid vector b are randomly determined, the output matrix H is uniquely determined, and the output weight β can be determined.

$$\hat{\beta} = H^{\dagger} T \tag{10}$$

In Formula (10), $H^{\dagger}$ is the Moore-Penrose generalized matrix of H.

$$Y = H' \hat{\beta} \tag{11}$$

Substitute the test set into Formula (10) to calculate the hidden layer output matrix H', and obtain the test set result.

3.4. Random Forest

Breiman first proposed random forest [37]. Random forest is based on a decision tree model. A more stable model is obtained by fusion of multiple decision trees, combining a random selection of features and integration ideas. The model randomly selects components and samples, so each tree has differences and similarities. Each tree predicts the pieces and obtains the final decision through voting [3].

3.5. Particle Swarm Optimization Algorithm

Particle swarm optimization is an evolutionary algorithm that imitates the foraging behavior of birds, first proposed by Kennedy and Eberhart [38,39]. Particle swarm optimization has the advantages of fast convergence speed and high optimization performance. Moreover, it will not fall into a local optimum.

The core of particle swarm optimization is that in the D-dimensional particle search space, there are n particles. All particles have a fitness value determined by the optimized function. Each particle's vector velocity determines the distance and direction they fly. The particles will follow the current optimal particle to search in the space, and finally, all converge to the vicinity of the optimal value [40].

3.6. Uncertainty Analysis Method

3.6.1. ROC Curve Precision Analysis

The ROC curve is drawn by taking the valid positive rate (sensitivity) as the ordinate and the false positive rate (1-specificity) as the abscissa. The closer the curve is to the upper left corner, the higher the accuracy, and the larger the area under the ROC curve, the better the effect. The ROC curve indicator is defined as:

Sensitivity:

$$SST = \frac{TP}{TP + FN} \tag{12}$$

Specificity:

$$SPF = \frac{TN}{TN + FP} \quad (13)$$

From the sample results, the data can be divided into two categories. For example, in this paper, the positive data are the sample data of landslides, and the negative data are the non-landslide sample data. (1) *TP*: Positive data predict the correct number. (2) *TN*: Negative data predict the correct number. (3) *FP*: Number of positive data prediction errors. (4) *FN*: Number of negative data prediction errors. (5) *SST*: The proportion of positive samples that are correctly classified. (6) *SPF*: The proportion of negative samples that are correctly classified.

3.6.2. Frequency Ratio

The frequency ratio reflects the distribution of factors of the class and can well explain the intrinsic relationship between factors and classes [41]. The formula for calculating the frequency ratio is:

$$FR = \frac{N_j/N}{S_j/S} \quad (14)$$

where N_j is the number of landslide grids in a cluster, N is the number of landslide grids in all groups, S_j is the number of units in the bunch, and S is the total number of grids shared by all clusters.

3.6.3. Root Mean Square Error Analysis

The root mean square error is the square root of the square ratio of the deviation from the observed value and the actual value and the number of observations n. *RMSE* is very sensitive to the large and small errors of measurement data, so *RMSE* can well reflect the accuracy of the measurement. The mathematical formula for the root mean square error is:

$$RMSE = \sqrt{\frac{1}{n}\sum_{i=1}^{n} (\hat{y}_i - y_i)^2} \quad (15)$$

where $\hat{y}_i$ is the actual value, y_i is the predicted value, and n is the number of observations.

4. Modeling of Landslide Susceptibility Assessment in Fu'an

4.1. Semi-Supervised Learning Framework Construction

Before making a landslide susceptibility prediction, high-trust non-landslide points were selected and high-trust landslide points were expanded to compensate for the lack of landslide data and the uncertainty caused by the random selection of non-landslide problems.

The flowchart of the semi-supervised learning framework is shown in Figure 4: (1) The data of the study area were organized into raster data. A total of 2,191,350 grid cells were obtained in the study area, with randomly selected 622 raster cell data from which landslide data and non-landslide data are 1:1, and the 622 data are clustered by the density peak clustering algorithm, as shown in Figure 5. Figure 5a is the cluster center selection diagram, the abscissa is the density of data points, and the ordinate is the distance from the point to the nearest higher density point. The density peak clustering algorithm selects a point with a higher density and no higher density nearby as the cluster center point. Therefore, according to Figure 5a, 489, 324, 367, 455, and 388 were selected as the cluster center points. Figure 5b shows that the remaining points are divided according to the five cluster center points. All the data are divided into five categories, and the cross symbols indicate the positions of the five cluster centers. (2) The categories calculated by the clustering algorithm were analyzed according to the frequency ratio method, and category a, with the most landslide data, and category b, with the most non-landslide data, were selected. The optimal condition is that both the proportion of landslide data in a and the ratio of non-landslide data in b exceed 0.7 (according to the existing research foundation and multiple experiments, it has been proved that 0.7 is the best threshold for experimental results; less than 0.7 is not ideal, and data with a threshold over 0.7 are prone to local redundancy). Otherwise, repeat step 1. The meaning of this step is to select two types of data from the five types of data, one of which has a landslide ratio higher than 0.7 and the other type of data whose non-landslide percentage is higher than 0.7. According to the density peak clustering algorithm, the same kind of data approximates in space if the proportion of landslide or non-landslide in a data class is the majority. It can be demonstrated that landslides or non-landslides in such data may be a standard feature. (3) The high-trust clustering data is used as training data on the random forest. Predict existing high-trust clustered data (the data for the first loop is itself). Assign prediction results to pseudo labels. In this step, the cluster data obtained in the above steps are added to the training data of the random forest model, and then the data are predicted and classified. Suppose the label of the predicted class is the same as the label of the previous step. The credibility of this piece of data will increase. (4) Add frequency labels. When the pseudo-labels and clustering labels are the same, the frequency label is increased by one. When the number of program loops gradually increases, the training data of the random forest will also gradually increase, which will cause the prediction results of the random forest model to fluctuate. As the number of loops increases, the larger the value of the frequency label, the more it can be proved that when the training data increase, the data have little effect on it. It is proved that the landslides (non-landslides) in the data have more in common with the data of multiple cluster classifications, indicating that the cluster labels of the data are more credible. (5) Determine whether the value of the frequency label reaches the threshold set by the end condition (when the frequency label value of a certain piece of data reaches 10, select all data with a frequency label greater than seven as high-confidence data). If not, return to step 1. Change the raster data selection method selected in step 1. A random selection of 622 cell data from all raster cells, regardless of the proportion of landslide and non-landslide data. Moreover, the data were compared with existing high-confidence data to remove redundant data.

The final high-trust data obtained are shown in Table 1. The more matching values, the higher the reliability of the data, and the smaller the number of matching values, which proves that the data fluctuate wildly and cannot be accepted as high-trust data.

Figure 6 shows the high-trust non-landslide points distribution in the study area. It can be seen from Figure 6a that a large part of the high-trust non-landslide point data are on the water surface. Figure 6b shows that most high-trust non-landslide point data are distributed over low-altitude areas, proving that it is advisable to adopt a semi-supervised framework to select high-trust non-landslide points.

Table 1. Highly trusted data (excerpt).

Grid Cell Number	Elevation (m)	Slope Direction (°)	Slope (°)	Distance from Water System (m)	Cluster Labels	Match Count
1,994,470	0	−1.00	0.00	0	No landslide	10
392,364	93	343.98	25.87	100	No landslide	10
1,160,375	30	282.52	4.27	200	No landslide	6
1,161,694	37	67.28	20.70	100	No landslide	5
153,813	478	109.13	11.87	500	Landslide	10
888,368	429	135.66	44.76	300	Landslide	10
1,784,541	271	203.08	28.26	100	Landslide	5

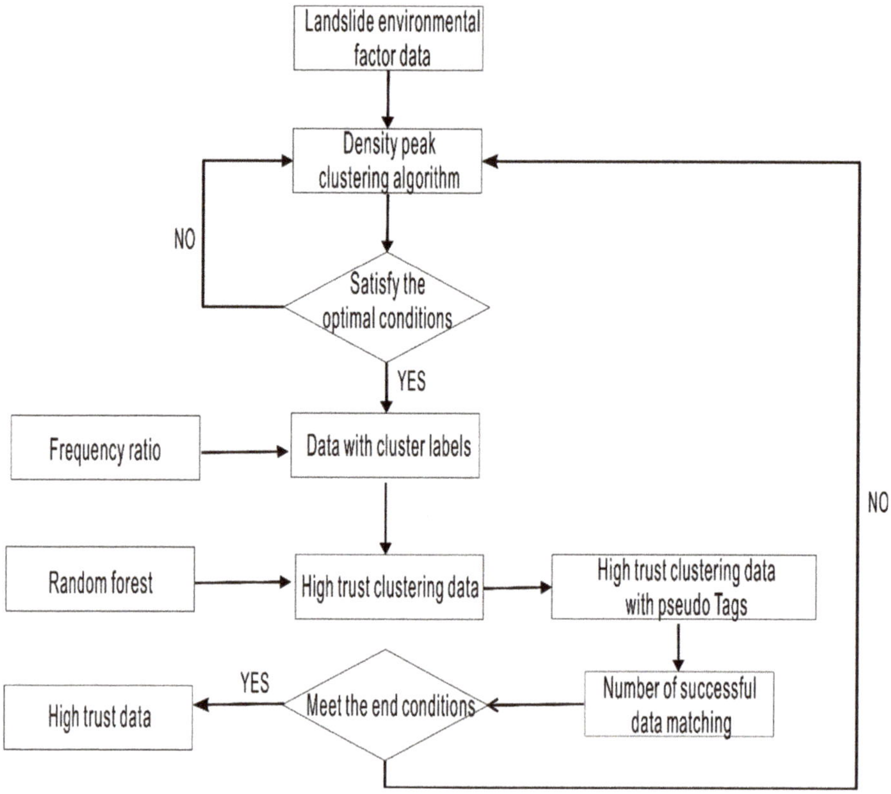

Figure 4. Flowchart of the semi-supervised learning framework.

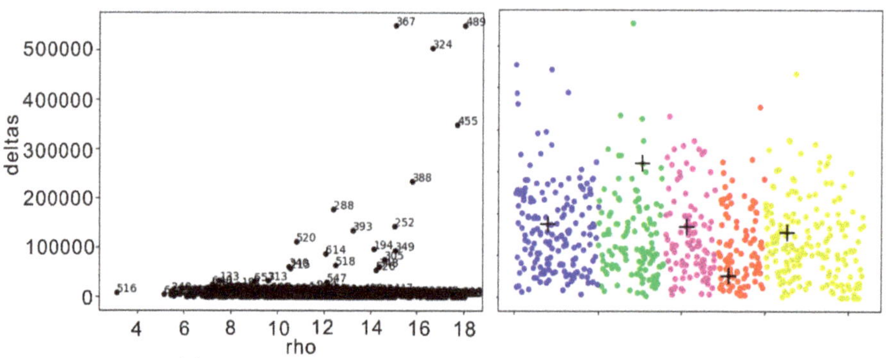

(**a**) (**b**)

Figure 5. Flowchart of the semi-supervised learning framework. (**a**) Cluster center distribution map. (**b**) Cluster distribution map.

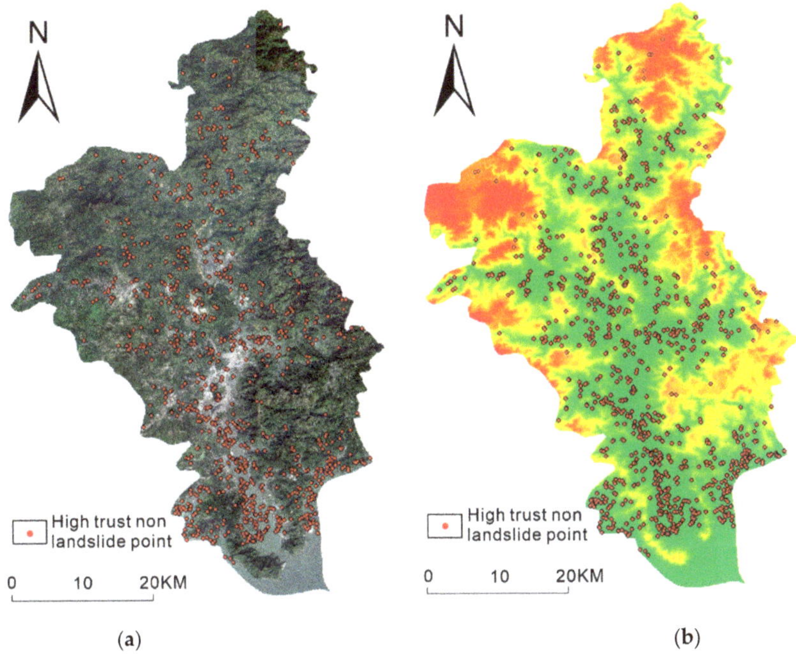

Figure 6. High-trust non-landslide location map. (**a**) High-trust non-landslide satellite imagery. (**b**) High-trust non-landslide elevation map.

4.2. Weight Determination Analysis

Based on the high-trust data obtained above, the maximum correlation minimum redundancy algorithm is used to calculate the weight of the landslide environmental factors. The mutual information on each environmental element and the landslide is shown in Figure 7, and the weights of environmental factors are shown in Figure 8. Mutual information represents the amount of information one random variable contains in another. Therefore, the higher the mutual information, the closer the relationship between the two variables. Figure 7 shows the mutual information between various environmental factors. It can be seen from the mutual information between each environmental element and landslide in Figure 7 that the mutual information between the slope aspect and landslide is the largest, with a value of 0.86. However, in the final weights shown in Figure 8, the influence of the slope direction on the landslide is ranked second. The slope direction and landslide have high mutual information, and the slope direction and other environmental factors also have high mutual information. Therefore, when the slope direction is used as the input for landslide prediction, if the weight of the slope direction is too high, it will lead to more redundancy and more significant prediction errors. Therefore, in the final weights calculated by the maximum correlation minimum redundancy algorithm, the weight of the slope direction is less than the weight of the elevation. This proves that it is feasible to calculate the importance of the landslide factor based on the maximum correlation minimum redundancy algorithm.

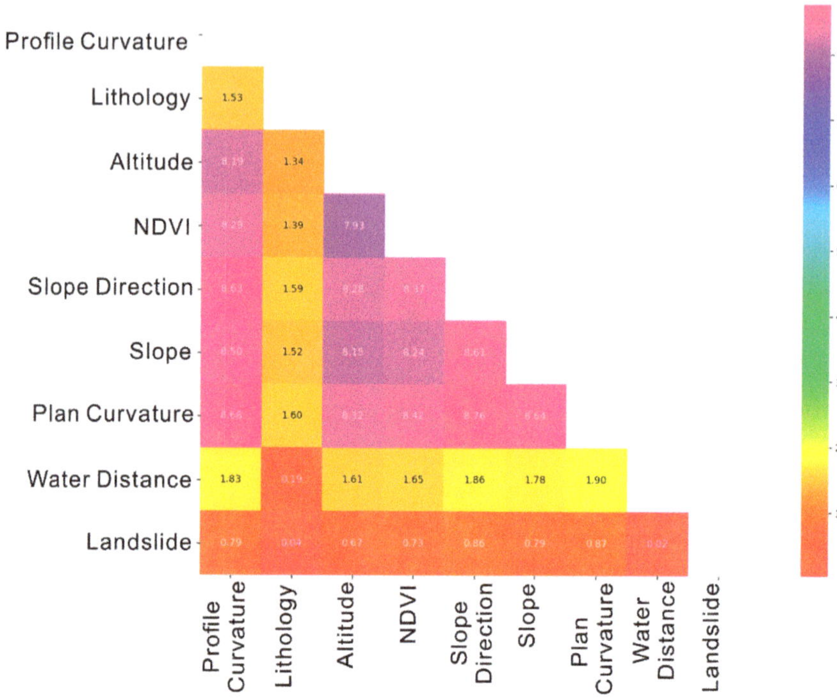

Figure 7. Environmental factor mutual information.

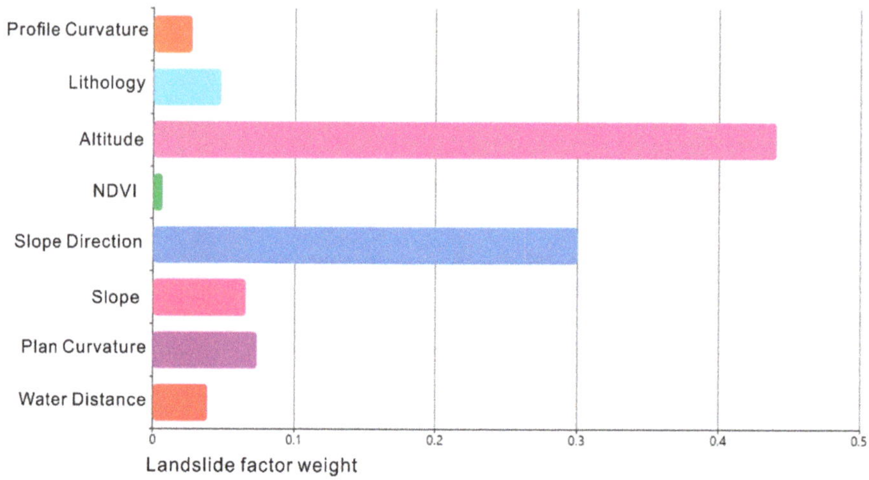

Figure 8. Environmental factor weight.

4.3. PSO-ELM Prediction Model

The model flow of the extreme learning machine optimized by the particle swarm optimization algorithm is shown in Figure 9: (1) To posit the velocity of the random particle swarm; (2) To evaluate the fitness value of all particles to get the optimal global position; (3) To update the velocity and position of each particle; (4) To evaluate the optimal fitness value of each particle of the previous iteration process, compare it with its own historical optimal fitness value, and select a better one; (5) To update the optimal global position—each particle moves towards the optimal global position and its optimal historical position; (6) To predict landslide susceptibility by assigning optimal parameters to an extreme learning machine.

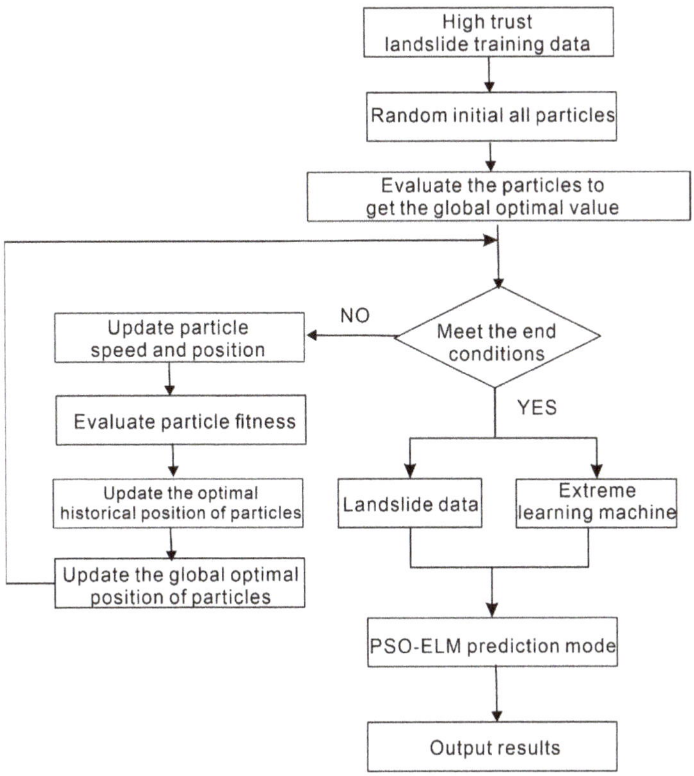

Figure 9. PSO-ELM flowchart.

4.4. Landslide Susceptibility Mapping

The landslide susceptibility mapping of the study area is shown in Figure 10. The natural discontinuity method divides landslide susceptibility into five zones: very low, low, medium, high, and very high. Figure 10 indicates that:

(1) The landslide points in the figure are landslide high-trust points expanded by the semi-supervised learning framework. Because the original landslide point may be accidental, it may be difficult for subsequent landslides to occur in this area over time. Therefore, this paper uses the expanded landslide high-confidence points to test the landslide susceptibility mapping.

(2) The results of the four models, SS-PSO-ELM, SS-ELM, PSO-ELM, and ELM, are shown in the figure. The high-trust landslide points all fall in the high-risk and very high-risk areas, proving that the four models can effectively predict landslides. However, in the PSO-ELM model and the ELM model, the high-risk and very high-risk areas account for a large proportion of the entire study area, which is inconsistent with reality. The SS-PSO-ELM model and the SS-ELM model are more realistic.

(3) In the northwest corner of the study area, the SS-PSO-ELM model and the SS-ELM model predicted a very high-risk area. The prediction results in the PSO-ELM and ELM models are low-risk and very low-risk areas. After data inspection and analysis, the reason is that the non-landslide points of the model without the semi-supervised

learning framework are randomly selected in the study area. However, randomly selected points within the study area do not guarantee that they are credible non-landslide points. As shown in this case, the area that was initially a high risk of the landslide was used as a sample to enter the training data into non-landslide points, resulting in a large discrepancy between the results and the actual results.

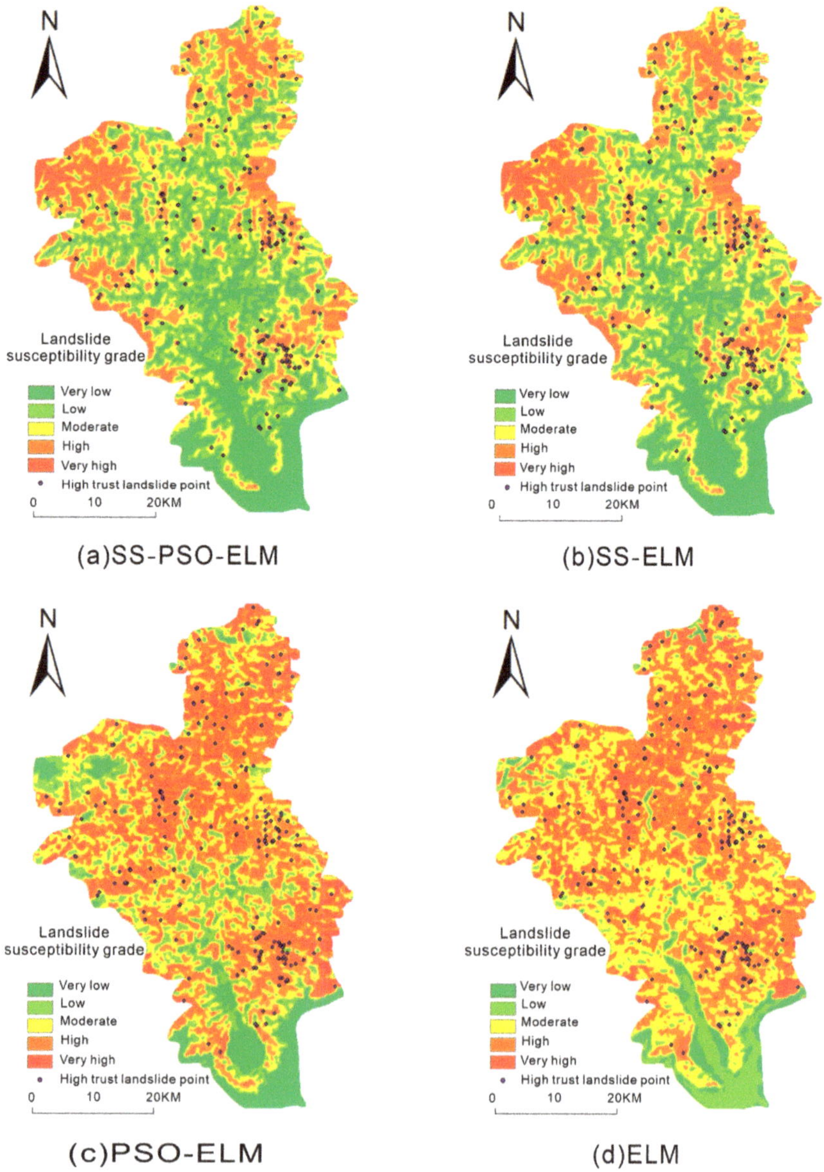

Figure 10. Landslide susceptibility map.

5. Modeling Uncertainty Analysis

5.1. ROC Accuracy Evaluation

As shown in Figure 11, the model's prediction accuracy is evaluated by the AUC area under the ROC curve. The AUCs of SS-PSO-ELM, SS-ELM, PSO-ELM, and ELM was 0.893,

0.867, 0.788, and 0.710, respectively. From the image, SS-PSO-ELM and SS-ELM have better prediction performance of landslide susceptibility. However, the curve of the SS-ELM model rises slowly in the later stage, and the prediction performance fluctuates wildly. Furthermore, this proves that the extreme learning machine model optimized by particle swarm optimization algorithm has higher accuracy and stability in landslide susceptibility prediction. The AUC accuracy of the SS-PSO-ELM model is 0.105 more increased than that of the PSO-ELM model without the semi-supervised learning framework. This shows that using a semi-supervised learning framework to screen non-landslide high-trust points can significantly improve the performance of landslide susceptibility prediction.

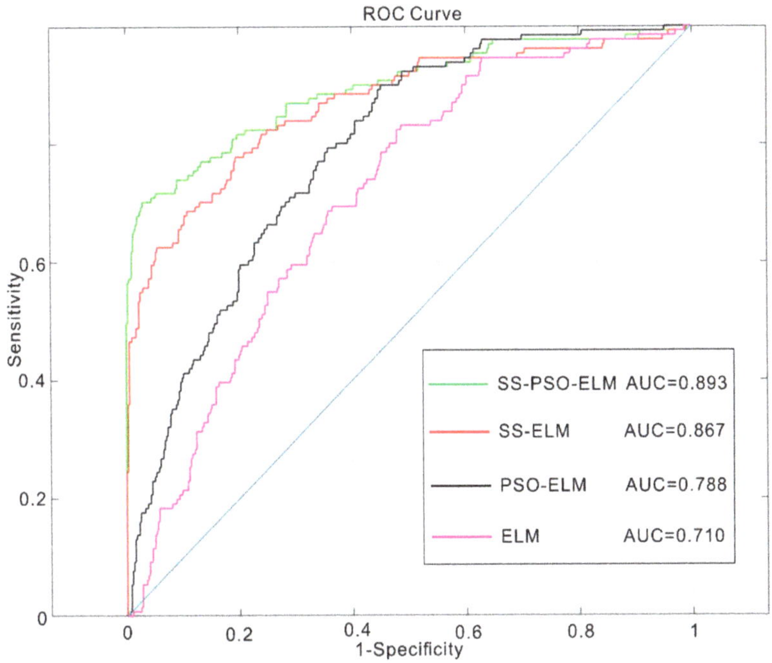

Figure 11. ROC accuracy plot.

5.2. Susceptibility Index Distribution

The distribution of the susceptibility index can visually observe the number of individuals in the specific susceptibility index interval in the study area. In practice, the range of landslide sites is much smaller than that of non-landslide sites. Therefore, we will focus on the very high-risk and very low-risk areas' scale when judging the model's performance in landslide susceptibility prediction. The larger the scale of the two areas, the better the model's ability to identify landslides. Therefore, the distribution of the susceptibility index can intuitively see the proportion of each risk area of the model and can more intuitively reflect the predictive performance of the model.

Figure 12 shows the susceptibility index distribution, showing the amount included in each landslide probability interval. The mean and standard deviation is shown in Figure 12 can better reflect the prediction level of the four models and the dispersion degree of the predicted landslide data. Figure 13 demonstrates the proportion of each landslide-prone zone in the study area. Both figures can show the stability of the model for landslide prediction and judge whether the model prediction is in line with the actual situation.

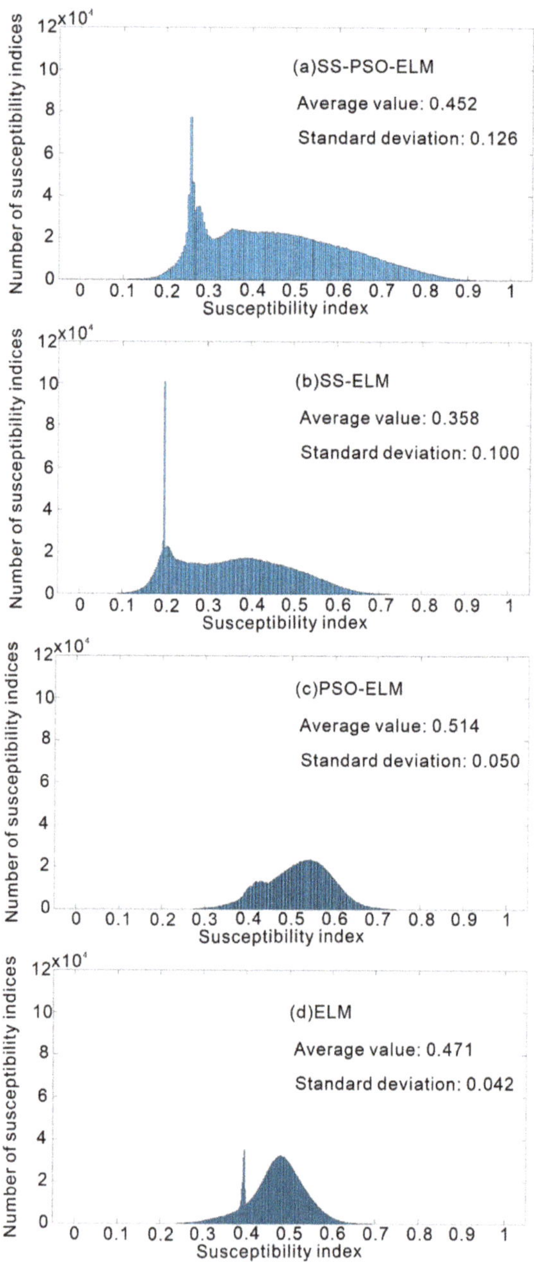

Figure 12. Distribution map of landslide susceptibility index.

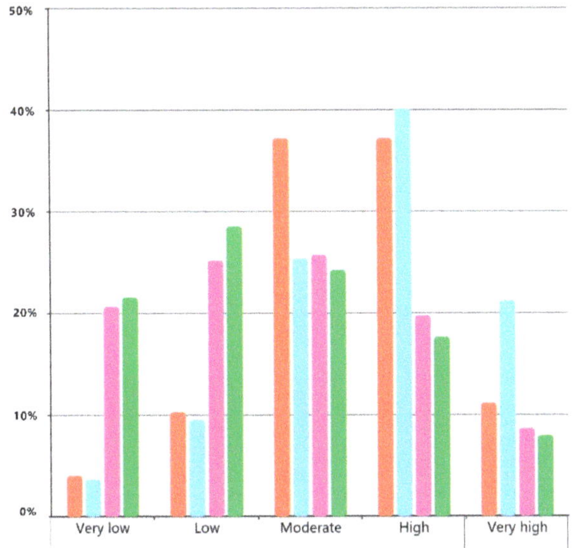

	Very low	Low	Moderate	High	Very high
ELM	4.09%	10.31%	37.22%	37.22%	11.16%
PSO-ELM	3.71%	9.56%	25.40%	40.09%	21.24%
SS-ELM	20.68%	25.19%	25.72%	19.76%	8.65%
SS-PSO-ELM	21.59%	28.56%	24.22%	17.67%	7.95%

Figure 13. Model classification ratio chart.

(1) The landslide risk areas of the SS-PSO-ELM model and the SS-ELM model are concentrated in low-risk and very low-risk areas and less in high-risk and very high-risk areas. The overall trend of landslide susceptibility is that the area from low risk to high risk gradually decreases, which is more in line with reality.

(2) The mean value of landslide occurrence probability of SS-PSO-ELM and SS-ELM models is smaller than that of the PSO-ELM model and ELM model. It is proved that the semi-supervised learning framework's prediction of landslide susceptibility is in line with reality, and the extremely low-susceptibility and low-susceptibility areas of landslides are the mainstream in the study area.

(3) In Figure 12, the standard deviations of the four models are compared from large to small, namely SS-PSO-ELM, SS-ELM, PSO-ELM, and ELM. The SS-PSO-ELM standard deviation is the largest, proving that the SS-PSO-ELM model can distinguish and identify landslides and better reflect the differences in landslide susceptibility to the study area. However, since the PSO-ELM and ELM models do not use high-trust non-landslide points as training data, the probability of landslides in most places is concentrated between 0.4 and 0.6, and there is no good ability to discriminate landslides. Furthermore, most of the predicted areas are in the high-risk prone regions to landslides, which is inconsistent with the actual situation.

5.3. Model Evaluation

The evaluation indexes of each model are shown in Table 2. The RMSE of the particle swarm optimized SS-PSO-ELM model is smaller than that of the PSO-ELM model. Furthermore, the small model volatility indicates that the extreme learning machine model optimized by the particle swarm optimization algorithm has significantly improved the landslide susceptibility prediction performance. The AUC values of the SS-PSO-ELM model and the SS-ELM model are higher than those of the PSO-ELM model and the ELM model, proving that the semi-supervised learning framework can significantly improve the performance of the model. Figures 12 and 13 and Table 2 indicate that SS-PSO-ELM has a higher performance in predicting landslide susceptibility, which is more in line with reality.

Table 2. Model evaluation metrics.

Model	Mean	Standard Deviation	AUC	RMSE
SS-PSO-ELM	0.452	0.126	0.893	0.370
SS-ELM	0.358	0.100	0.867	0.438
PSO-ELM	0.514	0.050	0.788	0.417
ELM	0.471	0.042	0.710	0.442

6. Conclusions

This paper takes Fu'an City, Fujian Province, as the research object and selects eight environmental factors: elevation, slope, NDVI, plane curvature, section curvature, water system distance, slope aspect, and lithology. A model evaluating landslide susceptibility is established with semi-supervised learning as the framework and the extreme learning machine of particle swarm optimization as the driving model. A comparative analysis was conducted with SS-ELM, PSO-ELM, and ELM as contrast models.

The SS-PSO-ELM model has the highest AUC accuracy, indicating that the model has the best performance in landslide susceptibility prediction. The mean value of SS-PSO-ELM is small, which is in line with the actual situation because the landslide area in the study area is much smaller than the non-landslide area. The standard deviation of SS-PSO-ELM is the largest, which proves that the landslide probability values of landslide sites are higher, the landslide probability values of non-landslide sites are lower, and they have better landslide identification ability. In addition, the RMSE of the SS-PSO-ELM model is the smallest, proving that the model is less volatile in landslide susceptibility prediction.

The high-trust landslide points and high-trust non-landslide points selected according to the semi-supervised learning framework can effectively improve the accuracy of landslide susceptibility prediction by the data-driven model. High-trust landslide points can delete occasional landslide points, avoiding the problem of many high-risk and very high-risk areas when data-driven models predict landslide susceptibility, which is more in line with this reality.

Because the purpose of the clustering algorithm is to cluster similar data together, it may lead to overfitting of the high-trust non-landslide data, and the selected high-trust non-landslide data may belong to an approximate area or a geographically similar area. In the next step, the research area can be divided into several areas to ensure that the number of grid cells in each area can meet the highest performance of the clustering algorithm and the diversity of data. The landslide sensitivity of the prediction accuracy can be further improved, and the drawing is more reasonable.

Author Contributions: Conceptualization, Yizhun Zhang and Qisheng Yan; Methodology, Yizhun Zhang; Software, Yizhun Zhang; Validation, Yizhun Zhang and Qisheng Yan; Formal Analysis, Yizhun Zhang; Investigation, Yizhun Zhang; Resources, Yizhun Zhang; Data Curation, Yizhun Zhang; Writing—Original Draft Preparation, Yizhun Zhang; Writing—Review and Editing, Qisheng Yan; Visualization, Yizhun Zhang; Supervision, Qisheng Yan; Project Administration, Qisheng Yan; Funding Acquisition, Qisheng Yan. All authors have read and agreed to the published version of the manuscript.

Funding: This research was funded by [National Natural Science Foundation of China] grant number [No: 71961001].

Institutional Review Board Statement: Not applicable.

Informed Consent Statement: Not applicable.

Data Availability Statement: Not applicable.

Acknowledgments: The authors would like to thank the anonymous reviewers for their helpful and valuable comments and suggestions.

Conflicts of Interest: The authors declare that they have no known competing financial interests or personal relationships that could have appeared to influence the work reported in this paper.

References

1. Guzzetti, F.; Carrara, A.; Cardinali, M.; Paola, R. Landslide hazard evaluation: A review of currenttechniques and their application in a multi-scale study, Central Italy. *Geomorphology* **1999**, *31*, 181–216. [CrossRef]
2. Nadim, F.; Kjekstad, O.; Peduzzi, P.; Herold, C.; Jaedicke, C. Global landslide and avalanche hotspots. *Landslides* **2006**, *3*, 159–173. [CrossRef]
3. Assilzadeh, H.; Levy, J.K.; Wang, X. Landslide catastrophes and disaster risk reduction:a GISframework for landslide prevention and management. *Remote Sens.* **2010**, *2*, 2259–2273. [CrossRef]
4. Guzzetti, F.; Reichenbach, P.; Ardizzone, F.; Cardinali, M.; Galli, M. Estimating the quality of landslide susceptibility models. *Geomorphology* **2006**, *81*, 166–184. [CrossRef]
5. Montgomery, D.R.; Dietrich, W.E. A physically based model for the topographic control onshallow landsliding. *Water Resour. Res.* **1994**, *30*, 1153–1171. [CrossRef]
6. Guzzetti, F.; Mondini, A.C.; Cardinali, M.; Fiorucci, F.; Santangelo, M.; Chang, K.T. Landslide inventory maps: New tools for an old problem. *Earth-Sci. Rev.* **2012**, *112*, 42–66. [CrossRef]
7. Aleotti, P.; Chowdhury, R. Landslide hazard assessment: Summary review and newperspectives. *Bull. Eng. Geol. Environ.* **1999**, *58*, 21–44. [CrossRef]
8. Ruff, M.; Czurda, K. Landslide susceptibility analysis with a heuristic approach in theEastern Alps (Vorarlberg, Austria). *Geomorphology* **2008**, *94*, 314–324. [CrossRef]
9. Lin, S.Y.; Lin, C.W.; Gasselt, S.V. Processing Framework for Landslide Detection Based on Synthetic Aperture Radar (SAR) Intensity-Image Analysis. *Remote Sens.* **2021**, *13*, 644. [CrossRef]
10. Reichenbach, P.; Rossi, M.; Malamud, B.; Mihir, M.; Guzzetti, F. A review of statistically-based landslide susceptibility models. *Earth-Sci. Rev.* **2018**, *180*, 60–91. [CrossRef]
11. Hai-Min, L.; Jack, S.; Arul, A. Assessment of Geohazards and Preventative Countermeasures Using AHP Incorporated with GIS in Lanzhou, China. *Sustainability* **2018**, *10*, 304.
12. Khan, H.; Shafique, M.; Khan, M.A.; Bacha, M.A.; Shah, S.U.; Calligaris, C. Landslide susceptibility assessment using Frequency Ratio, a case study of northern Pakistan. *Egypt. J. Remote Sens. Space Sci.* **2018**, *22*, 11–24. [CrossRef]
13. Bui, D.T.; Pradhan, B.; Lofman, O.; Revhaug, I.; Dick, O. BLandslide susceptibility mapping at Hoa Binh province (Vietnam) using an adaptive neuro-fuzzy inference system and GIS. *Comput. Geosci.* **2011**, *45*, 199–211.
14. Ghorbanzadeh, O.; Shahabi, H.; Crivellari, A.; Homayouni, S.; Blaschke, T.; Ghamisi, P. Landslide detection using deep learning and object-based image analysis. *Landslides* **2022**, *19*, 929–939. [CrossRef]
15. Ghorbanzadeh, O.; Blaschke, T.; Gholamnia, K.; Meena, S.R.; Tiede, D.; Aryal, J. Evaluation of Different Machine Learning Methods and Deep-Learning Convolutional Neural Networks for Landslide Detection. *Remote Sens.* **2019**, *11*, 196. [CrossRef]
16. Ghorbanzadeh, O.; Xu, Y.; Ghamis, P.; Kopp, M.; Kreil, D. Landslide4Sense: Reference Benchmark Data and Deep Learning Models for Landslide Detection. *arXiv* **2022**, arXiv:2206.00515.
17. Alb, A.; Frb, C.; Qbp, D.; Gigović, L.; Drobnjak, S.; Aina, Y.A.; Panahi, M.; Yekeen, S.T.; Lee, S. Spatial prediction of landslide susceptibility in western Serbia using hybrid support vector regression (SVR) with with GWO, BAT and COA algorithms. *Geosci. Front.* **2020**, *12*, 101104.
18. Depina, I.; Oguz, E.A.; Thakur, V. Novel Bayesian framework for calibration of spatially distributed physical-based landslide prediction models. *Comput. Geotech.* **2020**, *125*, 103660. [CrossRef]
19. Guo, Z.; Chen, L.; Gui, L.; Du, J.; Yin, K.; Do, H.M. Landslide displacement prediction based on variational mode decomposition and WA-GWO-BP model. *Landslides* **2019**, *17*, 567–583, (In Chinese with English abstract). [CrossRef]
20. Zhang, Y.G.; Tang, J.; Liao, R.P.; Zhang, M.F.; Zhang, Y.; Wang, X.M.; Su, Z.Y. Application of an enhanced BP neural network model with water cycle algorithm on landslide prediction. *Stoch. Environ. Res. Risk Assess.* **2021**, *35*, 1273–1291. [CrossRef]
21. Benbouras, M.A. Hybrid meta-heuristic machine learning methods applied to landslide susceptibility mapping in the Sahel-Algiers. *Int. J. Sediment Res.* **2022**, *37*, 601–618. [CrossRef]
22. Huang, F.M.; Pan, L.H.; Yao, C.; Zhou, C.; Huang, J.; Guo, Z. Prediction Model of Landslide Susceptibility Based on Semi-Supervised Machine Learning. *J. Zhejiang Univ. (Eng. Ed.)* **2021**, *55*, 1705–1713, (In Chinese with English abstract).
23. Huang, F.M.; Wang, Y.; Dong, Z.L.; Wu, L.Z.; Guo, Z.Z.; Zhang, T.L. Sensitivity Evaluation of Regional Landslide Based on Grey Relational Grade Model. *Earth Sci.* **2019**, *44*, 664–676, (In Chinese with English abstract).
24. Ito, R.; Nakae, K.; Hata, J.; Okano, H.; Ishii, S. Semi-supervised deep learning of brain tissue segmentation. *Neural Netw.* **2019**, *116*, 25–34. [CrossRef]
25. Huang, G.; Song, S.; Gupta, J.N.; Wu, C. Semi-supervised and unsupervised extreme learning machines. *IEEE Trans. Cybern.* **2014**, *44*, 2405–2417. [CrossRef]

26. Jin, G.; Liu, C.; Chen, X. Adversarial network integrating dual attention and sparse representation for semi-supervised semantic segmentation. *Inf. Processing Manag.* **2021**, *58*, 102680. [CrossRef]
27. Jian, C.; Yang, K.; Ao, Y. Industrial fault diagnosis based on active learning and semi-supervised learning using small training set. *Eng. Appl. Artif. Intell.* **2021**, *104*, 104365. [CrossRef]
28. Xu, L.; Cui, L.; Weise, T.; Li, X.; Wu, Z.; Nie, F.; Chen, E.; Tang, Y. Semi-Supervised Multi-Layer Convolution Kernel Learning in Credit Evaluation. *Pattern Recognit.* **2021**, *120*, 108125. [CrossRef]
29. Yao, J.; Qin, S.; Qiao, S.; Che, W.; Chen, Y.; Su, G.; Miao, Q. Assessment of Landslide Susceptibility Combining Deep Learning with Semi-Supervised Learning in Jiaohe County, Jilin Province, China. *Appl. Sci.* **2020**, *10*, 5640. [CrossRef]
30. Hu, H.; Wang, C.; Liang, Z.; Gao, R.; Li, B. Exploring Complementary Models Consisting of Machine Learning Algorithms for Landslide Susceptibility Mapping. *ISPRS Int. J. Geo-Inf.* **2021**, *10*, 639. [CrossRef]
31. Shahabi, H.; Rahimzad, M.; Tavakkoli Piralilou, S.; Ghorbanzadeh, O.; Homayouni, S.; Blaschke, T.; Lim, S.; Ghamisi, P. Unsupervised Deep Learning for Landslide Detection from Multispectral Sentinel-2 Imagery. *Remote Sens.* **2021**, *13*, 4698. [CrossRef]
32. Liang, L.H. Analysis of Influencing Factors of Geological Hazards in Fu'an City. *Fujian Geol.* **2012**, *31*, 185–190, (In Chinese with English abstract).
33. Rodriguez, A.; Laio, A. Clustering by fast search and find of density peaks. *Science* **2014**, *344*, 1492. [CrossRef]
34. Qian, L.X.; Wang, H.R.; Dang, S.Z.; Hong, M.; Zhao, Z.Y.; Deng, C.Y. A coupling Model of Water Resources Shortage Risk Assessment and its Application. *Syst. Eng.* **2021**, *41m*, 1319–1327, (In Chinese with English abstract).
35. Peng, H.; Long, F.; Ding, C. Feature selection based on mutual information criteria of max-dependency, max-relevance, and min-redundancy. *IEEE Trans. Pattern Anal. Mach. Intell.* **2005**, *27*, 1226–1238. [CrossRef]
36. Huang, G.B.; Zhu, Q.Y.; Siew, C.K. Extreme learning machine: Theory and applications. *Neurocomputing* **2006**, *70*, 489–501. [CrossRef]
37. Breiman, L. Random forests. *Mach. Learn.* **2001**, *45*, 5–32. [CrossRef]
38. Kennedy, J.; Eberhart, R.C. Particle Swarm Optimization. In Proceedings of the International Conference on Neural Networks (ICNN'95), Perth, Australia, 27 November–1 December 1995; Volume 4, pp. 1942–1948.
39. Eberhart, R.C.; Kennedy, J. A new optimizer using particle swarm theory. In Proceedings of the MHS95 Sixth International Symposium on Micro Machine and Human Science, Nagoya, Japan, 4–6 October 1995; pp. 39–43.
40. Zhang, S.; Qian, X.M.; Lou, P.H.; Wu, X.; Sun, C. Path Planning Optimization of Large-Scale Agv System Based on Improved Particle Swarm Optimization Algorithm. *Comput.-Integr. Manuf.* **2020**, *26*, 2484–2496, (In Chinese with English abstract).
41. Li, W.B.; Fan, X.M.; Huang, F.M.; Wu, X.L.; Yin, K.; Chang, Z. Uncertainty of Landslide Susceptibility Modeling Based on Different Environmental Factors Linkage and Prediction Models. *Earth Sci.* **2021**, *46*, 3777–3795, (In Chinese with English abstract).

International Journal of *Geo-Information*

MDPI

Article

Certainty Factor Analyses and Spatiotemporal Characteristics of Landslide Evolution: Case Studies in the Chishan River Watershed in Taiwan

Chunhung Wu

Department of Water Resources Engineering and Conservation, Feng Chia University, Taichung 40724, Taiwan; chhuwu@fcu.edu.tw

Abstract: The 1999 Chichi earthquake and Typhoon Morakot in 2009 caused two serious landslide events in the Chishan river watershed in southern Taiwan. In this study, certainty factor analysis was used to evaluate the effectiveness of landslide occurrence, and spatiotemporal hotspot analysis was used to explain the pattern and distribution of landslide hotspots. The Z-values from the Getis–Ord formula were used to assess the clustering strength of landslide evolution on different scales and with different landslide sizes in different time periods. The landslide-prone area had an elevation of 1000–1750 m, a slope of >40°, and hillslopes with N, NE, E, and SE aspects and was within 100 m of rivers. The main spatiotemporal hotspot patterns of landslide evolution during 1999–2017 were oscillating hotspots, intensifying hotspots, and persistent hotspots, and the three main hotspot patterns occupied 80.1–89.4% of all hotspot areas. The main spatiotemporal landslide hotspots were concentrated in the core landslide areas and the downslopes of riverbank landslide areas, especially in the upstream subwatersheds. The landslide clustered strength in the upstream watershed was 3.4 times larger than that in the Chishan river watershed, and that in large landslides was 2.4 and 6.6 times larger than those in medium and small landslides, respectively.

Keywords: landslide evolution; landslide recovery; spatiotemporal hotspot; certainty factor analysis

Citation: Wu, C. Certainty Factor Analyses and Spatiotemporal Characteristics of Landslide Evolution: Case Studies in the Chishan River Watershed in Taiwan. *ISPRS Int. J. Geo-Inf.* **2022**, *11*, 382. https://doi.org/10.3390/ijgi11070382

Academic Editors: Walter Chen and Fuan Tsai

Received: 30 April 2022
Accepted: 8 July 2022
Published: 10 July 2022

1. Introduction

Extreme rainfall and large earthquake events have caused severe landslides in several countries; examples of such events include the 1999 Chichi earthquake in Taiwan [1], the 2005 Kashmir earthquake in Pakistan [2,3], the 2008 Wenchuan earthquake in China [4–6], Typhoon Morakot in 2009 in Taiwan [7,8], the 2015 Gorkha earthquake in Nepal [9], and the 2017 Jiuzhaigou earthquake in China [10]. Consequently, the areas affected by such events have become fragile and susceptible to landslides in subsequent years. The landslide susceptibility of mountainous areas after extreme rainfall or large earthquakes changes with time and space. Watersheds that experience numerous earthquake- or extreme-rainfall-induced landslides usually require more than 10 years to recover from such landslides. Moreover, sediment generated by continuous landslides in fragile watersheds is transported to rivers and dominates the subsequent geomorphologic evolution of these rivers. The unstable sediment yield from numerous landslides induced by large earthquake or extreme rainfall events in mountainous regions can cause further sediment disasters in the following years [7,8,11].

Research on post-extreme rainfall and postearthquake landslide evolution is essential for disaster prevention and reduction [4–8]. Accordingly, the spatiotemporal characteristics of landslide evolution and the dominant factors affecting it must be determined to ensure effective watershed management in the years after severe landslide events. Several studies have analyzed the relationship between geomorphological factors—including elevation, slope, aspect, and distance to river—and annual landslide distributions in subsequent years to determine the dominant factors influencing landslide evolution [4,5] and determine the sensitivity of geomorphological factors affecting landslide occurrence [6]. Because of numerous landslide distributed in the mountainous regions, the thresholds of rainfall for triggering the debris flows in southern Taiwan sharply decreased [7,8]. Furthermore, the distribution and concentration of landslides in the years that follow extreme-rainfall- and large-earthquake-induced landslide events must be compared. Some spatial differences in the distribution of landslide evolution before and after large-earthquake- and extreme-rainfall-induced landslide events have been found [4,7,8]. In the years that follow large earthquake-induced landslide events, the subsequent landslides that occur are typically concentrated in areas with elevations of <2000 m, slopes of 30–50°, aspects identical to the thrust directions of earthquake-induced faults, and slope toes located near rivers [2,4,6]. Nevertheless, few studies have discussed the distribution and concentration of landslides in the years following extreme-rainfall-induced landslide events [12,13]. Landslide recovery time after large earthquake and extreme rainfall events is a key factor for watershed management and has been assessed by several methods. The landslide recovery time was estimated as 2 years by assessing the landslide activity after the 2005 Kashimir earthquake in northern Pakistan [2]; 3–4 years by observing the number and area of annual landslides after the 2008 Wenchuan earthquake in Sichuan, China [5,14]; and 5 years by assessing the ratios of the annual decline in landslide area after Typhoon Morakot in 2009 in southern Taiwan [7,8].

The emerging hotspot analysis technique for spatiotemporal analysis has been used by several scholars to investigate the transmission of the COVID-19 virus [15–17], explore temporal changes in the volume of capture fisheries [18], investigate the distribution of traffic accidents [19,20], track pollutant emissions [21], and analyze the relationship between urban growth and urban fire [22]. Moreover, this new technique has been used to explain the patterns and characteristics of the long-term landslide evolution after extreme-rainfall-induced landslide events in Taiwan [12,13]. Spatiotemporal analysis has been used to explore the spatiotemporal distribution of landslide hotspots [12] and analyze differences in landslide spatiotemporal distribution based on different bin sizes [13] in the landslide-prone watersheds in southern Taiwan. The traditional hotspot analysis approach considers the spatial distribution and clustering of landslides by using a specific time database; by contrast, the emerging hotspot analysis technique can consider the spatial distribution and clustering of landslides by using multiple time databases. The new technique can provide spatiotemporal plots of landslide hotspots, which can be used to explain changes in the spatial patterns and distributions of landslides over time. Accordingly, the spatiotemporal distribution of landslide clustering can be used to explain landslide recovery and watershed evolution in the years that follow severe landslide events.

Southwestern Taiwan experienced substantial rainfall during Typhoon Morakot in 2009. Consequently, severe landslides, including the well-known Xiaolin landslide, occurred in the Chishan river watershed [7,8]. The sediment yield induced by this typhoon in the Chishan river watershed was estimated to exceed 10^8 m^3 [23]. Therefore, this study investigated the effect of the typhoon-induced sediment yield on watershed evolution in the years that followed Typhoon Morakot. The spatial and temporal changes in landslide clustering intensity in the landslide-prone watershed have never been discussed. This study focused on explaining the spatial and temporal changes in landslide clustering intensity in the landslide-prone watershed in Taiwan by using spatiotemporal analysis and the Z-score of the Getis–Ord formula. The study also compared the landslide recovery after the 1999 Chichi earthquake (M_L = 7.2) with that after Typhoon Morakot in 2009 in the Chishan river watershed to determine the differences between and characteristics of the recovery processes after earthquake- and rainfall-induced landslides. The long-term observation of the evolution of landslides in the Chishan river watershed could provide valuable insights for understanding landslide recovery and evolution in southern Taiwan.

2. Research Area

The study area was the Chishan river watershed in southwestern Taiwan and covered an area of 612.8 km^2 (Figure 1). The elevation in the study area ranged from 40 to 3950 m above sea level, and the area with elevation <1000 m occupied 51.9% of the study area. The slope in the study area ranged from 0 to 85°, and more than 79.3% of the total study area had a slope <40°. Around 88.2% of the study area was covered by natural forest and agriculture. The Chishan river was the main river in the study area, and its average discharge was 30.1 m^3/day. The study area was surrounded by 11 faults (Figure 2), including 2 faults passing through the C10 watershed (Figure 1) and 9 faults passing through the C01 to C06 subwatersheds (Figure 1). The lithology in the study area was characterized by sandstone, shale, clay, slate, and interlayered sandstone and shale from the mid-Miocene to the Holocene. The weak geological formations in the study area lead to poor lithologic resistance to landslides and erosion, and the decadal erosion rate was estimated to be around 30 mm/year [24]. The mean suspended-sediment concentration measured at the Shanlin bridge in the downstream of the Chishan river watershed was 696 ppm, and the annual sediment yield was estimated to be around 1.06 Mt/year [24].

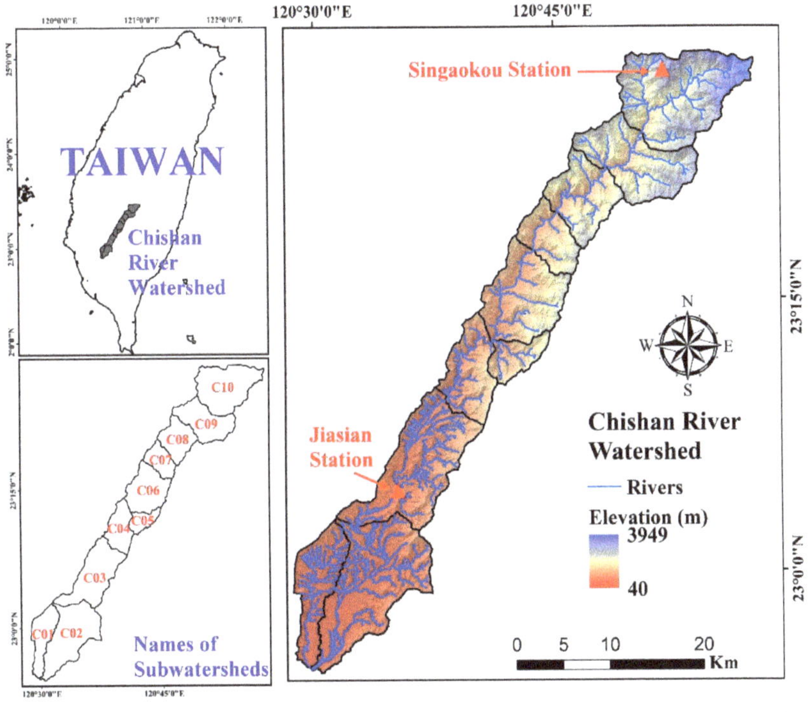

Figure 1. The distributions of elevation, rivers (main plot), and subwatersheds (left-down plot) in the Chishan river watershed.

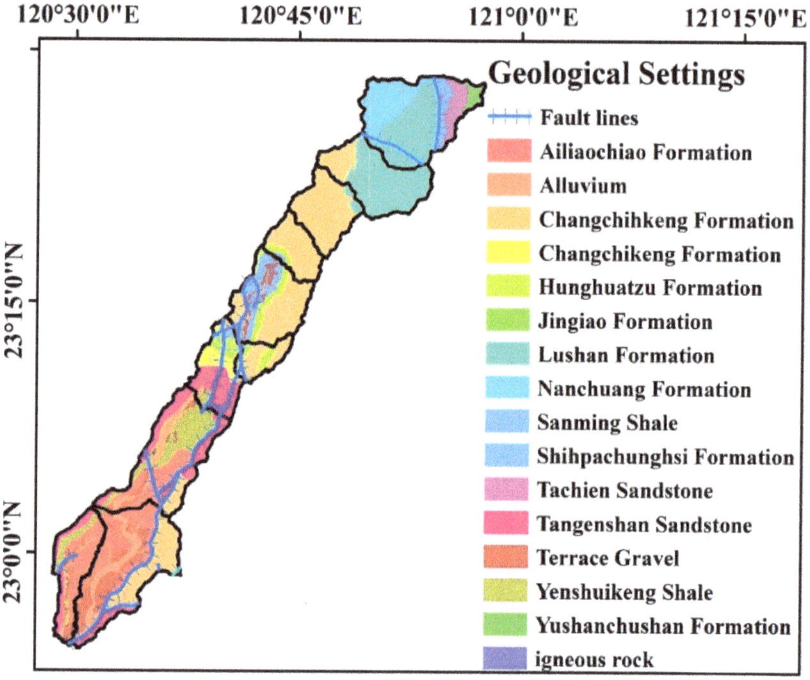

Figure 2. The geological settings in the Chishan river watershed.

The average annual rainfall at the Jiasian (low altitude, Figure 1) and Singaokou rainfall stations (high altitude, Figure 1) was around 2845 mm in 1951–2020 and 2908 mm in 1983–2020. The accumulated rainfall in the rainy season (from May to October) occupied >80% of the annual rainfall in the Chishan river watershed [18], and the heavy rainfall and typhoon events from June to September were the main reasons why. The Chishan river watershed is prone to natural disasters (e.g., floods, soil erosion, landslides, and debris flow). The two main geohazards in the Chishan river watershed are landslides and debris flow, and the main landslide-trigger event in 1999–2017 was Typhoon Morakot in 2009. Typhoon Morakot dumped record-breaking rainfall, over 2000 mm in 4 days, and caused very serious landslide disaster events in the Chishan river watershed [7]. As of 2021, the ratio and severity of extreme-rainfall-triggered landsliding from Typhoon Morakot in southern Taiwan were historical highs. Based on the rainfall record of the Jiasian rainfall station, the accumulated rainfall during Typhoon Morakot was 2142 mm, 72.3% of the annual rainfall in 2009. During Typhoon Morakot, there was 31.5 km^2 of landslide area, including 3.8 km^2 and 12.3 km^2 located in the upslope and downslope, respectively, in the Chishan river watershed, and the landslide ratio was estimated to be larger than 7.0% [7]. There were also 39 potential debris flow torrents in the Chishan river watershed in 2021.

3. Materials and Methods

3.1. Materials

This study selected the Jiasian and Singaokou rainfall stations (Figure 1) as the representative stations in the Chishan river watershed. The study used rainfall records collected at these two stations for the 1999–2017 period to analyze rainfall characteristics in the study area. The Central Geological Survey and Forestry Bureau of Taiwan provided annual landslide inventories during 1999–2017 in the Chishan river watershed. From these inventories, landslides that occurred during 1999–2002 were identified from SPOT 1 and 2 images with spatial resolutions of 10 m, and those that occurred during 2003–2017 were identified from Formosat-2 images with spatial resolutions of 2 m. The annual landslide inventories during

1999–2002 in Taiwan were produced by using supervised classification to delineate the landslide polygons, and those during 2003–2017 were produced by using the Formosat-2 automatic image processing system [25] and the Expert Landslide and Shaded Area Delineation System (ELSADS) [26] to delineate the landslide polygons. The Formosat-2 imagery during 2003–2017 was preprocessed by using the Formosat-2 automatic image processing system [25]. The Formosat-2 automatic image processing system was able to conduct band-to-band coregistration [25], automatic orthorectification [26], and multitemporal image geometrical registration [27], and it was the main system for producing the annual landslide inventories during 2003–2017 in Taiwan. The Expert Landslide and Shaded Area Delineation System (ELSADS) was used to delineate the polygons of landslides [28]. The landslide distribution maps during 1999–2017 and the landslide susceptibility maps after Typhoon Morakot in 2009 [23] are shown in Figure 3. The landslide cases were classified into three types for further analysis: large (area > 100,000 m^2), medium (area = 10,000–100,000 m^2), and small (area < 10,000 m^2) landslides. The study used a digital elevation model (DEM) with a spatial resolution of 5 m; moreover, a 5 m × 5 m grid was used as the unit of analysis. The study area was classified into 10 subwatersheds (C01–C10, as shown in Figure 1, lower left panel) on the basis of the delineation of subwatersheds in Taiwan.

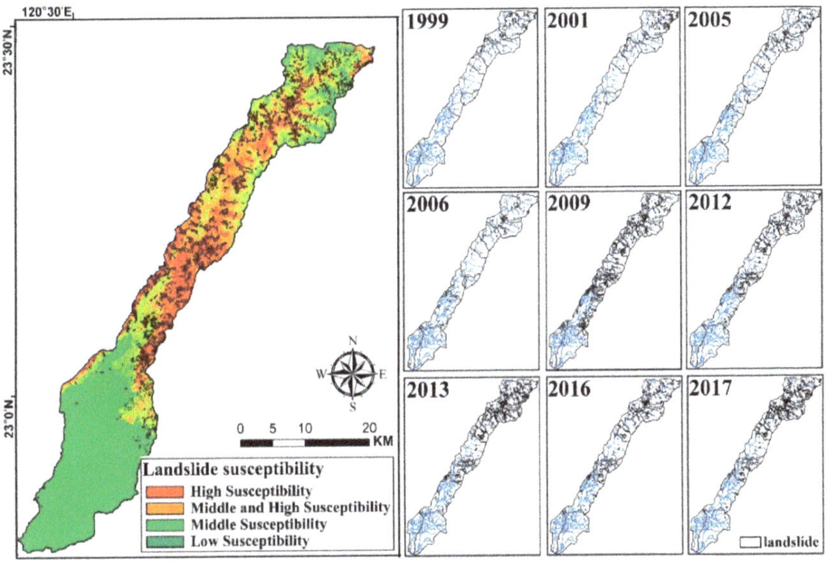

Figure 3. The landslide distribution during 1999–2017 (**right figures**) and the landslide susceptibility map (**left figure**) in the Chishan river watershed.

3.2. Certainty Factor Analyies

A certainty factor analysis and an effectiveness index were used to evaluate the effectiveness of factors on landslide occurrence in the study area. The certainty factor [29,30] can be used to analyze the effectiveness of different factors on the occurrence of events. This study used a certainty factor to identify the main factors influencing landslide occurrence and evolution after the 2008 Wenchuan earthquake in China [6,31]. The present study adapted the following certainty factor [32] for analysis:

$$\mathrm{CF} = \frac{P_a - P_S}{P_a \times (1 - P_s)} \text{ if } P_a \geq P_s; \ \frac{P_a - P_S}{P_S \times (1 - P_a)} \text{ if } P_a < P_S \tag{1}$$

where CF denotes the certainty factor, P_a denotes the ratio of the landslide area within a subcategory of a factor category to the landslide area in the factor category, and P_S denotes the ratio of the landslide area to the watershed area. The certainty factor value ranges from

−1 to 1, with positive and negative values indicating increasing and decreasing certainty of landslide occurrence.

This study also used an effectiveness index (E) to compare the effectiveness of various factors on landslide occurrence. The effectiveness index (E) can be defined as the difference between the maximum and minimum values of the certainty factor and can be used to quantify the influence of each factor on landslide occurrence and evolution.

3.3. Local Outlier Analysis

The study used the Anselin local Moran's I index [33] to analyze the clusters and outliers of landslide evolution in the Chishan river watershed. The Anselin local Moran's *I* index was used to identify local clusters and outliers of landslide in the space and time dimensions. Six kinds of patterns were defined: high–high cluster (HH), only high–low outlier (HL), only low–high outlier (LH), only low–low cluster (LL), multiple types (Mul.), and never significant (NS) based on the results of Anselin local Moran's I index. The low–low (LL) pattern revealed the bins that, along with their neighborhoods, had low landslide ratios or obvious landslide outliers, while the high–high (HH) pattern revealed the bins that, along with their neighborhoods, had high landslide ratios or obvious landslide clusters.

3.4. Spatiotemporal Hotspot Analysis

This study mainly used the Getis–Ord formula [34] and the emerging hotspot analysis technique to analyze the spatiotemporal trends of landslide evolution in the Chishan river watershed. The annual landslide inventories for the Chishan river watershed during 1999–2017 were used to establish a space–time cube model. Moreover, the annual landslide inventories during 1999–2008 were used to establish a space–time cube model to analyze the long-term evolution of landslides induced by the 1999 Chichi earthquake; this model comprised 797,912 bins. The annual landslide inventories during 2008–2017 were used to establish a space–time cube model to analyze the long-term evolution of landslides induced by Typhoon Morakot in 2009; this model comprised 1,946,705 bins. In each of these models, a bin represented a 5 m × 5 m grid that had been identified as landslide during 1999–2017. The x- and y-axes in the space–time cube model represented the longitude and latitude coordinate information of the bins, and the z-axis represented the years in which the bins were identified as being affected by landslides.

The Getis–Ord formula was used to explain the distribution of landslide clustering intensity and statistical significance, and the emerging hotspot analysis technique was used to analyze the spatiotemporal distribution and pattern of landslide hotspots. A Z-score was derived using the Getis–Ord formula; this measure indicated the strength and statistical significance of landslide clustering hotspots and cold spots (Table 1).

Table 1. Classification and statistical significance of hotspot and cold spots based on Z-scores.

Z-Score	Hotspot or Cold Spot	Confidence Level
>2.58	Hotspot	99% confidence level
1.96–2.58	Hotspot	95% confidence level
1.65–1.96	Hotspot	90% confidence level
−1.65–1.65	No statistical significance	
−1.96–−1.65	Cold spot	90% confidence level
−2.58–−1.96	Cold spot	95% confidence level
<−2.58	Cold spot	99% confidence level

The emerging hotspot analysis technique, available as a tool in the space–time pattern mining module of the Arc Pro software suite, was used to evaluate temporal evolution trends and characterize landslide clustering patterns. The spatial and temporal bins were set to 5 m × 5 m and 1 year, respectively, and the spatial and temporal search radii were set to 25 m and 5 years, respectively. The derived landslide clustering patterns were classified into 17 categories: 8 hotspot patterns, 8 cold spot patterns, and no pattern. The

8 hotspot and cold spot patterns were denoted as consecutive, diminishing, historical, intensifying, new, oscillating, persistent, and sporadic hotspots or cold spots, depending on the number of times the landslide bins were identified, the increases and decreases in landslide clustering intensity, and statistical significance (Table 2).

Table 2. The definition of spatiotemporal hotspot patterns by emerging hotspot analysis.

Pattern	Abbr. [1]	Definition
No pattern detected	—	A bin that did not fall into any of the hotspot or cold spot patterns defined below.
New hotspot (cold spot)	NHS (NCS)	A bin that was a statistically significant hotspot (cold spot) for the final year and had never been a statistically significant hotspot (cold spot) before.
Consecutive hotspot (cold spot)	CHS (CCS)	A bin with a single, uninterrupted run of statistically significant hotspot (cold spot) bins in the final year. The bin had never been a statistically significant hotspot (cold spot) prior to the final hotspot (cold spot) run, and less than 90% of all bins were statistically significant hotspots (cold spots).
Intensifying hotspot (cold spot)	IHS (ICS)	A bin that was a statistically significant hotspot (cold spot) for 90% of the year intervals, including the final year. In addition, the intensity of clustering of high counts in each year was increasing overall, and that increase was statistically significant.
Persistent hotspot (cold spot)	PHS (PCS)	A bin that was a statistically significant hotspot (cold spot) for 90% of the year intervals with no discernible trend indicating an increase or decrease in the intensity of clustering over time.
Diminishing hotspot (cold spot)	DHS (DCS)	A bin that was a statistically significant hotspot (cold spot) for 90% of the year intervals, including the final year. In addition, the intensity of clustering in each year was decreasing overall, and that decrease was statistically significant.
Sporadic hotspot (cold spot)	SHS (SCS)	A bin that was an on-again then off-again hot spot. Less than 90% of the year intervals were statistically significant hotspots (cold spot), and none of the year intervals were statistically significant cold spots (hotspots).
Oscillating hotspot (cold spot)	OHS (OCS)	A statistically significant hotspot (cold spot) for the final year that had a history of also being a statistically significant cold spot (hotspot) during a prior year. Less than 90% of the time-step intervals were statistically significant hotspots (cold spots).
Historical hotspot (cold spot)	HHS (HCS)	The most recent year was not hot (cold), but at least 90% of the year intervals were statistically significant hotspots (cold spots).

[1] Abbr. means the abbreviation for the pattern.

4. Results

4.1. Annual Rainfall and Landslide Records in 1999–2017

The annual rainfall data from the Jiasian and Singaokou stations were used to analyze the temporal changes in annual rainfall in the Chishan river watershed during 1999–2017. The rainfall characteristics of the Chishan river watershed during 1999–2017 are presented in Figure 4 and summarized in Table 3. At the Jiasian station, the average annual rainfall during 1999–2017 was higher than that during 1950–2020; moreover, at the Singaokou station, the average annual rainfall during 1999–2017 was higher than that during 1983–2020. The annual rainfall during 1999–2017 was higher than that during 1950–1998. The accumulated rainfall in the rainy season at the Jiasian station during 1999–2017 accounted for 92.2–93.8% of the average annual rainfall, and that at the Singaokou station accounted for 78.6–82.8%. Typhoon Morakot in 2009, a heavy-rainfall event with a return period of >200 years [7], was the strongest heavy-rainfall event during 1999–2017 in the Chishan river watershed. This study compared the rainfall characteristics before and after Typhoon Morakot.

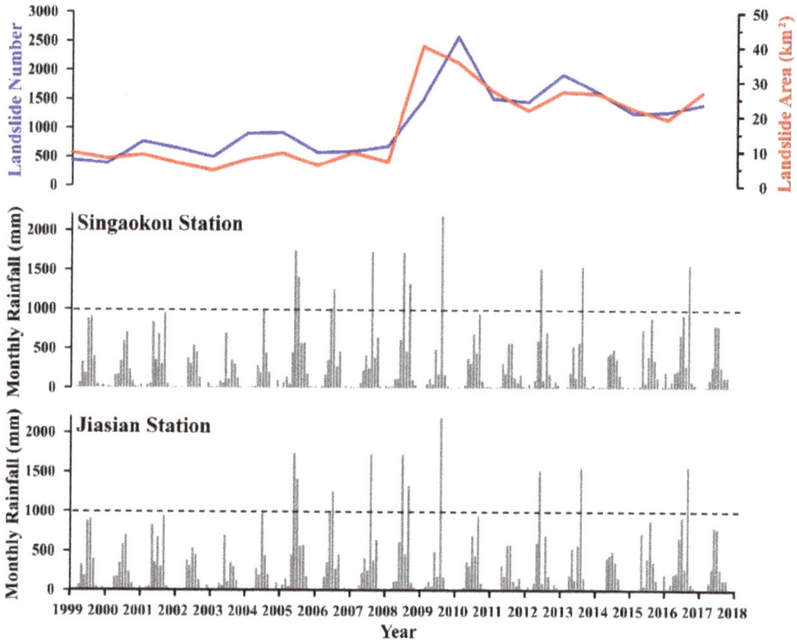

Figure 4. The distribution of monthly rainfall at the Jiasian (down plot) and Singaokou (middle plot) stations and the annual landslide number (blue line in upper plot) and area (red line in upper plot) during 1999–2017 in the Chishan river watershed.

Table 3. The rainfall characteristics at the Jiasian and Singaokou rainfall stations in 1999–2017 in the Chishan river watershed.

Time	Rainfall Characteristic	Jiasian	Singaokou
1999–2017	Average annual rainfall	3100 mm	3323 mm
	Average accumulated rainfall in rainy seasons	2886 mm	2683 mm
1999–2008	Average annual rainfall	3208 mm	3233 mm
	Average accumulated rainfall in rainy seasons	3010 mm	2678 mm
	Number of monthly rainfalls >1000 mm	7	6
	Number of daily rainfalls >652.8 mm [1]	1	1
2009–2017	Average annual rainfall	2980 mm	3423 mm
	Average accumulated rainfall in rainy seasons	2748 mm	2689 mm
	Number of monthly rainfalls >1000 mm	4	4
	Number of daily rainfalls >659.6 mm [1]	1	2

[1] The 652.8 mm and 659.6 mm measurements represent the daily rainfalls with return periods of 50 years at the Jiasian and Singaokou stations based on empirical equations from the regulations on soil and water conservation in Taiwan.

The accumulated rainfall in the rainy seasons, i.e., from May to October, and the number of monthly rainfalls >1000 mm at the Jiasian and Singaokou stations were used to describe the temporal distribution during 1999–2008 in the Chishan river watershed. The accumulated rainfall in the rainy seasons and the number of monthly rainfalls >1000 mm during 1999–2008 were greater than or close to those during 2009–2017. Using empirical equations provided in the regulations on soil and water conservation in Taiwan, this study estimated that the amounts of daily rainfall with return periods of 50 and 200 years were 601.3 and 704.3 mm, respectively, at the Jiasian rainfall station and 607.6 and 711.6 mm, respectively, at the Singaokou rainfall station. During 1999–2017, the numbers of daily rainfall events in which the amount of rainfall exceeded 601.3 mm at the Jiasian station and 607.6 mm at the Singaokou station were two and three, respectively. The two events recorded at the Jiasian station were Typhoon Kalmaegi (629.5 mm on 17 July 2008) and Typhoon Morakot (1072.0 mm on 8 August 2009), and the three events recorded at the Singaokou station were Typhoon Kalmaegi (609.5 mm on 9 June 2005), Typhoon Morakot (833.0 mm on 8 August 2009), and Typhoon Morakot (780.0 mm on 9 August 2009). The rainfall characteristics in the Chishan river watershed during 1999–2017 featured plenty of rainfall with high rainfall intensity, concentrated in the rainy seasons. These results indicated that the strengths of rainfall-induced landslides during 1999–2008 in the Chishan river watershed were greater than those during 2009–2017, except for the landslides induced by Typhoon Morakot.

Data on the numbers and areas of landslides during 1999–2017 in the Chishan river watershed were used to explain temporal changes in landslides. The results revealed obvious differences between the numbers and areas of landslides before and after Typhoon Morakot. Analysis of landslide disaster data during 1999–2017 showed that the considerable number of landslide disasters during 2008–2017 was larger than that during 1999–2008. The numbers of landslides were 385–927 during 1999–2008, 1494 after Typhoon Morakot, and 1258–2581 during 2010–2017. The landslide areas were 4.4–9.7 km^2 during 1999–2008, 40.4 km^2 after Typhoon Morakot, and 19.2–35.6 km^2 during 2010–2017. The numbers and areas of landslides during 2010–2017 were 2.8–3.3 and 3.7–4.4 times higher, respectively, than those during 1999–2008. The obvious increases in the numbers and areas of landslides demonstrated the considerable impact of Typhoon Morakot on the landslide susceptibility of the Chishan river watershed.

The frequency of landslides [13] was defined as the total number of landslides occurring in each grid during 1999–2017, as shown in Figure 5. This information was used to explain the spatial changes in landslide susceptibility in the Chishan river watershed. Areas with a landslide frequency of 0 accounted for 91.0% of the study area, and those with a landslide frequency of >5 accounted for 3.0% of the area. Areas with a landslide frequency of >5 were concentrated along the rivers and in the sources of rivers in the C04–C10 subwatersheds, especially in the C07–C10 subwatersheds. These results indicated that the landslide distributions during 1999–2017 in the Chishan river watershed were related to the distribution of rivers. The landslide distributions in 1999, 2003, 2005, 2008, 2009, 2012, 2013, 2016, and 2017 in the Chishan river watershed are illustrated in Figure 4. Landslides were distributed in the C08–C10 subwatersheds before 2009 and then distributed densely in the C03–C10 subwatersheds after Typhoon Morakot. The landslide distribution narrowed, but it was still densely spread in the C04–C05 and C07–C10 subwatersheds during 2010–2017. The upstream subwatersheds, including C07–C10, were always susceptible to landslides, but the midstream subwatersheds, including C04–C05, became susceptible to landslides after Typhoon Morakot.

Rainfall events primarily triggered the evolution of large-earthquake-induced landslides [1,4], but they were not the only triggers for the evolution of extreme-rainfall-induced landslides [7,8]. According to the spatial and temporal distributions of landslides during 1999–2017, differences existed between the landslide distributions before and after Typhoon Morakot in the Chishan river watershed. Despite the decreasing strength and numbers of rainfall-induced landslides after Typhoon Morakot, the landslide area tended to decrease continuously, but the number of landslides tended to fluctuate in the Chishan river watershed. This fluctuation indicates that recovery from landslides was difficult and that landslides were easily reinduced.

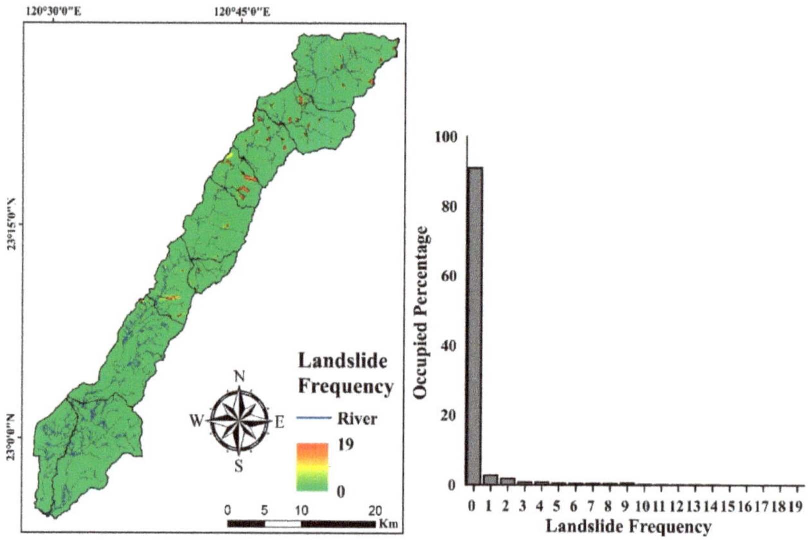

Figure 5. The distribution of landslide frequency from 1999 to 2017 in the Chishan river watershed.

4.2. Certainty Factor Analyses of Landslide Occurrence in 1999–2017

This study applied a certainty factor to analyze the effectiveness of four geomorphologic factors—namely elevation, slope, aspect, and distance to rivers—at affecting the occurrence of landslides during 1999–2017 in the Chishan river watershed by using the landslide inventories of 1999, 2001, 2005, 2006, 2009, 2010, 2012, 2013, 2015, and 2017. Most of the landslides in the Chishan river watershed were induced by rainfall events. The analysis results obtained in the Chishan river watershed by using the certainty factor were compared with those obtained in other earthquake-induced landslide areas [6,31]. Figure 6 presents the relationships between landslide occurrence and the four geomorphologic factors.

Positive certainty factor values were derived in areas with elevations of >1000 m during 1999–2008. Moreover, the certainty factor values in areas with elevations of >2000 m increased with time. In 2009, positive certainty factor values were obtained in areas with elevations of 500–1750 m, but the certainty factor values in the areas with elevations of >2000 m were negative. The certainty values in areas with elevations of <1000, 1000–1750, and >1750 m decreased, were positive, and increased, respectively. In areas with elevations of >2000 m, recovery from landslides was difficult, and landslides were easily reinduced, as determined on the basis of the long-term distribution of certainty factor values with respect to elevation during 1999–2017. The areas in which recovery from landslides was difficult and in which landslides were easily reinduced were determined to be valuable for the analysis of landslide evolution. The elevations of these areas in the Chishan river watershed after Typhoon Morakot were similar to the elevations observed after the 2008 Wenchuan earthquake in Sichuan, China [6].

The areas with positive certainty factor values during 1999–2008, 2009, and 2010–2017 in the Chishan river watershed had slopes of >40°. The certainty factor values in these areas during 2010–2017 increased gradually. The certainty factor values were higher in areas with steeper slopes. In these areas, recovery from landslides was difficult, and landslides were easily reinduced, as determined on the basis of the long-term distribution of the certainty factor values with respect to slope during 1999–2017. The slopes in these areas in the Chishan river watershed after Typhoon Morakot were the same as those observed after the 2008 Wenchuan earthquake in Sichuan, China [6].

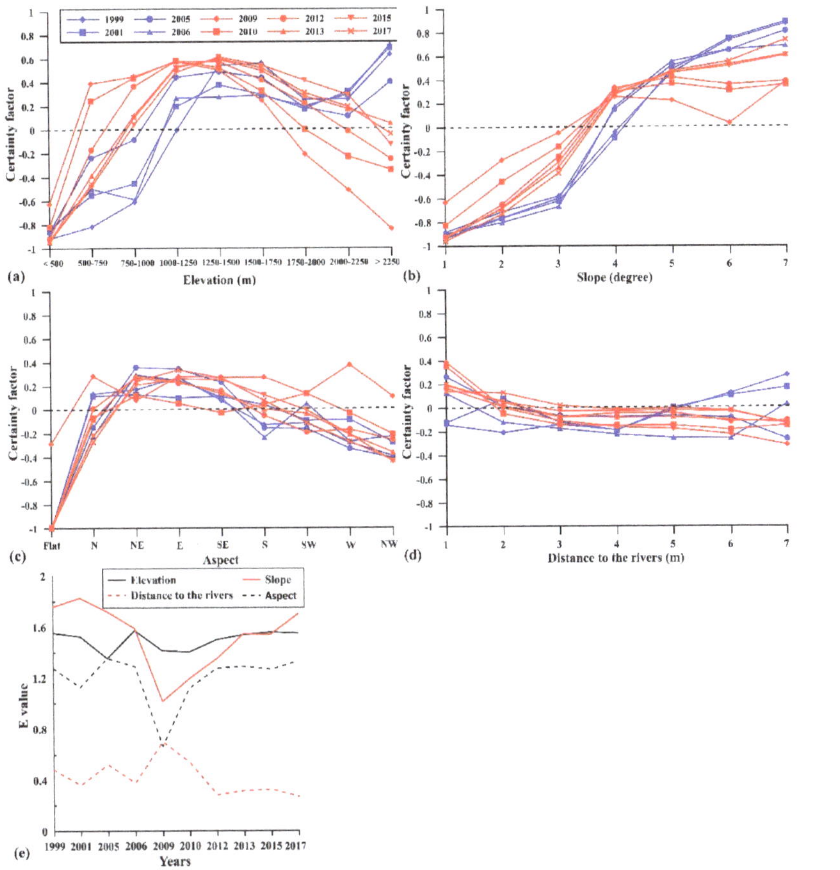

Figure 6. The certainty factor analysis in regard to elevation (**a**), slope (**b**), aspect (**c**), and distance to the rivers (**d**) and the *E* index values for these four factors (**e**) in the Chishan river watershed.

The slopes of the areas with positive certainty factor values during 1999–2008 had NE, E, and SE aspects. After Typhoon Morakot, the slopes of the areas with positive certainty factor values were in all aspects except for the flat aspect. Furthermore, the slopes of the areas with positive certainty factor values during 2009–2017 had N, NE, E, and SE aspects. In the Chishan river watershed, the slopes of the areas with high landslide susceptibility and difficulty recovering from landslides had N, NE, E, and SE aspects, as determined on the basis of the long-term distribution of certainty factor values with respect to aspect during 1999–2017. The aspects of slopes in the landslide-prone areas after large-earthquake-induced landslide events were similar to the directions of earthquake-causing faults [6], but the aspects of slopes in the landslide-prone areas in the Chishan river watershed after extreme-rainfall-induced landslide events were similar to windward side of the monsoon.

Studies have reported that the surroundings of rivers in upstream watersheds in Taiwan are susceptible to landslides [9,10]. This study revealed that the certainty factor values derived in areas within 400 m of rivers in the Chishan river watershed were negative after the 1999 Chichi earthquake event, indicating that the surroundings of rivers are not earthquake-induced landslide-prone areas. The certainty factor values derived in areas within 100 m of rivers were positive during 2005–2008. The certainty factor values reached their maximum during 1999–2017, 0.4, in areas within 100 m of rivers after Typhoon Morakot in the Chishan river watershed. The certainty factor values derived in areas within 100 m of rivers in the Chishan river watershed were greater than 0.15 during 2010–2017, and those derived in areas within 100–200 m of rivers were positive during 2012–2017. These results indicated that in terms of long-term landslide evolution after extreme rainfall or large earthquake events, areas around rivers in the Chishan river watershed had difficulty recovering from landslides, and landslides were easily reinduced in these areas.

The effectiveness index (*E*) was used to compare changes in the temporal effect of each factor on landslide occurrence during 1999–2017 in the Chishan river watershed (Figure 6e). The top two factors affecting landslide occurrence during 1999–2017 were elevation and slope, meaning that landslide location was a major factor affecting landslide evolution in the Chishan river watershed. Distance to rivers did not seem to have a clear effect on landslide evolution in the Chishan river watershed. Distance to rivers had the lowest effectiveness index (*E*) among all four factors, and its effectiveness index decreased rapidly after Typhoon Morakot.

4.3. Result of Local Outlier Analysis

The study conducted local outlier analysis to recognize the bins with landslide spatiotemporal statistical significance by using the local Anselin Moran's I index, and the statistical results are shown in Table 4. The multiple type dominated the spatiotemporal clustering pattern of landslide evolution during 1999–2008 and 2008–2017 in the Chishan river watershed. The areas of only high–high and low–low cluster patterns during 2008–2017 in the Chishan river watershed were obviously larger than those during 1999–2008. The distribution trend of the spatiotemporal clustering patterns of landslide evolution in the C10 subwatershed was similar to that in the Chishan river watershed. The distribution maps of local outlier analysis in the C10 subwatershed and Chishan river watershed are shown in Figure 7. The multiple types were widely spread in the C04 to C10 subwatersheds from 1999–2008 to 2008–2017. The only high–high clusters were recognized in the source areas of creeks and the neighborhoods of sinuous rivers, especially in the C10 subwatershed.

Table 4. The statistical results of local outlier analysis during 1999–2008 and 2008–2017 in the C10 subwatershed and the Chishan river watershed.

Patterns	Chishan River Watershed				C10 Subwatershed			
	1999–2008		2008–2017		1999–2008		2008–2017	
	A (ha) [1]	Per. [1]	A (ha)	Per.	A (ha)	Per.	A (ha)	Per.
Mul. [1]	1881.2	94.3	4073.7	83.7	703.0	94.8	798.3	83.1
NS. [1]	0.1	0.0	0.4	0.0	0.0	0.0	0.1	0.0
HH [1]	42.3	2.1	413.5	8.5	17.2	2.3	91.3	9.5
HL [1]	0.0	0.0	0.1	0.0	0.0	0.0	0.0	0.0
LH [1]	2.3	0.1	11.3	0.2	1.1	0.2	3.1	0.3
LL [1]	68.9	3.5	368.4	7.6	20.3	2.7	68.3	7.1

[1] Mul., NS., HH, HL, LH, and LL mean multiple types, never significant, only high–high cluster, only high–low cluster, only low–high cluster, and only low–low cluster, respectively. A and Per. mean area and percentage, respectively.

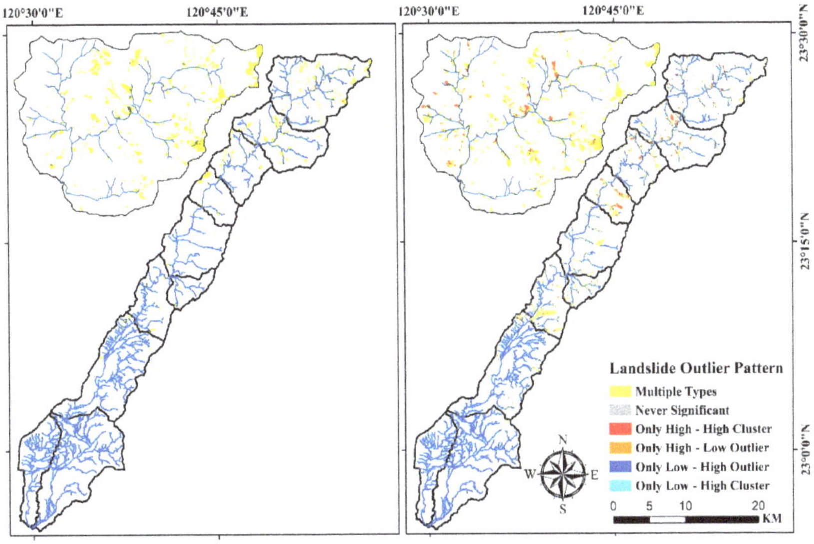

Figure 7. Local outlier analysis of landslide evolution during 1999–2008 (**left figure**) and 2008–2017 (**right figure**) in the Chishan river watershed.

4.4. Spatiotemporal Hotspot Distribution

Emerging hotspot analysis was used to explore the spatiotemporal hotspot distribution and patterns of landslide evolution during 1999–2017 in the Chishan river watershed. Previous studies have not examined the evolution trends of landslide clustering before and after Typhoon Morakot in the Chishan river watershed. A spatiotemporal analysis can provide a comprehensive view of landslide clustering locations. Accordingly, the present study performed the Mann–Kendall trend test to demonstrate changes in the spatial clustering of landslides in each grid during 1999–2008 and 2008–2017 (Figure 6). The spatial clustering distributions of landslides in the Chishan river watershed during 1999–2008 (Figure 6a) and 2008–2017 (Figure 6b) were remarkably different; areas with increasing landslide occurrence trends are clearly highlighted in Figure 6. The numbers of grids with no significant trend, declining trends, and increasing trends during 2008–2017 were 1.1, 3.1, and 8.9 times larger, respectively, than those during 1999–2008. These results imply that the trends of landslide clustering in the Chishan river watershed after Typhoon Morakot were clearer than those after the 1999 Chichi earthquake.

Moreover, grids with increasing trends were concentrated in the upstream subwatersheds (C07–C10) during 2008–2017. The numbers of grids with no significant trend, declining trends, and increasing trends during 2008–2017 in the upstream subwatersheds were 0.89, 1.07, and 9.06 times larger, respectively, than those during 1999–2008. The number of grids with increasing trends in the upstream subwatersheds during 2008–2017 accounted for 82.6% of those in the Chishan river watershed, meaning that the upstream subwatersheds represented the main areas with landslide clustering in the Chishan river watershed during 2008–2017. The spatial distribution of landslide clustering in C10 is illustrated in Figure 8. The grids with increasing trends in C10 were clustered along the rivers, in the source areas of the rivers, and in the surroundings of large landslides. In Taiwan, the locations of landslides have been reported to be related to geomorphological factors and rainfall distribution during heavy rainfall events [1,7,8,12]. However, the present study revealed that river systems and large landslide cases influenced the locations where landslides were difficult to recover from and easy to reinduce in the upstream of the Chishan river watershed during 2008–2017.

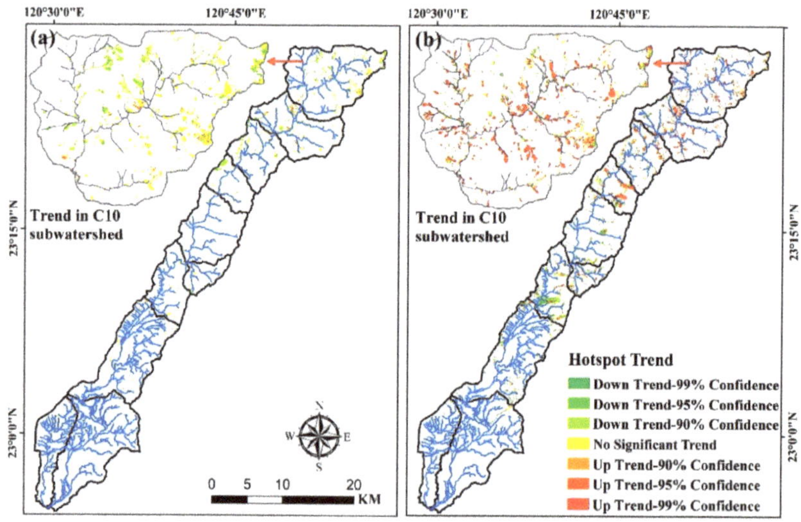

Figure 8. The spatial clustering distributions of landslides in 1999–2008 (**a**) and 2008–2017 (**b**) as assessed by using the Mann–Kendall trend test in the Chishan river watershed. The upper-left plots show the spatial clustering distribution of landslides in the C10 subwatershed in 1999–2008 (**a**) and 2008–2017 (**b**).

Figure 9 shows the differences in the spatiotemporal patterns and distributions of landslide hotspots during 1999–2008 and 2008–2017. The three main cold spot patterns during 1999–2008 were oscillating cold spots (OCS), persistent cold spots (PCS), and intensifying cold spots (ICS), and those during 2008–2017 were OCS, PCS, and sporadic cold spots (SCS). The three main hotspot patterns during 1999–2008 were oscillating hotspots (OHS), persistent hotspots (PHS), and intensifying hotspots (IHS), and those during 2008–2017 were OHS, HIS, and PHS. The characteristics of the main hotspot and cold spot patterns before and after Typhoon Morakot were oscillating and persistent, respectively.

During 1999–2017, the landslide hotspots were mainly distributed upstream of the Chishan river watershed. The main landslide hotspots in the upstream subwatersheds during 1999–2008 and 2008–2017 occupied 89.4% and 80.1%, respectively, of those in the Chishan river watershed. The main landslide hotspot in the upstream subwatersheds during 2008–2017 was 3.2 times larger than that during 1999–2008. OHS appeared in all upstream subwatersheds, but they were centralized in the surroundings of rivers and gullies and in the source areas of landslides. IHS and PHS were centralized in the core landslide areas and the downslopes of riverbank landslide areas.

The spatiotemporal patterns and distributions of the main landslide hotspots in the Chishan river watershed were related to the landslide area. The main landslide hotspot areas occupied 42.3%, 50.1%, and 59.1% of the large-, medium-, and small-landslide areas, respectively.

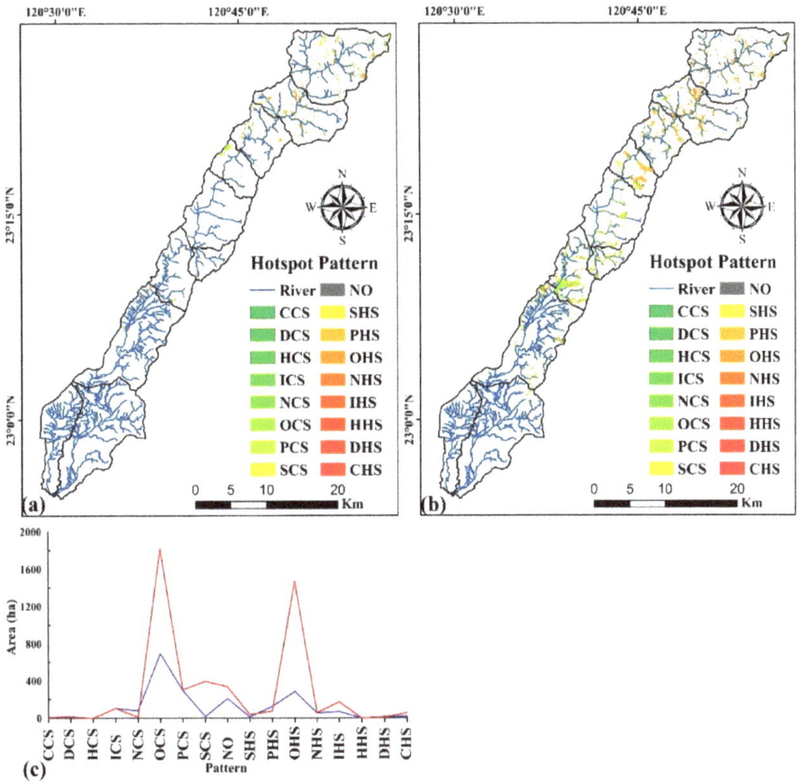

Figure 9. The patterns and distributions of landslide spatiotemporal hotspots in 1999~2008 (**a**) and 2008~2017 (**b**) in the Chishan river watershed. (**c**) shows the occupied percentage of each spatiotemporal hotspot pattern in 1999~2008 (blue line) and 2008~2017 (red line).

4.5. Clustering Strength of Small, Medium, and Large Landslides

In the Getis–Ord formula for hotspot analysis, the Z-score indicates the statistical significance of hotspots and cold spots; this study used it as an index to assess the strength of landslide clustering and estimate oscillation periods. A higher Z-score was considered to indicate more obvious landslide clustering, with a Z-score of 0 or a negative Z-score indicating the absence of landslide clustering. Figure 10 presents the average Z-scores during 1999–2008 and 2008–2017 in the Chishan river watershed.

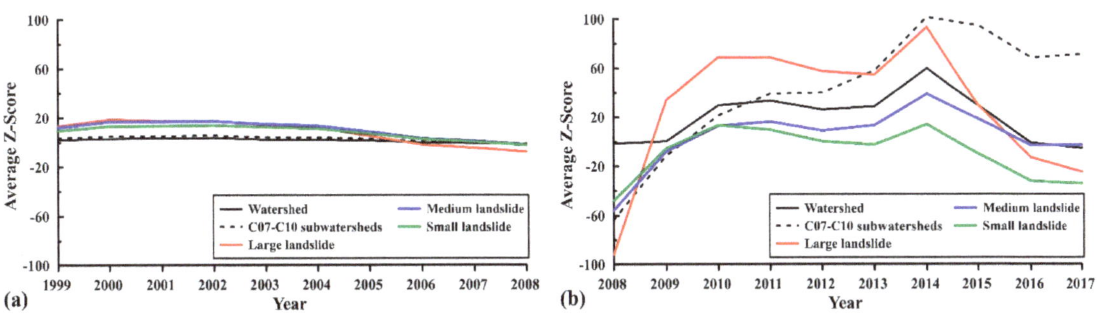

Figure 10. Temporal distribution of average Z-scores in the Chishan river watershed (black line), upstream subwatersheds (C07–C10, black dash line), large landslide cases (red line), medium landslide cases (blue line), and small landslide cases (green line) in 1999–2008 (**a**) and 2008–2017 (**b**).

The temporal distributions of the clustering strengths of large, medium, and small landslides during 1999–2008 were similar, and those of landslides during 2008–2017 were also similar. During 1999–2004, the landslides could be ordered as follows in terms of strength (in decreasing order): large, medium, and small landslides. Moreover, during 2005–2008, they could be ordered as follows in terms of strength (in decreasing order): medium, small, and large landslides. These results demonstrate that the clustering strength of large landslides in the Chishan river watershed was high only in the early years after the 1999 Chichi earthquake.

The distributions of the clustering strength of the large, medium, and small landslides during 2010–2014 fluctuated, and the strength decreased rapidly from 2015 onward. The clustering of small landslides induced by Typhoon Morakot was not obvious, and the average Z-score derived for small landslides was positive only during 2010–2012 and in 2014. The average Z-scores derived for large and medium landslides induced by Typhoon Morakot were positive in the subsequent 5 years, and they started to decrease rapidly from 2015 onward.

5. Discussion

The spatiotemporal hotspots and cold spots of landslides can indicate the clustering strength of landslides and the patterns of landslide locations. Moreover, the spatiotemporal distribution of landslide hotspots is directly related to the evolution and recovery of a watershed after extreme-rainfall-induced landslide events, and it exerts a strong negative effect on the sustainability of watershed management. The oscillation period [9] can be defined as the period during which the number or area of landslides fluctuates in the years that follow severe landslide events. The oscillation period can also be defined as the time needed for the watershed to stably recover from landslides. Landslide evolution research has used different methods to assess landslide recovery time and the oscillation period after large-earthquake- and extreme-rainfall-induced serious landslide events. The spatial distribution of landslide recovery has been assessed by estimating the landslide activity, and the temporal changes in landslide recovery have been assessed by observing the numbers and areas of landslides every year after serious landslide events, after the 2005 Kashimir earthquake [2] and the 2008 Wenchuan earthquake [4–6]. The oscillation period was estimated to be 3–5 years after the 2005 Kashmir earthquake and the 2008 Wenchuan earthquake and 5 years after Typhoon Morakot in southern Taiwan [9].

The 1999 Chichi earthquake induced 439 landslide cases covering a total area of 9.5 km^2 in the Chishan river watershed. The average Z-score in the Chishan river watershed (Figure 10) was 2.15 in 1999; it increased to 3.04–3.68 in 2000–2002 and then started to decrease in 2003 before finally dropping to −1.48 in 2008. Typhoon Morakot in 2009 induced 1494 landslide cases covering a total area of 40.3 km^2 in the Chishan river watershed. The average Z-score in the Chishan river watershed was 0.31 in 2009; it increased rapidly to 29.43 in 2010 and fluctuated between 26.02 and 29.78 during 2010–2014, after which it started to decrease in 2015 until reaching −5.37 in 2017. The average Z-score after the 1999 Chichi earthquake in the Chishan river watershed was 1.41 times greater than that before 1999, and the average Z-score after Typhoon Morakot was 94.94 times greater than that before 2009. These results demonstrate that the clustering strength of landslides induced by Typhoon Morakot in the Chishan river watershed was considerably greater than that of landslides induced by the 1999 Chichi earthquake. According to the temporal changes in the average Z-scores, the Chishan river watershed required 2 years to achieve stable recovery after the 1999 Chichi earthquake but 4 years after Typhoon Morakot. Accordingly, the oscillation period after Typhoon Morakot in the Chishan river watershed was twice as long as that after the 1999 Chichi earthquake. The results also indicated that recovery from landslides induced by a large earthquake (M_L = 7.3) in Taiwan was faster than from those induced by an extreme rainfall event with accumulated rainfall of >2000 mm.

The average Z-scores in the upstream subwatersheds during 1999–2008 ranged from −0.95 to 5.66, and they were marginally higher than those in the Chishan river watershed.

However, the average Z-scores in the upstream subwatersheds during 2008–2017 ranged from −63.70 to 101.38, and those after 2010 were higher than those in the Chishan river watershed. The average Z-score in the upstream subwatersheds during 2016–2017 was approximately 70.0, and that in the other area of the Chishan river watershed was negative. These results indicate that the landslide recovery period in the upstream subwatersheds was longer than that in the Chishan river watershed.

The spatial resolution of images that were used to identify the landslide was the major factor for the number of identified small landslides. The images used to identify landslides were Spot 1 and Spot 2 images with spatial resolution of 10 m during 1999–2002 and Formosat-2 images with spatial resolution of 2 m during 2003–2017. The number of small landslides ranged from 238 to 550 during 1999–2002, from 503 to 604 during 2003–2008, and from 1411 to 1930 during 2009–2017. The number of small landslides in the Chishan river watershed increased obviously after 2003, especially after Typhoon Morakot in 2009.

The number of small landslides was also the major reason explaining the temporal changes in Z-values. The number of small landslides was 54.2 to 71.8% of the total number of landslides during 1999–2002 and 70.2 to 80.4% during 2003–2017. The number of small landslide was 1494 in 2009, peaked at 1930 in 2014, and ended at 1411 in 2017. The fluctuation in the number of small landslides was similar to the temporal change in Z-values during 2009–2017. We suggest that the number of small landslides was the major factor for the average Z-value and that the location of medium and large landslides was the major factor for the distribution and patterns of spatiotemporal landslide hotspots based on the results on long-term landslide evolution in the Chishan river watershed.

The Z-values in the study provided quantified data on landslide clustering at different scales (watershed or subwatershed) and different landslide sizes (small, medium, and large) in different time periods. Using Z-values to explain the spatial distribution of and temporal changes in landslide recovery can be a convincing method for explaining landslide evolution.

Using spatiotemporal hotspot analysis to analyze landslide evolution in the study provided powerful theoretical evidence to explain the spatial and temporal distribution of landslide clustering. The patterns and distributions of landslide spatiotemporal hotspots can also explain the characteristics of watershed recovery after serious landslide events and can thus represent useful information for formulating policies for watershed recovery.

6. Conclusions

We analyzed the effectiveness of landslide occurrence and recovery and explained the patterns and distributions of landslide hotspots in the Chishan river watershed during 1999–2017. The numbers and areas of landslides in the Chishan river watershed after Typhoon Morakot in 2009 were around three times larger than those after the 1999 Chichi earthquake, and this result indicated that the impact of Typhoon Morakot on the landslide susceptibility in the Chishan river watershed was stronger and more obvious than that of the 1999 Chichi earthquake. We evaluated the effectiveness of landslide occurrence and recovery during 1999–2017 by using a certainty factor and compared the importance of four geomorphologic factors affecting the landslide occurrence. The concentration of landslide occurrence and evolution in the Chishan river watershed was more related to the factors of elevation and slope than to those of aspect and distance to the river. We built a spatiotemporal cube model by using the annual landslide inventories during 1999–2017 and explained the spatiotemporal characteristics of landslide hotspots by using emerging hotspot analysis. The main hotspot patterns in the Chishan river watershed were OHS, PHS, and IHS, while the main cold spot patterns were OCS, PCS, and SCS. We explained the temporal changes in landslide clustering by using the Z-score and found that the core landslide areas and the downslopes of riverbank landslide areas in the upstream of the Chishan river watershed were the areas of highest landslide susceptibility. The long-term monitoring of landslide occurrence and recovery is useful for understanding the recovery time after landslides and temporal changes in landslide susceptibility. The findings on the evolution of postearthquake and post-extreme rainfall landslides in this study are valuable for the prediction and disaster prevention of future geohazards.

Funding: This research was funded by Project Research in Feng Chia University (number: 20H00710 and 21H00710).

Institutional Review Board Statement: Not applicable.

Informed Consent Statement: Not applicable.

Data Availability Statement: Not applicable.

Conflicts of Interest: The author declares no conflict of interest.

References

1. Lin, C.W.; Liu, S.H.; Lee, S.Y.; Liu, C.C. Impacts of the Chi-Chi earthquake on subsequent rainfall-induced landslides in Central Taiwan. *Eng. Geol.* **2006**, *86*, 87–101. [CrossRef]
2. Shafique, M. Spatial and temporal evolution of co-seismic landslides after the 2005 Kashmir earthquake. *Geomorphology* **2020**, *362*, 107228. [CrossRef]
3. Saba, S.B.; van der Meijde, M.; van der Werff, H. Spatiotemporal landslide detection for the 2005 Kashmir earthquake region. *Geomorphology* **2010**, *124*, 17–25. [CrossRef]
4. Yang, W.; Qi, W.; Wang, M.; Zhang, J.; Zhang, Y. Spatial and temporal analyses of post-seismic landslide changes near the epicentre of the Wenchuan earthquake. *Geomorphology* **2017**, *276*, 8–15. [CrossRef]
5. Li, C.; Wang, M.; Liu, K. A decadal evolution of landslides and debris flows after the Wenchuan earthquake. *Geomorphology* **2018**, *323*, 1–12. [CrossRef]
6. Chen, M.; Tang, C.; Xiong, J.; Shi, Q.Y.; Li, N.; Gong, L.F.; Wang, X.D.; Tie, Y. The long-term evolution of landslide activity near the epicentral area of the 2008 Wenchuan earthquake in China. *Geomorphology* **2020**, *367*, 107317. [CrossRef]
7. Wu, C.H.; Chen, S.C.; Chou, H.T. Geomorphologic Characteristics of Catastrophic Landslides during Typhoon Morakot in the Kaoping Watershed, Taiwan. *Eng. Geol.* **2011**, *123*, 13–21. [CrossRef]
8. Wu, C.H.; Chen, S.C.; Feng, Z.Y. Formation, failure, and consequences of the Xiaolin landslide dam, triggered by extreme rainfall from Typhoon Morakot, Taiwan. *Landslides* **2014**, *11*, 357–367. [CrossRef]
9. Rosser, N.; Kincey, M.; Oven, K.; Densmore, A.; Robinson, T.; Pujara, D.S.; Shrestha, R.; Smutny, J.; Gurung, K.; Lama, S.; et al. Changing significance of landslide Hazard and risk after the 2015 Mw 7.8 Gorkha, Nepal Earthquake. *Prog. Disaster Sci.* **2021**, *10*, 100159. [CrossRef]
10. Chen, X.L.; Shan, X.J.; Wang, M.M.; Liu, C.G.; Han, N.N. Distribution Pattern of Coseismic Landslides Triggered by the 2017 Jiuzhaigou Ms 7.0 Earthquake of China: Control of Seismic Landslide Susceptibility. *ISPRS Int. J. Geo-Inf.* **2020**, *9*, 198. [CrossRef]
11. Huang, R.; Fan, X. The landslide story. *Nat. Geosci.* **2013**, *6*, 325–326. [CrossRef]
12. Wu, C.; Lin, C. Spatiotemporal hotspots and decadal evolution of extreme rainfall-induced landslides: Case studies in southern Taiwan. *Water* **2021**, *13*, 2090. [CrossRef]
13. Wu, C.H. Evaluating the landslide stability and vegetation recovery: Case studies in the Tsengwen reservoir watershed in Taiwan. *Water* **2021**, *13*, 3479. [CrossRef]
14. Zhang, S.; Zhang, L.M.; Glade, T. Characteristics of earthquake- and rain-induced landslides near the epicenter of Wenchuan earthquake. *Eng. Geol.* **2014**, *175*, 58–73. [CrossRef]
15. De Cos, O.; Castillo, V.; Cantarero, D. Differencing the risk of reiterative spatial incidence of COVID-19 using space–time 3D bins of Geocoded Daily Cases. *ISPRS Int. J. Geo-Inf.* **2021**, *10*, 261. [CrossRef]
16. Purwanto, P.; Utaya, S.; Handoyo, B.; Bachri, S.; Astuti, I.S.; Utomo, K.S.B.; Aldianto, Y.E. Spatiotemporal analysis of COVID-19 spread with emerging hotspot analysis and space–time cube models in east Java, Indonesia. *ISPRS Int. J. Geo-Inf.* **2021**, *10*, 133. [CrossRef]
17. Syetiawan, A.; Harimurti, M.; Prihanto, Y. A spatiotemporal analysis of COVID-19 transmission in Jakarta, Indonesia for pandemic decision support. *Geospat. Health* **2022**, *17*, 1042. [CrossRef]
18. Everett, B.I.; Fennessy, S.T.; van den Heever, N. Using hotspot analysis to track changes in the crustacean fishery off KwaZulu-Natal, South Africa. *Reg. Stud. Mar. Sci.* **2021**, *41*, 101553. [CrossRef]
19. Kang, Y.; Cho, N.; Son, S. Spatiotemporal characteristics of elderly population's traffic accidents in Seoul using space-time cube and space-time kernel density estimation. *PLoS ONE* **2018**, *13*, e0196845. [CrossRef]
20. Cheng, Z.; Zu, Z.; Lu, J. Traffic crash evolution characteristic analysis and spatiotemporal hotspot identification of urban road intersections. *Sustainability* **2019**, *11*, 160. [CrossRef]
21. Cheng, S.; Zhang, B.; Peng, P.; Yang, Z.; Lu, F. Spatiotemporal evolution pattern detection for heavy-duty diesel truck emissions using trajectory mining: A case study of Tianjin, China. *J. Clean. Prod.* **2020**, *244*, 118654. [CrossRef]
22. Zhang, X.; Yao, J.; Sila-Nowicka, K.; Jin, Y. Urban Fire Dynamics and Its Association with Urban Growth: Evidence from Nanjing, China. *ISPRS Int. J. Geo-Inf.* **2020**, *9*, 218. [CrossRef]

23. Wu, C.H. Landslide susceptibility based on extreme rainfall-induced landslide inventories and the following landslide evolution. *Water* **2019**, *11*, 2609. [CrossRef]
24. Yang, S.Y.; Jan, C.D.; Wang, J.S. Landslides Triggered by Typhoon Morakot in Taiwan. In *Environmental Risks*, 1st ed.; Mihai, F.C., Grozavu, A., Eds.; Intech Open: London, UK, 2018; pp. 13–43.
25. Liu, C.C.; Liu, J.G.; Lin, C.W.; Wu, A.M.; Liu, S.H.; Shieh, C.L. Image processing of formosat-2 data for monitoring south asia tsunami. *Int. J. Remote Sens.* **2007**, *28*, 3093–3111. [CrossRef]
26. Liu, C.C.; Chen, P.L. Automatic extraction of ground control regions and orthorectification of formosat-2 imagery. *Opt. Express* **2009**, *17*, 7970–7984. [CrossRef]
27. Liu, C.C.; Shieh, C.L.; Wu, C.A.; Shieh, M.L. Change detection of gravel mining on riverbeds from the multi-temporal and high-spatial-resolution formosat-2 imagery. *River Res. Appl.* **2009**, *25*, 1136–1152. [CrossRef]
28. Liu, C.C.; Ko, M.H.; Wen, H.L.; Fu, K.L.; Chang, S.T. Instability Index Derived from a Landslide Inventory for Watershed Stability Assessment and Mapping. *ISPRS Int. J. Geo-Inf.* **2019**, *8*, 145. [CrossRef]
29. Shortliffe, E.H.; Buchanan, B.G. A model of inexact reasoning in medicine. *Math. Biosci.* **1975**, *23*, 351–379.
30. Heckerman, D. Probabilistic interpretations for Mycin's certainty factors. *Mach. Intell. Patt. Rec* **1986**, *4*, 167–196.
31. Xu, C.; Dai, F.C.; Yao, X.; Chen, J.; Tu, X.B.; Cao, Y.B.; Xiao, J.Z. GIS based certainty factor analysis of landslide triggering factors in Wenchuan earthquake. *Chin. J. Rock Mech. Eng.* **2010**, *29*, P2972–P2981. (In Chinese)
32. Lan, H.X.; Zhou, C.H.; Wang, L.J.; Zhang, H.Y.; Li, R.H. Landslide hazard spatial analysis and prediction using GIS in the Xiaojiang watershed, Yunnan, China. *Eng. Geol.* **2004**, *76*, 109–128. [CrossRef]
33. Anselin, L. Local Indicators of Spatial Association—LISA. *Geogr. Anal.* **1995**, *27*, 93–115. [CrossRef]
34. Ord, J.K.; Getis, A. Local spatial autocorrelation statistics: Distributional issues and an application. *Geogr. Anal.* **1995**, *27*, 286–306. [CrossRef]

International Journal of *Geo-Information*

MDPI

Article

Ecological Impact Prediction of Groundwater Change in Phreatic Aquifer under Multi-Mining Conditions

Shenghui Zhou [1], Tingxi Liu [1,2,*] and Limin Duan [1,2]

1 Inner Mongolia Key Laboratory of Water Resource Protection and Utilization, Water Conservancy and Civil Engineering College, Inner Mongolia Agricultural University, Hohhot 010018, China; 2018202060045@emails.imau.edu.cn (S.Z.); duanlimin820116@163.com (L.D.)

2 Collaborative Innovation Center for Integrated Management of Water Resources and Water Environment in the Inner Mongolia Reaches of the Yellow River, Hohhot 010018, China

* Correspondence: txliu1966@163.com; Tel.: +86-471-430-9386

Abstract: In aeolian sandy grass shoal catchment areas that rely heavily on groundwater, mining-induced geological deformation and aquifer drainage are likely to cause irreversible damage to natural groundwater systems and affect the original circulation of groundwater, thus threatening the ecological environment. This study aimed to predict the impact of groundwater level decline on vegetation growth in the Hailiutu River Basin (HRB), which is a coal-field area. Based on remote-sensing data, the land use/cover change was interpreted and analyzed, and the central areas of greensward land in the basin were determined. Subsequently, the correlation between groundwater depth and grassland distribution was analyzed. Then, the groundwater system under natural conditions was modeled using MODFLOW, and the groundwater flow field in 2029 was predicted by loading the generalized treatment of coal mine drainage water to the model. The change in groundwater depth caused by coal mining and its influence on the grassland were obtained. The results show that coal mining will decrease the groundwater depth, which would induce degradation risks in 4 of the original 34 aggregation centers of greensward land that originally depended on groundwater for growth in HRB because they exceeded the groundwater threshold. The prediction results show that the maximum settlement of groundwater level can reach 5 m in the northern (Yinpanhao), 6 m in the eastern (Dahaize), and 10 m in the southern (Balasu) region of HRB. Attention should be paid to vegetation degradation in areas where groundwater depth exceeds the minimum threshold for plant growth.

Keywords: groundwater depth; greensward land; MODFLOW; coal mining; Hailiutu River Basin (HRB)

Citation: Zhou, S.; Liu, T.; Duan, L. Ecological Impact Prediction of Groundwater Change in Phreatic Aquifer under Multi-Mining Conditions. *ISPRS Int. J. Geo-Inf.* **2022**, *11*, 359. https://doi.org/10.3390/ijgi11070359

Academic Editors: Wolfgang Kainz and Fuan Tsai

Received: 19 April 2022
Accepted: 20 June 2022
Published: 23 June 2022

1. Introduction

As the transfer base of coal mining in China, the Inner Mongolia Autonomous Region and Shaanxi province are gaining importance [1,2]. Controlled by the burial characteristics, more than 90% of coal resources in China are extracted by underground mining [3]. However, high-intensity underground coal mining has brought about a series of water resource and ecological environmental problems [4,5]. Continuous development of underground coal mining would lead to the deformation and collapse of the overlying strata, resulting in fissures and subsidence, which will destroy the groundwater aquifer system [6], reduce the regional groundwater level, and affect the surface ecological environment. Unlike other regions, coal deposits in Inner Mongolia and Shaanxi are located in arid and semi-arid regions [7–9], which are areas of serious soil erosion in China, and their ecological environment is very fragile. Consequently, there is a major contradiction between groundwater loss through drainage by coal mining and the lack of water resources for vegetation growth. Therefore, it is meaningful to study the ecological changes caused by the decline of groundwater level due to coal mining.

Coal burial in China is characterized by strong geological reconstruction after coal formation due to which water inrush accidents occur frequently during coal mining [10]. Such accidents pose serious threats to the safety of miners and mine production equipment, hinder the development of coal industry and mining operations [11], and even bring a series of ecological and environmental problems. Therefore, it is necessary to study the groundwater system of coal mining areas. For example, Yu et al. [12] carried out field trials on coal mine geological engineering in the Bucun coal mine, China, based on theoretical calculation and numerical simulation, and proposed a combined controlling measure considering underground coal mining and the water environment. Andres and Jose [13] presented a conceptual and numerical model of two linked mines in Spain for the assessment of possible environmental risks following closure. Wu and Wang [14] presented a conceptual framework with multilayered groundwater flow systems for the characterization of water bursting and discharge in underground mines, based on the characteristics of regional geological conditions, and they analyzed mine water inrush and flooding events in the north China coal basin.

With the gradual improvement of governmental efforts and increase in people's awareness on the protection of the ecological environment [15], the impact of coal mining on the ecological environment is receiving increasingly more attention [4]. For example, in order to evaluate related effects in mining areas and non-mining areas from an ecological viewpoint, Xiao et al. [16] constructed an ecosystem service measurement and evaluation index system in a typical high-intensity coal mining area using remote sensing and geographic information system (GIS) and obtained good application results. Aiming to investigate the contamination level, distribution, and possible sources of polycyclic aromatic hydrocarbons (PAHs) in six coal mines in Xuzhou, China, and assess their ecological risks, Chen et al. [17] determined the concentrations of 16 PAHs in 26 underground samples, including coal, mine water, and underground sludge. Quaranta et al. [18] studied the influence of coal slurry injection in two underground coal mines on the potential water quality of surface water and groundwater in the central Appalachians and predicted the standard excess of surface water and groundwater using a dilution analysis model.

Recent studies have defined three stages of impact (damage phase, post-damage phase, and restoration phase) based on the effect of mining-induced groundwater changes on vegetation conditions [19]. Xie et al. [20] established a hydrogeological conceptual model and a groundwater flow numerical model to predict mining-induced changes in the groundwater flow field over five years (30 April 2015 to 30 April 2020). Based on the vegetation index determined using remote-sensing data, Zhang and Wang [21] reported that groundwater depth has a prominent effect on the spatial variation of vegetation distribution in the Ordos Plateau of China.

The border between Inner Mongolia and Shaanxi feature mega-coal fields, which are slated for mining in the next decade [1,22]. However, these coal fields occur below the Mu Us Sandy Land, which has the most vulnerable environment in China [23,24]. Therefore, changes in the groundwater and natural environment induced by coal mining are bound to affect the sustainable development of the region. These effects are mainly manifested in the reduction in groundwater, deterioration of vegetation that relies heavily on groundwater for growth, and regional vegetation shifting toward more drought-tolerant species [25], which are not suitable for grazing. More seriously, when groundwater shortage exceeds the basic threshold for maintaining vegetation growth, desertification will occur, making the area unsuitable for human habitation. Therefore, when groundwater discharge by coal mining exceeds natural recharge, groundwater resources will inevitably decrease sharply, ultimately threatening the healthy development of regional agriculture, animal husbandry, and industry. In particular, the Hailiutu River Basin (HRB) is a typical representative of this area. HRB has multiple planned coal fields, and its ecological vegetation and agricultural development mainly depend on groundwater recharge [26]. The relationship between the groundwater and ecology of HRB and its surrounding areas has been widely studied. In areas with groundwater depth less than 10 m within the HRB, the mean and standard deviation of normalized difference vegetation index (NDVI) values have been found to decrease with increasing groundwater depth [27]. Mata-González et al. [28] reported that vegetation growth and plant diversity are closely related to groundwater depth, especially in arid and semi-arid regions, and areas with shallow groundwater have higher vegetation coverage and plant diversity. Accordingly, the decline of groundwater level caused by coal mining will inevitably affect the status of vegetation growth, especially in the aeolian sand grass shoal catchment area, which is heavily dependent on groundwater [29]. Previous studies have investigated and predicted the influence of coal mining on groundwater mainly by establishing groundwater models [30,31]. In these models, the drainage by coal mining is used as input [32]. Subsequently, future changes in the groundwater flow field are predicted after model calibration and verification [33]. However, the prediction of the impact of groundwater change on the ecological environment has not been comprehensively analyzed thus far. In particular, detailed information on the specific impact of model predictions on the corresponding spatial distribution of a certain vegetation type after water level changes is lacking.

To address this gap, this study predicted the grassland degradation risk associated with groundwater decline induced by coal mining. Selecting HRB as a representative of the coal fields of Inner Mongolia and Shaanxi, mining-induced changes in groundwater circulation and their effects on the ecological environment were investigated. The groundwater depth of HRB was determined by measuring submersible wells, and the distribution of greensward land in the basin was interpreted using remote-sensing satellite images. Finally, changes in the groundwater flow field in the next 10 years (2020–2029), after the commencement of mining in first three mining areas in the basin, were simulated using a numerical model.

2. Materials and Methods

2.1. Study Area

HRB is located at the junction of Inner Mongolia and Shaanxi (Figure 1), which is an important coal mining base in northwest China. This basin belongs to the middle course of the Yellow River Basin. HRB is a small multiple bottomland basin in the transition area from the Ordos Plateau to the northern Shaanxi Loess Plateau [34]. The entire basin lies above the Mu Us Sandy Land, and the surface is mainly covered by Quaternary aeolian sand and lacustrine sand [35] (Figure 1). The basin covers an area of approximately 2600 km^2 and has a surface elevation ranging from 1020 m in the southeast to 1480 m above mean sea level in the northwest [36]. The perennial Hailiutu River flows from the northwest to the southeast of the basin and enters the Wuding River, the main tributary of the middle Yellow River. The Hanjiamao hydrological station is located at the outlet of watershed catchments, which had an average annual runoff of 0.8×10^6 m^3 from 1957 to 2014. HRB has a typical steppe environment with arid, semi-arid sandy land and beach land. It has a temperate continental monsoon climate, and atmospheric precipitation is the main source of water supply. The annual precipitation is 370 mm, and the annual average potential evaporation is 2000 mm [37]. The precipitation is mainly concentrated between July and September, and the wind direction is mainly northwest.

2.2. Geology and Hydrogeology

The study area has flat terrain, gentle strata, simple geological structure, and no distinct faults. The tectonic activity is not strong [38], and the watershed of HRB is composed of denudation loess ridges and high dunes [39]. Along at the river channel, the topography is steep, mainly due to river scouring and cutting of Cretaceous strata through flow erosion at the outlet of the basin. In addition to psammophytes, grass, and farmland, most of the land surface is covered by Quaternary sand dunes, with scattered spots of small ponds formed by groundwater. The Quaternary layer has various diagenetic features, but the thickness is generally small. Mesozoic and Cenozoic strata are widely exposed and generally water bearing, and the Lower Cretaceous Huanhe Formation of the Baoan Group is exposed only in some areas [40]. Within the basin, the Quaternary and Cretaceous aquifers exhibit distinct vertical variation, and their thickness varies widely.

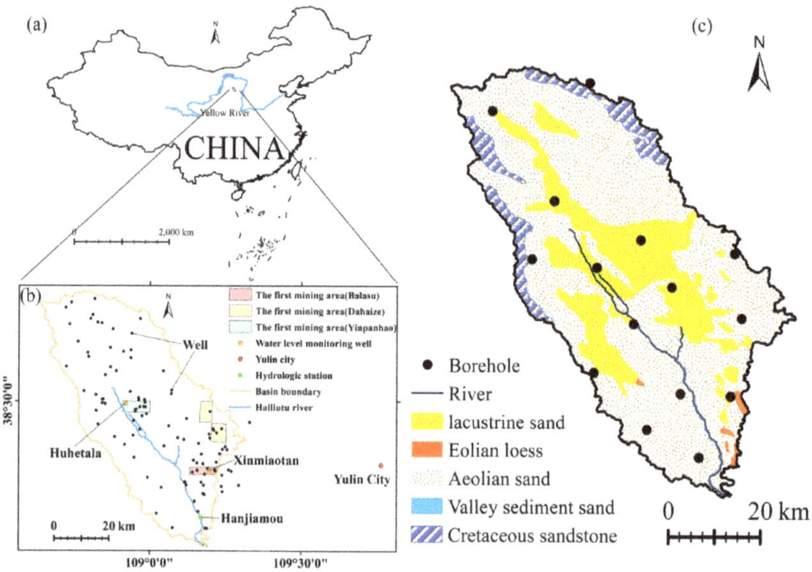

Figure 1. Location of the study area: (**a**) Location of the studied river basin in China; (**b**) Location of wells, hydrological stations, and first mining areas of coal mines in the basin; (**c**) Borehole drilling sites and geological background.

The average thickness of the upper Quaternary loose rock soil is approximately 50 m. The thickness of the Cretaceous sandstone aquifers is approximately 350 m. Between the main 2# coal seam and the Cretaceous floor is a special aquifer group formed by the alternate deposition of sandstone aquifers and impervious layers, which is the main source of mine water inrush [41]. The thickness of this special aquifer group is approximately 180 m. The Jurassic 2# coal seam in the study area is the first mining coal seam with an average thickness of 5 m (Figure 2). Restricted by the paleotopography during deposition, the thickness of each stratum in the basin varies greatly [42]. Generally, the strata are the thickest in the paleochannel and low-lying center, gradually thinning to both sides, and nip-out at the watershed [43]. However, the thickness of Quaternary strata gradually decreases from downstream to upstream, and the thickness of some beach land in the middle of the basin increases. The thickness of Cretaceous strata gradually increases from downstream to upstream, forming a large aquifer. Among them, there is no continuous and effective aquifuge between Quaternary and Cretaceous aquifers, but there are multiple aquifuge and aquifer alternations in the Jurassic strata [44]. Therefore, the aquifer above Cretaceous strata is regarded as a water-rich layer, and the aquifer below is regarded as a weak aquifer. Based on the above, the Quaternary and Cretaceous strata in the study area can be regarded as a unified phreatic aquifer group. The Quaternary and Cretaceous aquifer groups have different thicknesses, but they generally exhibit a continuous distribution. Stratigraphic integration is relatively gentle among the Jurassic strata, and the 2# coal seam is also evenly distributed in the basin because of the stress action of the upper strata [45].

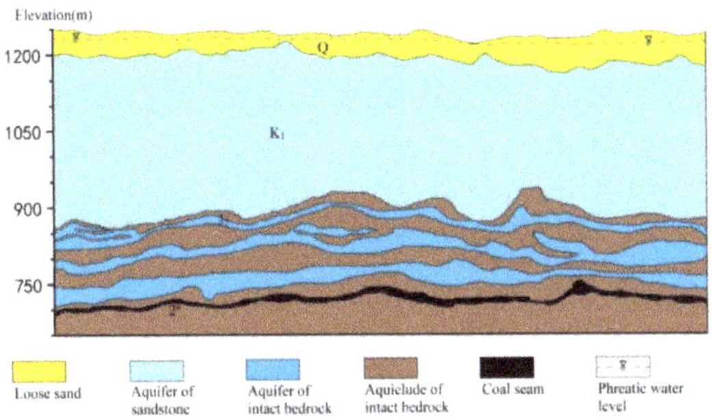

Figure 2. Schematic stratigraphic section of the study area.

2.3. Land Use/Cover Classification

Land use/cover change (LUCC) can provide crucial information for global environmental change and human environmental science [46]. LUCC can directly reflect the way humans utilize the land and the influence of human activities on the ecological environment [47]. In this study, we used Landsat images (TM/ETM+/OLI) of moderate resolution (30 m), remote-sensing data, and optical images from October 2019 acquired from the United States Geological Survey (USGS). We selected remote-sensing images with no or less cloud cover to improve the accuracy of land cover classification. For HRB, only one-scene images are required. Regarding the images, the cloud cover was 0.49%, and the Landsat path/row numbers were 128/33. After pretreatment, which involved radiometric calibration and atmospheric correction, land cover information was extracted using a supervised classification method [48]. In this study, land cover was classified into seven classes: greensward, psammophyte, water bodies, roads, buildings, farmland, and bare sand. The classification was performed according to the quantity and distribution characteristics in land use/cover types of the study area determined through field surveying and the research objectives of this study. We selected training samples from field investigations and Google Earth Engine for land use classification. A total of 287 samples were selected from evenly distributed sampling points in the watershed, including 58 greensward, 23 psmmophyte, 6 water bodies, 20 roads, 45 buildings, 68 farmlands, and 67 bare sandy lands. For pixel sample training, 70% of the samples were employed, and the remaining samples were used for testing (Table 1).

Table 1. Description of land use/cover classification.

LUCC	Training	Testing	Total	Description
Greensward	144,330	61,855	206,185	Emergence of green
Psammophyte	857,465	367,485	1,224,950	No obvious geometric features, mostly dark green or brown in color
Water bodies	3331	1427	4758	Lakes and reservoirs, and some wide rivers
Roads	7146	3063	10,209	Usually near buildings, in a regular straight line
Buildings	45,126	19,340	64,466	Regular geometry exists, concentrated distribution, with red or blue roofs
Farmland	348,413	149,320	497,733	Emergence of green, with clear geometric boundaries
Bare sand	622,805	266,916	889,721	Yellow, with the feeling of sand flowing

The distribution of greensward area in HRB was processed in the ENVI software, which was used to validate the classification results. Although greensward area accounts for only 7.1% of the basin, its sensitivity to changes in groundwater depth is the highest among all land use types.

2.4. Interpolation of Groundwater Depth

Spatial interpolation is an effective tool for studying groundwater distribution and has been widely used in the field of groundwater resources [49]. As only limited point-scale groundwater depth information can be obtained through monitoring stations in HRB, we used spatial interpolation to process the measured 100 wells and obtain spatially continuous groundwater depth data. To ensure that the wells were spatially representative, we selected 21 wells for continuous monitoring from July to early September (Figure 3). The monitoring showed that the average variation of groundwater level was only 0.1 m. To gain a more realistic view of groundwater depth in the watershed, we used the measured groundwater depth as the main variable and some small lakes and river node locations in the watershed as auxiliary variables, and we used the Co-kriging method of the ArcGIS software for interpolation [50]. Co-kriging is an extension of ordinary kriging, in which the best estimation method of regionalized variables is developed from a single attribute to two or more synergistic regionalized attributes, and one or more auxiliary variables are applied. These auxiliary variables, which are interpolated and estimated, are related to the main variables, and the correlations between the variables are assumed to be used to improve the accuracy of the main prediction target [51]. The interpolation of groundwater depth is mainly based on the following equation:

$$Z^*(x_0) = \sum_{i=1}^{n} \lambda_{1i} Z_1(x_i) + \sum_{j=1}^{m} \lambda_{2j} Z_2(x_j) \tag{1}$$

where x is variable location, $Z^*(x_0)$ is the predicted value for the x position, is the measured value of the primary variable, $Z_2(x_j)$ is the measured value of the covariate, λ is Co-kriging weights for Z, n is the number of the primary variable, and m is the number of the covariate.

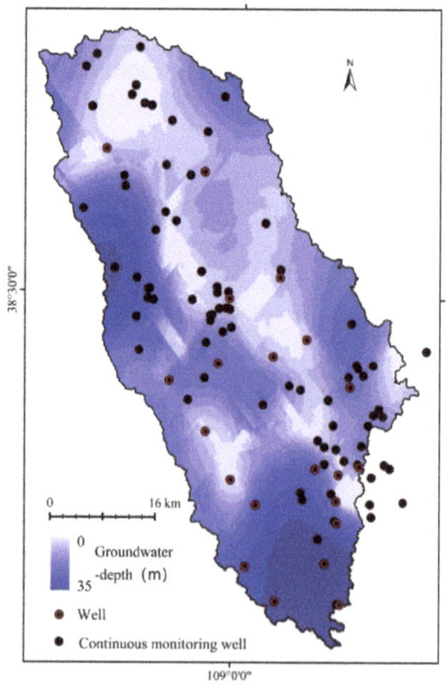

Figure 3. Groundwater depth interpolation and well location.

2.5. Model Description and Setup

Groundwater was simulated to predict changes in groundwater level in the basin under the combined exploitation of multiple underground mines and to study the change in groundwater resources using the simulated water level. The circulation of the groundwater system was simulated using the MODFLOW groundwater model, based on Darcy's law and the foundation of Dupuit, which are widely used in groundwater research [52,53]. MODFLOW was implemented in the GMS software platform to simulate the watershed. Through MODFLOW finite difference numerical simulation, HRB was subdivided into square meshes of 200 rows and 200 columns, on average. Finally, the watershed was divided into 20,986 grids. To establish the model, hydrological data, hydrogeological information, meteorological data, and groundwater level data were acquired from various sources. Geological drilling information and topographic elevation data are needed in the initial stage of the model. Hydrogeological parameters, water level, evaporation, and rainfall were applied as the driving forces of the hydrological cycle. For the initial groundwater flow field, important data were acquired through a groundwater survey from April to May 2019. Simulated and measured values were compared according to various evaluation metrics, and the model was optimized by fitting the actual values. The optimized model was finally used to predict the future conditions. In the groundwater simulation, the input factor values required adjustments within a certain range for adaptation to the ground conditions.

In order to establish an effective groundwater model, the aquifer system was generalized into three layers consisting of Quaternary aquifers, Cretaceous aquifers, and Jurassic strata above the 2[#] coal seam considering the strata lithology, groundwater occurrence, and hydrodynamic characteristics of the study area. In addition, the boundary conditions and internal structure of the study area were generalized according to the actual relationship between water supply and discharge (Figure 4). The phreatic aquifer in the basin is directly related to the atmosphere and vertical water exchange through surface processes such as precipitation infiltration, evaporation and drainage, and farmland irrigation. Horizontal groundwater movement is determined by the topography and geological characteristics of the study area. Based on the accurate description of the conceptual hydrogeological model of groundwater system, a mathematical model of the basin was established. This three-dimensional and finite difference groundwater flow model is based on the groundwater flow equation, combined with the three-dimensional mathematical model of non-steady flow. The model was visualized in GMS. The final mathematical model of the basin is as follows:

$$\frac{\partial}{\partial x}\left(K_h\frac{\partial h}{\partial x}\right)+\frac{\partial}{\partial y}\left(K_h\frac{\partial h}{\partial y}\right)+\frac{\partial}{\partial z}\left(K_z\frac{\partial h}{\partial z}\right)+W=\mu_s\frac{\partial h}{\partial t} \tag{2}$$

$$h|_{t=0} = h_0(x,y,z)\ (x,y,z) \in \Omega \tag{3}$$

$$K\frac{\partial h}{\partial n}\Big|_{\Gamma} = q(x,y,z,t)\ (x,y,z) \in \Omega \tag{4}$$

$$h(x,y,z,t)|_{\mathit{И}} = h\ (x,y,z) \in \Omega \tag{5}$$

$$\frac{K_r A}{M_r}(H_r-h)=Q_r \tag{6}$$

where μ_s is the water storage coefficient, h is the groundwater head or elevation (m), K_h is the hydraulic conductivity in the horizontal direction (m/d), K_z is the hydraulic conductivity in the perpendicular direction (m/d), W is the strength of source and sink (m^3/d), $q(x,y,z,t)$ is the inflow or outflow function from a unit area in a unit time of the second boundary condition, Γ is the flow boundary, $\mathit{И}$ is the head boundary, Ω is the scope of the model, K_r is the river permeability coefficient (m/d), M_r is the sedimentation thickness of the river channel (m), H_r is the water level in the river (m), and Q_r is the stream flow (m^3/d).

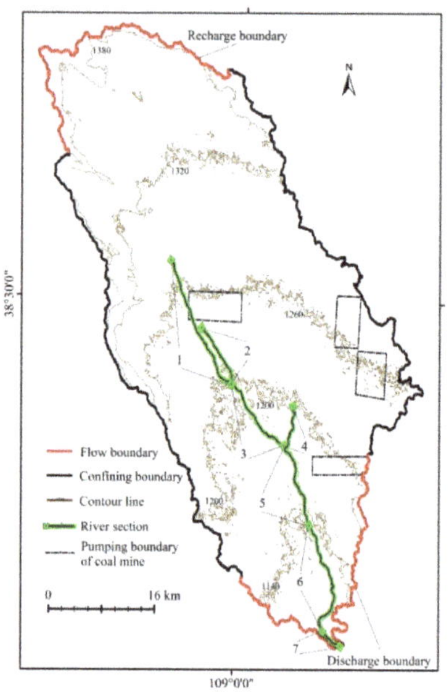

Figure 4. Division of recharge and discharge areas in the basin.

According to the characteristics of aquifers in the watershed and the distribution of the groundwater flow field, shallow water aquifers in the watershed were divided into flow boundary and closed boundary aquifers. Along the streamflow direction of groundwater, the north of the watershed is the recharge boundary, and the south is the discharge boundary. The Hailiutu River receives water from continuous groundwater discharge, and it was thus simulated as a drain [21,36]. In order to characterize the discharge process of the river in detail, the river course was divided into 7 sections, and the actual flow was measured in each section. The measured average flow was input into the model. The phreatic aquifer in HRB is directly recharged by atmospheric precipitation. As the sandy land is characterized by a high infiltration rate, scarce vegetation on the surface, and low evaporation capacity, the free surface of the phreatic aquifer is regarded as the boundary surface of vertical recharge. In the model setting, precipitation directly recharges groundwater, and evaporation is affected by groundwater depth. Moreover, farmland irrigation wells are evenly distributed in the basin, which can be regarded as the surface discharge of groundwater resources. For the simulation of recharge and evaporation processes, the Recharge Package and Evapotranspiration Package of MODFLOW were implemented [54], with data obtained from the measured precipitation and evaporation at the meteorological stations of Yulin City near the basin. The groundwater flow conditions of all aquifers in the study area are essentially similar, but the hydrogeological parameters are spatially different. The hydrogeological parameters were mainly determined through pumping tests, but the initial hydrogeological parameters simulated by the model may not closely match the actual situation. This is because the parameters obtained by the pumping test are only point data, while the model requires area data. Considering the characteristics of the groundwater flow field and hydrogeology data collected during the geological survey, based on hydrogeological parameters, the phreatic aquifer was divided into 4 zones (Quaternary) and 3 zones (Cretaceous), and the confined aquifer (Jurassic) was divided into 4 zones (Figure 5). Groundwater in the study area is mainly recharged by atmospheric precipitation, and the infiltration coefficient of precipitation is determined by the surface geotechnical properties. The infiltration coefficients of precipitation are 0.34 (aeolian sand), 0.26 (lacustrine sand), 0.29 (valley sediment sand), 0.07 (eolian loess), and 0.09 (Cretaceous sandstone) for the five zones of the top layer (Figure 1c). The depth limit of phreatic evaporation was defined as 4 m according to preliminary knowledge of the investigator. Water quotas for farmland irrigation were recorded from actual observations.

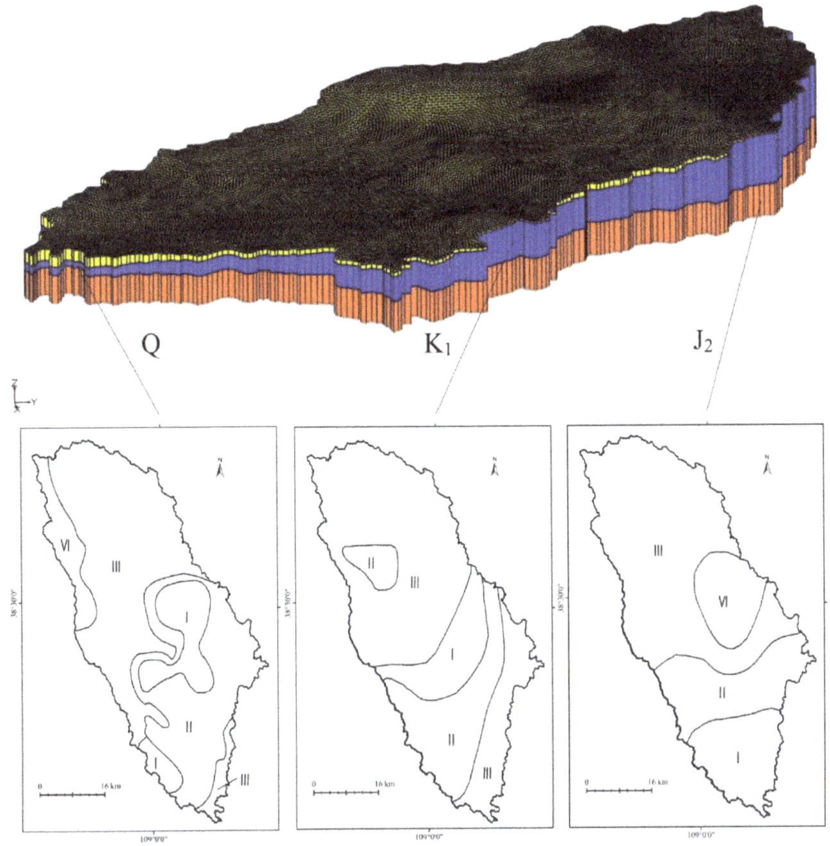

Figure 5. Three-dimensional hydrogeological solid model and hydrogeological parameter partitioning.

2.6. Model Calibration and Validation

Through the accurate calibration of the measured and calculated water levels, the groundwater flow in HRB was successfully simulated. The actual observed groundwater level and flow field were mainly fitted by trial-and-error calibration, such that the simulated groundwater levels are reasonably coordinated with the observed values. The observed groundwater level data from July 2019 to December 2019 (184 days in total) were used for model calibration. To ensure that the dynamic change in the simulated groundwater level was as close as possible to the observed data, the hydrogeological parameters were constantly adjusted [55]. At the end of the simulation (184 days), no significant difference was observed between the simulated non-pressurized aquifer flow field and the measured groundwater flow field.

As shown in Figures 6 and 7, the simulated groundwater level and the observed groundwater level were in line with the groundwater hydraulic properties of the basin. The groundwater levels of our long-term observation wells in Huhetala and Xinmiaotan are in good agreement with the simulated values. According to the centralized survey of the 21 selected wells at the beginning of each month from July to September, the water levels in the observation wells agree with the simulated values (Figure 7). Moreover, the maximum groundwater level error between the observed data and the simulated data of the two observation wells was 0.5 m, which is negligible compared to the natural variation of groundwater level. In the fitting process, hydrogeological parameters, as important data, were repeatedly adjusted manually to meet the actual situation, and the spatial distribution of hydrogeological parameters was finally determined. The hydrogeological parameters determined after model verification are shown in Table 2. In addition, as shown in Table 3, the discharge term in the basin was basically equal to the recharge term, indicating that the water cycle of HRB is in a dynamic equilibrium state, further indicating that the generalization of the water resources cycle in the study area is valid. Groundwater recharge mainly occurs through meteoric precipitation and lateral recharge, and drainage occurs mainly through surface evaporation, overflow to surface water, lateral discharge of groundwater, and artificial exploitation. The final verification results show that the groundwater model of HRB meets the accuracy requirements, which can reflect the hydraulic characteristics of the groundwater system in the basin and can be used to predict the change of groundwater level in the basin after coal mining.

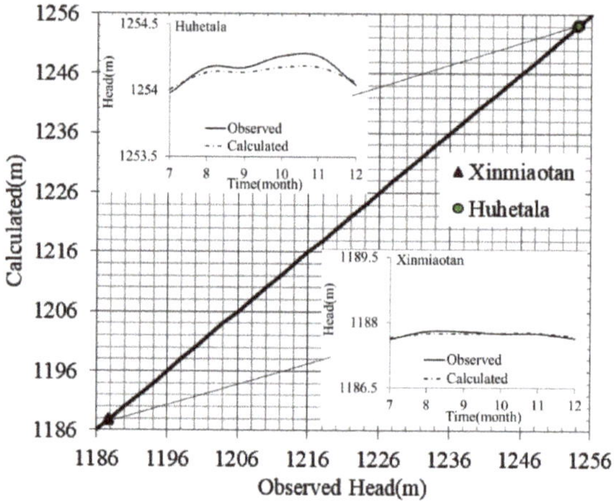

Figure 6. Comparison of observed head and calculated head.

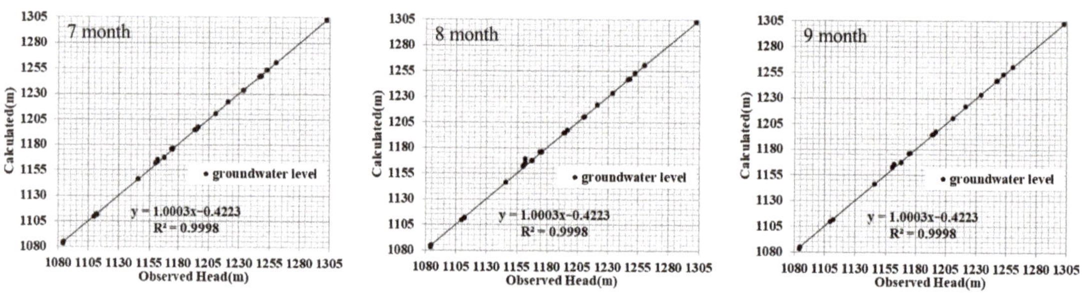

(a) (b) (c)

Figure 7. Comparison of observation head and calculated head of 21 monthly continuous monitoring wells.

Table 2. Hydrogeological parameters of aquifers.

Stratum	Area	Hydraulic Conductivity (m/d)	Specific Yield	Storage Coefficient (m^{-1})
Q	I	7	0.26	-
	II	3.5	0.2	-
	III	1.5	0.06	-
	VI	0.8	0.01	-
K_1	I	0.25	0.05	-
	II	0.3	0.05	-
	III	0.3	0.06	-
J_2	I	0.003	-	1.5×10^{-7}
	II	0.004	-	1.9×10^{-8}
	III	0.008	-	3.3×10^{-7}
	VI	0.007	-	4.0×10^{-8}

Table 3. Results of groundwater budget in the study area.

Balance Project		Annual Amount ($\times 10^6$ m^3/a)
Recharge	Precipitation	361.83
	Lateral recharge	3.23
Discharge	Evaporation	197.58
	Lateral discharge	4.17
	River discharge	81.99
	Artificial exploitation	80.43
	-	0.89

2.7. Mine Inflow

In order to ensure the safety of underground mining, groundwater in the upper aquifer of the coal seam and water inrush will be drained, which will significantly affect the groundwater flow field of the overlying aquifer and the natural circulation of groundwater resources in the region. Using the established numerical model of groundwater flow, the relationship between groundwater level and drainage in the mining area can be quantitatively described and predicted, and the possible impact can be analyzed. For the prediction, coal mine drainage should be preliminarily input to the model, while other hydrological data remain unchanged. After coal seam mining, the deformation and failure of the overlying strata would generate caved, fractured, and continuous deformation zones [6,56], which is the main reason for the change in the groundwater system. Considering that the study area has thick coal seams for mining, the combined thickness of caved and fractured zones corresponds to the major drained aquifers. Through the field investigation of the study area and surrounding coal mines, the water conductive fracture zone of the coal mine was found to develop only in the area above the working face, and the strata outside the mining face remained largely unaffected. Therefore, the effective thickness of the drainage layer is defined as the distance between the top boundary of the fractured zone with water flow and the bottom boundary of the coal seam. Considering the current situation of deep buried coal seams in HRB [45], with coal mining, the largest water-conducting layer would include only the Jurassic formation, and it would be separated from the Cretaceous bottom by a certain distance. The distance would be 100.87 m in the Yinpanhao (Yph) coal mine, 94.22 m in the Dahaize (Dhz) coal mine, and 110.9 m in the Balasu (Bls) coal mine. Pumping wells for simulating mine inflow were expanded to the drainage layer previously defined on the basis of the first mining area. According to the prediction report of the collected mine water inflows, the drainage water of the three coal mines was 13,549.8 m^3/d (Yph), 35,520 m^3/d (Dhz), and 37,000.8 m^3/d (Bls), and the planned mining volume of the three coal mines was 15 Mt/a (Yph), 15 Mt/a (Dhz), and 10 Mt/a (Bls). The drainage water of Bls was equal to that of Dhz, and the drainage water of Yph was the smallest. The drainage water increased gradually from upstream to downstream, according to the location of the coal mines in the basin.

3. Results and Discussion

3.1. Relationship between Vegetation and Groundwater Depth

Compared with the use of only groundwater depth as the interpolation factor, Co-kriging interpolation provides results with smoother spatial distribution without any sudden point aggregation characteristics, which can better reflect the spatial distribution of groundwater [57]. Therefore, during the field survey of 100 wells (Figure 3) in October 2019, the groundwater depth of the basin (Figure 8a) was acquired by Co-kriging interpolation. The results show that groundwater depth exhibits wide spatial variability across the study area. Specifically, groundwater depth is relatively large in the south and northwest regions, mainly distributed in the loess ridge and high dunes. On the basin scale, areas with suitable groundwater depth for vegetation are mainly distributed in the upper and middle reaches. According to the results of the land use type distribution (Figure 8b), vegetation coverage is relatively high in the areas with suitable groundwater depth, concentrated in the middle and upper reaches of the basin. Comparing Figure 8a,b, it can be ascertained that the location of shallow groundwater depth coincides with the dominant greensward area, which exhibits central aggregation points. Accordingly, the frequency graph of each pixel was plotted cumulatively, and the value of central aggregation points was determined based on the following conditions: greensward land use type with a resolution of 1500 m and dominant greensward pixels over other land use classifications. The dependency of greensward land on groundwater depth was investigated by organizing pair data of greensward land and groundwater depth on the same base map (Figure 8c). The diagram of the final results reveals 34 central aggregation points of greensward land, of which 24% are distributed in areas with groundwater depth less than 3 m and 11.17% in areas with groundwater depth more than 10 m.

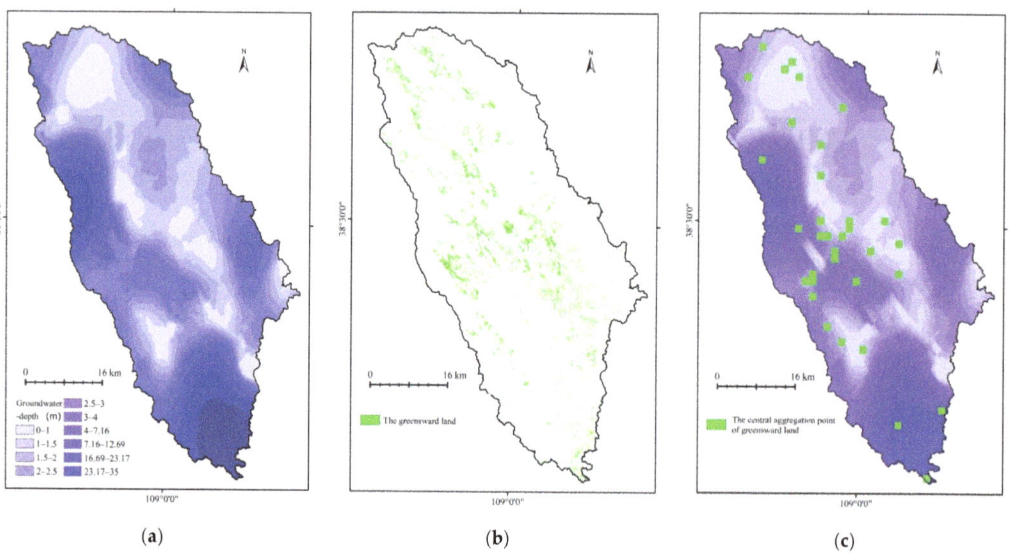

(a) (b) (c)

Figure 8. (**a**) Interpolation of groundwater depth in the watershed; (**b**) Distribution of greensward land use type in the watershed; (**c**) Location of the center of greensward land in the watershed (colors of the groundwater flow field are illuminated for contrast).

The dominant spatial pattern of the distribution of greensward land in HRB is presented in Figure 8b. Greensward land clearly occupies a relatively independent niche in space, which is related to groundwater depth, and indirectly reflects its different dependences on different groundwater depth conditions. The pixel number of greensward land was counted from LUCC data, and 82% of the greensward area was found to be distributed in areas with groundwater depth of 0–4 m (Figure 9). From the relationship between the distribution of groundwater depth and the number of greensward pixels, the groundwater depth of 4 m can be ascertained to be an important threshold. In aeolian sandy soil, grass can grow normally only when the deep roots of grass have access to groundwater. If the groundwater depth exceeds the root depth, the survivability of grass significantly declines. For example, greensward land is unlikely to develop in sand dunes because the groundwater level exceeds 8 m. The results suggest that grass growth is seriously limited in areas where the groundwater depth exceeds 4 m. It is noteworthy that groundwater irrigation also determines the coverage of green space. Figure 8c also confirms this, but this aspect is not the focus of our study.

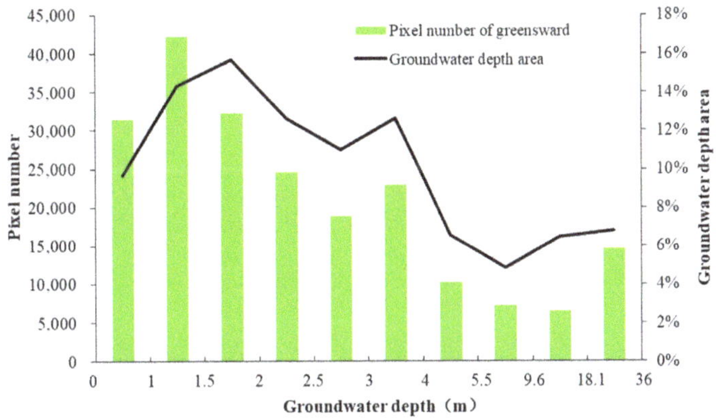

Figure 9. Distribution area of groundwater depth and the number of greensward pixels.

The relationship between groundwater and ecological vegetation has always been an important issue for research [58,59]. In particular, the relationship between vegetation and groundwater is very complex in arid and semi-arid areas, where plant roots have a strong response to changes in groundwater [60]. It is not sufficient to discuss the relationship between vegetation and groundwater depth by taking only a specific area of HRB as an example. To comprehensively present the characteristics of vegetation coverage in arid and semi-arid regions suitable for groundwater burial depth, we compared studies on vegetation and groundwater in similar regions to confirm the credibility of our conclusions [61–66]. As shown in Table 4, some arid and semi-arid regions around the world have a limited range of groundwater levels for the healthy growth of vegetation. If the minimum threshold of groundwater required for vegetation growth is exceeded, the growth of vegetation will be affected. Attributable to the difference in hydrogeology, different study areas exhibit distinct differences in the threshold. Nevertheless, comparing the conclusions of previous studies, the threshold of groundwater burial depth for the normal growth of herbs was found to be approximately 2–5 m, which is essentially the same as the results obtained in this study. This can be explained by the limited root systems of herbs.

Table 4. Relationship between vegetation growth and groundwater level in different areas.

Research Objectives	Number of Studies	The Link between Vegetation and Groundwater Levels
The Owens Valley of California, USA (1991–2004)	30 permanent monitoring sites	About 3.5 m is the groundwater threshold for normal vegetation coverage in most areas
The riparian area in southeastern Arizona, USA (2003)	Three sites	The average annual groundwater depth of the grassland is 2.6 m
The lower section of the Tarim River, China (2005)	Nine 50 m wide transects	When the groundwater depth is −3.14 m, the density and diversity of above-ground plants are significantly improved
The hinterland of the Badain Jaran Desert, China (2014)	10 long-term observation sites	In the area where the groundwater depth is 0–2 m, the growth of vegetation is more vigorous
Al Qunfudah City and the surrounding coastal plain in southwest Saudi Arabia (2018)	Six locations	Vegetation coverage is higher in areas where the groundwater depth is less than 5 m
The Manasi River riparian zone, China (2018–2019)	13 sites	The water-table depth appropriate for herbs is 1–1.5 m

3.2. *Groundwater Level Prediction*

According to the groundwater model established after verification, aquifer drainage caused by coal seam mining was separated, and mine gushing water was set as the output of the model. Finally, the model was used to predict future changes in groundwater induced by coal mining. In the model, wells were used to simulate mine water inflow. A simulation was performed for each grid of the first mining area, and the inflow was then interpolated within the effective mining drainage layer of the Jurassic strata. The value of total mine water inflow of the grids should be close to the value of the previously mentioned prediction report. Therefore, to ensure accurate simulation results, all pumping wells were set in the mesh grids. These wells, representing mine water inflow, were negative and were evenly distributed in the first mining area of Yph, Dhz, and Bls. In the model, the first mining area of the three coal mines was assumed to satisfy the designed production volume in the future. By running the model, the groundwater flow field of the aquifers in the next 10 years (2020–2029) could be predicted. Figure 10a shows the simulated flow of the phreatic aquifer in December 2029.

The change of the flow field curve represents the influence of coal mine drainage. Therefore, the flow field curve of the phreatic aquifer above the first mining area of the coal mine is different from that of other areas of the basin. It can be observed that the maximum drawdown may be as high as 5 m (Yph), 6 m (Dhz), and 10 m (Bls), indicating three cones of depression centered on the three mines. The cone of depression exerts a distinct influence in Yph, but the change of flow field in Dhz and Bls coal mines is more complicated because of the influence of groundwater watershed on the hydraulic gradient. Overall, mining activities affected groundwater in phreatic aquifers within a certain range [2]. The size of groundwater depression is closely related not only to the amount of drainage by the coal mine but also to the geographical location of coal mines in the basin. An obvious drop of groundwater levels could be observed for floor leakages in the first mining area of the three coal mines (Figure 10b), and the groundwater depth shows no significant change in the upstream of HRB because it is located far away from coal mining sites. It is essentially certain that, as coal mining continues, groundwater depression will continue to expand, thereby affecting the hydrological cycle of the basin.

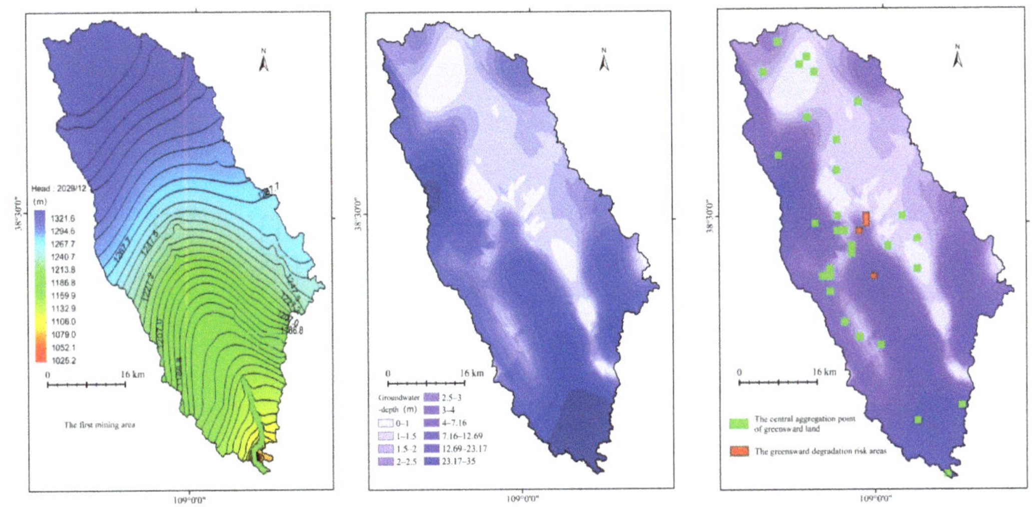

(**a**) (**b**) (**c**)

Figure 10. (**a**) Groundwater simulation flow field in the watershed; (**b**) Interpolation of simulated groundwater depth in the watershed; (**c**) Location of greensward aggregation centers in the watershed (colors of the groundwater flow field are illuminated for contrast).

3.3. Vegetation Change Prediction

Given the single vegetation structure and sensitive environment of the study area, the ecological function of the vegetation community is susceptible to interference from climate change and human activities, which may even completely destroy the original vegetation cover [21]. Because HRB is the recharge source area of the Wuding River Basin, and precipitation is the main source of groundwater recharge in the basin, groundwater resources are important ecological factors. The greensward land in the basin is mainly supported by groundwater. Therefore, groundwater depth is the decisive factor controlling the distribution of greensward land. Grass coverage and occurrence frequency are related to groundwater depth. With the increase in groundwater depth, the coverage, density, and population richness of grass decrease. The growth and development of grass is higher in areas with suitable groundwater depth, with higher frequency and coverage. On the contrary, in water deficit areas, the growth and development of grass are relatively poor, with low occurrence frequency and corresponding coverage. The most suitable groundwater level for grass growth in HRB is 0–4 m, and this groundwater depth range is also the key guarantee to maintaining the sustainable development of the ecosystem in arid and semi-arid greensward areas [67,68]. If the groundwater level continues to decline below 4 m, the growth and development of greensward vegetation in pastoral areas will be limited, and the number of biological populations will gradually decrease. As the groundwater depth continues to drop below 7 m, arid greensward areas will undoubtedly undergo irreversible desertification. Therefore, the upper limit for the stable survival of the existing greensward vegetation can be determined to be 0 m and the lower limit to be 4 m.

Using MODFLOW to predict the groundwater flow field in the basin after 10 years, we obtained the groundwater depth of the future basin, and central aggregation points of greensward land over the spatial distribution of groundwater depth were determined (Figure 10c). Compared with Figure 8b, the groundwater level in the first mining area of Yph decreased significantly, and the groundwater depth increased. The change of the flow field above the first mining area causes the increase in groundwater depth near the river, and this area is the central aggregation point of greensward land in the basin. The increase in groundwater depth will seriously affect the normal growth of grass vegetation. Therefore, these regions require serious attention. As mentioned earlier, the groundwater depth of 4 m is an important threshold. However, in order to take necessary measures before reaching the warning line, attention should be paid to affected areas when the groundwater depth drops below 3 m. Accordingly, the central aggregation point of greensward land in this region is marked as red to alert the government and local population of the degradation risk. By red tagging, four greensward degradation risk areas were found to be concentrated in the area and surroundings of Yph in the midstream of the basin. As clearly shown in Figure 11, the predicted groundwater depth in this area after coal mining exceeds 3 m, increasing the proportion central aggregation points with groundwater depth greater than 3 m to 35%. Although the greensward coverage near Dhz and Bls is small, and there is no major central aggregation point of the greensward land, attention should also be paid to the impact of groundwater level decline on ecological vegetation diversity. Finally, the prediction of groundwater and grassland coverage under multi-mining conditions shows that the overall ecological prospect of the basin is negative without the protection of the ecological environment. From the final research results, the change of groundwater in the next 10 years will affect only the growth of vegetation within a certain range. However, the influence of groundwater depth change caused by coal mining on natural vegetation should be carefully examined at the watershed scale.

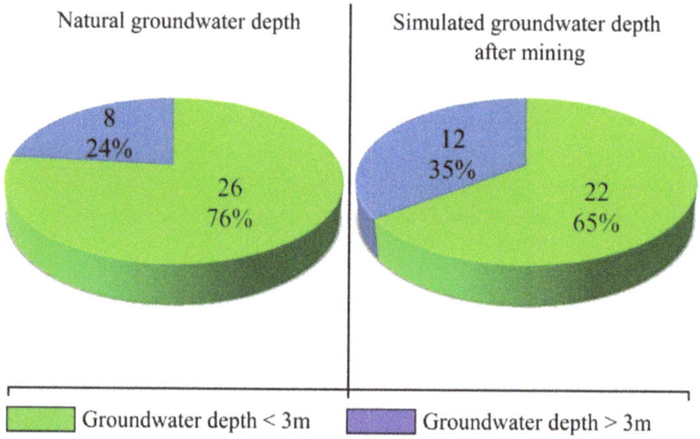

Figure 11. Proportion and number of greensward aggregation centers at different groundwater depths.

Although vegetation can adapt to changes in the environment, some vegetation types [69] require a certain amount of groundwater to maintain normal growth. With changes in groundwater, groundwater-sensitive plants may completely disappear from the corresponding area. Although the environmental crisis caused by coal mining is globally widespread [70], the coal industry is particularly important for regional economic development and national energy security in developing regions. Unless a breakthrough is achieved in the technology of renewable energy, and renewable energy is made affordable, coal mines will continue to operate, which will lead to the deterioration of the ecological environment after the drainage of groundwater [71]. Indeed, the best solution is to prevent mining and stop the destruction, but this solution is not feasible for developing countries and poor countries. We believe that coal mining activities in developed countries should be gradually discontinued, and clean energy measures should be increasingly implemented, such as in Germany [72]. For developing countries, the mining of coal resources should be limited by environmental protection. Under the condition of maintaining environmental sustainability, underground resources should be gradually exploited, and part of the profits from coal mining should be used for the management of hazards caused by coal mining, such as in China [73]. The contradiction between coal mining and environmental protection requires the preliminary intervention of government departments, for which the prediction of potential environmental changes caused by coal mining and subsequent active formulation of environmental protection policies are necessary.

4. Conclusions

In this study, variations in the flow field of phreatic groundwater caused by the joint action of multiple coal mines in a basin were explored, and the groundwater flow field in the basin in the next few years was predicted. The results show that the MODFLOW model exhibits good performance in simulating the variation law of the groundwater flow field. The law of groundwater movement in the natural state of the basin was first determined by simulating groundwater movement in the basin under non-mining conditions, and the model as well as the accuracy of the completed system were verified. The drainage attributable to underground coal mining was input to the established MODFLOW model, and potential changes in groundwater depth throughout the basin were simulated. Through the correlation between groundwater depth and distribution of greensward land obtained by remote-sensing interpretation, the influence of mining on the future distribution of greensward land was predicted. The following conclusions can be drawn:

1. The topographical features of HRB comprise arid and semi-arid dunes and denuded loess ridges. The distribution of greensward land is closely related to aquifer thickness and groundwater depth. Greensward land is mainly distributed in the upstream and midstream of the basin, especially in areas with shallow groundwater depth and thicker aquifers. The sporadic distribution of greensward land in some areas is related to artificial pumping irrigation. There are 34 central aggregation points of greensward land in HRB. Among them, 24% is distributed in areas with groundwater depth less than 3 m and only 8.82% in areas with groundwater depth more than 10 m. Furthermore, grassland accounts for only 11.11% of the total area in the downstream of the basin. The relationship between groundwater depth and central aggregation points of greensward land in the basin contributes to deeper understanding of the distribution of vulnerable vegetation in the Mu Us Sandy Land.
2. The combined action of multiple underground mines has a strong impact on the phreatic aquifer system in the basin, and coal mine drainage changes the flow field and increases the depth of groundwater. The MODFLOW simulation of the groundwater aquifer flow field in 2020–2029 suggests that coal mining will have distinct effects on the groundwater aquifer above the first mining area. However, variations in the groundwater hydraulic gradient induced by coal mining would not be prominent in the next 10 years compared with the hydraulic gradient caused by differences in terrain. The groundwater model reveals three cones of groundwater depression centered on three mines, and the maximum drawdown may be as high as 5 m (Yph), 6 m (Dhz), and 10 m (Bls).
3. According to the prediction of changes in groundwater depth induced by the combined mining of multiple underground mines in HRB, the groundwater level will prominently decrease around Yph, posing degradation risks to four central aggregation points of greensward land in the basin. It is necessary to implement timely measures to counter the changes. Although the coverage of greensward land is small in the first mining area of Dhz and Bls, the depth of groundwater decline is greater than that of Yph, and its impact on plant growth is more serious. This simulation approach provides a reference for the prediction of the distribution of vegetation with changes in groundwater depth.

Author Contributions: Shenghui Zhou developed the initial and final versions of this manuscript and analyzed the data. Tingxi Liu and Limin Duan contributed their expertise and insights, overseeing all of the analysis and supporting the writing of the final manuscript. All authors have read and agreed to the published version of the manuscript.

Funding: This research was funded by the National Key R&D Program of China (No. 2018YFC0406400), the National Natural Science Foundation of China (Nos. 51939006 and 51620105003), the Inner Mongolia Major science and technology projects (Nos. 2020ZD0009 and 2019ZD007), the Inner Mongolia Science and Technology Plan Project (Nos. 2020 and 2021GG0071), the Ministry of Education Innovative Research Team (No. IRT_17R60), the Innovation Team in Priority Areas Accredited by the Ministry of Science and Technology (No. 2015RA4013), the Natural Science Foundation of Inner Mongolia Autonomous Region of China (No. 2020JQ06).

Institutional Review Board Statement: Not applicable.

Informed Consent Statement: Informed consent was obtained from all subjects involved in the study.

Data Availability Statement: Please contact corresponding author.

Conflicts of Interest: The authors declare no conflict of interest.

References

1. Bai, X.; Ding, H.; Lian, J.; Ma, D.; Yang, X.; Sun, N.; Xue, W.; Chang, Y. Coal production in China: Past, present, and future projections. *Int. Geol. Rev.* **2018**, *60*, 535–547. [CrossRef]
2. Qu, S.; Shi, Z.; Wang, G.; Han, J. Application of multiple approaches to investigate hydraulic connection in multiple aquifers system in coalfield. *J. Hydrol.* **2021**, *595*, 125673. [CrossRef]
3. Tao, M.; Cheng, W.; Nie, K.; Zhang, X.; Cao, W. Life cycle assessment of underground coal mining in China. *Sci. Total Environ.* **2022**, *805*, 150231. [CrossRef] [PubMed]
4. Tao, S.; Fang, J.; Zhao, X.; Zhao, S.; Shen, H.; Hu, H.; Tang, Z.; Wang, Z.; Guo, Q. Rapid loss of lakes on the Mongolian Plateau. *Proc. Natl. Acad. Sci. USA* **2015**, *112*, 2281–2286. [CrossRef]
5. Xiao, W.; Lv, X.; Zhao, Y.; Sun, H.; Li, J. Ecological resilience assessment of an arid coal mining area using index of entropy and linear weighted analysis: A case study of Shendong Coalfield, China. *Ecol. Indic.* **2020**, *109*, 105843. [CrossRef]
6. Zhang, J.; Shen, B. Coal mining under aquifers in China: A case study. *Int. J. Rock Mech. Min.* **2004**, *41*, 629–639. [CrossRef]
7. Chen, W.; Li, H.; Wu, Z. Western China energy development and west to east energy transfer: Application of the Western China Sustainable Energy Development Model. *Energy Policy* **2010**, *38*, 7106–7120. [CrossRef]
8. Li, N.; Yan, C.; Xie, J. Remote sensing monitoring recent rapid increase of coal mining activity of an important energy base in northern China, a case study of Mu Us Sandy Land. *Resour. Conserv. Recycl.* **2015**, *94*, 129–135. [CrossRef]
9. Jin, J.; Yan, C.; Tang, Y.; Yin, Y. Mine Geological Environment Monitoring and Risk Assessment in Arid and Semiarid Areas. *Complexity* **2021**, *2021*, 3896130. [CrossRef]
10. Cao, D.; Zhan, W.; Li, H.; Li, X.; Liu, D.; Wei, Y. Tectonic setting and risk zoning of dynamic geologicl disasters in coal mines in China. *J. China Coal Soc.* **2020**, *45*, 2376–2388. (In Chinese)
11. Wu, C.; Wu, X.; Zhu, G.; Qian, C. Predicting mine water inflow and groundwater levels for coal mining operations in the Pangpangta coalfield, China. *Environ. Earth Sci.* **2019**, *78*, 130. [CrossRef]
12. Yu, S.; Xu, J.; Zhu, W.; Wang, S.; Liu, W. Development of a combined mining technique to protect the underground workspace above confined aquifer from water inrush disaster. *Bull. Eng. Geol. Environ.* **2020**, *79*, 3649–3666. [CrossRef]
13. Gonzalez-Quiros, A.; Fernández-Álvarez, J.P. Conceptualization and finite element groundwater flow modeling of a flooded underground mine reservoir in the Asturian Coal Basin, Spain. *J. Hydrol.* **2019**, *578*, 124036. [CrossRef]
14. Wu, Q.; Wang, M. Characterization of water bursting and discharge into underground mines with multilayered groundwater flow systems in the North China coal basin. *Hydrogeol. J.* **2006**, *14*, 882–893. [CrossRef]
15. Fang, Z.; Kong, X.; Sensoy, A.; Cui, X.; Cheng, F. Government's awareness of environmental protection and corporate green innovation: A natural experiment from the new environmental protection law in china. *Econ. Anal. Policy.* **2021**, *70*, 294–312. [CrossRef]
16. Xiao, W.; Zhang, W.; Ye, Y.; Lv, X.; Yang, W. Is underground coal mining causing land degradation and significantly damaging ecosystems in semi-arid areas? A study from an Ecological Capital perspective. *Land. Degrad. Dev.* **2020**, *31*, 1969–1989. [CrossRef]
17. Chen, D.; Feng, Q.; Liang, H.; Gao, B.; Alam, E. Distribution characteristics and ecological risk assessment of polycyclic aromatic hydrocarbons (PAHs) in underground coal mining environment of Xuzhou. *Hum. Ecol. Risk. Assess.* **2019**, *25*, 1564–1578. [CrossRef]
18. Quaranta, J.D.; Mack, B.; Van Aken, B.; Ducatman, A.; Ziemkiewicz, P. Practical application of dilution analysis for estimating groundwater quality effects due to coal slurry injection into underground mine voids. *Mine Water Environ.* **2014**, *33*, 353–361. [CrossRef]
19. Shen, Z.; Zhang, Q.; Chen, D.; Singh, V.P. Varying effects of mining development on ecological conditions and groundwater storage in dry region in Inner Mongolia of China. *J. Hydrol.* **2021**, *597*, 125759. [CrossRef]
20. Xie, P.; Li, W.; Yang, D.; Jiao, J. Hydrogeological model for groundwater prediction in the Shennan mining area, China. *Mine Water Environ.* **2018**, *37*, 505–517. [CrossRef]
21. Zhang, H.; Wang, X.-S. The impact of groundwater depth on the spatial variance of vegetation index in the Ordos Plateau, China: A semivariogram analysis. *J. Hydrol.* **2020**, *588*, 125096. [CrossRef]

22. Zhang, H.; Shen, L.; Zhong, S.; Elshkaki, A. Economic structure transformation and low-carbon development in energy-rich cities: The case of the contiguous area of shanxi and shaanxi provinces, and inner Mongolia Autonomous Region of China. *Sustainability* **2020**, *12*, 1875. [CrossRef]
23. Wang, X.; Chen, F.H.; Dong, Z.; Xia, D. Evolution of the southern Mu Us Desert in North China over the past 50 years: An analysis using proxies of human activity and climate parameters. *Land. Degrad. Dev.* **2005**, *16*, 351–366. [CrossRef]
24. Zhang, D.; Deng, H. Historical human activities accelerated climate-driven desertification in China's Mu Us Desert. *Sci. Total Environ.* **2020**, *708*, 134771. [CrossRef] [PubMed]
25. Orellana, F.; Verma, P.; Loheide, S.P.; Daly, E. Monitoring and modeling water-vegetation interactions in groundwater-dependent ecosystems. *Rev. Geophys.* **2012**, *50*, RG3003. [CrossRef]
26. Jin, X.; Guo, R.; Zhang, Q.; Zhou, Y.; Zhang, D.; Yang, Z. Response of vegetation pattern to different landform and water-table depth in Hailiutu River basin, Northwestern China. *Environ. Earth Sci.* **2014**, *71*, 4889–4898. [CrossRef]
27. Lv, J.; Wang, X.S.; Zhou, Y.; Qian, K.; Wan, L.; Eamus, D.; Tao, Z. Groundwater-dependent distribution of vegetation in Hailiutu River catchment, a semi-arid region in China. *Ecohydrology* **2013**, *6*, 142–149. [CrossRef]
28. Mata-González, R.; Averett, J.P.; Abdallah, M.A.; Martin, D.W. Variations in Groundwater Level and Microtopography Influence Desert Plant Communities in Shallow Aquifer Areas. *Environ. Manag.* **2022**, *69*, 45–60. [CrossRef]
29. Shao, G.; Zhang, D.; Guan, Y.; Xie, Y.; Huang, F. Application of SWAT model with a modified groundwater module to the semi-arid Hailiutu River Catchment, Northwest China. *Sustainability* **2019**, *11*, 2031. [CrossRef]
30. Xue, S.; Liu, Y.; Liu, S.; Li, W.; Wu, Y.; Pei, Y. Numerical simulation for groundwater distribution after mining in Zhuanlongwan mining area based on visual MODFLOW. *Environ. Earth Sci.* **2018**, *77*, 400. [CrossRef]
31. Wu, Q.; Liu, Y.; Zhou, W.; Li, B.; Zhao, B.; Liu, S.; Sun, W.; Zeng, Y. Evaluation of water inrush vulnerability from aquifers overlying coal seams in the Menkeqing coal mine, China. *Mine Water Environ.* **2015**, *34*, 258–269. [CrossRef]
32. Light, D.D.; Donovan, J.J. Mine-water flow between contiguous flooded underground coal mines with hydraulically compromised barriers. *Environ. Eng. Geosci.* **2015**, *21*, 147–164. [CrossRef]
33. Mu, W.; Wu, Q.; Xing, Y.; Qian, C.; Wang, Y.; Du, Y. Using numerical simulation for the prediction of mine dewatering from a karst water system underlying the coal seam in the Yuxian Basin, Northern China. *Environ. Earth Sci.* **2018**, *77*, 215. [CrossRef]
34. Pande, C.B.; Moharir, K.N.; Singh, S.K.; Elbeltagi, A.; Pham, Q.B.; Panneerselvam, B.; Varade, A.M.; Kouadri, S. Groundwater flow modeling in the basaltic hard rock area of Maharashtra, India. *Appl. Water Sci.* **2022**, *12*, 1–14. [CrossRef]
35. Yue, L.; Li, J.; Zheng, G.; Li, Z. Evolution of the Ordos Plateau and environmental effects. *Sci. China Ser. D: Earth Sci.* **2007**, *50*, 19–26. [CrossRef]
36. Yang, Z.; Zhou, Y.; Wenninger, J.; Uhlenbrook, S.; Wang, X.; Wan, L. Groundwater and surface-water interactions and impacts of human activities in the Hailiutu catchment, northwest China. *Hydrogeol. J.* **2017**, *25*, 1341–1355. [CrossRef]
37. Jing, T.; Fang, N.; Zeng, Y.; Huang, X.; Shi, Z. Catchment properties controlling suspended sediment transport in wind-water erosion crisscross region. *J. Hydrol. Reg. Stud.* **2022**, *39*, 100980. [CrossRef]
38. Zhou, S.; Liu, T.; Duan, L.; Zhang, W.; Ji, R.; Sun, L. Simulation and Prediction of Precipitation in the Hailiutu River Basin Based on Time Series. *Res. Soil Water Conserv.* **2021**, *28*, 88–94. (In Chinese)
39. Qu, S.; Wang, G.; Shi, Z.; Zhou, P.; Xu, Q.; Zhu, Z. Temporal changes of hydraulic properties of overburden aquifer induced by longwall mining in Ningtiaota coalfield, northwest China. *J. Hydrol.* **2020**, *582*, 124525. [CrossRef]
40. Russell, J.; Goodman, P.; Cai, F. From dust to dust: Quaternary wind erosion of the Mu Us Desert and Loess Plateau, China. *Geology* **2015**, *43*, 835–838.
41. Kapp, P.; Pullen, A.; Pelletier, J.D.; Jin, R.; Teng, X.; Li, X.; Si, Q.; Wang, W. Genesis of sandstone-type uranium deposits along the northern margin of the Ordos Basin, China. *Geosci. Front.* **2020**, *11*, 215–227.
42. Fang, G.; Liu, Y.; Liang, X.; Huang, H. Hydrogeological characteristics and mechanism of a water-rich coal seam in the Jurassic coalfield, northern Shaanxi Province, China. *Arab. J. Geosci.* **2020**, *13*, 1088. [CrossRef]
43. Zhang, W. The early Ordovician Majiagou reservoir of the Jingbian Field, northwest China: Karstic peritidal dolomite. *Carbonates Evaporites* **2004**, *19*, 93–106. [CrossRef]
44. Yang, J.; Dong, S.; Wang, H.; Li, G.; Wang, T.; Wang, Q. Mine water source discrimination based on hydrogeochemical characteristics in the northern Ordos Basin, China. *Mine Water Environ.* **2021**, *40*, 433–441. [CrossRef]
45. Chen, W.; Li, W.; He, J.; Qiao, W.; Wang, Q.; Yang, Y. Impact of mining-induced bed separation spaces on a cretaceous aquifer: A case study of the Yingpanhao coal mine, Ordos Basin, China. *Hydrogeol. J.* **2022**, *30*, 691–706. [CrossRef]
46. Hu, Y.; Zhen, L.; Zhuang, D. Assessment of land-use and land-cover change in Guangxi, China. *Sci. Rep.* **2019**, *9*, 2189. [CrossRef]
47. Lu, M.; Chen, J.; Tang, H.; Rao, Y.; Yang, P.; Wu, W. Land cover change detection by integrating object-based data blending model of Landsat and MODIS. *Remote. Sens. Environ.* **2016**, *184*, 374–386. [CrossRef]
48. Guo, L.; Xi, X.; Yang, W.; Liang, L. Monitoring Land Use/Cover Change Using Remotely Sensed Data in Guangzhou of China. *Sustainability* **2021**, *13*, 2944. [CrossRef]
49. Antonakos, A.; Lambrakis, N. Spatial Interpolation for the Distribution of Groundwater Level in an Area of Complex Geology Using Widely Available GIS Tools. *Environ. Process.* **2021**, *8*, 993–1026. [CrossRef]
50. Nikroo, L.; Kompani-Zare, M.; Sepaskhah, A.R.; Fallah Shamsi, S.R. Groundwater depth and elevation interpolation by kriging methods in Mohr Basin of Fars province in Iran. *Environ. Monit. Assess.* **2010**, *166*, 387–407. [CrossRef]

51. Madani, N.; Emery, X. A comparison of search strategies to design the cokriging neighborhood for predicting coregionalized variables. *Stoch. Environ. Res. Risk Assess.* **2019**, *33*, 183–199. [CrossRef]
52. Serrano-Hidalgo, C.; Guardiola-Albert, C.; Heredia, J.; Elorza Tenreiro, F.J.; Naranjo-Fernández, N. Selecting Suitable MODFLOW Packages to Model Pond–Groundwater Relations Using a Regional Model. *Water* **2021**, *13*, 1111. [CrossRef]
53. Sreekanth, J.; Crosbie, R.; Pickett, T.; Cui, T.; Peeters, L.; Slatter, E.; Northey, J.; Merrin, L.E.; Davies, P.; Miotlinski, K. Regional-scale modelling and predictive uncertainty analysis of cumulative groundwater impacts from coal seam gas and coal mining developments. *Hydrogeol. J.* **2020**, *28*, 193–218. [CrossRef]
54. Vasilevskiy, P.; Wang, P.; Pozdniakov, S.; Wang, T.; Zhang, Y.; Zhang, X.; Yu, J. Simulating River/Lake–Groundwater Exchanges in Arid River Basins: An Improvement Constrained by Lake Surface Area Dynamics and Evapotranspiration. *Remote Sens.* **2022**, *14*, 1657. [CrossRef]
55. Sun, W.; Wu, Q.; Liu, H.; Jiao, J. Prediction and assessment of the disturbances of the coal mining in Kailuan to karst groundwater system. *Phys. Chem. Earth.* **2015**, *89*, 136–144. [CrossRef]
56. Shi, Z.; Hu, X.; Wang, C.-Y. Hydro-mechanical coupling in the shallow crust–Insight from groundwater level and satellite radar imagery in a mining area. *J. Hydrol.* **2021**, *594*, 125649. [CrossRef]
57. Niu, T.; Yu, J.; Yue, D.; Yu, Q.; Hu, Y.; Long, Q.; Li, S.; Mao, X. Differential Law and Influencing Factors of Groundwater Depth in the Key Agricultural and Pastoral Zones Driven by the Minimum Hydrological Response Unit. *Appl. Sci.* **2020**, *10*, 7105. [CrossRef]
58. Eamus, D.; Zolfaghar, S.; Villalobos-Vega, R.; Cleverly, J.; Huete, A. Groundwater-dependent ecosystems: Recent insights from satellite and field-based studies. *Hydrol. Earth Syst. Sci.* **2015**, *19*, 4229–4256. [CrossRef]
59. Niu, G.Y.; Troch, P.A.; Paniconi, C.; Scott, R.L.; Durcik, M.; Zeng, X.; Huxman, T.; Goodrich, D.; Pelletier, J. An integrated modelling framework of catchment-scale ecohydrological processes: 2. The role of water subsidy by overland flow on vegetation dynamics in a semi-arid catchment. *Ecohydrology* **2014**, *7*, 815–827. [CrossRef]
60. Fan, Y.; Miguez-Macho, G.; Jobbágy, E.G.; Jackson, R.B.; Otero-Casal, C. Hydrologic regulation of plant rooting depth. *Proc. Natl. Acad. Sci. USA* **2017**, *114*, 10572–10577. [CrossRef]
61. McLendon, T.; Hubbard, P.J.; Martin, D.W. Partitioning the use of precipitation-and groundwater-derived moisture by vegetation in an arid ecosystem in California. *J. Arid Environ.* **2008**, *72*, 986–1001. [CrossRef]
62. Scott, R.L.; Huxman, T.E.; Williams, D.G.; Goodrich, D.C. Ecohydrological impacts of woody-plant encroachment: Seasonal patterns of water and carbon dioxide exchange within a semiarid riparian environment. *Glob. Chang. Biol.* **2006**, *12*, 311–324. [CrossRef]
63. Xu, H.-L.; Mao, Y.; Li, J.-M. Changes in groundwater levels and the response of natural vegetation to transfer of water to the lower reaches of the Tarim River. *J. Environ. Sci.* **2007**, *19*, 1199–1207. [CrossRef]
64. Zhang, W.; Zhao, L.; Yu, X.; Zhang, L.; Wang, N.A. Estimation of Groundwater Evapotranspiration Using Diurnal Groundwater Level Fluctuations under Three Vegetation Covers at the Hinterland of the Badain Jaran Desert. *Adv. Meteorol.* **2020**, *2020*, 8478140. [CrossRef]
65. Alshehri, F.; Sultan, M.; Karki, S.; Alwagdani, E.; Alsefry, S.; Alharbi, H.; Sahour, H.; Sturchio, N. Mapping the distribution of shallow groundwater occurrences using Remote Sensing-based statistical modeling over southwest Saudi Arabia. *Remote Sens.* **2020**, *12*, 1361. [CrossRef]
66. Wang, Z.; Wang, W.; Zhang, Z.; Hou, X.; Duan, L.; Yao, D. Assessment of the effect of water-table depth on riparian vegetation along the middle and lower reaches of the Manasi River, Northwest China. *Hydrogeol. J.* **2021**, *29*, 579–589. [CrossRef]
67. Jia, L.; Jiao, R.; Liao, Z.; Long, Y. The Ecological Quality Present Situation and Water Requirement Research of Grassland Vegetation in rid Pastorl Areas. *China Rural. Water Hydropower* **2013**, *6*, 49–52+56. (In Chinese)
68. Fan, L.; Hou, G.; Tao, Z. Characteristics of Groundwater and Its Relation with Vegetation Distribution in Sara Wusu Formation of Mu US Desert. *J. Soil Water Conserv.* **2018**, *32*, 151–157. (In Chinese)
69. Niu, G.Y.; Fang, Y.H.; Chang, L.L.; Jin, J.; Yuan, H.; Zeng, X. Enhancing the Noah-MP ecosystem response to droughts with an explicit representation of plant water storage supplied by dynamic root water uptake. *J. Adv. Model. Earth Syst.* **2020**, *12*, e2020MS002062. [CrossRef]
70. Currell, M.; Irvine, D.; Werner, A.; McGrath, C. Science sidelined in approval of Australia's largest coal mine. *Nat. Sustain.* **2020**, *3*, 644–649. [CrossRef]
71. Ma, Q.; Wu, J.; He, C.; Fang, X. The speed, scale, and environmental and economic impacts of surface coal mining in the Mongolian Plateau. *Resour. Conserv. Recycl.* **2021**, *173*, 105730. [CrossRef]
72. Kessler, T.; Mugova, E.; Jasnowski-Peters, H.; Rinder, T.; Stemke, M.; Wolkersdorfer, C.; Hilberg, S.; Melchers, C.; Struckmeier, W.; Wieber, G. Groundwater in former German coal mining areas—a scientific perspective on mine floodings. *Grundwasser* **2020**, *25*, 259–272. [CrossRef]
73. Xu, J.; Lv, C.; Zhang, M.; Yao, L.; Zeng, Z. Equilibrium strategy-based optimization method for the coal-water conflict: A perspective from China. *J. Environ. Manag.* **2015**, *160*, 312–323. [CrossRef] [PubMed]

International Journal of *Geo-Information*

MDPI

Article

Continuous Monitoring of the Surface Water Area in the Yellow River Basin during 1986–2019 Using Available Landsat Imagery and the Google Earth Engine

Qingfeng Hu [1], Chongwei Li [1], Zhihui Wang [2,3,*], Yang Liu [2] and Wenkai Liu [1]

1 College of Surveying and Geo-Informatics, North China University of Water Resources and Electric Power, Zhengzhou 450046, China; huqingfeng@ncwu.edu.cn (Q.H.); 201401014@stu.ncwu.edu.cn (C.L.); liuwenkai@ncwu.edu.cn (W.L.)
2 Key Laboratory of Soil and Water Conservation on the Loess Plateau of Ministry of Water Resources, Yellow River Institute of Hydraulic Research, Yellow River Conservancy Commission, Zhengzhou 450003, China; liuyang@hky.yrcc.gov.cn
3 Henan Key Laboratory of Ecological Environment Protection and Restoration of the Yellow River Basin, Zhengzhou 450003, China
* Correspondence: wangzhihui@hky.yrcc.gov.cn; Tel.: +86-133-5380-6578

Abstract: The Yellow River Basin (YRB) has been facing severe water shortages; hence, the long-term dynamic monitoring of its surface water area (SWA) is essential for the efficient utilization of its water resources and sustainable socioeconomic development. In order to detect the changing trajectory of the SWA of the YRB and its influencing factors, we used available Landsat images from 1986 through to 2019 and a water and vegetation indices-based method to analyze the spatial–temporal variability of four types of SWAs (permanent, seasonal, maximum and average extents), and their relationship with precipitation (Pre), temperature (Temp), leaf area index (LAI) and surface soil moisture (SM).The multi-year average permanent surface water area (SWA) and seasonal SWA accounted for 46.48% and 53.52% in the Yellow River Basin (YRB), respectively. The permanent and seasonal water bodies were dominantly distributed in the upper reaches, accounting for 70.22% and 48.79% of these types, respectively. The rate of increase of the permanent SWA was 49.82 km^2/a, of which the lower reaches contributed the most (34.34%), and the rate of decrease of the seasonal SWA was 79.18 km^2/a, of which the contribution of the source region was the highest (25.99%). The seasonal SWA only exhibited decreasing trends in 13 sub-basins, accounting for 15% of all of the sub-basins, which indicates that the decrease in the seasonal SWA was dominantly caused by the change in the SWA in the main river channel region. The conversions from seasonal water to non-water bodies, and from seasonal to permanent water bodies were the dominant trends from 1986 to 2019 in the YRB. The SWA was positively correlated with precipitation, and was negatively correlated with the temperature. Because the permanent and seasonal water bodies were dominantly distributed in the river channel region and sub-basins, respectively, the change in the permanent SWA was significantly affected by the regulation of the major reservoirs, whereas the change in the seasonal SWA was more closely related to climate change. The increase in the soil moisture was helpful in the formation of the permanent water bodies. The increased evapotranspiration induced by vegetation greening played a significant positive role in the SWA increase via the local cooling and humidifying effects, which offset the accelerated water surface evaporation caused by the atmospheric warming.

Keywords: Yellow River Basin; surface water area; Google Earth Engine; spatio-temporal change; influencing factors

Citation: Hu, Q.; Li, C.; Wang, Z.; Liu, Y.; Liu, W. Continuous Monitoring of the Surface Water Area in the Yellow River Basin during 1986–2019 Using Available Landsat Imagery and the Google Earth Engine. *ISPRS Int. J. Geo-Inf.* **2022**, *11*, 305. https://doi.org/10.3390/ijgi11050305

Academic Editors: Walter Chen and Fuan Tsai

Received: 17 March 2022
Accepted: 7 May 2022
Published: 10 May 2022

1. Introduction

Water resources are vital to human economic prosperity, production development, the maintenance of ecosystem functions, and the promotion of sustainable development [1].

As an essential component of water resources, surface water bodies (i.e., lakes, reservoirs, rivers, streams, and ponds) provide a series of ecosystem services, such as water supply and regulation, climate regulation, and food production [2,3]. Therefore, the spatial distribution of surface water and its changes over time are central to many agricultural, environmental, and ecological issues, and are important factors for human socioeconomic development [4–6]. Global climate change and anthropogenic activities can have dramatic impacts on the inter-annual and intra-annual variations in surface water bodies, which can have profound influences on human society and natural ecosystems [6–8]. Therefore, monitoring the spatial–temporal dynamics of the Surface Water Area (SWA) using remote sensing technology is crucial for scientific research, as well as for adaptive and sustainable ecosystem management and social development [9–11].

Due to the convenient data acquisition, simple preprocessing, large range of observations, and relatively high accuracy of the automatic detection of water bodies, optical remote sensing imagery has become the primary data source for the long-term dynamic monitoring of surface water in large areas, especially on global and continental scales. The single-band surface reflectance threshold method [12,13] was first employed to automatically extract surface water bodies. Although it is simple and straightforward to use, it has difficulty extracting small surface water bodies, and the separation of shadows and water bodies is problematic [14]. Subsequently, Mcfeeters et al. [15] introduced the Normalized Difference Water Index (NDWI), and the threshold approach based on this index considerably enhances water body extraction accuracy; however, it is still ineffective when the water body exists in a background with buildings. In order to address this issue, Xu et al. [16] proposed the modified NDWI (mNDWI), which has become one of the most extensively used and effective methods for the delineation of open water bodies using the Landsat Thematic Mapper™ green (band 2) and short-wave infrared (band 5) channels, which can effectively suppress the signal from built-up land noise [17]. The mNDWI still makes mistakes when distinguishing between water bodies and vegetation [18,19]. Wetland vegetation is the main factor leading to classification mistakes due to the mixed distributions of water and grasses in wetlands [20].

This has become one of the most commonly used water indexes. Although the water index and threshold-based water body mapping approaches are computationally efficient, there is much uncertainty in the determination of the optimal threshold for classifying water and non-water areas. Although supervised classification algorithms—such as the Support Vector Machine (SVM), Maximum Likelihood (ML), and Random Forest (RF) algorithms [21,22]—can also be used for water body classification, the computational processes of these machine learning algorithms are time-consuming, and the accuracy of the identified water body is strongly influenced by the training samples. Recently, Zou et al. [23] combined the mNDWI and vegetation indices, including the Enhanced Vegetation Index (EVI) and the Normalized Difference Vegetation Index (NDVI), to map water bodies, which did not require a unique threshold value for the water index. Recently, it has been proposed that combining the mNDWI, NDVI, and EVI could produce better results and would be more stable than using the individual indexes [24–26], especially for wetlands [27]. This method has been used predominantly in the mapping of surface water bodies on the global and continental scales [24].

With the rapid development of cloud computing platform technology (e.g., Google Earth Engine) in recent years, it has become more practical and more efficient to map land cover types at the global and continental scales using dense time-series remote sensing data stacked at a medium spatial resolution (e.g., Landsat and Sentinel) [25,28]. Using the Google Earth Engine cloud platform, Pekel et al. [29] created the first global remote sensing continuous surface water dynamics product with a 30-m resolution. The Google Earth Engine has also been used to reveal the spatial–temporal dynamics of the SWA and its driving factors in Australia [30], Inner Mongolia Plateau [31], and China [32].

The Yellow River Basin (YRB), which only accounts for 2.2% of China's total runoff, is responsible for supplying water for 15% of the arable land and 12% of the population in

China [33,34]. The contradiction between the supply and demand for the water resources in the YRB has become increasingly prominent under the influences of climate change and human activities, and the YRB is becoming one of the regions with the most serious water resource shortages in China [35,36]. Because surface water resources are very significant for the preservation of ecosystem stability and for economic and social sustainable development in the YRB, it is critical to monitor the continuous changes in the SWA using satellite remote sensing observations. However, previous studies have primarily focused on sub-regions of the YRB, such as the Hetao irrigation area [37], Hongjiannao Lake [38], and the headwater region [39], and no study has continuously tracked the changes in the SWA in the entire YRB over the last few decades. Therefore, in this study, we used surface water body time-series data with a 30-m resolution for 1986–2019 based on the Google Earth Engine to reveal the historical variation characteristics of the different types of surface water bodies in the different sub-regions of the YRB, and we analyzed the relationships between the SWA and climate and vegetation changes.

2. Materials and Methods

2.1. Study Area and Data Processing

The Yellow River is the second-longest river in China, with a total length of 5464 km. The YRB is mostly located in the semi-arid to semi-humid regions, where the mean annual precipitation ranges from 300 to 700 mm, and the average multi-year temperatures range from −4 to 14 °C. The multi-year average runoff volume and sediment discharge in the main river channel are 58 billion m^3 and 1.6 billion tons [40], respectively. The top parts of the river are where the Yellow River flow originates, whereas the middle reaches are where the Yellow River sediment originates. In the upper and intermediate reaches, historic climate change and intensive human activities have resulted in substantial ecosystem deterioration. It has been reported that the underlying surface conditions have been dramatically changed, and that the ecological environment and the soil and water conservation capacity significantly improved after 2000, when extensive ecological restoration projects were implemented [41]. In addition, several large reservoirs have been gradually built on the main river channel by the national government for flow and sediment regulation and the optimal allocation of water resources in the YRB. In this study, the YRB was divided into six sub-regions based on the geographic locations of six key hydrological stations, including the Tangnaihai (TNH), Qingtongxia (QTX), Toudaoguai (TDG), Longmen (LM), Huayuankou (HYK), and Lijin (LJ) stations, located along the main river channel (Figure 1). The area above TNH is the source region of the YRB, and the area above TDG is the upper reaches of the YRB. The QTX-TDG sub-region is a typical windy desert area, and an irrigation area. The area between TDG and HYK is the middle reaches of the YRB, and the TDG-LM section is the main sediment formation area. The area between HKY and LJ is the upper reaches of the YRB. The upper reaches of the YRB are the main source area of the river runoff, and the middle reaches are the main source area of the sediments.

All of the available Landsat satellite images with a 30-m resolution (Landsat5 TM, Landsat7 ETM+ and Landsat8 OLI) covering the entire YRB from 1986 to 2019 were used in this study based on the Google Earth Engine platform (https://earthengine.google.com/, accessed on 16 March 2022), for which the Landsat ecosystem disturbance adaptive processing system (LEDAPS) [42] was employed to produce the surface reflectance of each Landsat pixel, and an F-mask algorithm [26,43] was then used to identify the contaminated pixels, such as clouds, cloud shadows and snow cover. The spatial distribution of the frequency of clear Landsat observations over the YRB from 1986 to 2019 is shown in the Figure 2, in which the effective number of observations in the northern region is larger than that in the southern region with more cloud coverage; the high values of observations occur in the area of overlapping satellite observations, and the low values are mainly distributed in the headwater region. Figure 3 depicts the interannual variation of the area ratio of the frequency with different levels (0–4, 5–10, 10–20, 20–40, 40–70, and 70–100) of clear Landsat observations over the YRB. It can be seen that as the Landsat satellite sensors were updated, the number of clear Landsat observations increased gradually. In addition, meteorological parameters including the daily precipitation (pre) and temperature (temp) during 1986–2019 of 295 stations located in the YRB were acquired from the China Meteorological Science Data Sharing Service (http://data.cma.cn/, accessed on 16 March 2022), and the raster data of the precipitation and temperature at the 1-km scale were produced using a spatial interpolation algorithm in the AUSPLINE software [44].

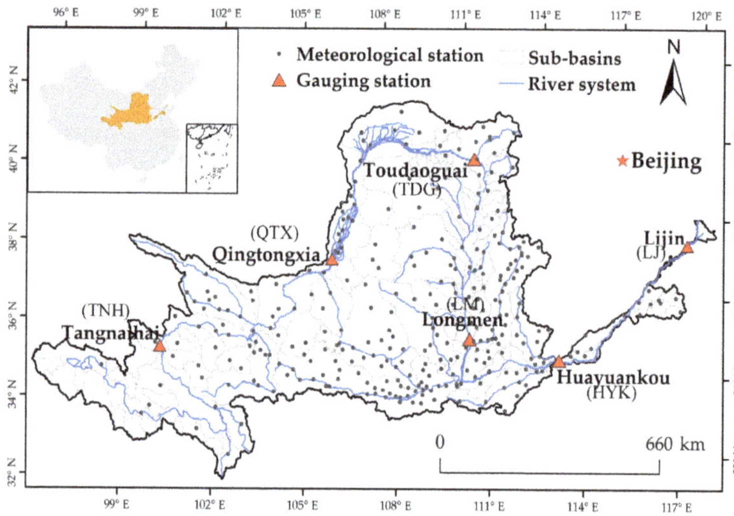

Figure 1. Spatial distribution of the meteorological and gauging stations in the YRB. The gauge stations are TNH, QTX, TDG, LM, HYK and LJ from upstream to downstream, respectively.

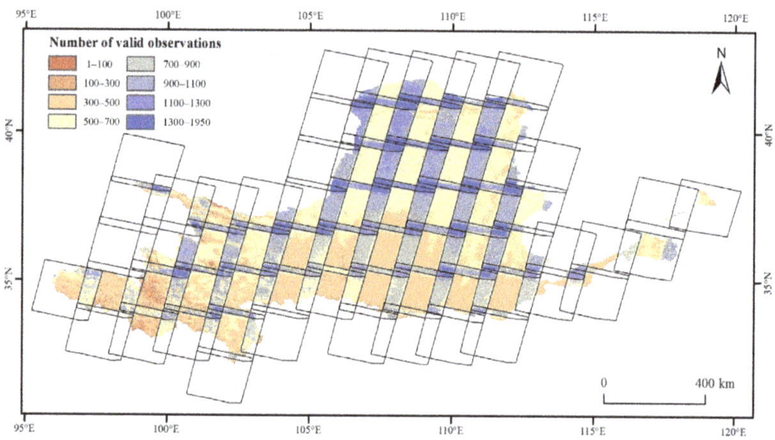

Figure 2. Spatial distribution of the frequency of clear Landsat observations in the YRB from 1986 to 2019.

The global land surface satellite (GLASS) (http://www.glass.umd.edu/, accessed on 16 March 2022) [45] leaf area index (LAI) product with 8-day and 1-km resolution was selected to characterize the vegetation structure change during 1986–2019. Daily evapotranspiration (ET) and surface soil moisture (SM) data at 0.25° during 1986–2019 were derived from the Global Land Evaporation Amsterdam Model (GLEAM) (https://www.gleam.eu/, accessed on 16 March 2022) [46] v3.5a product. Finally, the daily meteorological parameters, 8-day LAI, daily ET and SM were temporally upscaled into annual total Pre and ET, and the annual average temp, LAI and SM.

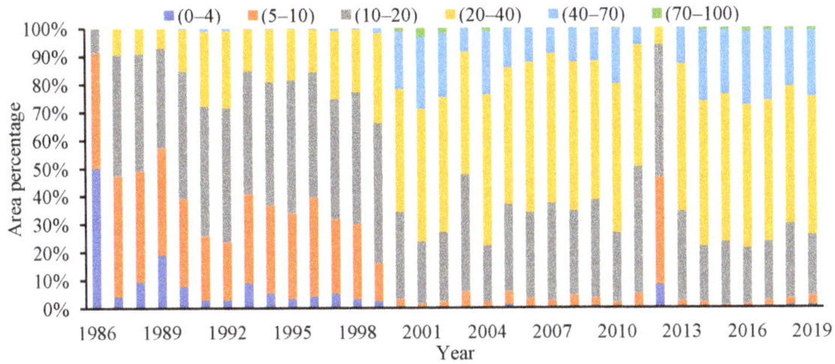

Figure 3. Interannual variation in the area ratio of the frequency with different levels of clear Landsat observations in the YRB.

2.2. Surface Water Body Mapping Algorithm

The water and vegetation index and the threshold-based method proposed by Zou et al. [23] were used in this study. The algorithms for the water and vegetation indices and threshold rules are as follows:

$$\mathrm{NDVI} = \frac{\rho_{NIR} - \rho_{red}}{\rho_{NIR} + \rho_{red}}, \tag{1}$$

$$\mathrm{MNDWI} = \frac{\rho_{green} - \rho_{SWIR1}}{\rho_{green} - \rho_{SWIR1}}, \tag{2}$$

$$\mathrm{EVI} = 2.5 \times \frac{\rho_{NIR} - \rho_{red}}{\rho_{NIR} + 6 \times \rho_{red} - 7.5 \times \rho_{biue} + 1}, \tag{3}$$

$$\mathrm{EVI} < 0.1 \text{ and} (\mathrm{mNDWI} > \mathrm{NDVI} \text{ or } \mathrm{mNDWI} > \mathrm{EVI}) \tag{4}$$

where ρ_{red}, ρ_{green}, ρ_{blue}, ρ_{NIR}, and ρ_{SWIR1} are the surface reflectance of the red, green, blue, near-infrared, and short-wave infrared wavelengths, respectively.

Because there are serious commission errors in the classified water bodies in shaded hills, the rule of terrain slopes of <8° was also employed in order to improve the identification accuracy of the water bodies in the hilly and mountainous areas [47]. Then, the Water Inundation Frequency (WIF) was calculated by dividing the number of identified water bodies by the number of clear Landsat observations in a year (Equation (4)). Water bodies with 25% ≤ WIF ≤ 75% were defined as seasonal water bodies, and water bodies with WIF > 75% were defined as permanent water bodies based on previous studies [48,49].

$$\mathrm{WIF} = \frac{W}{N} \times 100\%, \tag{5}$$

where N is the total number of valid observations in a year, and W is the total number of times the water body was detected.

Except for the SWA of the seasonal and permanent water bodies, the annual maximum SWA was calculated based on the water bodies with WIF ≥ 25%, and the annual average SWA was defined as the sum of all of the effective areas (900 m^2) in one Landsat pixel multiplied by the WIF [23,24]. Therefore, the inter-annual time series of the four types of SWA (permanent, seasonal, annual maximum, and annual average water body areas) for the different sub-regions and different sub-basins was derived in order to analyze the spatial–temporal changes.

2.3. Accuracy Assessment

In order to assess the classification accuracy of the water bodies derived in this study, we selected 995 samples containing typical types of water bodies (rivers, lakes, reservoirs, and check dams) via artificial visual interpretation based on the high spatial resolution Google Earth images. Because the high-resolution Google Earth images were all acquired during 2019–2020, we only evaluated the accuracy of the water body classification in 2019 using the Google Earth Engine. In addition, we compared the accuracies of the water bodies identified using different WIF thresholds (5%, 10%, 15%, 25%, 35%, 50%, and 75%).

2.4. Linear Slope Calculation

In order to reveal the temporal variation characteristics of the SWA over time in the YRB, we used the slope of the linear regression model to characterize the interannual rate of change of the SWA during 1986–2019 in the YRB. The slope was calculated as follows:

$$\text{Slope} = \frac{n \times \sum_{i}^{n} i \times X_i - \sum_{i=1}^{n} i \sum_{i=1}^{n} X_i}{n \times \sum_{i=1}^{n} i^2 - \left(\sum_{i=1}^{n} i\right)^2} \tag{6}$$

where X_i is the SWA in year i, and n is the total number of years. When the slope is >0, the SWA is increasing; and when the slope is <0, the SWA is decreasing.

Then, the slopes of the seasonal, permanent, annual maximum, and annual average SWAs in the entire YRB were analyzed, and the contributions of the SWAs in the different sub-regions and different sub-basins were compared.

2.5. Partial Correlation Analysis

Rainfall is the main source of surface water bodies, an increased temperature can increase the evaporation from surface water surfaces, vegetation is linked to evapotranspiration from surface water surfaces, and soil moisture influences temperature and rainfall. In order to explore the impacts of the natural water supply, atmospheric warming, vegetation greening, and SM on the SWA, we conducted a partial correlation analysis to examine the relationships between the SWA and the climatic parameters, vegetation status, and water storage status of the surface soil. Partial correlation analysis measures the strength and direction of a linear relationship between two variables while the effect of one or more other variables is controlled. In this way, the individual roles of the influencing factors in the SWA dynamics can be determined. The partial correlation coefficient was calculated as follows:

$$r_{i,j \cdot l_1 l_2 \ldots\ldots l_n} = \frac{r_{i,j \cdot l_1 l_2 \ldots\ldots l_{n-1}} - r_{i,l_n \cdot l_1 l_2 \ldots\ldots l_{n-1}} \cdot r_{j,l_n \cdot l_1 l_2 \ldots\ldots l_{n-1}}}{\sqrt{\left(1 - r^2_{i,l_n \cdot l_1 l_2 \ldots\ldots l_{n-1}}\right) \cdot \left(1 - r^2_{j,l_n \cdot l_1 l_2 \ldots\ldots l_{n-1}}\right)}} \tag{7}$$

where $r_{i,j \cdot l_1 l_2 \ldots\ldots l_n}$ is the nth (n = k − 2) order partial correlation coefficient between i and j when $l_1, l_2 \ldots l_n$ are controlled, and k is the total number of variables. $r^2_{i,l_n \cdot l_1 l_2 \ldots\ldots l_{n-1}}$ and $r^2_{j,l_n \cdot l_1 l_2 \ldots\ldots l_{n-1}}$ are the (n − 1)th order partial correlation coefficient. When $r_{i,j \cdot l_1 l_2 \ldots\ldots l_n} > 0$, the relationship between i and j is positive; otherwise, it is negative.

We also examined the significance of this correlation by performing hypothesis testing.

$$t = \frac{\sqrt{m - \text{k} - 2} \cdot r_{i,j \cdot l_1 l_2 \ldots\ldots l_n}}{\sqrt{1 - r^2_{i,j \cdot l_1 l_2 \ldots\ldots l_{n-1}}}} \tag{8}$$

where t is the t-test value, m is the sample size (here m = 34), and k is the degree of freedom. When t < 0.05, the partial correlation is statistically significant (at the 95% statistical significance level).

3. Results

3.1. Surface Water Body Classification Results and Accuracy Validation

The WIF in the YRB was calculated for each year from 1986 to 2019, and the WIFs of six typical surface water bodies in the YRB are presented in Figure 4. The SWA of Hongjiannao Lake gradually decreased, whereas the SWA of Longyangxia Reservoir increased. The SWA of the Xiaolangdi Reservoir increased abruptly due to reservoir storage after it began operating. The SWA of Wuliangsuhai Lake fluctuated sharply between years due to rainfall and artificial water diversion. It can be seen that the SWA in the check dam increased gradually, which demonstrates that the check dam constructed in the 1980s not only intercepted large amounts of sediment but also served as a water storage facility for local irrigation. It can also be seen that the downstream river regime in Kaifeng changed dramatically, and the river regime became relatively stable after 2015. It can be reasonably concluded that the temporal variation in the WIF can effectively reflect the dynamics of the different surface water bodies in the different regions.

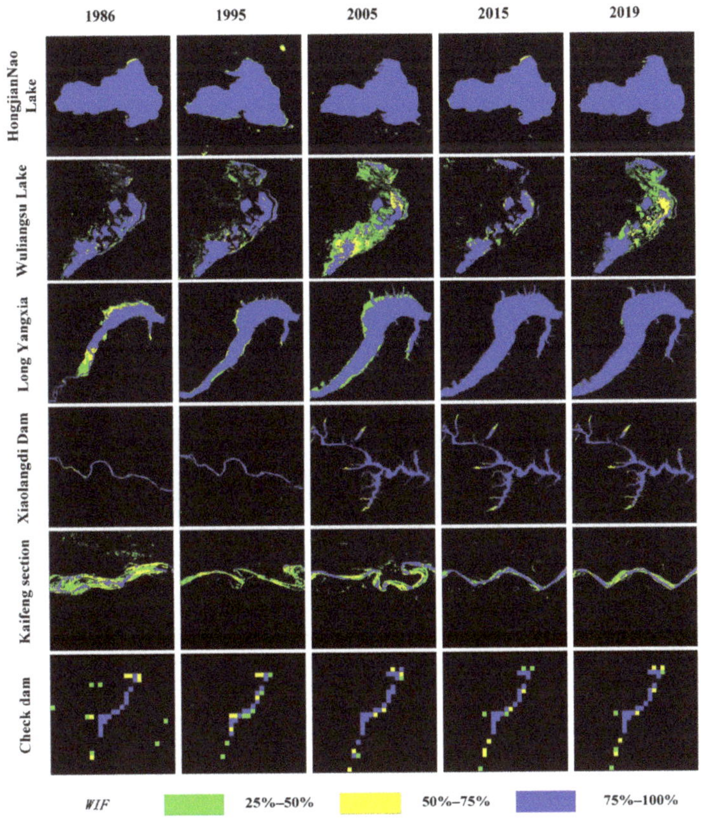

Figure 4. Variations in the WIFs of different types of typical water bodies in the YRB during 1986–2019.

The classification accuracy of the surface water bodies derived using different WIF thresholds was validated based on selected samples of typical types of water bodies. The results show that the classification accuracies of the water bodies using WIF thresholds of 5%, 10%, 15%, 25%, 35%, 50%, and 75% were 97.33%, 97.23%, 97.17%, 97.12%, 97.00%, 96.86%, and 96.38%, respectively (Figure 5). The average value of the classification accuracy was as high as 97.01%, indicating that the surface water identification algorithm used in this study is very accurate and robust.

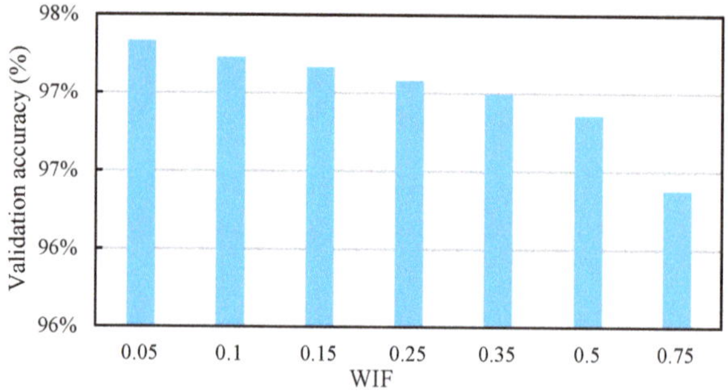

Figure 5. Validation of the accuracy of the water body classification in 2019 using different WIF thresholds.

3.2. Spatial Distribution of the Surface Water Bodies in the YRB

The spatial distribution of permanent and seasonal water bodies derived from the multi-year average WIF during 1986–2019 is shown in Figure 6. It can be seen that the seasonal water bodies always existed in the transition areas between the permanent water bodies and the non-water bodies, and there were more seasonal water bodies near the river channel.

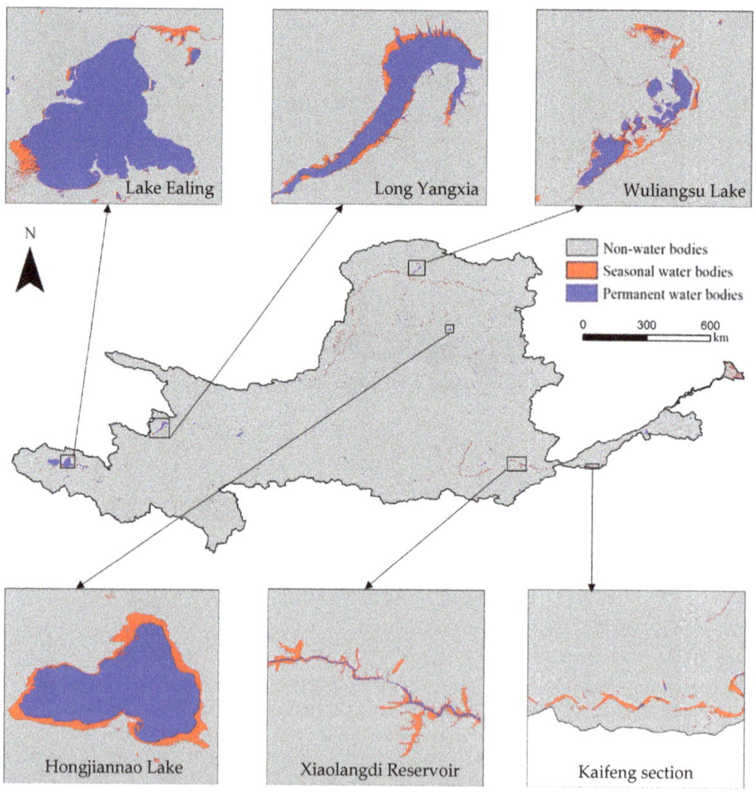

Figure 6. The spatial distribution of the surface water bodies derived from the multi-year average WIF during 1986–2019.

The different types of SWAs in all of the sub-regions of the YRB were calculated (Figure 7). For the entire YRB, the permanent water bodies covered an area of 4048.76 km^2, and the seasonal water bodies covered an area of 4661.69 km^2, accounting for 46.48% and 53.52% of the surface water bodies in the YRB, respectively. The maximum water body area was 8710.45 km^2, and the annual average water body area was 5853.17 km^2. The SWA in the source area of the upper reaches and the upstream area of the Yellow River (above Toudaoguai) contained 58.75% of the water bodies in the YRB, while the Huayuankou-Lijin segment contained 18.86%. The Toudaoguai–Longmen and Longmen–Huayuankou intervals contained 9.74% and 12.86% of the maximum water bodies in the YRB, respectively. These results indicate that the YRB was dominated by seasonal water bodies. However, it can be seen that the Tangnaihai section had far more permanent water bodies than seasonal water bodies. In the source region, the ratio of permanent to seasonal water bodies was 3.24, whereas the ratios in the Tangnaihai–Qingtongxia, Qingtongxia–Toudaoguai, Toudaoguai–Longmen, Longmen–Huayuankou, and Huayuankou–Lijin sections were 1.55, 0.27, 0.70, 0.33, and 0.55, respectively.

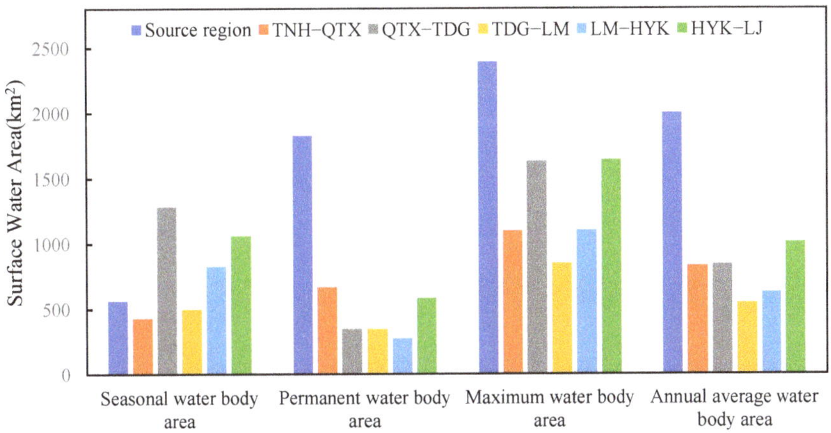

Figure 7. The areas of the different types of surface water bodies derived from the multi-year average WIF during 1986–2019 in the YRB.

Figure 8 shows the statistical areas of the different types of water bodies in the selected 86 sub-basins in the YRB. The Wei River (101.18 km^2) was the largest seasonal water body, and the Xianchuan River (0.07 km^2) was the smallest. The Shushui River (58.96 km^2) was the largest permanent water body, and the Mangla River was the smallest (0 km^2). In terms of the major water bodies, the Wei River had the largest area (135.63 km^2), and the Mangla River had the smallest area (0.08 km^2). For the annual average water bodies, the Shushui River (80.94 km^2) had the largest area, and the Mangla River (0.03 km^2) had the smallest area.

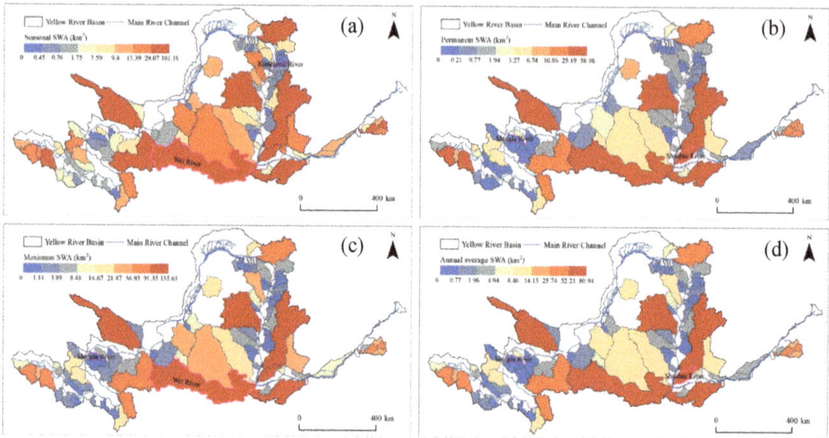

Figure 8. Different types of water bodies in each sub-basin derived from the multi-year average WIF during 1986–2019. Panels (**a**–**d**) are for the seasonal, permanent, maximum, and annual average water bodies, respectively.

3.3. Changes in the SWA in the YRB from 1986 to 2019

Figure 9 shows the continuous dynamic changes in the different types of water bodies—including the seasonal, permanent, maximum, and annual average water bodies—in the entire YRB during 1986–2019. The maximum and seasonal water body areas exhibited typical decreasing trends from 1986 to 2019, but the permanent and annual water areas exhibited slightly increasing trends. The maximum water area varied from 7984.02 km^2 to 13,145.55 km^2, i.e., 35.75% higher and 18.48% lower than the average area of 9683.99 km^2, respectively. The permanent water area varied from 4547.8 km^2 to 7109.7 km^2, i.e., 16.95% lower and 29.84% higher than the average value (5475.95 km^2), respectively. The seasonal water area varied from 2624.57 km^2 to 7595.80 km^2 during a single year, i.e., 37.63% lower and 80.51% higher than the average value (4208.04 km^2), respectively. Because the average water area was based on the pixel level of the water bodies, it reflects the changes in the water bodies in one year the best. The annual water area varied from 5708.17 km^2 to 8344.84 km^2, i.e., 16.8% lower and 21.64% higher than the average value (6860.40 km^2), respectively. As is shown in Figure 6, the statistics of the water bodies reveal decreasing trends in the maximum water body area ($p < 0.01$) and the seasonal water body area ($p < 0.01$) from 1986 to 2019. The permanent water area ($p < 0.01$) and the annual water body area ($p < 0.01$) exhibited increasing trends. The total SWA of the YRB initially decreased and then increased during the study period. According to the linear regression model, the annual water area in the YRB increased by 27.2 km^2 per year from 1986 to 2019.

Figure 10 depicts the changes in the different types of water bodies in six sub-regions in the YRB. The seasonal water bodies exhibited decreasing trends in all six sub-regions. The source area had the largest rate of change of -20.66 km^2/a, accounting for 25.99%. The permanent water bodies exhibited increasing trends in all six sub-regions. The Huayuankou–Lijin section had the largest rate of change of 17.1 km^2/a, accounting for 34.34%. Among the remaining subdivisions, in section Huayuankou–Lijin, the largest water body exhibited an increasing trend (9.33 km^2/a); the source area had the largest negative trend (16.85 km^2/a), accounting for 34.87%. Except for in the source region and the Toudaoguai–Longmen section, which exhibited minor decreasing trends of roughly -1 km^2/a, the Huayuankou–Lijin section continued to have the strongest increasing trend in terms of the annual mean water bodies (13.85 km^2/a), accounting for 44.48% in the entire YRB.

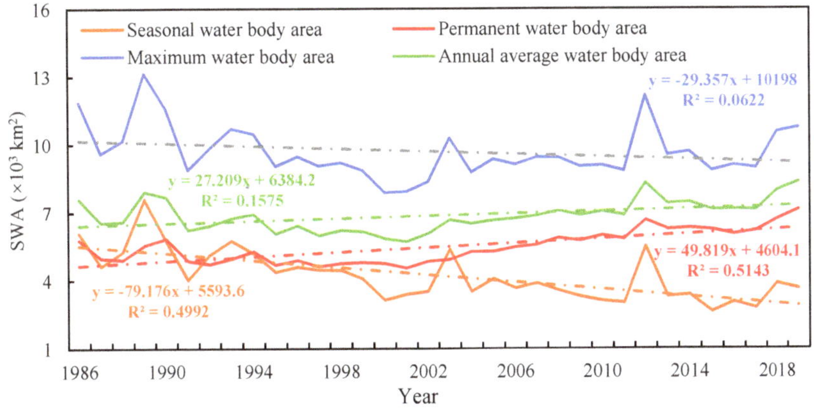

Figure 9. Continuous dynamic changes in the different types of water bodies—including the seasonal, permanent, maximum, and annual average water bodies—in the entire YRB during 1986–2019.

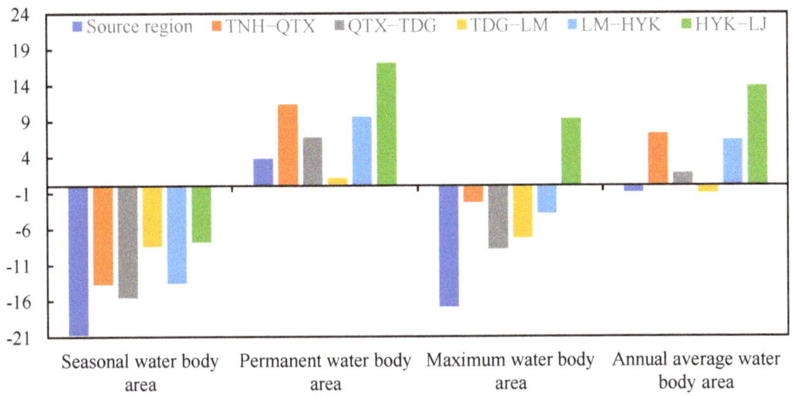

Figure 10. Interannual rates of change for the different types of water bodies in six sub-regions from the upper to the lower reaches in the YRB during 1986–2019.

Figure 11 depicts the rates of change of the various types of water bodies in the 86 sub-basins of the YRB from 1986 to 2019. The Dawen River basin had the highest permanent SWA growth rate (1.63 km^2/a), whereas the Kequ River basin had the highest permanent SWA reduction rate (−0.24 km^2/a). The Yiluohe River (0.54 km^2/a) had the highest seasonal SWA growth rate, and the Wei River (−5.56 km^2/a) had the highest seasonal SWA reduction rate. The Yiluohe River (1.89 km^2/a) had the highest maximum SWA growth rate, and the Wei River (−4.99 km^2/a) had the highest maximum SWA reduction rate. The Dawen River (1.63 km^2/a) had the highest annual average SWA growth rate, and the Huangshui (−0.74 km^2/a) had the highest reduction rate.

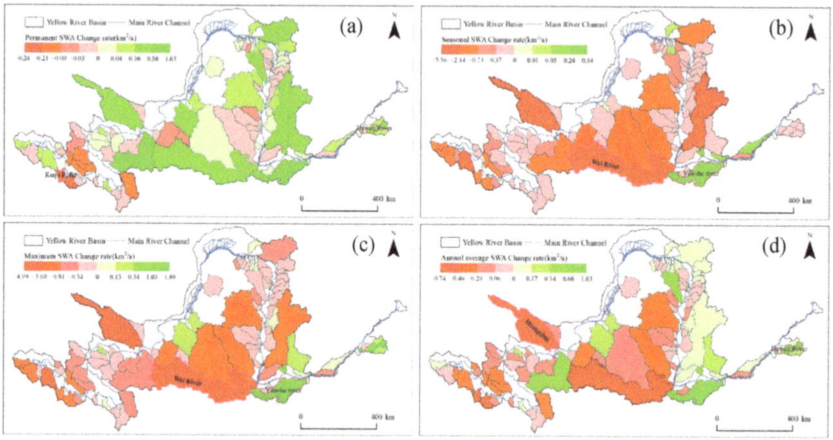

Figure 11. Changes in the different types of water bodies in each sub-basin in the YRB during 1986–2019. Panels (**a**–**d**) are for the permanent, seasonal, maximum, and annual average SWAs, respectively.

3.4. Conversions of Different Types of Surface Water Bodies

The types of conversions between different water bodies from 1986 to 2019 were divided into non-water to seasonal, non-water to permanent, seasonal to non-water, seasonal to permanent, permanent to non-water, and permanent to seasonal conversions. The spatial pattern of the different conversion types is shown in Figures 12 and 13. The most common conversion in the source region was from seasonal to non-water bodies, with a conversion area of 357.05 km^2, while the least common conversion was from non-water bodies to permanent water bodies, with a conversion area of 39.7 km^2. With a conversion area of 241.87 km^2, the conversions from seasonal to non-water bodies were the most prevalent in the Tangnaihai–Qingtongxia section. Conversely, the conversions from non-water bodies to seasonal water bodies were the least common, with a conversion area of 10.03 km^2. The transition from non-water bodies to permanent water bodies was the least common in the Qingtongxia–Toudaoguai section, whereas the transition from seasonal to non-water bodies was the most common, with a conversion area of 414.35 km^2. In the Toudaoguai–Longmen section, the most common transformation was from seasonal to permanent water bodies, with a conversion area of 104.5 km^2, while the least common transformation was from non-water bodies to seasonal water bodies, with a conversion area of 19.62 km^2. Hua's long scenario is comparable to that in the Qingtongxia–Toudaoguai section, in which the least common change was from non-water bodies to permanent water bodies, with a conversion area of 12.2 km^2, while the most frequent transformation was from seasonal to non-water bodies, with a conversion area of 173.75 km^2.

Figure 14 further shows the SWAs of the different water body conversion types in each sub-basin. It can be seen that the conversion of water bodies in the YRB exhibited strong spatial heterogeneity. The largest conversion of non-water bodies to permanent water bodies occurred in the Dawen River (2.58 km^2), and the smallest occurred in Jiaqu (0 km^2). The largest conversion of non-water bodies to seasonal water bodies occurred in the Duoqu (11.23 km^2) and the smallest occurred in Jiaqu (0 km^2). The largest conversion of seasonal water bodies to non-water bodies occurred in Huangshui (122.51 km^2) and the smallest occurred in Yangjiachuan (0 km^2). The largest conversion of seasonal water bodies to permanent water bodies occurred in Weihe (18.60 km^2) and the smallest occurred in Hulouhe (0 km^2). The largest conversion of permanent water bodies to non-water bodies occurred in Huangshui (22.31 km^2) and the smallest occurred in the Pianguan River (0 km^2). The largest conversion of permanent water bodies to seasonal water bodies occurred in Huangshui (13.96 km^2) and the smallest occurred in the Quchan River (0 km^2).

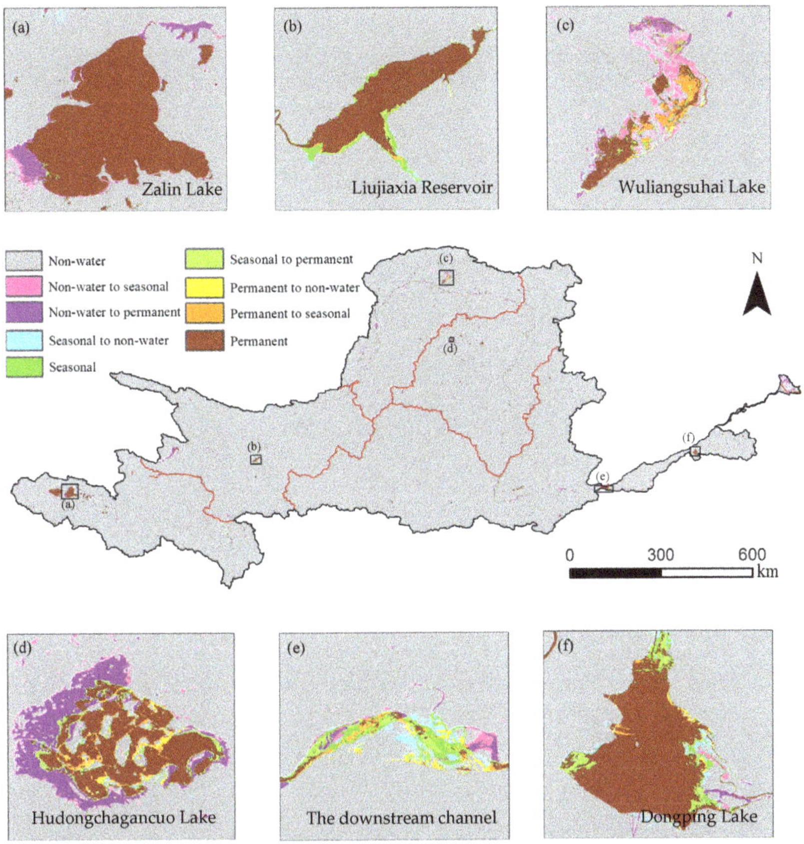

Figure 12. The spatial pattern of the conversion of surface water bodies from 1986 to 2019 in the YRB.

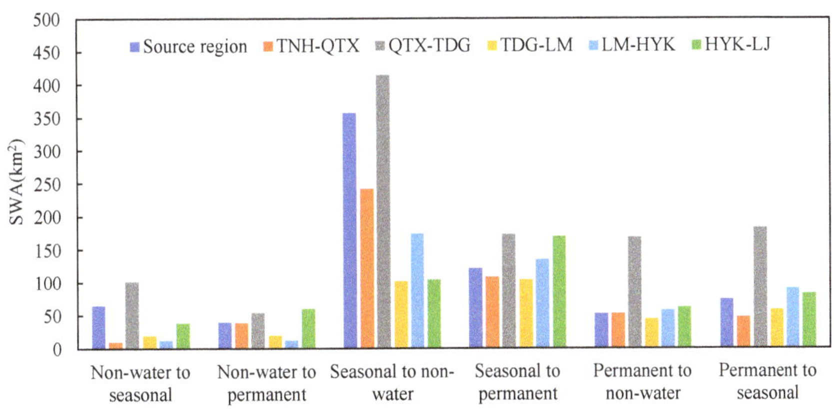

Figure 13. The SWA of the different surface water body conversion types from 1986 to 2019 in each sub-region in the YRB.

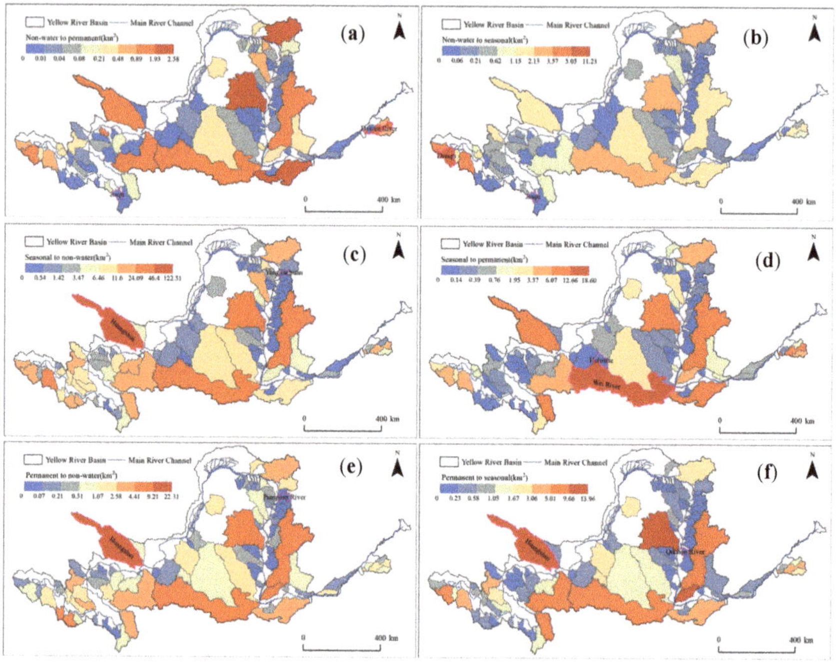

Figure 14. The different water body conversion types from 1986 to 2019 in the YRB. Panels (**a–f**) are for non-water to permanent, non-water to seasonal, seasonal to non-water, seasonal to permanent, permanent to non-water, and permanent to seasonal conversions, respectively.

3.5. Relationship between SWA and Environmental Factors

In order to explore the impacts of the water supply, atmospheric warming, vegetation greening, and surface SM on the SWA, we conducted partial correlation analysis to derive the correlations between the SWA and the Pre, Temp, LAI, and SM. During 1986–2019, the Pre, LAI, Temp, and SM in the entire YRB all increased, with the rate of change of Pre was 1.69 km^2/a, and the rates of change of LAI, Temp, and SM were not as large (Figure 15a). The partial correlation coefficients between the different types of water bodies and the Pre, Temp, LAI, and SM are shown in Figure 15b. The different types of water bodies were negatively correlated with temperature, while the LAI was positively correlated with the different types of water bodies. The creation of permanent water bodies was aided by an increase in the surface SM, which led to a small decrease in the seasonal water bodies. None of the correlations between Pre and the different types of water bodies were significant.

Divergent correlations between the SWA and the Pre, Temp, LAI, and SM were observed in the 86 sub-basins (Figures 16 and 17). In terms of the permanent water bodies, in 62.8% of the sub-basins, which were primarily located in the source area, the correlation between Pre and SWA was negative, and the Longwu River had the highest negative correlation coefficient. The increase in temperature reduced the SWA in 76.7% of the watersheds, mostly in the source area and the upper and middle reaches, and the Fen River had the largest negative correlation coefficient. The correlation between the LAI and SWA was positive in 86% of the watersheds, spanning the entire YRB, and the Hantai River had the highest positive correlation coefficient. The correlation between the SM and SWA was positive in 70.9% of the watersheds, mostly in the middle reaches, and the Kuye River had the highest positive correlation coefficient. The correlations between Pre and SWA were essentially not significant, whereas the associations between the SWA and the temperature, LAI, and SM were all significant to varying degrees.

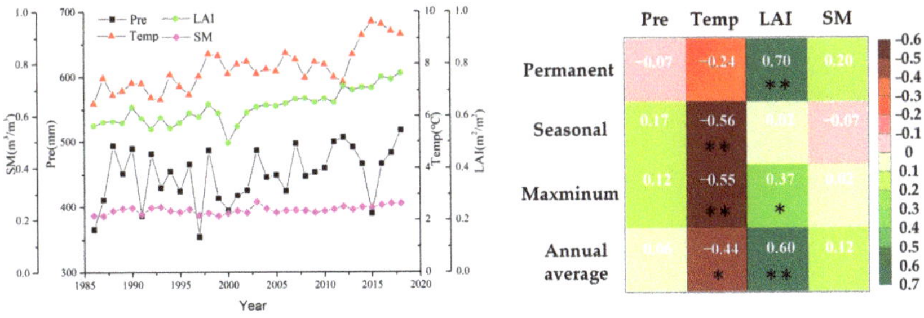

(**a**) (**b**)

Figure 15. (**a**) Interannual variations in the environmental factors, and (**b**) the correlation coefficients between the environmental factors and the different types of water bodies in the entire YRB. ** denotes $p < 0.01$; * denotes $p < 0.05$.

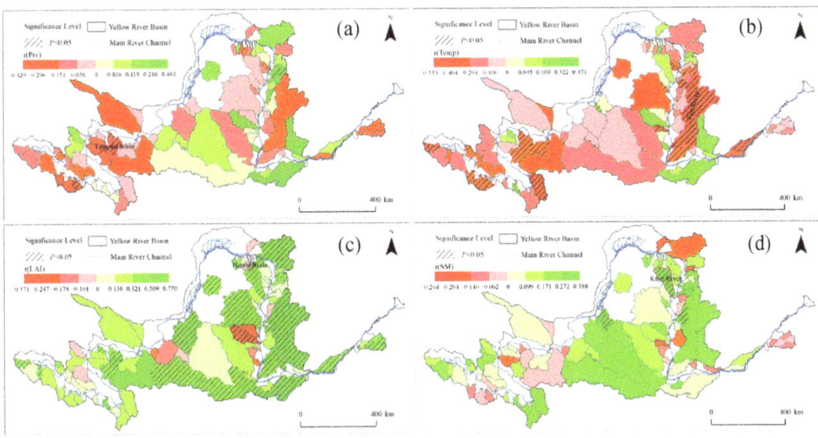

Figure 16. Correlation coefficients between the SWA and the Pre (**a**), Temp (**b**), LAI (**c**), and SM (**d**) for the permanent water bodies in each sub-basin.

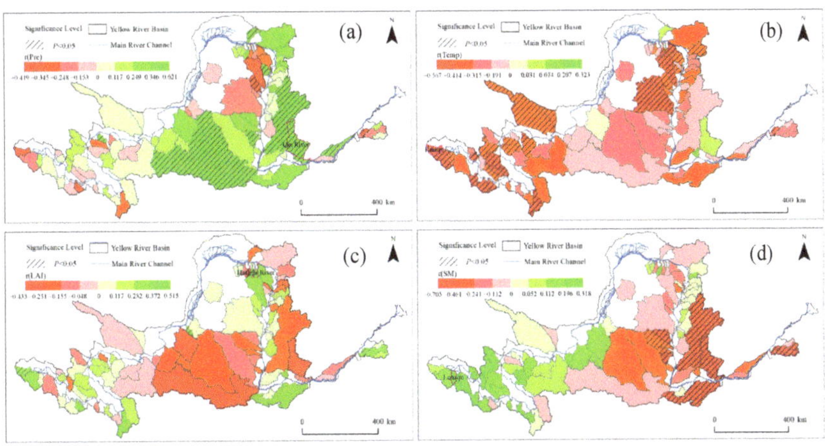

Figure 17. Correlation coefficients between the SWA and the Pre (**a**), Temp (**b**), LAI (**c**), and SM (**d**) for the seasonal water bodies in each sub-basin.

In terms of the seasonal water bodies, the correlation between Pre and SWA was positive in 67.4% of the sub-basins, mostly in the middle reaches, and the Qin River had the highest positive correlation coefficient. The SWA decreased with the increasing temperature in 94.2% of the watersheds in the YRB, and the negative correlation coefficient of Kariqu was the highest. In 55.8% of the watersheds, LAI was positively correlated with SWA, primarily in the middle reaches, and the Hashila River had the largest positive effect on the SWA. In 55.8% of the watersheds, SM was positively correlated with the SWA, primarily in the source and upstream areas, and Lenaqu had the highest positive correlation coefficient. LAI was insignificantly correlated with the SWA, but the correlations between the SWA and the Pre, Temp, and SM were all significant at different levels.

4. Discussion

4.1. Potential Influence Mechanism of the Environmental Factors on the SWA

Natural rainfall is the main water resource of the surface water bodies in the YRB; however, insignificant correlations between Pre and the different types of water bodies were observed (Figure 15), which indicate that Pre was not the dominant factor influencing the change in the SWA in the YRB during 1986–2019. An increase in temperature can increase the potential evaporation and the atmospheric vapor pressure deficit, accelerating the actual evaporation of surface water bodies [23]; Xia et al. came to the same conclusion in their study of the Huai River [26].

According to studies, a considerable increase or decrease in plant cover can alter surface water patterns [50]. An increase in the LAI can lower the land surface albedo, increase the solar radiation absorbed by the vegetation canopy, and increase the transpiration and interception evaporation from the vegetation [51] (Figure 18a). An increase in the latent heat can lead to a decrease in the land surface temperature and an increase in the air humidity, resulting in a decrease in the evaporation from surface water bodies [52]. Therefore, an increase in the vegetation LAI contributes to an increase in SWA [53,54], and Zeng et al. found the same pattern in studies in the Northeast China and Loess Plateau regions [55]. In addition, vegetation improves the surface runoff impedance and greatly enhances the trapped ET and soil water storage [56], which may alter the development of seasonal water bodies. Thus, seasonal water bodies are not strongly correlated with the LAI.

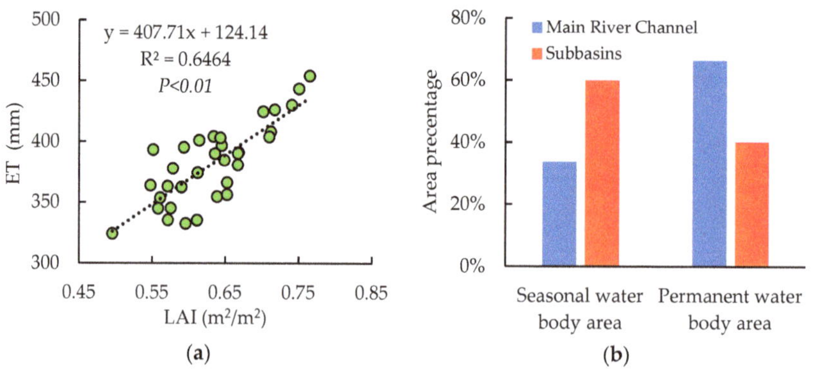

Figure 18. (**a**) Relationship between ET and LAI in the entire YRB, and (**b**) the water body area percentage in the main river channel region and the sub-basins for the permanent and seasonal water bodies.

Soil moisture is an important variable in the water and energy cycle because it impacts how rainfall is divided into runoff, surface storage, and infiltration components, and how entering solar and long-wave radiation is divided into outgoing long-wave radiation, latent heat, and so on. Soil moisture is an important variable in the water and energy cycle. The surface temperature, on the other hand, is a key component in establishing the land surface's heat and moisture balance. The surface temperature determines the incoming long-wave, sensible heat, and surface heat flow. Generally, the surface temperature shows an increase that corresponds to a decrease in the soil moisture [57]. However, the greater the surface SM is, the more easily the water storage soil layer becomes saturated, the longer the surface water body may persist after formation, and the more likely it is that a permanent water body will be established. As is illustrated in Figures 15 and 16, the correlation between the SM and the SWA was positive for the permanent water bodies and negative for the seasonal water bodies.

Due to the definition of permanent and seasonal water bodies, a rise in permanent water bodies equals a loss in seasonal water bodies, with opposite trends in their changes and thus major variations in the impact of influencing variables on both. It should be noted that compared with the seasonal water bodies, the SWA of the permanent water bodies was less strongly correlated with the Pre and Temp, and was even negatively correlated with Pre (Figures 15 and 16). This is mainly because most of the permanent water bodies are distributed in the main channel area (Figure 18b), and the storage and regulation effects of the large reservoirs located on the main channel have significantly influenced the SWA change in the channel area.

In contrast, the seasonal water bodies were generally located in the sub-basins where the effect of small-size reservoirs on SWA was less than that from large reservoirs located in the main channel; thus, the SWA change of the seasonal water bodies was more dependent on the natural precipitation and hydrological processes. However, the massive expansion of vegetation and water conservation engineering measures have significantly altered the precipitation redistribution through canopy interception, litter, and soil absorption [58,59], resulting in an insignificant correlation between the precipitation and surface water area [60].

4.2. Uncertainties

Despite the F-mask method having removed some clouds and cloud shadows from the remote sensing imagery, undetected thin clouds or mountain shadow areas still had an impact on the water body extraction in this investigation. Although the adopted hybrid index rule set for the water extraction had a high degree of accuracy, there were still some recognition mistakes based on comparisons with the Google Earth high-resolution photographs. The WIF thresholds used to distinguish between permanent and seasonal water bodies in this study were based on the findings of a vast number of previous studies, but these thresholds may vary with the location. In this study, the correlations between the SWA and various influencing factors were calculated using the partial correlation coefficient method, which only statistically analyzed the degree of correlation between the two changes, and did not quantitatively calculate the effects of the various influencing factors on the SWA. However, this study has taken into account the effects of the interactions between several contributing factors, which could reduce some uncertainties of the influence analysis. It should be also noted that due to the lack of groundwater observations, the effect of the groundwater supply on the SWA was neglected in this study.

5. Conclusions

In this study, we analyzed the spatial–temporal variability of the SWAs of four types of water bodies (permanent, seasonal, maximum, and average extents) and their relationships with precipitation (Pre), temperature (Temp), LAI, and surface SM using all of the available Landsat images acquired from 1986 to 2019 and a water and vegetation index-based method. The results demonstrate that the accuracy of the water body identification method used in this study was 97%, according to an accuracy verification based on high-resolution Google Earth images.

The YRB's permanent SWA increased at a rate of 49.82 km^2/y, with the largest proportion—34.34%—occurring in the downstream area, while the seasonal SWA decreased at a rate of 79.18 km^2/y, with the largest proportion—25.99%—occurring in the source area. In 50 of the 86 sub-basins, the SWA of the permanent water bodies increased, and these sub-basins were mainly located in the middle and lower reaches. The SWA of the seasonal water bodies increased in five of the sub-basins, accounting for just 5.8% of the total, indicating that the changes in the SWA in the sub-basin areas were the primary driver of the decrease in the SWA of the seasonal water bodies. The conversions from seasonal water bodies to non-water and from seasonal to permanent water bodies were the dominant trends from 1986 to 2019 in the YRB.

For the permanent water bodies, the SWA was less strongly correlated with Pre and Temp, and it was negatively correlated with Pre (Figures 15 and 16) because most of the permanent water bodies were distributed in the main channel area (Figure 18a), and the storage and regulation effects of the large terrace reservoirs located on the main channel significantly influenced the changes in the SWA in the channel area.

The seasonal water bodies were primarily located in the sub-basins, and the surface water was mostly controlled by natural hydrological processes. Thus, the SWA was more dependent on the climatic factors. An increase in the surface SM can lead to the establishment of permanent water bodies, as well as a reduction in seasonal water bodies. By modifying the local climatic conditions, a considerable increase in the ET due to increased vegetation may decrease the evaporation of the surface water, partially compensating for the faster evaporation of the water surface caused by the increased air temperature.

Author Contributions: All of the authors contributed to the design and development of this manuscript. Chongwei Li performed the data processing and wrote the first draft of the manuscript. Zhihui Wang gave constructive suggestions on the design and modification of the manuscript. Qingfeng Hu and Yang Liu also helped process the data in this paper. Wenkai Liu helped edit the manuscript prior to submission. All authors have read and agreed to the published version of the manuscript.

Funding: This research was funded by the Joint Funds of the National Natural Science Foundation of China (U21A20109), Young Elite Scientist Sponsorship from CAST (2017QNRC023).

Data Availability Statement: The data presented in this study are available on request from the corresponding author for research purposes.

Conflicts of Interest: The authors declare no conflict of interest.

References

1. Amprako, J.L. The United Nations World Water Development Report 2015: Water for a Sustainable World. *Future Food J. Food Agric. Soc.* **2016**, *4*, 64–65.
2. Costanza, R.; d'Arge, R.; De Groot, R.; Farber, S.; Grasso, M.; Hannon, B.; Limburg, K.; Naeem, S.; O'Neill, R.V.; Paruelo, J.; et al. The value of the world's ecosystem services and natural capital. *Nature* **1997**, *387*, 253. [CrossRef]
3. Wood, E.F.; Roundy, J.K.; Troy, T.J.; van Beek, L.P.H.; Bierkens, M.F.P.; Blyth, E.; de Roo, A.; Döll, P.; Ek, M.; Famiglietti, J.; et al. Hyperresolution global land surface modeling: Meeting a grand challenge for monitoring Earth's terrestrial water. *Water Resour. Res.* **2011**, *47*, W05301. [CrossRef]
4. Shevyrnogov, A.P.; Kartushinsky, A.V.; Vysotskaya, G.S. Application of Satellite Data for Investigation of Dynamic Processes in Inland Water Bodies: Lake Shira (Khakasia, Siberia), A Case Study. *Aquat. Ecol.* **2002**, *36*, 153–164. [CrossRef]
5. Rabus, B.; Eineder, M.; Roth, A.; Bamler, R. The Shuttle Radar Topography Mission—A New Class of Digital Elevation Models Acquired by Spaceborne Radar. *ISPRS J. Photogramm. Remote Sens.* **2003**, *57*, 241–262. [CrossRef]
6. Hall, J.W.; Grey, D.; Garrick, D.; Fung, F.; Brown, C.; Dadson, S.J.; Sadoff, C.W. Coping with the curse of freshwater variability. *Science* **2014**, *346*, 429–430. [CrossRef]
7. Tulbure, M.G.; Broich, M.; Stehman, S.V.; Kommareddy, A. Surface water extent dynamics from three decades of seasonally continuous Landsat time series at subcontinental scale in a semi-arid region. *Remote Sens. Environ.* **2016**, *178*, 142–157. [CrossRef]
8. Huang, C.; Chen, Y.; Zhang, S.; Wu, J. Detecting, extracting, and monitoring surface water from space using optical sensors: A review. *Rev. Geophys.* **2018**, *56*, 333–360. [CrossRef]
9. Cole, J.J.; Prairie, Y.T.; Caraco, N.F.; McDowell, W.H.; Tranvik, L.J.; Striegl, R.G.; Duarte, C.M.; Kortelainen, P.; Downing, J.A.; Middelburg, J.J. Plumbing the Global Carbon Cycle: Integrating Inland Waters into the Terrestrial Carbon Budget. *Ecosystems* **2007**, *10*, 172–185. [CrossRef]
10. Carroll, M.L.; Townshend, J.R.G.; DiMiceli, C.M.; Loboda, T.; Sohlberg, R.A. Shrinking Lakes of the Arctic: Spatial Relationships and Trajectory of Change. *Geophys. Res. Lett.* **2011**, *38*, L20406. [CrossRef]

11. Craglia, M.; de Bie, K.; Jackson, D.; Pesaresi, M.; Remetey-Fülöpp, G.; Wang, C.; Annoni, A.; Bian, L.; Campbell, F.; Ehlers, M. Digital Earth 2020: Towards the Vision for the next Decade. *Int. J. Digit. Earth* **2012**, *5*, 4–21. [CrossRef]
12. Jianbo, L.; Changda, D. The application of TM image in reservoir situation monitoring. *Natl. Remote Sens. Bull.* **1996**, *1*, 54–58.
13. Bi, H.Y.; Wang, S.Y.; Zeng, J.Y.; Zhao, Y.; Wang, H.; Yin, H. Comparison and analysis of several common water extraction methods based on TM image. *Remote Sens. Inf.* **2012**, *27*, 77–82.
14. Jiaju, L.; Shihong, L. Improvement of the techniques for distinguishing water bodies from TM data. *J. Remote Sens.* **1992**, *1*, 17–23.
15. McFeeters, S.K. The use of the Normalized Difference Water Index (NDWI) in the delineation of open water features. *Int. J. Remote Sens.* **1996**, *17*, 1425–1432. [CrossRef]
16. Xu, H. A study on information extraction of water body with the modified normalized difference water index (MNDWI). *J. Remote Sens.* **2005**, *9*, 589–595.
17. Xu, H. Modification of normalised difference water index (NDWI) to enhance open water features in remotelysensed imagery. *Int. J. Remote Sens.* **2006**, *27*, 3025–3033. [CrossRef]
18. Lei, J.I.; Zhang, L.I.; Wylie, B. Analysis of Dynamic Thresholds for the Normalized Difference Water Index. Photogramm. *Eng. Remote Sens.* **2009**, *75*, 1307–1317.
19. Verpoorter, C.; Kutser, T.; Tranvik, L. Automated Mapping of Water Bodies Using Landsat Multispectral Data. *Limnol. Oceanogr. Methods* **2015**, *10*, 1037–1050. [CrossRef]
20. Santoro, M.; Wegmueller, U.; Lamarche, C.; Bontemps, S.; Defoumy, P.; Arino, O. Strengths and weaknesses of multi-year Envisat ASAR backscatter measurements to map permanent open water bodies at global scale. *Remote Sens. Environ.* **2015**, *171*, 185–201. [CrossRef]
21. Gómez, C.; White, J.C.; Wulder, M.A. Optical remotely sensed time series data for land cover classification: A review. *ISPRS J. Photogramm. Remote Sens.* **2016**, *116*, 55–72. [CrossRef]
22. Khatami, R.; Mountrakis, G.; Stehman, S.V. A meta-analysis of remote sensing research on supervised pixel-based land-cover image classification processes: General guidelines for practitioners and future research. *Remote Sens. Environ.* **2016**, *177*, 89–100. [CrossRef]
23. Zou, Z.H.; Dong, J.W.; Menarguez, M.A.; Xiao, X.M.; QIN, Y.W.; Doughty, R.B.; Hooker, K.V.; Hambright, K.D. Continued decrease of open surface water body area in Oklahoma during 1984–2015. *Sci. Total Environ.* **2017**, *595*, 451–460. [CrossRef]
24. Zou, Z.H.; Xiao, X.M.; Dong, J.W.; Qin, Y.W.; Doughty, R.B.; Menarguez, M.A.; Zhang, G.L.; Wang, J. Divergent trends of open-surface water body area in the contiguous United States from 1984 to 2016. *Proc. Natl. Acad. Sci. USA* **2018**, *115*, 3810–3815. [CrossRef]
25. Liu, X.P.; Hu, G.H.; Chen, Y.M.; Li, X.; Xu, X.C.; Li, S.Y.; Pei, F.S.; Wang, S.J. High-resolution multi-temporal mapping of global urban land using Landsat images based on the Google Earth Engine Platform. *Remote Sens. Environ.* **2018**, *209*, 227–239. [CrossRef]
26. Xia, H.M.; Zhao, J.Y.; Qin, Y.C.; Yang, J.; Cui, Y.M.; Song, H.Q.; Ma, L.Q.; Jin, N.; Meng, Q.M. Changes in water surface area during 1989–2017 in the Huai river basin using landsat data and google earth engine. *Remote Sens.* **2019**, *11*, 1824. [CrossRef]
27. Zhou, Y.; Dong, J.W.; Xiao, X.M.; Xiao, T.; Yang, Z.Q.; Zhao, G.S.; Zou, Z.H.; Qin, Y.W. Open surface water mapping algorithms: A comparison of water-related spectral indices and sensors. *Water* **2017**, *9*, 256. [CrossRef]
28. Dong, J.W.; Xiao, X.M.; Menarguez, M.A.; Zhang, G.L.; Qin, Y.W.; Thau, D.; Biradar, C.; Moore III, B. Mapping paddy rice planting area in northeastern Asia with Landsat 8 images, phenology-based algorithm and Google Earth Engine. *Remote Sens. Environ.* **2016**, *185*, 142–154. [CrossRef]
29. Pekel, J.F.; Cottam, A.; Gorelick, N.; Belward, A.S. High-resolution mapping of global surface water and its long-term changes. *Nature* **2016**, *540*, 418–422. [CrossRef]
30. Mueller, N.; Lewis, A.; Roberts, D.; Ring, S.; Melrose, R.; Lymburner, L.; Mclntyre, A.; Tan, P.; Curnow, S. Water observations from space: Mapping surface water from 25 years of Landsat imagery across Australia. *Remote Sens. Environ.* **2016**, *174*, 341–352. [CrossRef]
31. Zhou, Y.; Dong, J.W.; Xiao, X.M.; Liu, R.G.; Zou, Z.H.; Zhao, G.S.; Ge, Q.S. Continuous monitoring of lake dynamics on the Mongolian Plateau using all available Landsat imagery and Google Earth Engine. *Sci. Total Environ.* **2019**, *689*, 366–380. [CrossRef]
32. Wang, X.X.; Xiao, X.M.; Zou, Z.H.; Dong, J.W.; Doughty, R.B.; Menarguez, M.A.; Chen, B.Q.; Wang, J.B.; Ye, H. Gainers and losers of surface and terrestrial water resources in China during 1989–2016. *Nat. Commun.* **2020**, *11*, 3471. [CrossRef]
33. Tang, Q.H.; Liu, X.C.; Zhou, Y.Y.; Wang, J.; Yun, X.B. Cascading impacts of Asian water tower change on downstream water systems. *Bull. Chin. Acad. Sci.* **2019**, *34*, 1306–1312.
34. Jia, S.; Liang, Y. Suggestions for strategic allocation of the Yellow River water resources under the new situation. *Resour. Sci.* **2020**, *42*, 29–36. [CrossRef]
35. Xia, J.; Peng, S.M.; Wang, C.; Hong, S.; Chen, J.; Luo, X. Impact of climate change on water resources and adaptive management in the Yellow River Basin. *Yellow River* **2014**, *36*, 15.
36. Liu, C.M.; Tian, W.; Liu, X.M. Analysis and understanding on runoff variation of the Yellow River in recent 100 years. *Yellow River* **2019**, *41*, 11–15.
37. Wang, R.M.; Xia, H.M.; Qin, Y.C.; Niu, W.H.; Pan, L.; Li, R.M.; Zhao, X.Y.; Bian, X.Q.; Fu, P.D. Dynamic monitoring of surface water area during 1989–2019 in the Hetao plain using landsat data in google earth engine. *Water* **2020**, *12*, 3010. [CrossRef]
38. Liang, K.; Li, Y.Z. Changes in lake area in response to climatic forcing in the endorheic Hongjian lake basin, China. *Remote Sens.* **2019**, *11*, 3046. [CrossRef]

39. Luo, D.L.; Jin, H.J.; Du, H.Q.; Li, C.; Ma, Q.; Duan, S.Q.; Li, G.S. Variation of alpine lakes from 1986 to 2019 in the Headwater Area of the Yellow River, Tibetan Plateau using Google Earth Engine. *Adv. Clim. Chang. Res.* **2020**, *11*, 11–21. [CrossRef]
40. Liu, X. *Causes of Sharp Decrease in Water and Sediment in Recent Years in the Yellow River*; Science Press: Beijing, China, 2016.
41. Li, J.; Peng, S.; Li, Z. Detecting and attributing vegetation changes on China s Loess Plateau. *Agric. For. Meteorol.* **2017**, *247*, 260–270. [CrossRef]
42. Ju, J.C.; Roy, D.P.; Vermote, E.; Masek, J.; Kovalskyy, V. Continental-scale validation of MODIS-based and LEDAPS Landsat ETM+ atmospheric correction methods. *Remote Sens. Environ.* **2012**, *122*, 175–184. [CrossRef]
43. Zhu, Z.; Wang, S.X.; Woodcock, C.E. Improvement and expansion of the Fmask algorithm: Cloud, cloud shadow, and snow detection for Landsats 4-7, 8, and Sentinel 2 images. *Remote Sens. Environ.* **2015**, *159*, 269–277. [CrossRef]
44. Liu, Z.; Li, L.; Tim, R.M.; Vanniel, T.G.; Li, R. Introduction of the professional interpolation software for meteorology data: ANUSPLINN. *Meteorol. Mon.* **2008**, *34*, 92–100.
45. Martens, B.; Miralles, D.G.; Lievens, H.; van der Schalie, R.; de Jeu, R.A.M.; Fernández-Prieto, D.; Beck, H.E.; Dorigo, W.A.; Verhoest, N.E.C. GLEAM v3: Satellite-based land evaporation and root-zone soil moisture. *Geosci. Model Dev.* **2017**, *10*, 1903–1925. [CrossRef]
46. Xiao, Z.; Liang, S.; Wang, J.; Chen, P.; Yin, X.; Zhang, L.; Song, J. Use of general regression neural networks for generating the GLASS leaf area index product from time-series MODIS surface reflectance. *IEEE Trans. Geosci. Remote Sens.* **2014**, *52*, 209–223. [CrossRef]
47. Zhang, X.; Liu, L.Y.; Wang, Y.J.; Hu, Y.; Zhang, B. A SPECLib-based operational classification approach: A preliminary test on China land cover mapping at 30 M. *Int. J. Appl. Earth Obs. Geoinf.* **2018**, *71*, 83–94. [CrossRef]
48. Deng, X.Y.; Song, C.Q.; Liu, K.; Ke, L.H.; Zhang, W.S.; Ma, R.H.; Zhu, J.Y.; Wu, Q.H. Remote sensing estimation of catchment-scale reservoir water impoundment in the upper Yellow River and implications for river discharge alteration. *J. Hydrol.* **2020**, *585*, 124791. [CrossRef]
49. Wang, C.; Jia, M.M.; Chen, N.C.; Wang, W. Long-term surface water dynamics analysis based on landsat imagery and the google earth engine platform: A case study in the middle Yangtze River basin. *Remote Sens.* **2018**, *10*, 1635. [CrossRef]
50. Egginton, P.; Beall, F.; Buttle, J. Reforestation–Climate change and water resource implications. *For. Chron.* **2014**, *90*, 516–524. [CrossRef]
51. Wang, Z.; Cui, Z.; He, T. Attributing the Evapotranspiration Trend in the Upper and Middle Reaches of Yellow River Basin Using Global Evapotranspiration Products. *Remote Sens.* **2021**, *14*, 175. [CrossRef]
52. Chen, X.Z.; Liu, L.Y.; Su, Y.X.; Yuan, W.P.; Liu, X.D.; Liu, Z.Y.; Zhou, G.Y. Quantitative association between the water yield impacts of forest cover changes and the biophysical effects of forest cover on temperatures. *J. Hydrol.* **2021**, *600*, 126529. [CrossRef]
53. Zhou, G.Y.; Sun, G.; Wang, X.; Zhou, C.Y.; McNulty, S.G.; Vose, J.M.; Amatya, D.M. Estimating forest ecosystem evapotranspiration at multiple temporal scales with a dimension analysis Approach1. *JAWRA J. Am. Water Resour. Assoc.* **2008**, *44*, 208–221. [CrossRef]
54. Wang, S.; Fu, B.J.; He, C.S.; Sun, G.; Gao, G.Y. A comparative analysis of forest cover and catchment water yield relationships in Northern China. *For. Ecol. Manag.* **2011**, *262*, 1189–1198. [CrossRef]
55. Zeng, Y.; Yang, X.K.; Fang, N.F.; Shi, Z.H. Large-scale afforestation significantly increases permanent surface water in China's vegetation restoration regions. *Agric. For. Meteorol.* **2020**, *290*, 108001. [CrossRef]
56. van den Hurk, B.J.J.M.; Viterbo, P.; Los, S.O. Impact of leaf area index seasonality on the annual land surface evaporation in a global circulation model. *J. Geophys. Res. Atmos.* **2003**, *108*, D6. [CrossRef]
57. Lakshmi, V.; Jackson, T.J.; Zehrfuhs, D. Soil moisture–temperature relationships: Results from two field experiments. *Hydrol. Processes* **2003**, *17*, 3041–3057. [CrossRef]
58. Helvey, J.D.; Patric, J.H. Canopy and litter interception of rainfall by hardwoods of eastern United States. *Water Resour. Res.* **1965**, *1*, 193–206. [CrossRef]
59. Jian, S.Q.; Zhao, C.Y.; Fang, S.M.; Yu, K. Effects of different vegetation restoration on soil water storage and water balance in the Chinese Loess Plateau. *Agric. For. Meteorol.* **2015**, *206*, 85–96. [CrossRef]
60. Wu, D.H.; Zhao, X.; Liang, S.L.; Zhou, T.; Huang, K.C.; Tang, B.J.; Zhao, W.Q. Time-lag effects of global vegetation responses to climate change. *Glob. Chang. Biol.* **2015**, *21*, 3520–3531. [CrossRef]

International Journal of *Geo-Information*

Article

Investigating Relationships between Runoff–Erosion Processes and Land Use and Land Cover Using Remote Sensing Multiple Gridded Datasets

Cláudia Adriana Bueno da Fonseca [1,2], Nadhir Al-Ansari [3,*], Richarde Marques da Silva [1,4], Celso Augusto Guimarães Santos [1,5], Bilel Zerouali [6], Daniel Bezerra de Oliveira [5] and Ahmed Elbeltagi [7]

1 Graduate Program in Geography, Federal University of Paraíba, João Pessoa 58051-900, Brazil; claudia.fonseca@ueg.br (C.A.B.d.F.); richarde@geociencias.ufpb.br (R.M.d.S.); celso@ct.ufpb.br (C.A.G.S.)
2 Course of Geography, Campus Cora Coralina, Goiás State University, Goiás 76600-000, Brazil
3 Department of Civil, Environmental and Natural Resources Engineering, Lulea University of Technology, 97187 Lulea, Sweden
4 Department of Geosciences, Federal University of Paraíba, João Pessoa 58051-900, Brazil
5 Department of Civil and Environmental Engineering, Federal University of Paraíba, João Pessoa 58051-900, Brazil; daniel.oliveira2@academico.ufpb.br
6 Vegetal Chemistry-Water-Energy Research Laboratory, Department of Hydraulic, Faculty of Civil Engineering and Architecture, Hassiba Benbouali, University of Chlef, B.P. 78C, Ouled Fares, Chlef 02180, Algeria; b.zerouali@univ-chlef.dz
7 Agricultural Engineering Department, Faculty of Agriculture, Mansoura University, Mansoura 35516, Egypt; ahmedelbeltagy81@mans.edu.eg
* Correspondence: nadhir.alansari@ltu.se

Citation: Fonseca, C.A.B.d.; Al-Ansari, N.; Silva, R.M.d.; Santos, C.A.G.; Zerouali, B.; Oliveira, D.B.d.; Elbeltagi, A. Investigating Relationships between Runoff–Erosion Processes and Land Use and Land Cover Using Remote Sensing Multiple Gridded Datasets. *ISPRS Int. J. Geo-Inf.* **2022**, *11*, 272. https://doi.org/10.3390/ijgi11050272

Academic Editors: Walter Chen, Fuan Tsai and Wolfgang Kainz

Received: 24 February 2022
Accepted: 15 April 2022
Published: 19 April 2022

Abstract: Climate variability, land use and land cover changes (LULCC) have a considerable impact on runoff–erosion processes. This study analyzed the relationships between climate variability and spatiotemporal LULCC on runoff–erosion processes in different scenarios of land use and land cover (LULC) for the Almas River basin, located in the Cerrado biome in Brazil. Landsat images from 1991, 2006, and 2017 were used to analyze changes and the LULC scenarios. Two simulations based on the Soil and Water Assessment Tool (SWAT) were compared: (1) default application using the standard model database (SWATd), and (2) application using remote sensing multiple gridded datasets (albedo and leaf area index) downloaded using the Google Earth Engine (SWATrs). In addition, the SWAT model was applied to analyze the impacts of streamflow and erosion in two hypothetical scenarios of LULC. The first scenario was the optimistic scenario (OS), which represents the sustainable use and preservation of natural vegetation, emphasizing the recovery of permanent preservation areas close to watercourses, hilltops, and mountains, based on the Brazilian forest code. The second scenario was the pessimistic scenario (PS), which presents increased deforestation and expansion of farming activities. The results of the LULC changes show that between 1991 and 2017, the area occupied by agriculture and livestock increased by 75.38%. These results confirmed an increase in the sugarcane plantation and the number of cattle in the basin. The SWAT results showed that the difference between the simulated streamflow for the PS was 26.42%, compared with the OS. The sediment yield average estimation in the PS was 0.035 ton/ha/year, whereas in the OS, it was 0.025 ton/ha/year (i.e., a decrease of 21.88%). The results demonstrated that the basin has a greater predisposition for increased streamflow and sediment yield due to the LULC changes. In addition, measures to contain the increase in agriculture should be analyzed by regional managers to reduce soil erosion in this biome.

Keywords: agricultural data; geoinformation; LULC changes; modeling; observation; SWAT

1. Introduction

Land use and land cover changes (LULCC) caused by the advance of agriculture have been causing severe environmental problems worldwide, mainly in Brazil [1]. Some of the LULCC are caused by climate variability that is independent of human activity [2]; however, in the Cerrado biome in Brazil, especially in an environment like the Almas River basin, the LULCC have been caused by the intense advancement of agriculture (e.g., sugarcane) [3]. LULCC lead to decreased fauna and flora biodiversity [4], and they affect streamflow and sediment yield [5]. This study investigates the relationships between LULCC and runoff–erosion processes using remote sensing multiple gridded datasets. In recent years, the relationship between LULC, climate, streamflow, and sediment yield has attracted the attention of society and researchers [6]; however, there is a lack of data to create a scientific basis for subjects such as the properties of the streamflow and sediment yield in the Savanna biome (e.g., the Cerrado biome of Brazil). Knowing these data is crucial to control erosion and sedimentation effectively because plans made without being based on scientific evidence can cause greater expense.

Although the importance of studying the relationships between the runoff–erosion process and LULCC using remote sensing multiple gridded datasets is well-understood, determining the spatial distribution of the runoff–erosion process is an essential prerequisite for the establishment of erosion management plans in any catchment. The advancement of agriculture and the influence of different LULC scenarios has been significantly studied [7–10]; however, research involving the impacts of LULC on runoff–erosion processes using estimated satellite data and runoff–erosion models in some regions of the planet, such as Brazil, is still scarce [11–13]. In addition, the published studies did not carry out estimates of runoff and sediment yield considering different LULC scenarios at watershed scales. In this sense, this study can be used in other hydrologically homogeneous regions because the methodology used can be easily replicated in other regions with the same type of data used in this study.

In Brazil, LULCC have impacted the quality and quantity of water in the basins [1]. This change is due to deforestation for the sale of wood and the increase in agricultural activities [14]. Such a change intensified from the 1990s onwards, causing a reduction in the area occupied with native vegetation cover. The expansion of cattle ranching played an essential role in the historical process of occupation of this biome, as it has transformed cattle raising in recent decades into one of the main economic activities within this biome [15–17]. Since the 1960s, the Cerrado biome has been marked by constant tax incentives and investments in agriculture, which favored the increase in agricultural activity and pastures [18]. This biome has been occupied due to an agricultural model focused on agribusiness without worrying about environmental preservation, which occurs in large parts of Brazil [19,20]. In recent years, extensive areas of native vegetation have been deforested because of LULCC, with the conversion of native vegetation into agricultural spaces and pastures [21]. The intense pace of deforestation in the Cerrado biome has caused several environmental impacts, such as ecosystem fragmentation, reduced soil quality, increased water erosion, siltation of water bodies, and increased sediment yield [22]. On the other hand, ignoring any historical LULCC and climatic variations within the Cerrado area means ignoring the cause-and-effect relationships of the hydrological cycle and the physical characteristics of a river basin, which can lead to numerous environmental problems.

The problem of the impacts of agricultural expansion and its implications on runoff–erosion processes in the Cerrado biome has been widely studied [23–26]; however, studies involving the flow behavior and the sediment yield in hypothetical LULC scenarios at a basin-scale in this Brazilian biome are still scarce [27–29]. For these reasons, the impacts of LULCC on streamflow and sediment yield still need to be further investigated in the Cerrado biome, which is of extreme importance for water resources and electrical production in Brazil [30]. Understanding the runoff–erosion behavior of this basin is vital for good planning of the service life of the Serra da Mesa hydroelectric power plant for energy generation. This hydroelectric plant totals 1275 MW and is strategic for the development of Brazil, as it produces electricity for all Brazilian regions [31]; therefore, knowing the contribution of sediments and inflow is essential for decision makers of water resources to estimate the reservoir service life and plan the water supply and electric energy generation.

This study also explores the applicability of remote sensing in ungauged basins to contribute to new research avenues on data-scarce regions, such as the Cerrado biome in Brazil. The availability of input parameter data for physically-based models is one of the most significant challenges for applying hydrological models today. This paper demonstrates how LULC, soil parameters, albedo, and leaf area index (LAI), obtained from remote sensing datasets, can successfully calibrate distributed hydrological models. In addition, this study seeks to analyze the satellite-estimated data quality for use as input data in hydrological modeling to estimate runoff and sediment yield at a basin scale [32,33]. This application would open up many possibilities in this biome where hydrological information is scarce, and it would help to improve the simulation accuracy. In this study, we choose the Almas River basin, which is representative of a typical humid tropical basin in the Cerrado biome in Brazil.

The Soil and Water Assessment Tool (SWAT) model has already been widely applied to basins worldwide [34–37]; however, this model performs poorly for tropical areas using the standard model dataset's soil parameters (e.g., albedo and LAI). Thus, many improvements to the SWAT model have been developed, such as SWAT-T [36]. In this study, two simulations based on the SWAT model are compared: (1) default application using the standard model database (SWATd); and (2) application using remote sensing multiple gridded datasets (albedo and LAI), downloaded using the Google Earth Engine (SWATrs). Thus, the objective of this study is to analyze the relationships between runoff–erosion processes and LULC under agricultural shift, comparing two simulations of the SWAT model, with and without remote sensing multiple gridded datasets, in a typical river basin of the Cerrado biome in Brazil.

2. Materials and Methods

2.1. Study Area

The Almas River basin has an area of 18,838 km^2 and is within the Cerrado biome. This basin is located between latitudes 14°37′00″ S and 16°15′00″ S and longitudes 50°08′00″ W and 48°49′00″ W (Figure 1). The Cerrado biome occupies an area of approximately 2,036,448 km^2 (24%) of the territory of Brazil. It is the second-largest biome in South America [5], recognized for the variability of the phytophysiognomy and the biodiversity of its flora and very rich fauna, with numerous species of plants and animals [38]. From a hydrological point of view, the Cerrado biome plays a fundamental role in producing water that flows into the main Brazilian river basins, such as Tocantins-Araguaia, Amazonia, Paraná, Paraguay, and São Francisco [39]. The Almas River basin is fundamental for energy generation, as is the Serra da Mesa Hydroelectric Power Plant, located in the basin outlet. This hydroelectric plant was inaugurated in March 1997 with a water volume capacity equivalent to 54.4 billion m^3 [40]. This plant is essential for power generation, supporting South, Southeast, Midwest, and North Brazil [41]. This basin has species of heterogeneous vegetation with arboreal and forest, herbaceous-shrubby, and herbaceous-grassy strata, with spaced and gnarled trees, which are generally endowed with thick bark deep roots [42]. This basin is represented by several phytophysiognomies, such as savannah formations (Cerradão/Forest and typical Savanna), grassy formations (grassy field, clean field, and rupestrian field), and forest formations (riparian forest, gallery forest, dry forest, and Cerradão).

According to the Köppen classification, the region's climate is Aw type (warm sub-humid tropical), and the average annual rainfall is approximately 1800 mm [43]. This region is marked by two well-defined seasons, a rainy season from October to April, with an average monthly rainfall of 250 mm, and a dry season from May to September with an average monthly of 10 mm [44]. The primary meteorological phenomenon that influences rainfall in the region during the rainy season is the South Atlantic Convergence Zone, formed from the arrival of subtropical fronts in central Brazil and is associated with moisture from the Amazon region, favoring the occurrence of rainfalls with large volumes [45]. Temperatures in the basin range from 17 °C to 34 °C, with an average relative humidity of approximately 80% [46].

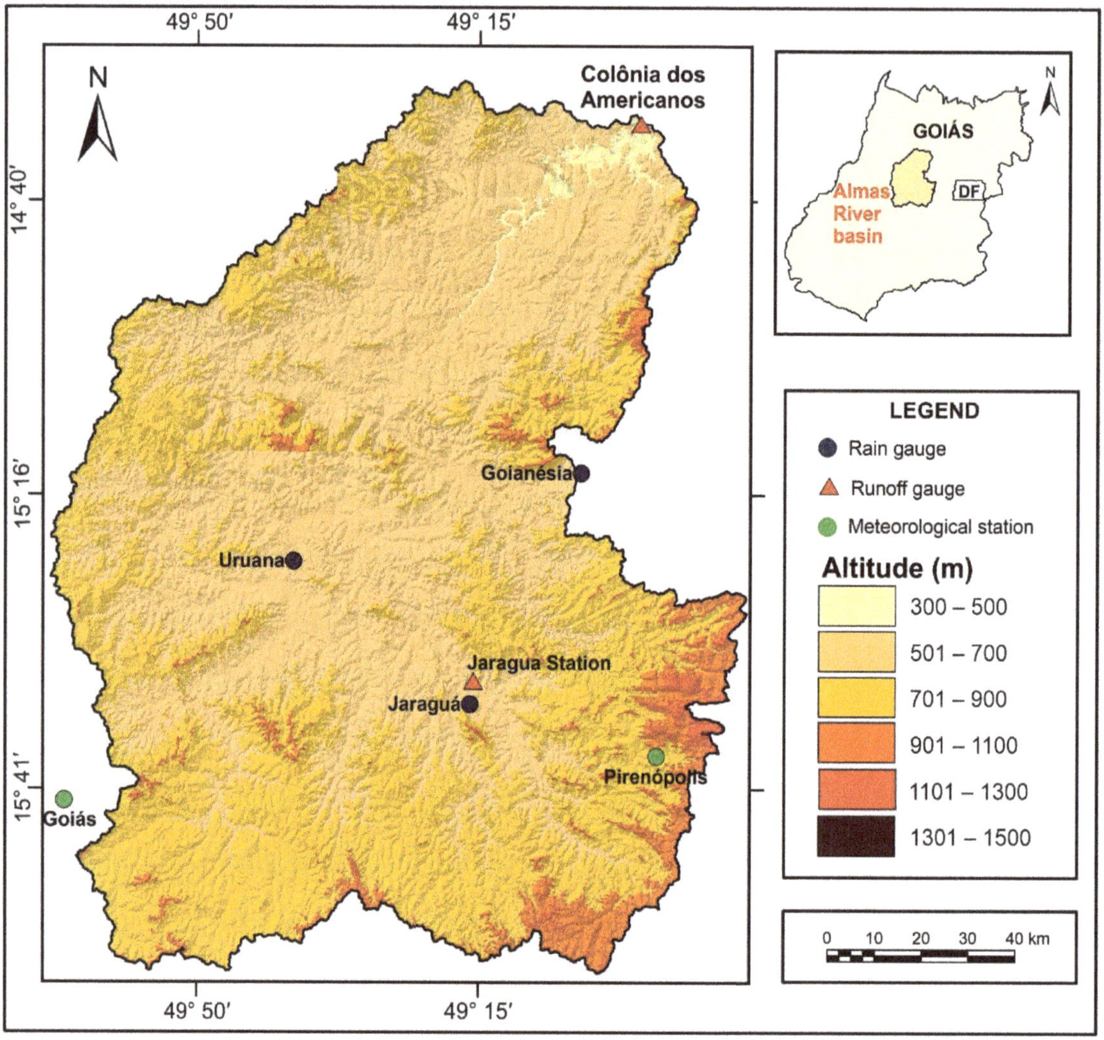

Figure 1. The geographic location of the Almas River basin in Brazil, Goiás State, and the federal district (DF), altimetry of the basin, rain gauges, and streamflow stations used in this study.

The population in this basin is approximately 729,108 inhabitants, being mainly composed of an urban population (89%) [47]. The demographic structure of the basin has undergone an intense transformation since 1970, when the population changed from rural to urban. This phenomenon directly results from the change in economic and production base that this region has gone through. The expansion of the industrial park, notably that of agribusiness, and the strengthening of the service sector, boosted the local economy, attracting immigrants [48].

This basin has a great diversity of habitats. Since the 1970s, this region has suffered several environmental impacts on flora due to fragmentation and habitat loss, which affect the region's fauna [49]. These modifications cause a disturbance and dispersion of the gene flow, influencing population density and genetic diversity, and occasionally causing local extinctions [50]. In addition to the local and regional effects of agricultural activities, global change is another critical factor that can impact the diversity and distribution of animal species such as the Quenquém (*Acromyrmex diasi*) and Ground-web Spider (*Anapistula guyri*) [49], and vegetation such as the Baru tree (*Dipteryx alata Vogel*) [50]. According to Ref. [49], there are currently more than 130 species of amphibians, birds, aquatic invertebrates, terrestrial invertebrates, mammals, fish, and reptiles in the Cerrado biome that are threatened with extinction.

There are large extensions of crops with intense mechanization and significant investments in technology and inputs in the Almas River basin. The region's crops are also characterized by the diversification of products, such as rice, sugarcane, beans, corn, soy, and sorghum. Agricultural production in the region is geared towards meeting regional particularities and commercial prospects as the demand for the products in international markets increases [51,52].

2.2. Evolution of Agriculture in the Region

This study collected data on the planted area of temporary agriculture of rice, sugarcane, beans, corn, soybean, and sorghum for 1991, 2000, 2006, 2011, and 2017 [53]. These years were chosen due to the advance of LULCC for agriculture and livestock. The data are available on the Automatic Recovery System (SIDRA)/Municipal Agricultural Production platform [54]. In addition, the data from the Municipal Cattle Raising Survey for 1991, 2000, 2006, 2011, 2016, and 2017 were also used to analyze the impact of changes in LULC arising from cattle ranching. These databases are the only official agricultural data sources in Brazil [53–57].

2.3. Hydrometeorological and Sediment Yield Data

Several data were used, such as the daily data of maximum and minimum air temperatures (°C), incident solar radiation (MJ/m^2/day), wind speed (m/s), and relative air humidity (%) from the Pirenópolis and Goiás meteorological stations. These data are from 1971 to 1994 and were collected from the Meteorological Database for Teaching and Research platform [58]. Those data were used in the modeling to analyze the behavior of the hydrological processes within the basin. For the rainfall time series, daily data from five rain gauges from 1971 to 1994 were used: Jaraguá (ID #01549003), Uruana (ID #01549009), HPP Serra da Mesa Fazenda Cajupira (ID #01449005), Goianésia (ID #01549001), and HPP Serra da Mesa Ceres (ID #01549000) (Figure 1). In addition, streamflow data were acquired for the following stream stations: Colônia dos Americanos (ID #20490000) and Jaraguá (ID #20100000), for the period from 1974 to 1994. Rainfall and streamflow data were obtained from the website of the National Water Agency [59].

The validation of the SWAT model was performed by comparing calculated and observed sediment yield data. The estimated sediment yield (T_S), in ton/ha/year, was determined according to:

$$T_S = 0.0864 \times Q \times C_{SS} \quad (1)$$

where Q is the water discharge (m^3/s), and C_{SS} is the suspended sediment concentration (mg/L). After calculating the suspended sediment discharge for each measurement, the sediment rating curve for each station was then established. Two criteria were used to evaluate the sediment rating curve quality: (a) the first was that the R^2 value must be higher than 60%, and (b) the second involved a visual assessment of how closely the exponential form of the generated curve followed the measured points.

The annual sediment transported by the Almas River basin was calculated, taking into account the discharge curve and the daily water flow dataset, the latter of which was obtained from the National Water Agency [59]. To develop this curve, total solids in the water and the respective discharge were collected between 2000 and 2019 in the São Félix do Araguaia gauging station (code 26350000), located near the study area, more precisely between coordinates latitude 11°37′02″ S and longitude 50°40′10″ W. Measured C_{SS} data were collected, which generated a good correlation curve between the flow data and the measured suspended sediment. After 2019, data were not used because the monitoring at the gauging station was discontinued after this date. The relationship between T_S and observed discharge was obtained, which presented an R^2 greater than 0.95 (Figure 2). In addition, the results obtained were discussed and compared to other studies, i.e., [60–62].

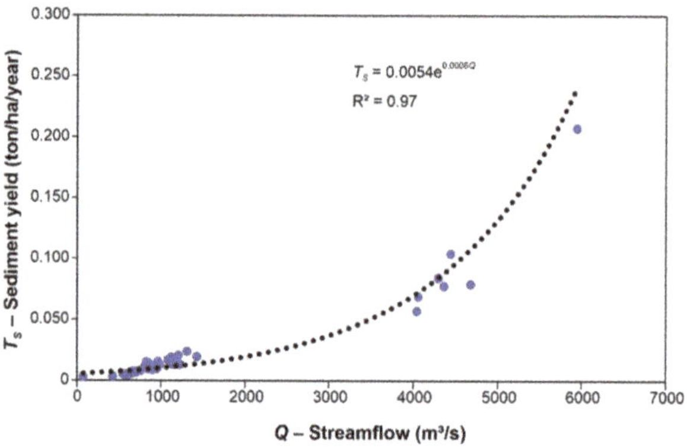

Figure 2. Correlation between annual sediment yield and water discharge.

2.4. SWAT Model

In the SWAT, the land phase of the streamflow process, the driving force behind the movement of sediments, nutrients, or pesticides, was examined. In the SWAT model, the water balance is based on the following equation:

$$SW_t = SW_0 + \sum_{i=1}^{t} \left(P_i - Q_i - E_{T_i} - R_i - Q_{G_i}\right) \tag{2}$$

where SW_t is the final soil water content (mm), SW_0 is the initial soil water content on day i (mm), t is the time (days), P is the rainfall depth for the day i (mm), Q is the amount of daily streamflow on day i (mm), E_T is the amount of evapotranspiration on day i (mm), R is the amount of water entering the vadose zone from the soil profile on day i (mm), and Q_G is the amount of return flow on day i (mm).

The streamflow was estimated using the Soil Conservation Service (SCS) curve number (CN) method. The amount of daily streamflow is given as:

$$Q = \frac{(R - I_a)^2}{(R - I_a + S)} \tag{3}$$

where I_a is the initial abstractions, including surface storage, interception, and infiltration prior to runoff (mm), and S is the retention parameter (mm). The retention parameter is defined as:

$$S = 25.4 \times \left(\frac{1000}{CN} - 10\right) \tag{4}$$

where CN is the applicable curve number for the day. The initial abstractions, I_a, is commonly approximated as $0.2 \times S$; hence, Equation (3) can be given as:

$$Q = \frac{(R - 0.2 \times S)^2}{(R - 8 \times S)} \tag{5}$$

The peak streamflow rate, which is the maximum runoff rate that occurs with a given rainfall event, is an indicator of the erosive power of a storm. It is used to calculate the sediment loss from the unit. SWAT calculates the peak runoff rate with a modified rational method, which is given as:

$$q_{peak} = \frac{C \times I \times A}{3.6} \quad (6)$$

where q_{peak} is the peak runoff rate (m^3/s), C is the runoff coefficient, I is the rainfall intensity (mm/h), A is the sub-catchment area (km^2), and 3.6 is a unit conversion factor from (mm/h) (km^2) to m^3/s.

The SWAT model uses the soil evaporation compensation factor (ESCO) to estimate the evaporation distribution better. The ESCO parameter must be between 0.01 and 1.0 and is used to adjust the depth distribution for evaporation from the soil to account for the effect of capillary action, crusting, and cracks. Calibrating this parameter is considered critical since it may vary from one catchment to another, even within the same geographical area. As the value for ESCO is reduced, the model can extract more of the evaporative demand from lower levels. ESCO coefficient is a calibration parameter and not a property that can be directly measured.

The SWAT model calculates sediment yield for each sub-basin using the Modified Universal Soil Loss Equation (MUSLE) [63]. MUSLE is a modified version of the Universal Soil Loss Equation (USLE) [64]. The MUSLE is given as:

$$S_Y = 11.8 \times (Q \times q_p \times A_h)^{0.56} \times K \times C \times P \times LS \times CFRG \quad (7)$$

where S_Y is the sediment yield on a given day (t), Q is the surface runoff volume (mm), q_p is the peak runoff rate (m^3/s), A_h is the area of the hydrologic response units (HRU) in ha, K is the soil erodibility factor (t·ha/MJ/mm), C is the USLE cover and management factor (dimensionless), P is the USLE support practice factor (dimensionless), LS is the USLE topographic factor, and $CFRG$ is the coarse fragmentation factor (dimensionless).

The SWAT allows simultaneous computations in each sub-basin and routes the water, sediment, and nutrients from the sub-basin outlets to the basin outlet. The routing model consists of two components, deposition and degradation, which operate simultaneously. The amount of sediment finally reaching the basin outlet, S_{out}, is given as:

$$S_{out} = S_{in} - S_d + D_t \quad (8)$$

where S_{in} is the sediment entering the last or final reach, S_d is the sediment deposited, and D_t is the total degradation. The total degradation is the sum of re-entrainment and bed degradation components, and it is given as:

$$D_T = (D_r + D_B) \times (1 - D_R) \quad (9)$$

where D_r is the sediment re-entrained, D_B is the bed material degradation component, and D_R is the sediment delivery ratio. Detailed theoretical documentation for the model is given by Neitsch [65]. More information about SWAT's equations can be founded in Arnold et al. [66], Silva et al. [67], Gassman et al. [68], and Neitsch et al. [69].

2.5. Application of the SWAT Model and Performance Indices

The Soil and Water Assessment Tool (SWAT) model [66] simulated the streamflow and sediment yield using different LULC scenarios for the Almas River basin. SWAT is a semi-distributed and continuous over time model that simulates the streamflow and sediment yield processes for long periods. The digital elevation model (DEM) used for the SWAT application was the Shuttle Radar Topographic Mission (SRTM) 1 Arc-Second Global, with a resolution of 30 m × 30 m. This DEM was used to determine the sub-basins (Figure 3a) and the slopes within the basin (Figure 3b). In this study, the LULC used in the modeling was obtained from Landsat 5/TM images (Figure 3c) path 222, and rows 070 and 071, downloaded from the USGS platform [70].

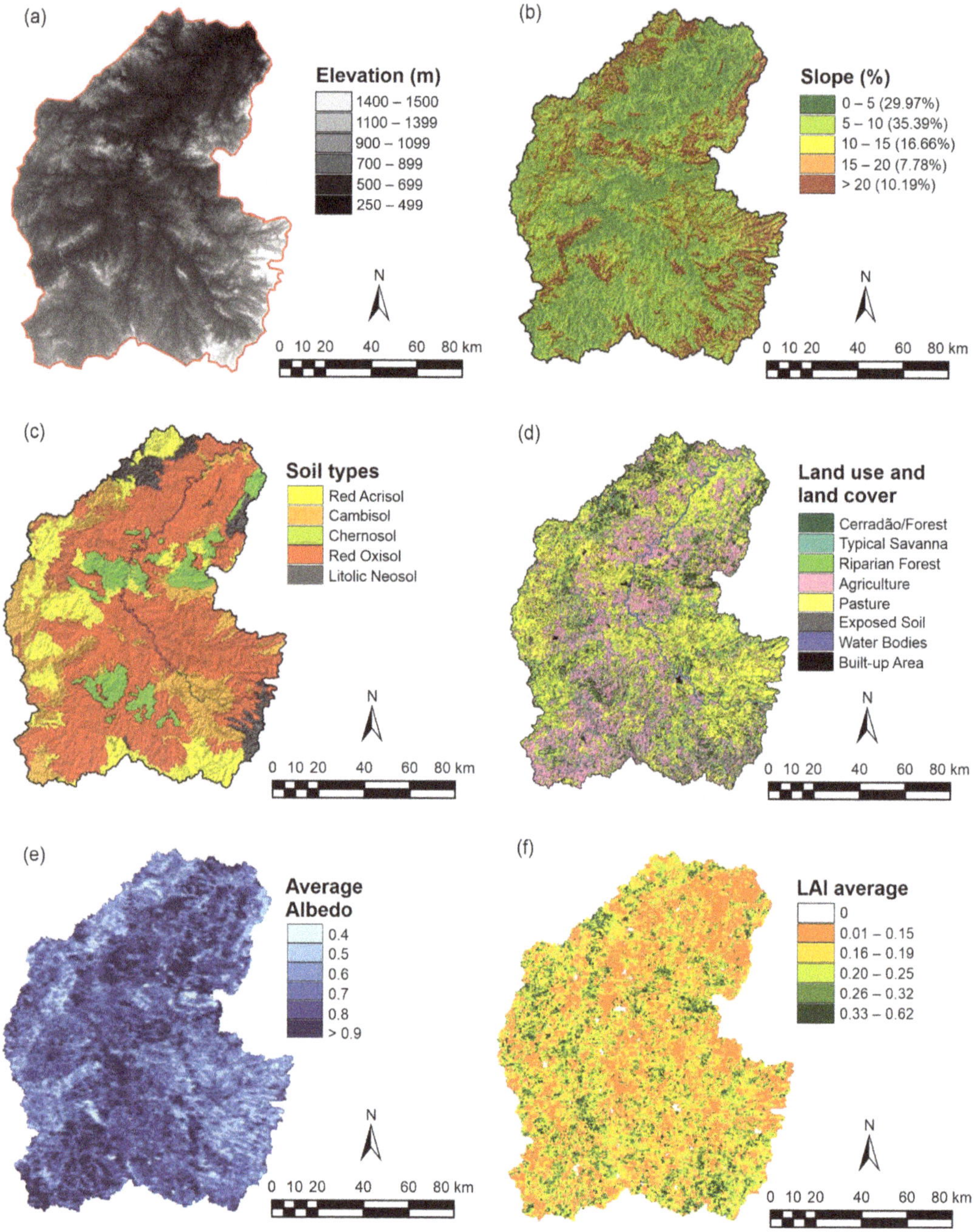

Figure 3. (**a**) Digital elevation model (DEM), (**b**) slopes, (**c**) soil types, (**d**) LULC, (**e**) albedo, and (**f**) LAI.

In this study, we chose to define the scenarios using Landsat image classification based on the research team's expertise in the chosen method and its knowledge of the study area. The classification validation process was based on the confusion matrix, using the user's accuracy, producer's accuracy, omission, and commission measures [71]. Fieldwork in the basin was carried out during research development when data on LULC were collected to check the errors and successes of the classification. The classified map was statistically tested with random validation samples collected from orbital imagery and samples verified in the field. An independent collection of points of each LULC class was used to validate the classified classes that remained unchanged in the analyzed image. A group of 800 samples was randomly selected after image fusion and checked in the field.

The accuracy statistics for the classification and image commission, and the omission results, showed that the accuracy ranged from 81.5% to 84.6%, and the kappa coefficient ranged from 86.6% to 89.9%. The overall kappa coefficient and overall accuracy calculated for the entire image were 89.3% and 79.7%, respectively. The results of commission and omission show that all classes had suitable adjustments in the classification. To analyze the accuracy of image classification, the kappa index was used. This test is a discrete multivariate measure of actual concordance minus the concordance due to chance [1,3] (i.e., it is a measure of the consistency between the classification and the reference data). The kappa index (κ) can be calculated by:

$$\kappa = \frac{D_o - D_e}{1 - D_e} \tag{10}$$

where D_o represents the accuracy of the observed classifications, and D_e represents the accuracy of the expected classifications.

The soil map (Figure 3d) used was on a 1:250,000 scale [72]. The albedo data were obtained from the MCD43A3 V6 Albedo Model dataset (Figure 3d), a product used daily for 16 days, with spatial resolution of 500 m, for the 2000–2018 period [73]. The LAI was obtained using the MCD15A3H V6 level 4, a product from a 4-day composite dataset with spatial resolution of 500 m (Figure 3e). For this product, the algorithm chooses the best pixel available from all the acquisitions of both MODIS sensors located on NASA's Terra and Aqua satellites within 4 days [74]. Albedo and LAI data were used for simulations using the SWAT model with grids at 500 m. All spatial bases were processed using ArcGIS 10.2® software.

2.6. Calibration, Validation, and Sensitivity Analysis

The Nash–Sutcliffe (NS) efficiency coefficient [75], the Pearson coefficient of determination (R^2), and the BIAS percentage (PBIAS) were used to evaluate the efficiency of the simulated data in the SWAT model. In addition, the performance of the calibration and validation results of the SWAT model was assessed based on the criteria recommended by Moriasi et al. [76]. These criteria establish guidelines for evaluating the model's performance by comparing observed and simulated values. A perfect simulation, which is unlikely to happen, would have NS = 1, R^2 = 1, and PBIAS = 0%. The calibrated parameters and initial intervals are summarized in Table 1.

Table 1. Parameters and ranges of variation used in the model calibration.

Parameter	Description	Inferior Limit		Upper Limit		Initial Simulation		Best Found Value		Adjusted Value		Method
		SWATd	SWATrs	SWATd	SWATrs	SWATd	SWATrs	SWATd	SWATrs	SWATd	SWATrs	
ALPHA_BF	Baseflow alpha factor (days)	0	0	1	1	0.048	0.048	0.7871	0.7871	0.7871	0.7871	–
CN2	Initial SCS runoff curve number for moisture condition II	−1	−1	100	100	79	79	−0.407249	−0.407249	47	47	×
SOL_K	Saturated hydraulic conductivity (mm/h)	−0.8	−0.8	100	100	2.3	2.3	15.49457	15.49457	38	38	×
ESCO	Soil evaporation compensation factor	0.5	0.5	1.0	1.0	0.95	0.95	0.7195	0.7195	0.7195	0.7195	–
GW_DELAY	Aquifer recharge time (days)	−30	−30	450	450	31	31	168.5042	168.5042	199.5042	199.5042	+

Table 1. *Cont.*

Parameter	Description	Inferior Limit		Upper Limit		Initial Simulation		Best Found Value		Adjusted Value		Method
		SWATd	SWATrs	SWATd	SWATrs	SWATd	SWATrs	SWATd	SWATrs	SWATd	SWATrs	
SURLAG	Delay coefficient of runoff (dimensionless). Smaller values represent greater delay in runoff	0	0	24	24	2	2	20.7109	20.7109	20.7109	20.7109	–
SOL_AWC	Available water capacity of the soil layer (mm water /mm soil)	−0.25	−0.25	1	1	0.18	0.18	0.189402	0.189402	0.218	0.218	×
CH_N2	Manning's n value for the main channel	0	0	0.3	0.3	0.014	0.014	0.176768	0.176768	0.176768	0.176768	–
GWQMN	Threshold depth of water in the shallow aquifer required for return flow to occur (mm water)	0	0	1000	1000	1000	1000	826.7252	826.7252	826.7252	826.7252	–
RCHRG_DP	Deep aquifer percolation fraction	0.1	0.1	1	1	0.05	0.05	0.767757	0.767757	0.088	0.088	×
GW_REVAP	Groundwater coefficient	0	0	0.2	0.2	0.02	0.02	0.1473	0.1473	0.1473	0.1473	–
CANMX	Maximum water storage in the vegetative canopy (mm)	0	0	100	100	0	0	52.6664	52.6664	52.6664	52.6664	–
SOL_ALB	The ratio of the amount of solar radiation reflected by a body to the amount incident upon its soil albedo	—	0.10	—	0.80	—	0.10	—	0.70	—	0.70	–
BLAI	Potential maximum of leaf area index for the plant	—	0	—	7	—	0	—	3	—	3	–

Values: substitution (−), addition (+), and multiplication (×).

The calibration of the SWAT model was performed using the observed streamflow data from the Jaraguá and Colônia dos Americanos streamflow stations for the period from 1 January 1974 to 31 December 1980. The period for the validation process was from 1 January 1985 to 31 December 1994. The SWAT model possesses many parameters that can be used; thus, the most sensitive parameters were initially analyzed during the calibration process. This procedure was possible using the SWAT calibration and uncertainty program—SWAT-CUP [77]. To determine the parameter values in the calibration and the uncertainty of hydrological modeling, the Sequential Uncertainty Fitting (SUFI-2) algorithm [78] was used. Two sensitivity analysis methods were performed (i.e., the Latin hypercube and the one-factor-at-a-time methods [77]). A sensitivity analysis was performed using these two methods, based on observed and simulated streamflow data. The percentage of measured data bracketed by the 95% prediction boundary (p-factor) was used to quantify all the uncertainties associated with the SWAT model [79]. In this study, a sensitivity analysis of the SWAT model parameters was performed using t-stat and p-value [61]. The t-stat was used to provide a sensitivity measurement, and the higher its value is, the more sensitive the parameter would be. After this step, 19 parameters were selected for further calibration.

The SWAT model was applied based on two datasets: (1) without RS data and (2) using RS data obtained using GEE. The RS data corresponded to soil albedo and LAI. These parameters are highly complex and challenging to obtain in the field. RS-estimated values can improve the calibration of physically-based models, such as the SWAT model, for ungauged or poorly gauged basins; thus, this study involves essential RS products obtained by the MODIS sensor using advanced techniques in the GEE environment. Both products and techniques used in this study are of great interest to the geo-information user community, which focuses on hydrological modeling. Finally, the data were downloaded and organized into the standard SWAT input format.

Landsat and SRTM data have the exact spatial resolution. They are imported directly into the SWAT model, which discretizes the basin into portions that possess unique land use/management/soil attributes, called HRU. The MODIS data, which have a spatial resolution different from the others, were treated and organized in a regular grid of 10 km. These datasets were imported into the SWAT model in tabular format, representing a regular mesh of stations. In this study, the sediment yield was divided into classes to represent better the spatialization of the results obtained. As described in Table 2, the data were classified to represent the spatialization of the sediment yield in the study area.

Table 2. Classes for sediment yield used in this study.

Number	Class	Sediment Yield (ton/ha/Year)
1	Very low	<0.01
2	Low	0.01–0.05
3	Moderate	0.06–0.10
4	High	0.11–0.15
5	Very high	0.16–0.20
6	Extremely high	>0.20

2.7. Recent Changes in LULC and Future LULC Scenarios

The years 1991, 2006, and 2017 were analyzed to assess changes in LULC. These years were selected because they contain dates with available images without clouds and with the most prolonged time interval to analyze changes in LULC for this basin. The LULC classification was performed using the maximum likelihood unsupervised classification method. The mappings used in this study were (a) LULC for 1991 (S1) using images from the TM/LANDSAT-5 sensor dated 13 June 1991, (b) LULC for 2006 (S2) using images of the TM/LANDSAT-5 sensor dated 13 June 2006, and (c) LULC for 2017 (S3) using images from the OLI/LANDSAT-8 sensor dated 13 June 2017. For the Almas River basin, the LULC identified were the following classes: cerradão/forest, typical Savanna, riparian forest, agriculture, pasture, and built-up area.

Two hypothetical LULC scenarios were proposed to evaluate the runoff–erosion processes in the basin, the (a) optimistic scenario (OS) and the (b) pessimistic scenario (PS). Based on the Brazilian forest code, the OS is considered the ideal LULC and was developed based on LULC S3 and the hypothetical recovery of permanent preservation areas close to watercourses, hilltops, and mountains. The future PS was simulated based on land use transformations that follow a historical trend in the basin, such as increased deforestation and growth in agricultural activities. The scenarios OS and PS were compared with observed and calibrated streamflows, the natural streamflow data measured at the streamflow stations, and simulated streamflow using the SWAT model based on the S1, OS, and PS scenarios. In addition, the OS maintained the existing native vegetation classes and estimated an increase in the remnant areas of the Cerrado biome (Cerradão/forest, typical Savanna, and riparian forest). The hypothetical PS is based on increased deforestation and the growth of agricultural activities, based on recent transformations of LULC that have taken place in recent years. The OS and PS scenarios were used as input data in the SWAT model, along with parameter values and meteorological data used in the calibration period. These simulations made it possible to compare the streamflow and sediment yield that occurred in these two scenarios with the simulations that took place in S1. Thus, the impacts of LULC changes on runoff–erosion processes are analyzed.

The different products used in this study aimed to provide the best historical representation of the analyzed processes. Unfortunately, the various datasets used do not have the same period. This limitation did not influence the methodology since the different products allowed for analyzing the phenomena separately. The integrated analysis of different products allowed a study in different stages: (a) LULCC, (b) simulation of LULC scenarios, (c) calibration and validation of the SWAT model with the longest existing time series, (d) validation of the sediment yield using the largest amount of data available, and (e) simulation of the runoff–erosion process in different LULC scenarios. Table 3 shows the period and source from which each product was obtained.

Table 3. Datasets, periods, and sources used in this study.

Dataset	Period	Source
Land use and land cover	1991–2017	https://earthexplorer.usgs.gov (accessed on 12 March 2021)
Agricultural production data	1991–2017	https://sidra.ibge.gov.br/ (accessed on 13 February 2021)
Weather data	1971–1994	http://www.inmet.gov.br/projetos/rede/pesquisa (accessed on 11 November 2021)
Hydrometeorological data	1971–1994	http://www.snirh.gov.br/hidroweb (accessed on 01 October 2021)
Albedo data	2000–2018	https://lpdaac.usgs.gov/products/mcd43a3v006 (accessed on 20 November 2021)
Leaf area index data	2000–2018	https://lpdaac.usgs.gov/products/mod15a2hv006 (accessed on 21 December 2021)
Total solid discharge	2000–2019	http://www.snirh.gov.br/hidroweb (accessed on 15 January 2022)

3. Results

3.1. Changes in LULC between 1991 and 2017

Figure 4a–c shows the spatial distribution of LULC in S1, S2, and S3. It is noticed that the agriculture and pasture classes occur in all portions of the basin and that in S3, there is an increase in these classes compared with S1 and S2. These classes predominate in the basin, and the agriculture class has constantly been increasing, whereas the pasture class showed a small oscillation. The results show that in S1, the agriculture and pasture classes accounted for 56.41% of the basin area. An advance can be seen in the agriculture and pasture classes, which influenced the growth of agriculture in the basin, mainly due to the increase in sugarcane being planted and cattle being raised in the area. Food crops of corn, beans, and rice were of lesser importance in the agricultural class of the region [20,21].

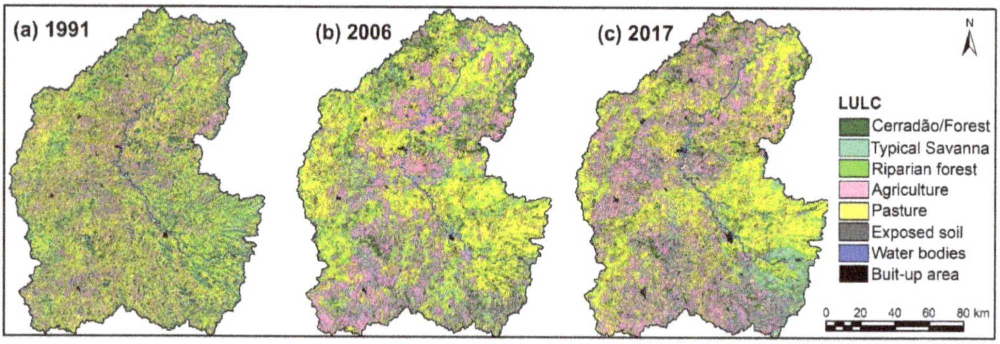

Figure 4. Land use and cover mapping of the Almas River basin in (**a**) 1991, (**b**) 2006, and (**c**) 2017.

The results show that in S3, the agriculture and pasture classes occupy approximately 71% of the basin area. In comparison, the Cerradão/forest and typical Savanna classes occupy 21%, whereas the other classes occupy only 8% of the basin's total area (Table 4). It should be noted that the pasture areas have expanded over the gently undulating and moderately undulating relief areas.

Table 4. Classified area and temporal variation of LULC for the Almas River basin.

LULC	S1 (1991)		S2 (2006)		Variation S1–S2 (%)	S3 (2017)		Variation S1–S3 (%)
	Area (km^2)	Area (%)	Area (km^2)	Area (%)		Area (km^2)	Area (%)	
Pasture	5427.32	28.8	6364.46	33.8	17.27	6791.82	36.1	25.14
Agriculture	3620.07	19.2	5790.45	30.7	59.95	6348.93	33.7	75.38
Cerradão/Forest	2366.42	12.6	2028.36	10.8	−14.29	2008.49	10.7	−15.13
Typical Savanna	1386.46	7.4	1936.24	10.3	39.65	1937.62	10.3	39.75
Riparian Forest	1566.73	8.3	1254.00	6.7	−19.96	956.39	5.1	−38.96
Exposed soil	1667.71	8.9	648.65	3.4	−61.11	402	2.1	−75.90
Water bodies	95.80	0.5	112.24	0.6	17.16	104.94	0.6	9.54
Urban area	40.61	0.2	55.78	0.3	37.36	86.64	0.5	113.35

The results show that in S1, there was a decline in corn, bean, and rice areas, and a steady decline from S2 onwards. In the period analyzed, the planted area of sorghum remained practically unchanged due to it being used in conjunction with soybean and corn. The relationships concerning the runoff–erosion process are similar to those obtained in various studies [22,23], which analyzed the streamflow in river basins in the Cerrado biome. The results show that uncertainties were high because the basin has a significant heterogeneity of soil physical parameters; however, the statistical results obtained show a good fit between the observed and estimated data in the basin.

The transformations in the basin landscape between S1 and S3 occurred with the association of the crop–pasture system (i.e., a production system that prioritized some commodity crops, such as soybeans and sugarcane). In addition, it strengthened pasture areas that are cultivated with incorporated agriculture, which made the use of pasture in confinement more profitable. The increase in cropland reduced occupied areas with exposed soil, riparian forest, and cerradão/forest by −76%, −38.96%, and −15.13%, respectively (Table 4). The results show that the agriculture and pasture classes significantly increased (75.38% and 25.14%, respectively), while the typical savanna class increased by 39.75% (Table 4). The classes of native vegetation (Cerradão/forest, typical Savanna, and riparian forest) decreased by 7.8%. The change between S1 and S3 represented the scenario of the advance of agriculture in the region. The comparison between S1 and S3 is helpful to understand the total changes over the entire period studied.

3.2. Advances in Agriculture between 1991 and 2017

Figure 5a–c shows the number of cattle in S1, S2, and S3 for each municipality within the basin. The results show the municipalities that have the highest concentration of cattle within the basin, which are Pirenópolis, Goiás, Goianésia, Itapuranga, Itaberaí, Uruaçu, Jaraguá, and Barro Alto. Together, these municipalities have a total of 1,095,085 cattle (58.8% of the basin's total). The results show that these municipalities grew by 18.6% during the study period, and that there was a more significant increase in the number of cattle in the municipalities in the southern portion of the basin (Figure 5c). It can also be highlighted that the municipalities of Santa Rosa de Goiás, Petrolina de Goiás, and Pilar de Goiás had a growth rate of 113.84%, 102.02%, and 101.66%, respectively. The cattle herd has expanded the number of cattle over the years, and the results show that the livestock area increased by 31% in the period. Figure 5d shows the areas planted with temporary crops and livestock between 1991 and 2017. As can be seen, the planted area data shows that soybean and sugarcane crops grew by 287% and 650%, respectively, whereas corn, bean, and rice crops showed a more significant decrease in the period analyzed. After 2000, the area planted with soybean crops predominated in the basin; therefore, such areas did not suffer reductions, even with the drop in the price of soybeans on the international market and the climate variations between 2006 and 2017. The same happened with the area planted with sugarcane because this crop had an appreciation in the international market during the same period. In addition, with tax incentives from the Government of Goiás

State, new sugar and alcohol plants were reactivated and built, and improvements in the region's agro-industrial complex enabled the expansion of sugarcane within the basin [37].

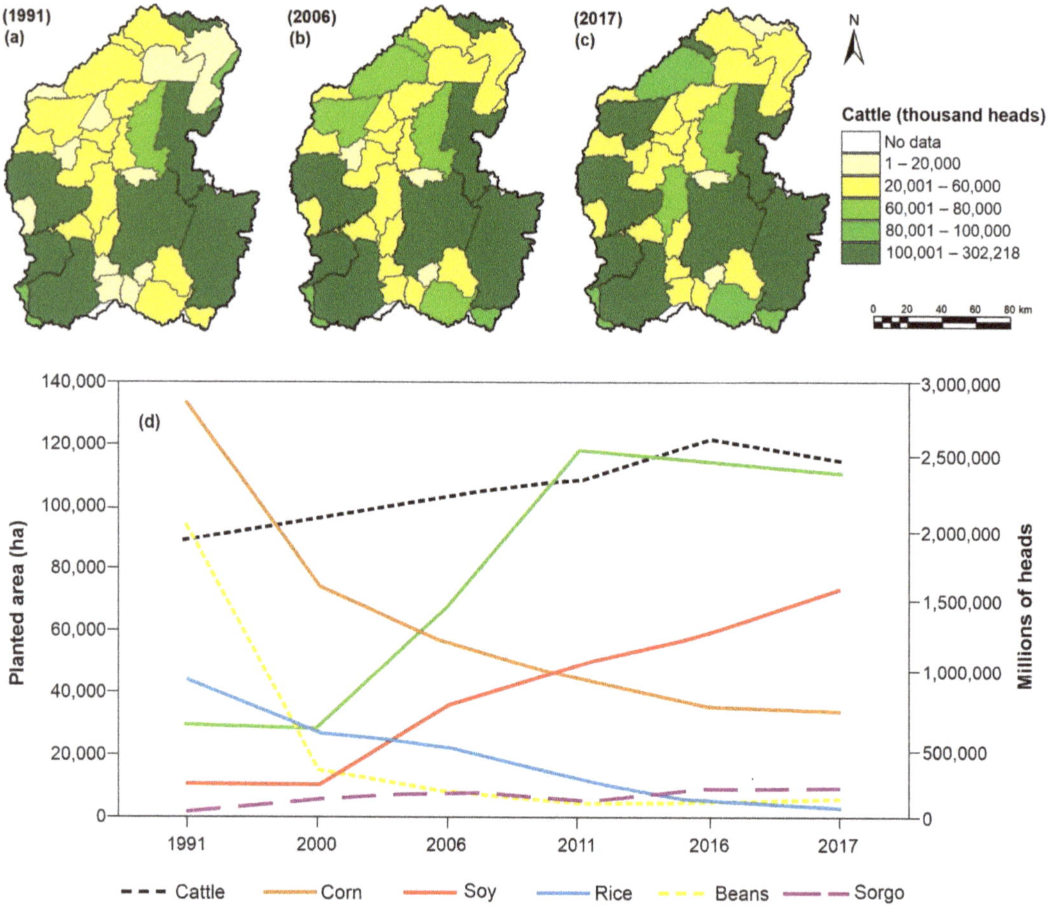

Figure 5. The number of cattle per municipality in (**a**) 1991, (**b**) 2006, (**c**) 2017, and (**d**) planted area of temporary crops and livestock between 1991 and 2017.

Figure 6a–f shows the spatial variation of the area planted with rice, sugarcane, beans, corn, soybean, and sorghum between 1991 and 2017 for each municipality within the basin. Figure 6a shows that most municipalities showed a decrease in the area planted with rice, and it is notable that the area planted with rice decreased by 96.9% between 1991 and 2017. The sugarcane planted area data show a significant increase in the period analyzed, and the incorporation of more municipalities in sugarcane production (28,972 ha in 1991 and 111,681 ha in 2017) represents an increase of approximately four times the area planted in the basin (Figure 6b). The results show that the area planted with beans within the basin showed a reduction of 95.4% (Figure 6c).

Figure 6d shows the spatiotemporal variation of the area planted with corn between 1991 and 2017. The results show a reduction of 75.4% in the period analyzed, which was similar to the cultivation of beans. On the other hand, the area planted with soybeans (Figure 6e) showed a significant increase (694.5%). Figure 6f shows the temporal variation from 1991 to 2017 in the area planted with sorghum. The results show an increase in the area with sorghum. The crop was spread in different portions of the basin (north and south), whereas the other areas did not show crop production, or they kept their planted area unchanged in the period analyzed.

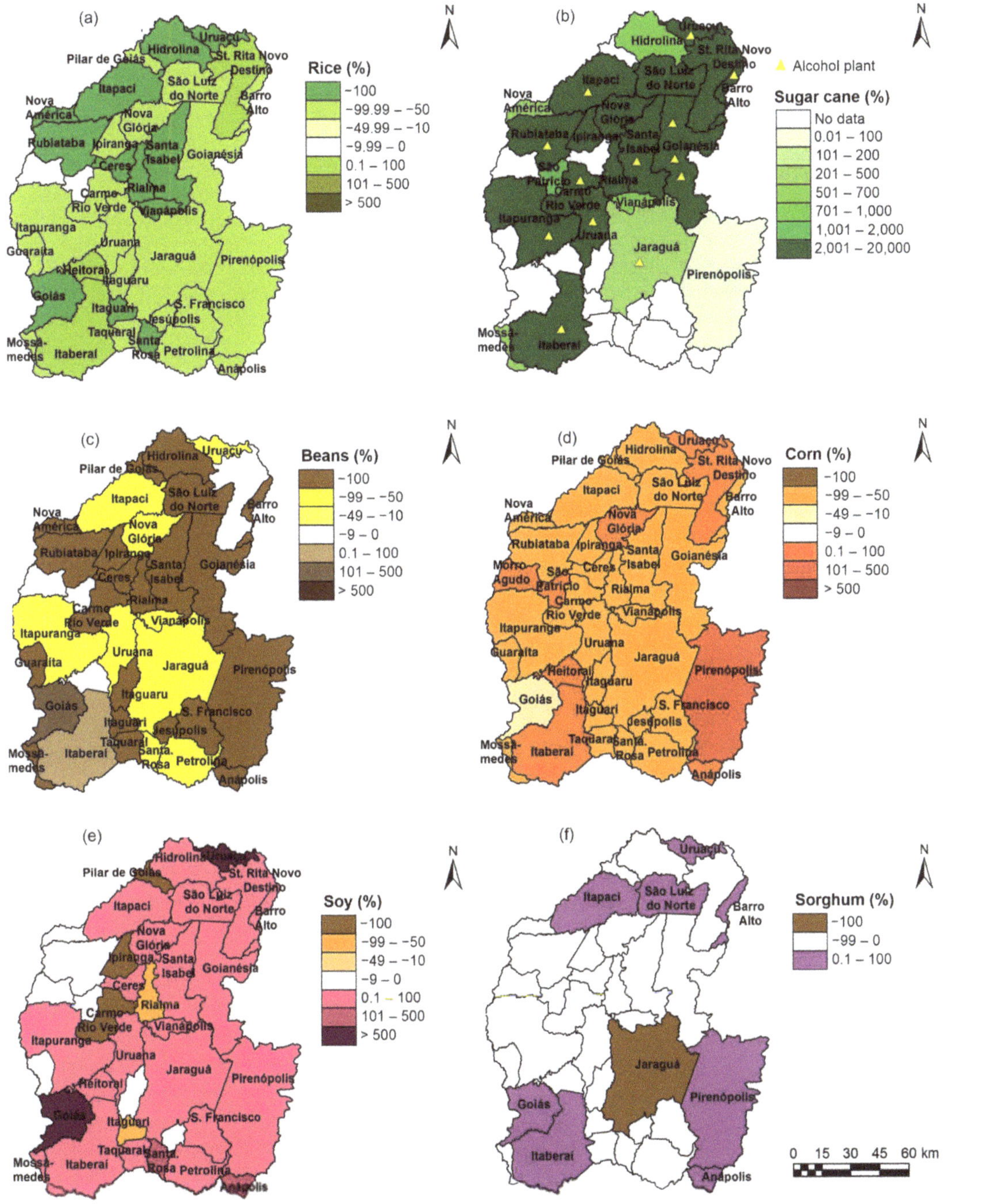

Figure 6. Spatial variation of (**a**) rice, (**b**) sugarcane and location of alcohol plants, (**c**) beans, (**d**) corn, (**e**) soybean, and (**f**) sorghum crops per cultivated area between 1991 and 2017.

3.3. Runoff–Erosion Modeling

3.3.1. Sensitivity Analysis

The results of the sensitivity of the best parameters assigned by SWAT-CUP are shown in Figure 7. The most sensitive parameters based on the *p*-value are grouped in descending order according to their greatest significance (i.e., closer to one). On the other hand, the t-stat is used to identify the relative significance of each parameter, estimating how changes in the value of a given parameter influence the results of the objective function. These two tests are used to analyze the sensitivity of the modeling (i.e., how the uncertainty in the modeling results can be attributed to different parameters that deal with the behavior of water in the basin, in such a way that it considers the entire amplitude of variation in the input data). Thus, the higher the *p*-value and the lower the t-stat value, the greater the sensitivity of the parameter in the modeling; therefore, it is not possible to group them by category. According to this figure, the most sensitive parameters in the modeling were CN2 and SOL_K. The other parameters that were also sensitive in the streamflow simulation were GWQMN, SOL_AWC, RCHRG_DP, GW-DELAY, SURLAG, CAMIX, CH_N2, ALFA_BF, GW_REVAP, and ESCO. The results show that the parameters which were considered more sensitive and influential for streamflow calibration are related to streamflow (CN2, SURLAG, and CH_N2), evapotranspiration (ESCO and CAMIX), soil water, and soil physical characteristics (SOL_K and SOL_AWC). It should also be noted that the groundwater parameters (ALFA_BF, GWQMN, GW_REVAP, GW_DELAY, and RCHRG_DP) were relevant in the modeling.

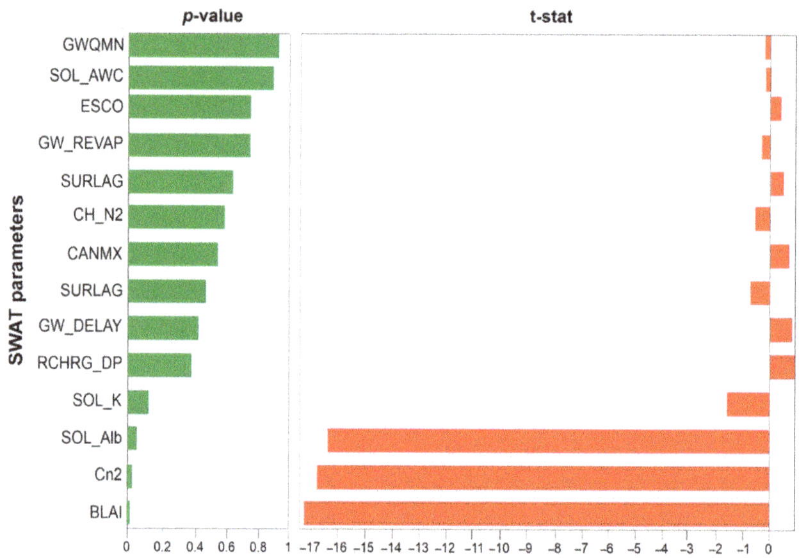

Figure 7. Sensitivity analysis of the SWAT model parameters used in the modeling for the Almas River basin.

3.3.2. Calibration and Validation

Figure 8a,b shows the observed and simulated streamflow time series after model calibration for the Jaraguá and Colonia dos Americanos stations. The calibration results for the Jaraguá station showed a satisfactory fit between the observed and simulated monthly streamflow (R^2 = 0.8 and NS = 0.61) and in the validation period (R^2 = 0.76 and NS = 0.5). The annual average of the observed streamflow was 35.44 m^3/s, whereas the simulated streamflow was 41.48 m^3/s, a difference of 17%. The PBIAS value for the Jaraguá station in the calibration period was −20.3% and −28.5% in the validation, indicating an overestimation bias.

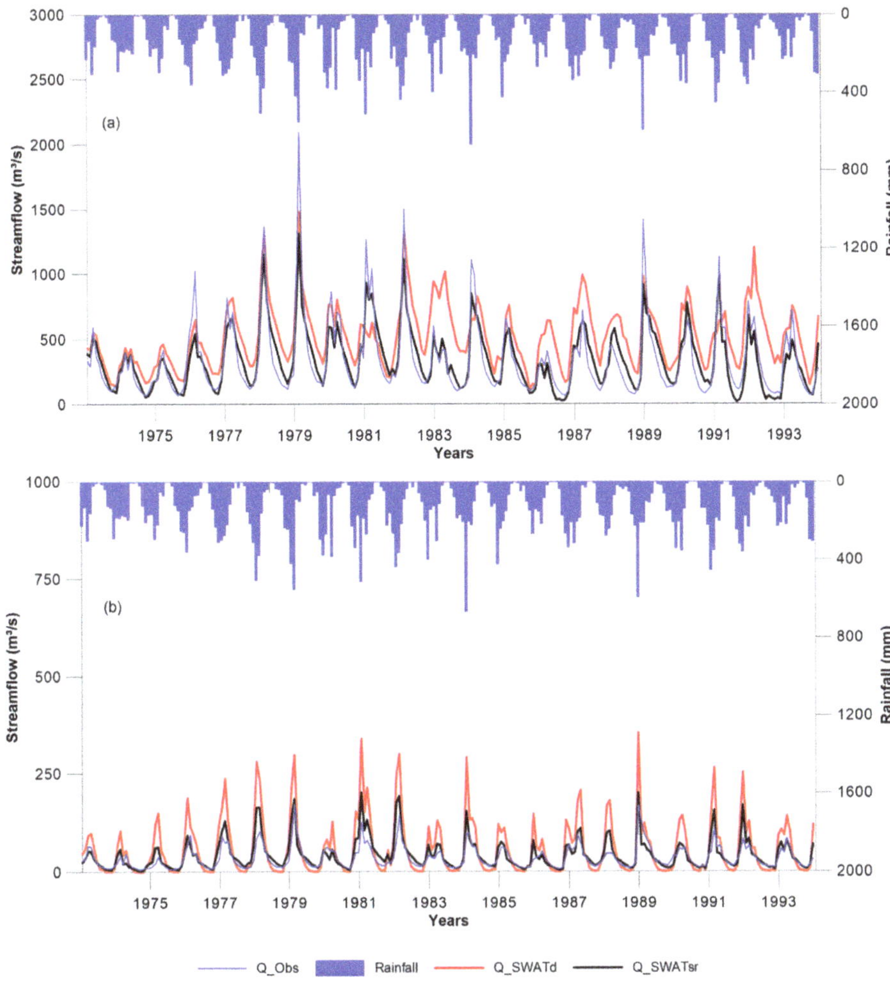

Figure 8. Comparison of calibration and flow validation in the SWAT model for (**a**) Colonia dos Americanos and (**b**) Jaraguá stations.

For the Colonia dos Americanos station, the results presented a very good performance in the calibration and validation, presenting R^2 = 0.85, NS = 0.82, and PBIAS = 0.9% for the calibration, whereas in the validation period, the values were R^2 = 0.84, NS = 0.80, and PBIAS = −15.5%. The Colonia dos Americanos station results also showed an overestimation bias to the observed values. The average observed streamflow was 337.80 m^3/s, and the simulated streamflow was 360.60 m^3/s, an increase of 6.74%, which can be considered low between the measured and simulated streamflow.

3.4. Estimate Sediment Yield

Figure 9a shows the spatial distribution of sediment yield in the Almas River basin between 1974 and 1994. Figure 9a also shows that the sediment yield is very variable and that the most significant amount of sediment occurs in the elevated regions, which are moderately wavy. The results show that the sediment yield in the sub-basins varied between 0.01 and 0.2 ton/ha/year. It is notable that the most significant volume of sediment occurred in areas with agriculture, pasture, exposed soil, and types of cambisols and red clay soils located in the eastern portion of the basin. In contrast, the smallest volumes of sediment occurred in areas with natural vegetation cover. The basin areas with agriculture, pasture, exposed soil, and red oxisol type soils, with slopes varying between 0% and 5%, had sediment yields between 0.01 and 0.12 ton/ha/year.

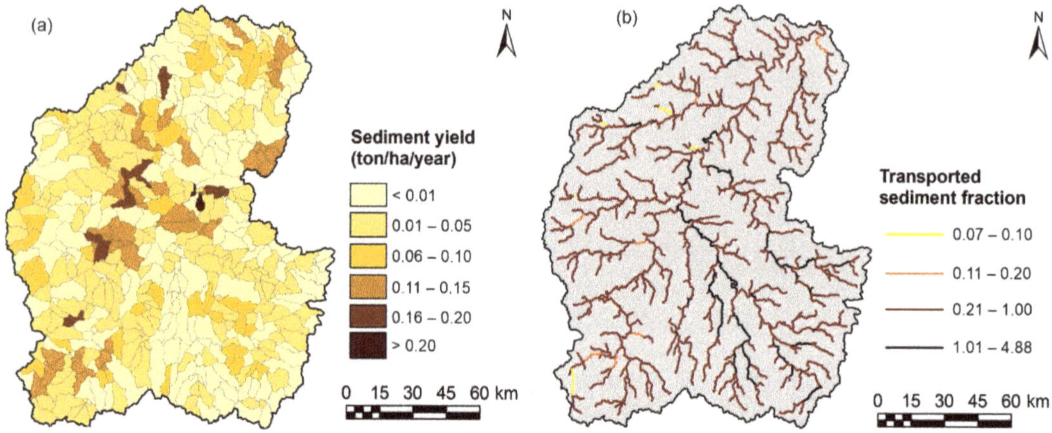

Figure 9. (**a**) Spatial distribution of sediment yield per sub-basins, and (**b**) sediment fraction transported per stretch between 1974 and 1994.

Figure 9b shows the sediment fraction that each segment of the drainage network transports to the subsequent channel stretch. Again, the pattern of the sediment fraction of river stretches can be seen located in the upper and middle portions of the basin, which show more significant sediment deposition. In contrast, the sub-basins close to the basin boundary had little or no sediment deposition.

Table 5 shows statistics of sediment yield and estimation errors between observed and calculated data. The results highlight that the calculated sediment yield underestimated the observed data by 22.42%. The curve fitting for the relationship between sediment yield and observed discharge presented R^2 equal to 0.97. This relationship can be considered very good due to the uncertainties in estimating sediments in rivers with a large volume of suspended sediments and non-continuous data collection. Furthermore, as the station used to measure the observed data is downstream of the Almas River basin (i.e., it has a water catchment area more extensive than the studied basin), it was expected that the data collected in situ would present an overestimation. This sediment measurement station is close to the basin outlet and past the Serra da Mesa hydroelectric plant; thus, the contribution area chosen as the study area comprises the catchment area up to the hydroelectric plant and the results of the SWAT model can be considered satisfactory.

Table 5. Observed and calculated sediment yield and estimation errors for the study area.

Years	Sediment Yield (ton/ha/Year)		
	Observed	Calculated	Estimation Error (%)
Average	0.032	0.025	−21.88
Standard deviation	0.041	0.016	
Mean deviation	0.031	0.013	
Coefficient of variation	1.288	0.646	

3.5. Hypothetical Land Use Scenarios and Simulation of Runoff–Erosion Processes

Table 6 and Figure 10a,b show the classes of hypothetical scenarios of optimistic and pessimistic LULC. The OS simulation results showed that the natural vegetation classes represent 63% of the basin's total area (i.e., 23.25% Cerradão/forest, 24.77% typical Savanna, and 14.32% riparian forests). The agriculture, pasture, and urban area classes represent 17.12%, 20.21%, and 0.34% of the basin's total area, respectively. The spatial distribution of the simulated LULC for the PS is shown in Figure 10b. The PS shows an intense change in land use, in which the natural vegetation was entirely replaced by agriculture (70.07%), pasture (6.34%), and urban area (0.46%), which represents 76.86% of the total area of the basin. The other LULC occupied 23.14% of the basin area (i.e., a reduction rate of 62.88% for the OS scenario). Changes in the PS indicate less protection against the direct impact of rainwater drops in the soil, which favors the runoff and the detachment and transport of sediment particles.

Table 6. Land use class data for the two hypothetical land-use scenarios, simulated with the SWAT model.

LULC	Optimistic LULC (OS)		Pessimistic LULC (PS)	
	Area (km^2)	Area (%)	Area (km^2)	Area (%)
Typical Savanna	4615.64	24.77	4312.03	23.14
Cerradão/Forest	4333.38	23.25	–	–
Pasture	3765.72	20.21	1181.39	6.34
Agriculture	3190.74	17.12	13,058.11	70.07
Riparian forest	2668.76	14.33	–	–
Urban area	62.59	0.34	85.31	0.46

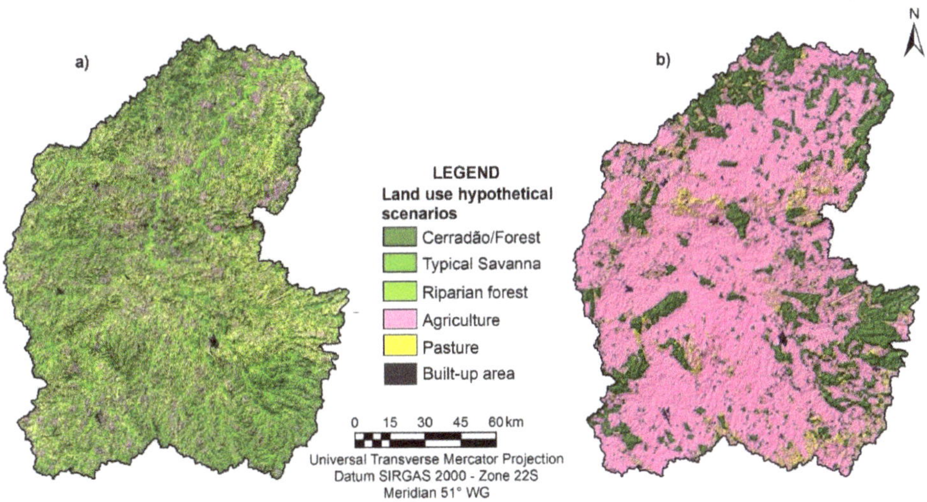

Figure 10. Proposed hypothetical scenarios: (**a**) OS and (**b**) PS.

To assess the efficiency of the SWAT model, we compared simulated monthly average streamflow data based on hypothetical scenarios with the observed data. Figure 11 shows the SWAT model simulations for the OS and PS scenarios for the Colonia dos Americanos stations. Table 7 presents the comparison of the observed, calibrated, and simulated streamflows for the OS and PS scenarios. The results of the monthly average streamflow for the OS showed that the simulated streamflow was 337.80 m^3/s and the observed value was 455.88 m^3/s, a difference of −25.9% for the Jaraguá station. For the Colonia dos Americanos station, the simulated streamflow was 504.17 m^3/s and the observed value was 514.17 m^3/s, a difference of −1.9%. The comparison between the observed and simulated streamflow using the PS at the Jaraguá station show that the streamflow was 500.44 m^3/s, presenting an increase of approximately 9.8%. For the Colonia dos Americanos station, the simulated streamflow was 547.23 m^3/s (i.e., a difference of 6.4%). The obtained coefficients of determination confirmed that the simulated flows accurately replicated the measured flows in the present research. The statistical indicators of the calibration and validation phases show the satisfactory performance of the model throughout the basin, mainly upstream and downstream. The obtained results follow other studies, which emphasize the good performance of the SWAT model in modeling Cerrado basins (e.g., [21,27,28,30,31,71]); therefore, the impacts of LULCC on the streamflow time series can be assessed using the calibrated and validated SWAT model.

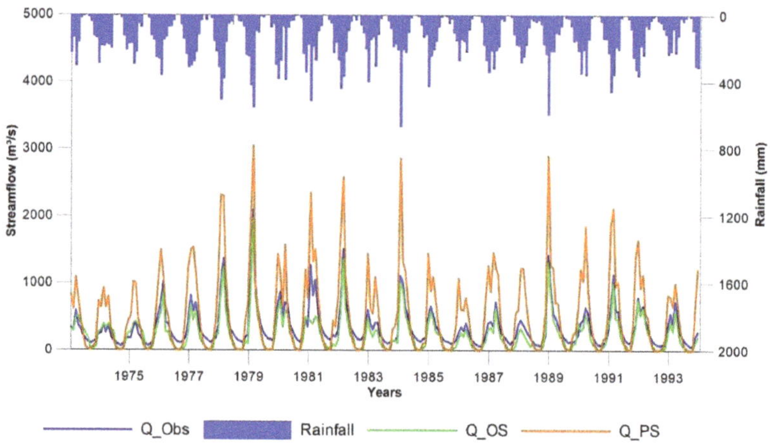

Figure 11. Comparison of observed streamflow and pessimistic and optimistic scenarios simulated using a SWAT model for Colonia dos Americanos station.

Table 7. Statistical comparison of mean streamflow and sediment yield for S1, OS, and PS scenarios.

Statistics	Rainfall (mm)	Streamflow (m^3/s)			Sediment Yield (ton/ha/Year)		
		S1	OS	PS	S1	OS	PS
Mean	1612.23	514.17	504.17	547.23	0.026	0.023	0.035
Maximum	2245.62	766.42	622.64	839.94	0.049	0.039	0.063
Minimum	1075.84	323.84	297.96	359.21	0.010	0.011	0.020
Standard deviation	285.70	133.66	104.56	140.74	0.011	0.007	0.011

Figure 12 compares annual differences in sediment yield for scenarios S1, OS, and PS. As can be seen, there is a significant discrepancy between the results of the three simulations. The results show an intensification of soil erosion when comparing the PS and OS, by approximately 54%. The results also show that the sediment yield in the OS scenario decreased by around 11% compared to the S1 scenario. LULC variation in the OS seems to cause less erosion and much more deposition than the S1 and PS variation. Comparing the S1 scenario with the OS, it is expected that native vegetation can reduce erosion, as it directly changes the infiltration parameters and especially the protection given to the soil against the direct impact of raindrops and increased surface roughness, as reported by [60].

Figure 13 shows the spatial distribution of sub-basin sediment yield classification in hypothetical optimistic (Figure 13a) and pessimistic (Figure 13b) land use scenarios. The results show that both LULC scenarios significantly influence the sediment yield since there are mainly occurrences of very low and low classes in the OS compared to PS. The total impact of the PS on the increase in the sediment yield is much more significant than in other scenarios; however, the differences comparing the OS and PS are not linear, mainly due to the different geographic locations occupied by native vegetation in both scenarios. As in OS, the LULC that predominated in steep areas was native forest vegetation, which acted as a barrier to sediments and areas with very low sediment yields. When the runoff passes from a sugarcane area to a native vegetation area, the flow velocity decreases because of the high surface roughness due to the vegetation. This characteristic of LULC reduces the sediment transport capacity, preventing them from reaching the drainage network due to the early sediment deposition. Considering that the riparian forest in the Almas River basin is in a good state of preservation, we can highlight how this LULC acts as a protective barrier to sediments, mainly reducing the flow rate and retaining sediments.

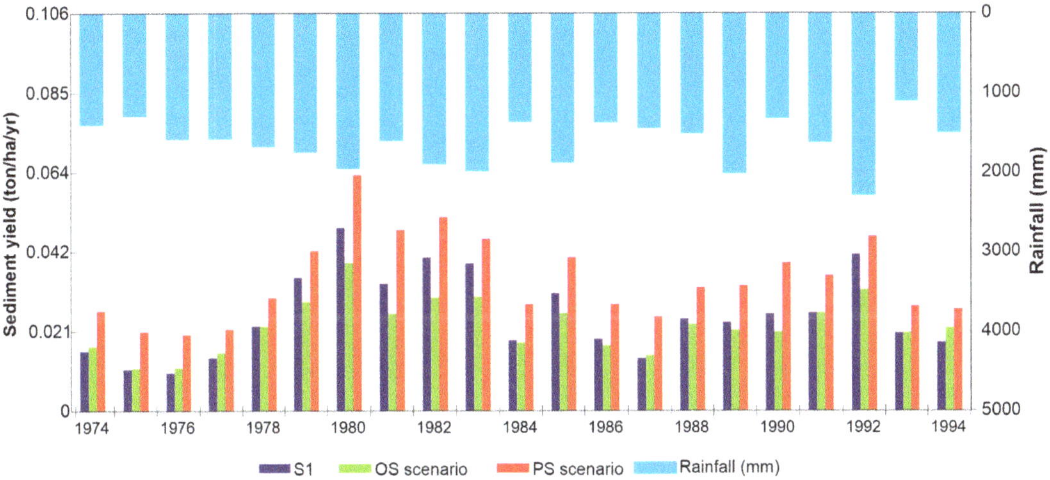

Figure 12. Comparison between estimated mean annual sediment yield for S1, OS, PS, and mean annual rainfall.

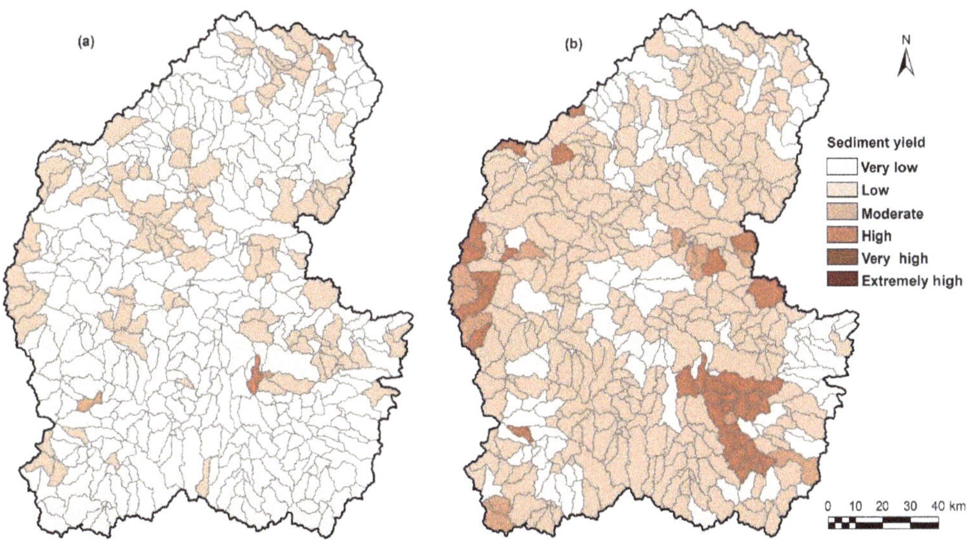

Figure 13. Classification of sub-basin sediment yield losses for the hypothetical scenarios (**a**) OS and (**b**) PS.

The results show that in the OS, there was a predominance of sub-basins classified as very low risk to erosion (88% of the basin), which can be due to the hypothetical regeneration of vegetation cover and linear corridors of riparian forest, which acted as a physical barrier to retain sediment from the slopes. The agricultural class occupied the sub-basins with greater erosion susceptibility with cambisol soils and slopes ranging from 0 to 5%. The erosion results in the PS were very variable, and the moderate erosion risk class was the one with the most significant predominance (71%). The results for the PS scenario showed that 50 sub-basins (10.3% of the basin area) presented sediment yield in the moderate class (i.e., such sub-basins are areas susceptible to the erosion process). When comparing the areas susceptible to sediment yield among the hypothetical scenarios, the OS had a significant increase of 1725.21% compared to the OS. The hypothetical simulated land use scenarios had significant differences. The optimistic scenario presented a very low to low risk of sheet erosion. In contrast, the other pessimistic scenario was unfavorable, presenting a high risk of erosion susceptibility and a high predisposition for the sediments to be transported to the drainage channels. These results highlight the importance of land cover in protecting the soil against erosion processes.

4. Discussion

Knowing the influence that changes in the LULC can have on the quantity and quality of sediments, and how streamflow can affect energy generation, ecosystems in the basin, and impact freshwater availability for human consumption and agro-industrial production, changes in the LULC influence streamflow and sediment yield behavior, as demonstrated in the simulations of the two LULC scenarios. The study highlighted that the increase in agricultural and pasture areas and the decrease in native vegetation cover caused severe environmental impacts, reinforcing the need to manage the LULC at a basin-scale in a biome such as the Cerrado. This methodology can be tested in ungauged or data-poor watersheds as it uses freely available datasets and consolidated and widely used methods. In addition, the applicability of this study allows the simulation of LULC future scenarios at a low cost, and it gives an estimation of streamflow and sediment yield time series.

The results of this study can help decision makers understand the changes to the landscape in recent decades and allow them to make future predictions about public policies for environmental preservation or the liberation of areas from pasture or agricultural activities. As a result, the OS and PS scenarios were proposed, and the streamflow and sediment yield behavior results were analyzed.

The calibration and validation results show that the LULCC in this region severely influence the streamflow pattern. The results of this modeling are similar to the results obtained in the Cerrado area by [29,30,32,33]. As expected, the sediment yield and streamflow results show that the highest values occurred in the PS, whereas a significant decrease was observed in the OS, considering the S1 scenario. The average annual sediment yield for the OS was 0.023 ton/ha/year, whereas for the PS, it was 0.035 ton/ha/year, representing a difference of 21.88% (Table 5). These results show that LULC greatly influences runoff–erosion processes in the region [67]. Vegetation cover plays a fundamental role in water conservation and supply, nutrient cycling, soil protection against erosion, temperature regulation, water cycling, and returning water to the atmosphere by evapotranspiration. For this reason, one may say that deforestation and LULCC are two of the world's leading environmental concerns, especially in Brazil, which is currently the country that devastates its native vegetation most (e.g., the Cerrado biome).

The estimate of sediment yield in S1 shows a reduction of 10.96% in the OS, and when compared to the PS, it shows an increase of 37.4% (Table 5). These results highlight the influence of LULC as one of the main controlling factors of hydrological processes, as it was possible to compare the results of streamflow and sediment yield with the same amount of rainfall but with different conditions of LULC.

Changes in areas of the pasture class by native vegetation reduced the erosion process. According to Falcão [12], grazing under adequate conditions usually does not increase sediment in water bodies after heavy rains. Nevertheless, intensive grazing on sloping terrain and fragile soils can cause severe erosion problems. In addition, according to the authors, the sediment yield increases when riparian areas are used as pasture, which leads to erosion of the riverbanks and deposition directly on the bed. There is still no accurate data on erosion from cultivated areas in Brazil. According to USDA [80], for instance, in the United States, erosion generated in cultivated areas is approximately 38%, while pasture erosion accounts for 26% of the sediments reaching water bodies. According to Santos et al. [61] surface roughness is the main factor in reducing surface streamflow, and consequently, the sediment yield.

The hypothetical land use scenarios of the Almas River basin alerted possible future situations in a river basin for issues related to runoff–erosion processes. Taking rigorous measures to preserve the vegetation cover and reforestation implies reducing environmental impacts and sediment yield within the basin. In this context, the methodology adopted to generate these hypothetical scenarios allowed us to satisfactorily show that the hydrological processes associated with land use and management play a fundamental role in understanding the water and sediment yield within the river basin.

Despite the SWAT model's many qualities, its limitations must be further discussed and analyzed. The SWAT model was developed for rural watersheds, and therefore, there is a need for parameter calibration; thus, identifying the parameters that have or do not have a significant influence on the model simulation is fundamental not only to reducing the modeling uncertainty but also to reduce the number of excessive parameters in the model calibration process, which can harm the physical representation of the basin in the model. In this regard, please see [61,67], which provide more details on SWAT's capabilities and limitations.

5. Conclusions

This study evaluated the impacts of historical LULCC on hydrological processes using the SWAT model and remote sensing multiple gridded datasets for a humid tropical basin in the Cerrado biome in Brazil. With the calibration of the SWAT model, it was possible to observe that some parameters are more influenced by the runoff–erosion process than others, providing the conditions to improve the simulation in the basin. After validation, the hydrological simulation satisfactorily represented the streamflow variability and the estimated sediment yield during the period studied. The temporal evolution of the changes in the LULC increased the mean streamflow and the sediment yield. This study highlighted that the LULCC in the study area play an essential role in the runoff–erosion process in the Cerrado biome, and consequently, impacts various human activities such as agribusiness, livestock, energy production, food security, and public water supply. The purpose of the study was to simulate the influence of LULCC on the amount of streamflow and sediment yield in the basin in different scenarios. Nevertheless, the water quality in the basin was not analyzed due to the methodology tested. The results alert decision makers about the importance of proper LULC management in the streamflow and sediment yield in the basin.

This study discussed the LULCC due to agricultural advances that caused a shift in the runoff–erosion dynamics, exploring the applicability of remote sensing in an ungauged basin in the Cerrado biome in Brazil that underwent intense modification in LULC. The analysis of the LULCC for 1991, 2006, and 2017, and the agricultural census data, allowed us to understand the reconfiguration of the basin's landscape over the twenty-six years, which proved to be fast and progressive in the process of expansion of the economic activity. This complexity involves replacing food grains (rice and beans) to incorporate crops in an area planted with sugarcane and soy and the expansion of cattle ranching. Reconciling the pressure of agribusiness with the preservation of natural areas is a challenge for environmental planning and management of water resources. The changes in LULC and deforestation interfere with the hydrological cycle, causing a reduction in water infiltration into the soil and increasing the streamflow, which affects the fluvial dynamics and erosion process. In addition, this paper demonstrated how LULC, soil parameters, albedo, and LAI obtained from RS datasets could successfully calibrate distributed hydrological models such as the SWAT model. This research showed that the influence of LULC on the runoff–erosion process using estimated satellite data and runoff–erosion models in the Cerrado biome is still scarce in Brazil. We can conclude that the current simulations are classified as good according to Moriasi [76].

The runoff–erosion modeling allowed us to understand the runoff–erosion process, helping the future planning and territorial management of water resources in this basin. This modeling also helps define public policies to control deforestation and preserve, maintain, and recover the Cerrado biome. From these future perspectives of land use in the hypothetical scenarios in different landscapes, it allowed us to analyze the responses in terms of the effects of anthropic action on the runoff–erosion processes within the basin.

The continuous agricultural activity in the basin permeates the confrontation and pressure from agribusiness on land regulation, the control of burning in the area in the Cerrado biome, and the lack of inspection and regulation of the forest code. Given the data from the pessimistic scenario simulated in the model, the trend is clear for the growth of social and environmental practices such as deforestation, climate change, water use for agricultural irrigation, water erosion, siltation of watercourses, and sediment yield, among others. It can be concluded that the parameters calibrated in this study are valid and correspond with all types of landscape and land use based on the performance of the SWAT model, and after comparing observed and calculated streamflow and sediment yield data. It can be concluded that estimated values of soil parameters obtained by remote sensing slightly improved the model's calibration. These results can be significantly valuable to governmental agencies as a communication model for better water resource management and energy generation. Furthermore, these results are highly relevant to the sustainable management of water resources within the region, as such obtained results allow decision makers to observe how water variables behave with changes in LULC caused by human actions; thus, managers can know in advance in which sub-basins this conditioning is more prominent, especially in areas with remnants of forests, or areas characterized by the advance of agriculture in recent years.

Author Contributions: Cláudia Adriana Bueno da Fonseca: data curation, writing—original draft. Nadhir Al-Ansari: writing—review, and editing. Richarde Marques da Silva: conceptualization, methodology, visualization, writing—review and editing. Celso Augusto Guimarães Santos: writing—review and editing. Bilel Zerouali: writing—review and editing. Daniel Bezerra de Oliveira: data curation, writing—review and editing. Ahmed Elbeltagi: writing—review and editing. All authors have read and agreed to the published version of the manuscript.

Funding: This research received no external funding.

Data Availability Statement: The data supporting this study's findings are available upon reasonable request from the corresponding author [CAGS].

Acknowledgments: This study was also financed in part by the Brazilian Federal Agency for the Support and Evaluation of Graduate Education (Coordenação de Aperfeiçoamento de Pessoal de Nível Superior-CAPES)—Finance Code 001, the National Council for Scientific and Technological Development, Brazil—CNPq (Grant Nos. 313358/2021-4 and 309330/2021-1).

Conflicts of Interest: The authors declare no conflict of interest.

References

1. Da Cunha, E.R.; Santos, C.A.G.; Silva, R.M.; Panachuki, E.; De Oliveira, P.T.S.; Oliveira, N.S.; Falcão, K.S. Assessment of current and future land use/cover changes in soil erosion in the Rio da Prata basin (Brazil). *Sci. Total Environ.* **2022**, *807*, 151811. [CrossRef]
2. Tirupathi, C.; Shashidhar, T. Investigating the impact of climate and land-use land cover changes on hydrological predictions over the Krishna river basin under present and future scenarios. *Sci. Total Environ.* **2020**, *721*, 137736.
3. Cunha, E.R.; Santos, C.A.G.; Silva, R.M.; Bacani, V.M.; Pott, A. Future scenarios based on a CA-Markov land use and land cover simulation model for a tropical humid basin in the Cerrado/Atlantic forest ecotone of Brazil. *Land Use Policy* **2021**, *100*, 105141. [CrossRef]
4. Verstegen, J.A.; van der Laan, C.; Dekker, S.C.; Faaij, A.P.C.; Santos, M.J. Recent and projected impacts of land use and land cover changes on carbon stocks and biodiversity in East Kalimantan, Indonesia. *Ecolog. Indicat.* **2019**, *103*, 563–575. [CrossRef]

5. Dos Santos, G.L.; Pereira, M.G.; Delgado, R.C.; Magistrali, I.C.; da Silva, C.G.; de Oliveira, C.M.M.; Laranjeira, J.P.B.; da Silva, T.P. Degradation of the Brazilian Cerrado: Interactions with human disturbance and environmental variables. *Forest Ecol. Manag.* **2021**, *482*, 118875. [CrossRef]
6. Korkanç, S.Y. Effects of the land use/cover on the surface runoff and soil loss in the Niğde-Akkaya Dam Watershed, Turkey. *Catena* **2018**, *163*, 233–243. [CrossRef]
7. Grecchi, R.C.; Gwyn, Q.H.J.; Bénié, G.B.; Formaggio, A.R.; Fahl, F.C. Land use and land cover changes in the Brazilian Cerrado: A multidisciplinary approach to assess the impacts of agricultural expansion. *Appl. Geogr.* **2014**, *55*, 300–312. [CrossRef]
8. Beuchle, R.; Grecchi, R.C.; Shimabukuro, Y.E.; Seliger, R.; Eva, H.D.; Sano, E.; Achard, F. Land cover changes in the Brazilian Cerrado and Caatinga biomes from 1990 to 2010 based on a systematic remote sensing sampling approach. *Appl. Geogr.* **2015**, *58*, 116–127. [CrossRef]
9. Ferreira, F.L.V.; Rodrigues, L.N.; da Silva, D.D. Influence of changes in land use and land cover and rainfall on the streamflow regime of a watershed located in the transitioning region of the Brazilian Biomes Atlantic Forest and Cerrado. *Environ. Monit. Assess.* **2021**, *193*, 16. [CrossRef] [PubMed]
10. Cunha, E.R.; Santos, C.A.G.; Silva, R.M.; Bacani, V.M.; Teodoro, P.E.; Panachuki, E.; Oliveira, N.S. Mapping LULC types in the Cerrado-Atlantic Forest ecotone region using a Landsat time series and object-based image approach: A case study of the Prata River Basin, Mato Grosso do Sul, Brazil. *Environ. Monit. Assess.* **2021**, *192*, 547–567. [CrossRef]
11. De Oliveira, V.A.; de Mello, C.R.; Beskow, S.; Viola, M.R.; Srinivasan, R. Modeling the effects of climate change on hydrology and sediment load in a headwater basin in the Brazilian Cerrado biome. *Ecolog. Eng.* **2019**, *133*, 20–31. [CrossRef]
12. Falcão, K.S.; Panachuki, E.; Monteiro, F.N.; Menezes, R.S.; Rodrigues, D.B.B.; Sone, J.S.; Oliveira, P.T.S. Surface runoff and soil erosion in a natural regeneration area of the Brazilian Cerrado. *Int. Soil Water Conser. Res.* **2020**, *8*, 124–130. [CrossRef]
13. Serrão, E.A.O.; Silva, M.T.; Ferreira, T.R.; de Ataide, L.C.P.; dos Santos, C.A.; de Lima, A.M.M.; Silva, V.P.R.; de Sousa, F.A.Z.; Gomes, D.J.C. Impacts of land use and land cover changes on hydrological processes and sediment yield determined using the SWAT model. *Int J. Sedim Res.* **2022**, *37*, 54–69. [CrossRef]
14. De Arruda, M.R.; Slingerland, M.; Santos, J.Z.L.; Giller, K.E. Agricultural land use change and associated driving forces over the past 180 years in two municipalities of the Brazilian Cerrado. *GeoJournal* **2019**, *84*, 555–570. [CrossRef]
15. Alves, W.S.; Martins, A.P.; Pôssa, É.M.; de Moura, D.M.B.; Morais, W.A.; Ferreira, R.S.; dos Santos, L.N.S. Geotechnologies applied in the analysis of land use and land cover (LULC) transition in a hydrographic basin in the Brazilian Cerrado. *Remote Sens. Applic. Soc. Environ.* **2021**, *22*, 100495. [CrossRef]
16. Parente, L.; Nogueira, S.; Baumann, L.; Almeida, C.; Maurano, L.; Affonso, A.G.; Ferreira, L. Quality assessment of the PRODES Cerrado deforestation data. *Remote Sens. Applic. Soc. Environ.* **2021**, *21*, 100444. [CrossRef]
17. Cunha, E.R.; Bacani, V.M.; Panachuki, E. Modeling soil erosion using RUSLE and GIS in a watershed occupied by rural settlement in the Brazilian Cerrado. *Nat. Hazards* **2017**, *85*, 851–868. [CrossRef]
18. Hunke, P.; Mueller, E.N.; Schröder, B.; Zeilhofer, P. The Brazilian Cerrado: Assessment of water and soil degradation in catchments under intensive agricultural use. *Ecohydrology* **2015**, *8*, 1154–1180. [CrossRef]
19. Hunke, P.; Roller, R.; Zeilhofer, P.; Schröder, B.; Mueller, E.N. Soil changes under different land-uses in the Cerrado of Mato Grosso, Brazil. *Geoderma Reg.* **2015**, *4*, 31–43. [CrossRef]
20. Anache, J.A.A.; Flanagan, D.C.; Srivastava, A.; Wendland, E.C. Land use and climate change impacts on runoff and soil erosion at the hillslope scale in the Brazilian Cerrado. *Sci. Total Environ.* **2018**, *622–623*, 140–151. [CrossRef]
21. Lopes, T.R.; Moura, L.B.; Nascimento, J.G.; Fraga Junior, L.S.; Zolin, C.A.; Duarte, S.N.; Folegatti, M.V.; Santos, O.N.A. Priority areas for forest restoration aiming at the maintenance of water resources in a basin in the Cerrado/Amazon ecotone, Brazil. *J. South Amer. Earth Sci.* **2020**, *101*, 102630. [CrossRef]
22. Serrão, E.A.O.; Silva, M.T.; Ferreira, T.R.; Ferreira, T.R.; de Ataide, L.C.P.; Wanzeler, R.T.S.; da Silva, V.P.R.; de Lima, A.M.; de Sousa, F.A.S. Large-Scale hydrological modelling of flow and hydropower production, in a Brazilian watershed. *Ecohydrol. Hydrobiol.* **2021**, *21*, 23–35. [CrossRef]
23. Oliveira, P.T.S.; Nearing, M.A.; Moran, M.S.; Goodrich, D.C.; Wendland, E.; Gupta, H.V. Trends in water balance components across the Brazilian Cerrado. *Water Res. Res.* **2014**, *50*, 7100–7114. [CrossRef]
24. Lamparter, G.; Nobrega, R.L.B.; Kovacs, K.; Amorim, R.S.; Gerold, G. Modelling hydrological impacts of agricultural expansion in two macro-catchments in Southern Amazonia, Brazil. *Reg. Environ. Change* **2018**, *18*, 91–103. [CrossRef]
25. Senent-Aparicio, J.; Blanco-Gómez, P.; López-Ballesteros, A.; Jimeno-Sáez, P.; Pérez-Sánchez, J. Evaluating the Potential of GloFAS-ERA5 River Discharge Reanalysis Data for Calibrating the SWAT Model in the Grande San Miguel River Basin (El Salvador). *Remote Sens.* **2021**, *13*, 3299. [CrossRef]
26. Jepsen, S.M.; Harmon, T.C.; Guan, B. Analyzing the Suitability of Remotely Sensed ET for Calibrating a Watershed Model of a Mediterranean Montane Forest. *Remote Sens.* **2021**, *13*, 1258. [CrossRef]
27. Zhou, S.; Zhang, W.; Wang, S.; Zhang, B.; Xu, Q. Spatial–Temporal Vegetation Dynamics and Their Relationships with Climatic, Anthropogenic, and Hydrological Factors in the Amur River Basin. *Remote Sens.* **2021**, *13*, 684. [CrossRef]
28. Salles, L.A. Calibração e Validação do Modelo SWAT Para Predição de Vazão na Bacia do Ribeirão Pipiripau. Ph.D. Dissertation, Universidade de Brasília, Brasília, Brasil, 2012; 114p.
29. Carvalho, F.H. Uso do Modelo SWAT na Estimativa da Vazão e da Produção de Sedimentos em Bacia Agrícola do Cerrado Brasileiro. Ph.D. Dissertation, Universidade de Brasília, Brasília, Brasil, 2014; 154p.

30. Narsimlu, B.; Gosain, A.K.; Chahar, B.R.; Singh, S.K.; Srivastava, P.K. SWAT model calibration and uncertainty analysis for streamflow prediction in the Kunwari River Basin, India, using Sequential Uncertainty Fitting. *Environ. Process.* **2015**, *2*, 79–95. [CrossRef]
31. FURNAS. Usina Hidrelétrica de Serra da Mesa. Available online: https://www.furnas.com.br/subsecao/129/usina-de-serra-da-mesa---1275-mw?culture=pt (accessed on 23 February 2022).
32. Ferreira, R.S. Análise da Produção da Carga Líquida na Bacia do Ribeirão do Gama-DF Através do Modelo SWAT. Ph.D. Dissertation, Universidade de Brasília, Brasília, Brasil, 2016; 126p.
33. Veiga, A.M. Calibração Hidrossedimentológica do Modelo SWAT na Bacia Hidrográfica do Córrego Samambaia, Goiânia—GO. Ph.D. Dissertation, Universidade Federal de Goiás, Goiânia, Brasil, 2014; 130p.
34. Aznarez, C.; Jimeno-Sáez, P.; López-Ballesteros, A.; Pacheco, J.P.; Senent-Aparicio, J. Analysing the Impact of Climate Change on Hydrological Ecosystem Services in Laguna del Sauce (Uruguay) Using the SWAT Model and Remote Sensing Data. *Remote Sens.* **2021**, *13*, 2014. [CrossRef]
35. Viana, J.F.S.; Montenegro, S.M.G.L.; da Silva, B.B.; da Silva, R.M.; Srinivasan, R.; Santos, C.A.G.; dos Santos Araujo, D.C.; Tavares, C.G. Evaluation of gridded meteorological datasets and their potential hydrological application to a humid area with scarce data for Pirapama River basin, northeastern Brazil. *Theor. Appl. Climatol.* **2021**, *145*, 393–410. [CrossRef]
36. Strauch, M.; Volk, M. SWAT plant growth modification for improved modeling of perennial vegetation in the tropics. *Ecol. Model.* **2013**, *269*, 98–112. [CrossRef]
37. Hernandes, T.A.D.; Scarpare, F.V.; Seabra, J.E.A. Assessment of the recent land use change dynamics related to sugarcane expansion and the associated effects on water resources availability. *J. Cleaner Product.* **2018**, *197*, 1328–1341. [CrossRef]
38. MMA—Ministério do Meio Ambiente. *Mapeamento do Uso e Cobertura da Terra do Cerrado: Projeto TerraClass Cerrado*; MMA: Brasília, Brazil, 2015; 69p.
39. Valente, C.R. Caracterização geral e composição florística do Cerrado. In *(Org.). Natureza viva Cerrado: Caracterização e Conservação*; Guimarães, L.D., Silva, M.A.D., Anacleto, T.C., Eds.; UCG: Goiânia, Brazil, 2006; pp. 21–44.
40. Cunha, E.R.; Bacani, V.M.; Facincani, E.M.; Sakamoto, A.Y.; Luchiari, A. Remote sensing and GIS applied to geomorphological mapping of the watershed stream Indaiá, MS, Brazil. In Proceedings of the 8th IAG International Conference on Geomorphology, Paris, France, 27–31 August 2013.
41. Pereira, R.C.G.; Braga, C.C.; Paz, R.L.F. Estudo da pluviometria no Estado de Goiás. In Proceedings of the XVI Congresso Brasileiro de Meteorologia, Belém, Brazil, 13–17 September 2010.
42. Fonseca, C.A.B. Análise Espaço-Temporal do Uso e Ocupação do Solo e Simulação dos Processos Hidrossedimentológicos Em uma Bacia do Bioma Cerrado. Ph.D. Thesis, Universidade Federal da Paraíba, João Pessoa, Brazil, 2020; 225p.
43. ANA—Agência Nacional de Águas. Sistema de Acompanhamento de Reservatórios. 2021. Available online: https://www.ana.gov.br/sar (accessed on 12 March 2021).
44. MME—Ministério das Minas e Energia. FURNAS—Usina de Serra da Mesa. 2020. Available online: https://www.furnas.com.br/subsecao/129/usina-de-serra-da-mesa (accessed on 22 January 2021).
45. SEGPLAN-IMB—Secretaria de Estado de Gestão e Planejamento. Anuário Estatístico do Estado de Goiás—2017. Instituto Mauro Borges de Estatísticas e Estudos Socioeconômicos. 2018. Available online: https://www.imb.go.gov.br/bde (accessed on 10 December 2021).
46. SEGPLAN-IMB—Secretaria de Estado de Gestão e Planejamento. Anuário Estatístico do Estado de Goiás—2015. Instituto Mauro Borges de Estatísticas e Estudos Socioeconômicos. 2016. Available online: https://www.imb.go.gov.br/bde (accessed on 10 December 2021).
47. Rocha, E.C.; Silva, J.; da Silva, P.T.; Araújo, M.S.; Castro, A.L.S. Medium and large mammals in a Cerrado fragment in Southeast Goiás, Brazil: Inventory and immediate effects of habitat reduction on species richness and composition. *Biota Neotrop.* **2019**, *19*, e20180671. [CrossRef]
48. Barbosa, F.F.; Godoy, B.S.; Oliveira, L.G. Trichoptera Kirby (Insecta) immature fauna from Rio das Almas Basin and Rio Paranã, Goiás State, Brazil, with new records for some genera. *Biota Neotrop.* **2011**, *11*, 21–25. [CrossRef]
49. Costa, H.A.O.; Stürmer, S.L.; Ragonezi, C.; Grazziotti, P.H.; Grazziotti, D.C.F.S.; Silva, E.B. Species richness and root colonization of arbuscular mycorrhizal fungi in Syngonanthus elegans, an endemic and threatened species from the Cerrado domain in Brazil. *Ciência Agrotec.* **2016**, *40*, 326–336. [CrossRef]
50. Nabout, J.C.; Soares, T.N.; Diniz-Filho, J.A.F.; De Marco Júnior, P.; Telles, M.P.C.; Naves, R.V.; Chaves, L.J. Combining multiple models to predict the geographical distribution of the Baru tree (*Dipteryx alata Vogel*) in the Brazilian Cerrado. *Braz. J. Biol.* **2010**, *70*, 911–919. [CrossRef]
51. Fischer, J.; Lindenmayer, D.B. Landscape modification and habitat fragmentation: A synthesis. *Glob. Ecol. Biogeog.* **2007**, *16*, 265–280. [CrossRef]
52. Alves, G.L.F. Expansão Canavieira e Seus Efeitos na Violência em Goianésia. Ph.D. Dissertation, Universidade Federal de Goiás, Goiânia, Brazil, 2012; 103p.
53. IBGE—Instituto Brasileiro de Geografia e Estatística. *Census of Agriculture 1995/1996*; IBGE; 1996. Available online: https://ftp.ibge.gov.br/Censo_Agropecuario/Censo_Agropecuario_1995_96/Goias/ (accessed on 13 February 2021).
54. IBGE—Instituto Brasileiro de Geografia e Estatística. *Census of Agriculture 2006*; IBGE; 2006. Available online: https://ftp.ibge.gov.br/Censo_Agropecuario/Censo_Agropecuario_2006/Censo_Agropecuario_2006.zip (accessed on 13 February 2021).

55. IBGE—Instituto Brasileiro de Geografia e Estatística. Produção da Pecuária Municipal. Brasil. 2016. Diretoria de Agropecuária, Recursos Naturais e Geografia. 2016. Available online: https://biblioteca.ibge.gov.br/visualizacao/periodicos/84/ppm_2016_v44_br.pdf (accessed on 23 March 2021).
56. IBGE—Instituto Brasileiro de Geografia e Estatística. *Produção da Pecuária Municipal Brasil. Ministério da Agricultura. Pecuária*; Diretoria de Agropecuária, Recursos Naturais e Geografia; 2017. Available online: https://biblioteca.ibge.gov.br/visualizacao/periodicos/84/ppm_2017_v45_br_informativo.pdf (accessed on 23 March 2021).
57. IBGE—Instituto Brasileiro de Geografia e Estatística. Pesquisa da Pecuária Municipal. 2018. Available online: https://biblioteca.ibge.gov.br/visualizacao/periodicos/84/ppm_2018_v46_br_informativo.pdf (accessed on 23 March 2021).
58. INMET—Instituto Nacional de Meteorologia. Banco de Dados Meteorológicos do INMET. 2021. Available online: http://www.inmet.gov.br/projetos/rede/pesquisa (accessed on 11 November 2021).
59. ANA—National Water Agency. Rede Hidrometeorológica Nacional. 2021. Available online: http://www.snirh.gov.br/hidroweb (accessed on 11 November 2021).
60. Barreto, L.; Schoorl, J.M.; Kok, K.; Veldkamp, T.; Hass, A. Modelling potential landscape sediment delivery due to projected soybean expansion: A scenario study of the Balsas sub-basin, Cerrado, Maranhão state, Brazil. *J. Environ. Managem.* **2013**, *115*, 270–277. [CrossRef]
61. Santos, J.Y.G.; Montenegro, S.M.G.L.; Silva, R.M.; Santos, C.A.G.; Quinn, N.W.; Xavier, A.P.C.; Ribeiro Neto, A. Modeling the impacts of future LULC and climate change on runoff and sediment yield in a strategic basin in the Caatinga/Atlantic forest ecotone of Brazil. *Catena* **2021**, *202*, 558–578. [CrossRef]
62. Silva, R.M.; Santos, C.A.G.; dos Santos, J.Y.G. Evaluation and modeling of runoff and sediment yield for different land covers under simulated rain in a semiarid region of Brazil. *Int. J. Sediment Res.* **2018**, *33*, 117–125. [CrossRef]
63. Williams, J.R.; Berndt, H.D. Sediment Yield Prediction Based on Watershed Hydrology. *Trans. ASABE.* **1977**, *20*, 1100–1104. [CrossRef]
64. Wischmeier, W.H.; Smith, D.D. *Predicting Rainfall-Erosion Losses from Cropland East of the Rocky Mountains, Agriculture Handbook*; US Department of Agriculture, Agriculture Research Service, 282: Washington, DC, USA, 1965.
65. Neitsch, S.L.; Arnold, J.G.; Kiniry, J.R.; Williams, J.R. *Soil and Water Assessment Tool—Theoretical Documentation*; Version 2005; United States Department of Agriculture: Temple, TX, USA, 2005.
66. Arnold, J.G.; Srinivasan, R.; Muttiah, R.S.; Williams, J.R. Large area hydrologic modeling and assessment: Part I: Model development. *J. Amer. Water Res. Assoc.* **1998**, *34*, 73–89. [CrossRef]
67. Silva, R.M.; Santos, C.A.G.; Dantas, J.C.; Beltrão, J.A. Hydrological simulation in a tropical humid basin in the Cerrado biome using the SWAT model. *Hydrol. Res.* **2018**, *49*, 908–923. [CrossRef]
68. Gassman, P.W.; Reyes, M.R.; Green, C.H.; Arnold, J.G. The Soil and Water Assessment Tool: Historical Development, Applications, and Future Research Directions. *Trans. ASABE* **2007**, *50*, 1211–1250. [CrossRef]
69. Neitsch, S.L.; Arnold, J.G.; Kiniry, J.R. *Soil and Water Assessment Tool: Theoretical Documentation*; Version 2009; United States Department of Agriculture: Temple, CA, USA, 2009.
70. USGS—United States Geological Survey. Satellite Images. 2021. Available online: https://earthexplorer.usgs.gov (accessed on 12 March 2021).
71. Lima, C.E.S.; Costa, V.S.O.; Galvíncio, J.D.; Silva, R.M.; Santos, C.A.G. Assessment of automated evapotranspiration estimates obtained using the GP-SEBAL algorithm for dry forest vegetation (Caatinga) and agricultural areas in the Brazilian semiarid region. *Agric. Water Manag.* **2021**, *250*, 106863. [CrossRef]
72. EMBRAPA—Empresa Brasileira de Pesquisa Agropecuária. Mapa de Solos do Brasil. 2020. Available online: http://geoinfo.cnps.embrapa.br/layers (accessed on 2 June 2021).
73. Chrysoulakis, N.; Mitraka, Z.; Gorelick, N. Exploiting satellite observations for global surface albedo trends monitoring. *Theor. Appl. Climatol.* **2019**, *137*, 1171–1179. [CrossRef]
74. Myneni, R.; Knyazikhin, Y.; Park, T. *MOD15A2H MODIS Leaf Area Index/FPAR 8-Day L4 Global 500m SIN Grid V006*; Level-1 and Atmosphere Archive & Distribution System Distributed Active Archive Center, NASA: Houston, TX, USA, 2015.
75. Nash, J.E.; Sutcliffe, J.V. River flow forecasting through conceptual models part I: A discussion of principles. *J. Hydrol.* **1970**, *10*, 282–290. [CrossRef]
76. Moriasi, D.N.; Arnold, J.G.; Van Liew, M.W.; Bingner, R.L.; Harmel, R.D.; Veith, T.L. Model evaluation guidelines for systematic quantification of accuracy in watershed simulations. *Trans. ASABE* **2007**, *50*, 885–900. [CrossRef]
77. Abbaspour, K.C.; Rouholahnejad, E.; Vaghefi, S.A.; Srinivasan, R.; Yang, H.; Kløve, B. A continental-scale hydrology and water quality model for Europe: Calibration and uncertainty of a high-resolution large-scale SWAT model. *J. Hydrol.* **2015**, *524*, 733–752. [CrossRef]
78. Van Griensven, A.; Meixner, T.; Grunwald, S.; Bishop, T.; Di Luzio, M.; Srinivasan, R. A global sensitivity analysis tool for the parameters of multi-variable catchment models. *J. Hydrol.* **2006**, *324*, 10–23. [CrossRef]
79. Franz, C.; Abbt-Braun, G.; Lorz, C.; Roig, H.L.; Makeschin, F. Assessment and evaluation of metal contents in sediment and water samples within an urban watershed: An analysis of anthropogenic impacts on sediment and water quality in Central Brazil. *Environ Earth Sci.* **2014**, *72*, 4873–4890. [CrossRef]
80. USDA—United States Department of Agriculture. *Riparian Forest Buffers: Function and Design for Protection and Enhancement of Water Resources*; National Agroforestry Center, U.S. Department of Agriculture: Washington, DC, USA, 1991; 24p.

International Journal of *Geo-Information*

Article

Mapping of Rill Erosion of the Middle Volga (Russia) Region Using Deep Neural Network

Artur Gafurov

Institute of Environmental Sciences, Kazan Federal University, 420008 Kazan, Russia; amgafurov@kpfu.ru

Abstract: Soil erosion worldwide is an intense, poorly controlled process. In many respects, this is a consequence of the lack of up-to-date high-resolution erosion maps. All over the world, the problem of insufficient information is solved in different ways, mainly on a point-by-point basis, within local areas. Extrapolation of the results obtained locally to a more extensive territory produces inevitable uncertainties and errors. For the anthropogenic-developed part of Russia, this problem is especially urgent because the assessment of the intensity of erosion processes, even with the use of erosion models, does not reach the necessary scale due to the lack of all the required global large-scale remote sensing data and the complexity of considering regional features of erosion processes over such vast areas. This study aims to propose a new methodology for large-scale automated mapping of rill erosion networks based on Sentinel-2 data. A LinkNet deep neural network with a DenseNet encoder was used to solve the problem of automated rill erosion mapping. The recognition results for the study area of more than 345,000 sq. km were summarized to a grid of 3037 basins and analyzed to assess the relationship with the main natural-anthropogenic factors. Generalized additive models (GAM) were used to model the dependency of rill erosion density to explore complex relationships. A complex nonlinear relationship between erosion processes and topographic, meteorological, geomorphological, and anthropogenic factors was shown.

Keywords: rill; soil erosion; neural networks; remote sensing; relationship modeling

Citation: Gafurov, A. Mapping of Rill Erosion of the Middle Volga (Russia) Region Using Deep Neural Network. *ISPRS Int. J. Geo-Inf.* **2022**, *11*, 197. https://doi.org/10.3390/ijgi11030197

Academic Editor: Walter Chen

Received: 25 December 2021
Accepted: 12 March 2022
Published: 15 March 2022

1. Introduction

Soil erosion became the main factor in degrading a fertile layer of agricultural lands. Intensive land use combined with natural factors creates conditions for increased development of soil erosion, including rills, erosion furrows, and ephemeral gullies. Often, these erosion forms turn into permanent gullies, completely removing the area from agricultural turnover. Therefore, it is not surprising that researchers worldwide pay special attention to the study of soil erosion and, in particular, rill erosion.

Both field and mathematical methods are the main methods for studying soil erosion. The field method is very accurate in estimating the volume of erosion changes on local areas, which somewhat complicates the spatial interpretation of the results at the region or landscape levels. Such methods include the classical methods of reference sites, landmarks, microprofiling, and measuring the length and width of washouts with a tape measure. Reference sites allow estimating the direct volume of soil washed away from the territory, providing an opportunity to artificially change the conditions of the "environment" [1]. Based on the data from such sites, the data for creating mathematical models of soil erosion assessment, which will be discussed below, were obtained. The method of landmarks and microprofiling makes it possible to accurately estimate the depth of washout in representative sites [2]. Studies using these techniques helped find a strong correlation between the width of washout and its depth, which, in turn, estimates the volumes of soil washed away from the territory [3,4]. At present, geodetic methods based on relief reconstruction using scanning data or photogrammetry, both ground and airborne, are being developed to assess erosion. The most accurate results can be achieved using ground-based laser scanning data using reference points [5]. Researchers from all over the world manage to record in this way, not only gully [6] or rill [7] soil erosion but also micro-rill and sheet soil erosion [8], the changes of which are within the first millimeters. Unfortunately, it is problematic to use such technologies for direct monitoring in areas larger than 1 hectare. This problem is attempted to be solved using airborne laser scanning; however, this technology has its limitations related to the positioning of the scanning equipment for the subsequent alignment of the repeated survey. This fact immediately precludes the use of airborne scanning to evaluate the dynamics of micro-rill and sheet erosion, but it does allow for more or less automated estimation of the total length and width of rill and gully erosion. Different approaches are used for this. For example, the most common one is based on the threshold value of the number of digital elevation model (DEM) cells, from which runoff into neighboring cells is possible. Depending on the resolution of the initial DEM obtained from scanning data, this approach allows recognizing rills depths from 5 cm. However, airborne laser scanning is not widely used due to the high cost of the scanning equipment. Nowadays, more affordable scanning sensors appeared installed on manned aircraft and unmanned aerial vehicles (UAV). Nevertheless, the applicability of such devices for solving the problem of soil erosion assessment has yet to be discovered—the analysis of the existing experience has demonstrated that the most significant relevance of such devices is in forestry [9]. Despite the current trend toward cheaper scanning systems, they are still not widely available to most researchers. Combining these factors has led to the most widespread use of photosensors as payloads for UAVs [10]. UAV photogrammetry makes it possible to achieve comparable point cloud densities with scanning systems' competitive accuracy of the resulting models and allows one of the end products to be an ultra-high-resolution orthophotoplane. The use of UAVs makes it possible to provide geomorphological surveys with up-to-date information about the condition of the survey area cheaply, quickly, and accurately enough. Unfortunately, despite the overall high productivity, the use of UAVs does not allow for making a continuous survey of large regions [11,12]; however, it will enable verification model data.

Model data obtained from office studies make it possible to estimate the intensity of exogenous processes at both local and global levels. A whole range of mathematical models has been developed to assess soil and gully erosion, subdivided into empirically and physically based [13]. The most well-known and widely used soil erosion models are USLE [14], RUSLE [15], MUSLE [16], CREAMS [17], LISEM [18], and so on. WEPP [19], RillGrow [20], GUEST [21,22], and EUROSEM [23] models are used to estimate rill erosion. All these models have their advantages and disadvantages, and the latter does not allow using these models everywhere. The main limitations are related to the lack of publicly available large-scale remote sensing data [24], such as terrain models [25] and detailed comparable climate data [26,27], which does not allow the mathematical simulations needed to map streamflow paths. The lack of detailed remote sensing data also limits the possibility of applying object-oriented recognition of rill erosion [28] and stochastic modeling of erosion exposure [29,30].

The data obtained by erosion models in the form of erosion and accumulation balances often do not reflect the actual development of rill erosion processes. For example, in the catchments of the European part of Russia, it is not uncommon that the rate of accumulation in the bottoms of dry valleys tends to decrease [31–33], while the number, total length, and, accordingly, the density of rill network increases from year to year [34]. This is often due to the peculiarities of plowing with the creation of special sloughs that retain runoff from the fields, corrupting the results obtained by erosion models.

Considering the existing limitations, approaches based on manual extraction of washouts from remote sensing data, namely space images [35,36], are applied to solve the problem of rill erosion mapping. Remote sensing mapping of rill erosion was carried out using Landsat-5 (30 m) [37,38], QuickBird (0.6 m) and SPOT (5 m) satellite imagery [39], GeoEye-1 stereo pair [40], and UAV data [41]. Sentinel-2 (10 m) images were also used to create NDVI-images during the predictive model development [42], but the mapping has not been carried out directly based on them before. Such solutions allow for achieving the best accuracy but are labor consuming and low productive. To solve this problem, the development of approaches based on the automation of rills washout extraction is necessary, which is the goal of this work. The source data used as a cartographic basis sets a list of possible approaches that can be applied to solve the purpose. The simplest of them is object recognition based on a thresholding approach, where the threshold defines the boundary of the reflectivity of the spectral data characteristic of the study object. Such methods are suitable for identifying different land-use types [43,44]. However, the result will be unpredictable when recognizing rills using a similar approach on different types of soils. Machine learning methods can be used to consider the spatial variability of environmental factors, such as the highly proven Random Forest or Support Vector Machine method [45,46]. However, such approaches only give a probability of soil erosion in a particular pixel, while not distinguishing the erosions themselves into a "tree-like pattern" in the landscape. In addition, the methods are very sensitive to the amount of input data—the more information used for analysis, the more consistent the results will be. On small watersheds, such approaches can be successfully applied; for large areas, their applicability is questionable.

Recently, there has been a rapid increase in the number of works related to deep neural networks (DNNs) for the semantic segmentation of remote sensing data. This has been facilitated by the increased quality of remote sensing data, user-friendly deep-learning framework availability, and the increased computing power available to researchers. At present, DNNs enable successful automated interpretation of anthropogenic objects [47–49], shorelines [50,51], land use [52–54], vegetation cover dynamics [55,56], and exogenous processes [57,58]. In all cases, the authors note the higher recognition accuracy of the objects of interest than other methods and emphasize scaling the trained models. Deep neural networks have not been applied to the recognition and mapping of rill erosion. Nevertheless, the importance of the problem under study and the promising use of artificial intelligence for solving this problem determines the necessity of developing an appropriate methodology.

Given the intensification of rill erosion, the necessity of mapping it, and the limitations of field survey and modeling, this study aims to develop and apply a deep neural network-based methodology for automated rill erosion detection from remote sensing data.

2. Study Area

The study area is the Middle Volga region of Russia. The territory is located in the central part of the East European Plain [59]. The Middle Volga region's agro-industrial complex (AIC) is of all-Russian importance. The region holds one of the top places in the country to produce grain, including the valuable grain crop—wheat. Agriculture is characterized by high efficiency due to very favorable natural conditions. At the same time, the agricultural sector of the territory has a substantial impact on natural-territorial complexes, primarily on the soil cover. Given the intensity of plowing, cattle grazing, and natural conditions conducive to the development of exogenous processes, this area is highly affected by soil erosion. This situation is not new; it is not accidental that the territory of the Middle Volga region is historically called an erosion belt. At the same time, anti-erosion measures in recent decades are of a point and haphazard nature, mainly due to the poor study of the rate and development direction of the process, primarily rill erosion.

The study area is 345,477 sq. km. The macroregion includes Mari El, Tatarstan, Chuvashia, Saratov, Samara, Ulyanovsk, and Penza regions. The Middle Volga region is located in the forest landscape zone in the north, the forest-steppe landscape zone in the center, and the steppe landscape zone in the south (Figure 1).

Figure 1. The study area. The 0—Saratov region, 1—Penza region, 2—Republic of Chuvashia, 3—Samara region, 4—Republic of Tatarstan, 5—Republic of Mari El, and 6—Ulyanovsk region.

The relief of the Middle Volga region has a spotted asymmetry of slopes: the average height—146 m, maximum elevation—316.8 m, and minimum—13.2 m [60]. The region's climate is moderately continental and continental in the south, varying wildly from region to region. Average annual precipitation ranges from 471 to 549 mm in the north to 264–424 mm in the south.

The steepness of the slopes of the territory, in general, is favorable for agriculture; a quarter of the region is steep to 1 degree, and a significant part of the territory is characterized by steepness in the range of 1–2.5 degrees, rising to 5 degrees in the area of Bugulminsko-Belebeyevskaya upland [60]. Erosion risks are not uniform and vary from region to region [61]. Soil-forming rocks are not homogeneous and are mainly clayey and heavy loam. There are sandy soils and medium-loam in the north of the Republic of Mari El region and light loam in the forest-steppe zone (Penza Oblast). Soils vary; chernozems of various subtypes predominate, mostly leached (Haplic Chernozems), as well as ordinary (Calcic Chernozems), typical (also Haplic Chernozems), and southern (Haplic Chernozems too)(steppe soils). The remaining types have a clearly expressed zonal pattern, where the gray forests (Haplic Greyzems) are typical for the forest territories and north of the forest-steppe zone [60].

3. Materials and Methods

The development and implementation of a methodology for automatic rill erosion recognition involve several interrelated points:

1. Selection and preparation of remote sensing data from space;
2. Preparation of training and test sets;
3. Training the neural network and evaluating the quality of rill recognition;
4. Implementation of the neural network for the entire study area and vectorization of recognition results;
5. Calculation of rill erosion length as a measure of rill erosion density index in the basins;
6. Analysis of the obtained results.

3.1. Remote Sensing Data

Data from the Sentinel-2 satellite constellation was used for recognition. Unlike other sources of multispectral remote sensing data, such as Landsat, MODIS, SPOT, RapidEye and so on, Sentinel-2 imagery provides the best resolution in the target bands (10 m) at no charge. The cloudless composite data of the Sentinel-2 mission for the spring period, April-June, were used as input data to develop the rill erosion recognition technique. Near-infrared images with baseline resolution (10 m) were used for recognition. The composite was created using the Google Earth Engine (GEE) [62], product "Sentinel-2 MSI: MultiSpectral Instrument, Level-2A" (COPERNICUS/S2_SR). GEE allows the processing of big remote sensing data and facilitates some routine operations. A total of 2988 scenes from both S2A and S2B satellites were used to create the composite. To create the composite, the 2018–2019 images were filtered by date, images were cleaned from clouds and shadows, median pixel brightness values were calculated, and imagery was clipped to the boundaries of the study area and reprojected into the WGS 84/Pseudo-Mercator (EPSG:3857) projection coordinate system. A cloud probability raster was used to clear the imagery from the clouds. The S2 cloud probability is created using the sentinel2-cloud-detector library (LightGBM [63]). All bands are upsampled using bilinear interpolation to 10 m resolution before the gradient boost base algorithm is applied. The resulting 0 to 1 floating point probability is scaled to 0–100 and stored as a UINT8. Areas missing any or all of the bands are masked out. Higher values are more likely to be clouds or highly reflective surfaces (e.g., roof tops or snow). A 15% probability of having clouds on the scene was used as the cloud threshold.

3.2. Preparation of the Training Dataset

In the resulting spring composite, an area with visually the densest rill network and the presence of major land-use classes of water, anthropogenic, forest, and agricultural features was selected and clipped. For this area of more than 2500 square kilometers, continuous manual rill erosion recognition was performed to create a set of reference data samples (Ground truth, GT) (Figure 2). Manual recognition was made by QGIS tools using Sentinel-2 data and ultra-high resolution (UHR) images (Bing, Google, and so on, available as a substrate at no charge). The main selection was made on the S2 images. In controversial cases, the UHR images were used. They were also used for the non-selection of permanent gullies. Due to the difficulty of classifying erosion forms on such a vast territory with different natural conditions, the differentiation of erosion forms was not carried out. Ephemeral gullies" and "rills" were defined as all erosion forms that can be interpreted on satellite images but do not belong to permanent gullies. Considering the resolution of the satellite image, rills are the most minor erosion form, which transforms into an ephemeral gully.

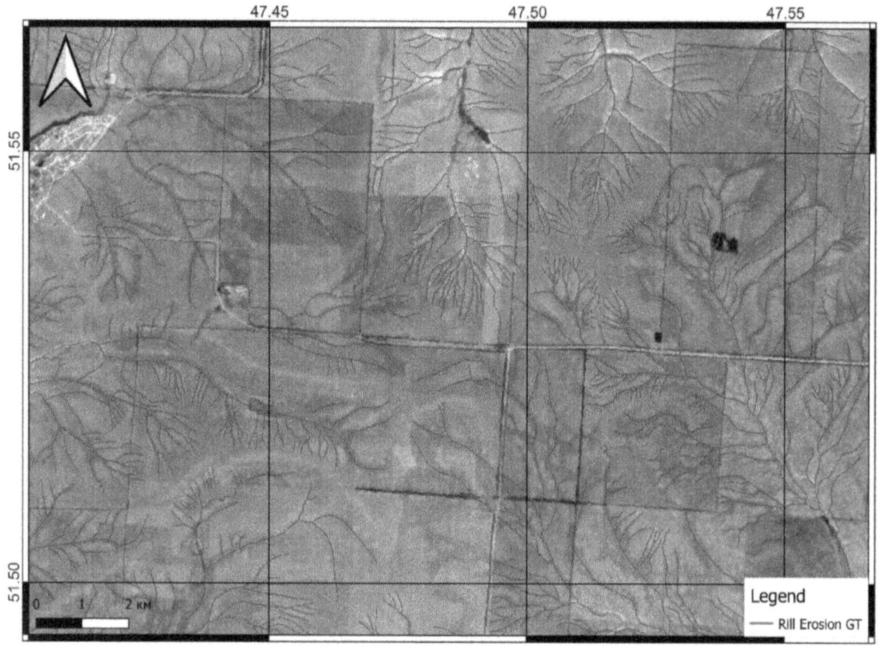

Figure 2. Fragment of a rill erosion ground truth data (GT).

The resulting sample was rasterized and reduced to the resolution of the satellite image fragment, and then both images were cut into 256 × 256 pixels patches. In total, 10,933 satellite image-binary mask pairs have been obtained this way (Figure 3). The resulting rasters were further randomly transformed to augment the number of rasters by three times artificially. The resulting dataset was divided into training and test sets in the ratio of 1/5.

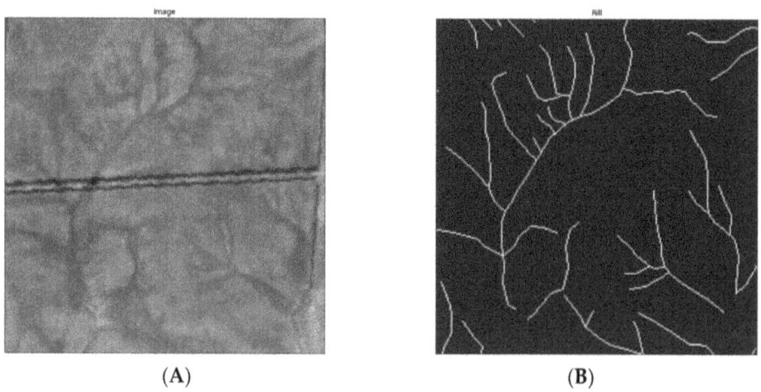

Figure 3. Example of a satellite image patch (**A**) and a binary mask (**B**) of rill erosion.

3.3. Training a Deep Neural Network

A convolutional neural network is a class of artificial neural network that uses convolutional layers to filter inputs for useful information. The convolution operation involves combining input data (feature map) with a convolution kernel (filter) to form a transformed feature map. The filters in the convolutional layers (conv layers) are modified based on learned parameters to extract the most useful information for a specific task.

Convolutional networks adjust automatically to find the best feature based on the task. Applications of convolutional neural networks include various image (image recognition, image classification, video labeling, text analysis) and speech (speech recognition, natural language processing, and text classification) processing systems, along with state-of-the-art AI systems, such as robots, virtual assistants, and self-driving cars.

LinkNet (Figure 4), a fully convolutional neural network for semantic image segmentation, was chosen as a neural network architecture [64]. In contrast to the well-known and well-proven U-NET architecture [57], which has been used for semantic segmentation of satellite images in geomorphological studies [65], LinkNet uses residual blocks instead of convolutional blocks in its encoder and decoder networks. The LinkNet deep neural network architecture efficiently exchanges the information received by the encoder with the decoder after each downsampling block. This turns out to be better than using pooling indices in the decoder, or just using fully convolutional networks in the decoder. This feature transfer technique not only gives good accuracy, but also allows a small number of parameters to be used in the decoder.

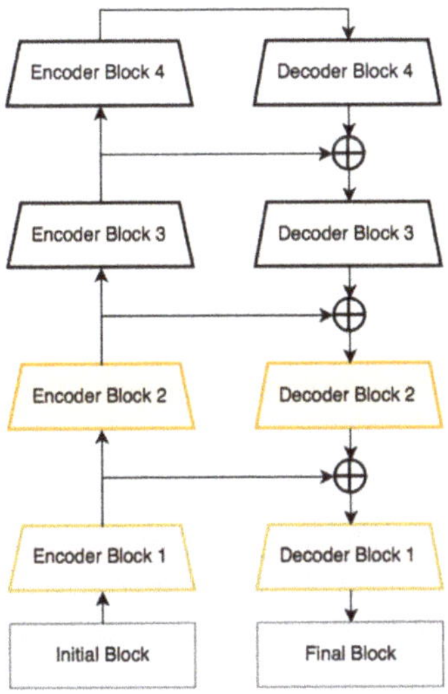

Figure 4. DNN LinkNet architecture scheme [64].

The initial block contains a convolution layer with a kernel size of 7 × 7 and a stride of 2, then a max-pool layer with a window size of 2 × 2 and a stride of 2. Similarly, the last block performs a full convolution, taking feature maps from 64 to 32, then a 2D convolution. Finally, a full convolution is used as a classifier with a kernel size of 2 × 2. As practice has shown, this allows for better semantic aggregation, including dendro-like constructions (tree-like patterns on the landscape) and rill networks. By trial-and-error method, it became clear that the best results in decoding rills can be achieved using transfer learning, a deep neural network learning method that allows using the knowledge obtained about one deep learning problem and applying it to another, but with a similar problem. The encoders used for DenseNet image classification [66] were used in our case. In contrast to similar models, the features are not summed up but concatenated (channel-wise concatenation) into a single tensor before passing them to the next layer. The model learning and implementation algorithm was performed in the Python 3.7 programming environment using the Tensorflow library. The input of the neural network was a stack of image-mask pairs prepared in the previous step. EarlyStopping monitoring was used to prevent overtraining of the model, and the IOU Score metric, Jaccard's coefficient, was used to check the model's learning capability [67].

3.4. Statistical Analysis of the Obtained Results

Further research was conducted to analyze the influence of various natural–anthropogenic factors on the intensity of erosion processes in the study area. Classical correlation, regression analysis, and analysis of variance (AOV) (one-way [68] and multivariate [69]) were used to analyze the relationship between natural and anthropogenic factors on the density of the rill network. The AOV was performed using the Statsmodel package for Python 3. To analyze complex relationships, an analysis based on generalized additive models was used. The generalized additive model (GAM) is a generalized linear model in which a linear predictor depends linearly on unknown smooth functions of some predictor variables, and interest is focused on inference about these smooth functions. GAM uses spline functions, functions that can be combined to approximate arbitrary functions. GAM introduces penalties for the weights so that they remain close to zero, which effectively reduces the flexibility of the splines and reduces the possibility of overfitting. The smoothness parameter, which is normally used to control curve flexibility, is then adjusted using cross-validation. The natural–anthropogenic factors presented in Table 1 [60] were used for the analysis.

Table 1. Natural–anthropogenic factors of rill erosion selected for analysis [60].

Name	Description
AREA	Basin area (sq. km)
SLOPE	Average slope of the basin (degrees)
HMEAN	Average elevation in the basin (m)
HMAXMIN	Elevation range in the basin (m)
TMEAN	Mean annual air temperature (degrees C) in the basin
TMAX	Mean annual maximum temperature (degrees C) in the basin
TMIN	Mean annual minimums temperature (degrees C) in the basin
TAMP	Mean annual variation of air temperature (degrees C) in the basin
T1MEAN	Mean air temperature in January (deg. C) in the basin
TAKT	Sum of active temperatures (degrees C) in the basin
RMEAN	Mean annual precipitation in the basin (mm)
R58	Mean May-August precipitation (mm) in the basin
RCOLD	Mean precipitation for the cold period of the year (mm) in the basin
RWARM	Mean precipitation for warm period of the year (mm) in the basin
RVC	Mean annual precipitation variation coefficient (%) in the basin
GTK	Mean value of the hydrothermal coefficient in the basin
PARENT1	Predominant type of soil-forming rocks
SOIL0	Predominant soil type
LES_PROC	Forest cover of the basin (%)
LAND_COD	Landscape subtype
PLT_RANGE	Population density (people/square km)
ANTR1	Anthropogenic load (score)
X	Longitude of the basin centroid
Y	Latitude of the basin centroid

4. Results and Discussion

The trained RECNN (rill erosion convolutional neural network) was tested on an independent test dataset. The recognition accuracy was 0.62, F1-measure was 0.76, and loss-function was 0.27. Qualitatively analyzing the obtained evaluation results (Figure 5), we can note a relatively high level of recognition of the rill erosion network. It is crucial that not a single case of detection of gullies or ground roads, which are abundant in the study area, instead of rills, was recorded.

Figure 5. An example of the results of applying RECNN to a test dataset.

The trained and tested model was decided to apply to the entire study area (Figure 6). For the obtained geometry, the length was calculated, which was aggregated to a grid of basins [60] (Figure 7). Analyzing the obtained maps, the presence of specific clusters becomes evident. The first cluster—the highest density of rill erosion—is located in the left-bank part of the Volga River in the Saratov region. The second cluster—the area of medium erosion—characterizes the right bank of the Volga in the Saratov region and Zakamye of the Republic of Tatarstan. The Republics of Mari El and Chuvashia territories are characterized mainly by low rill erosion, forming the third cluster.

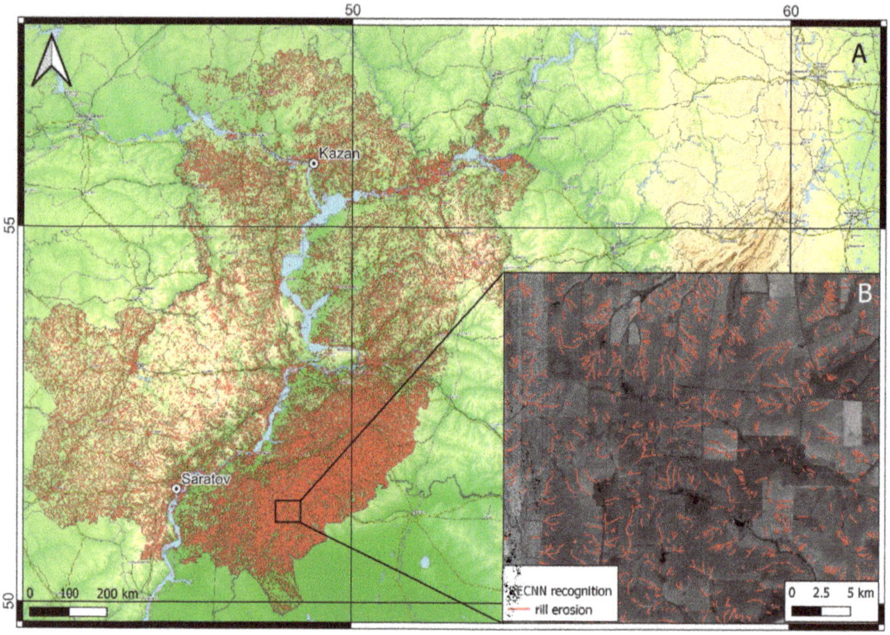

Figure 6. Results of rill erosion recognition over the entire study area (**A**) and enlarged area with recognized streams in Sentinel 2 image (**B**).

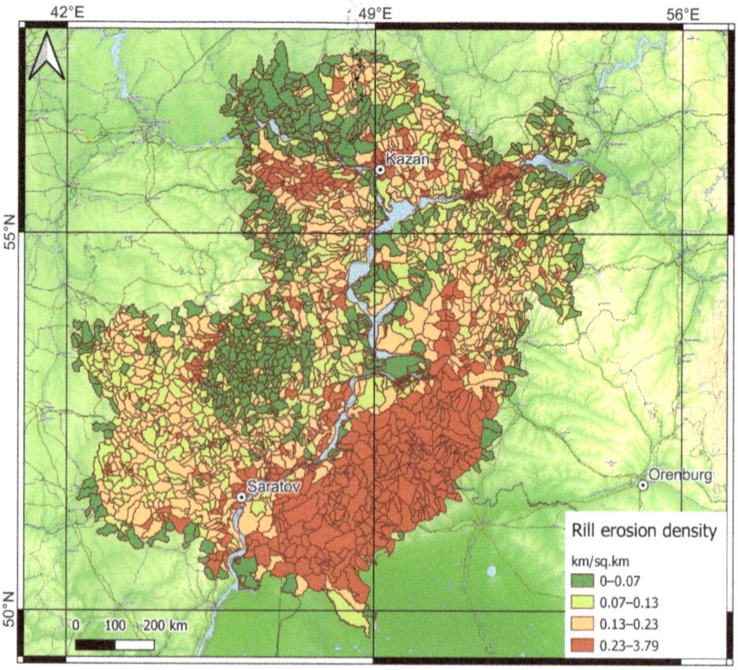

Figure 7. Map of rill erosion density. Quartiles calculate intervals.

In total, information in the study area is aggregated on 3037 basins with an average area of 117.83 sq. km. The average density of rill erosion over the entire study area is 0.23 km/sq. km, and the maximum density is 3.79 km/sq. km. In general, the distribution of the total length of rill erosion in the study area is lognormal; that is, most of the measurements contain, in general, relatively small values of the total length (Figure 8).

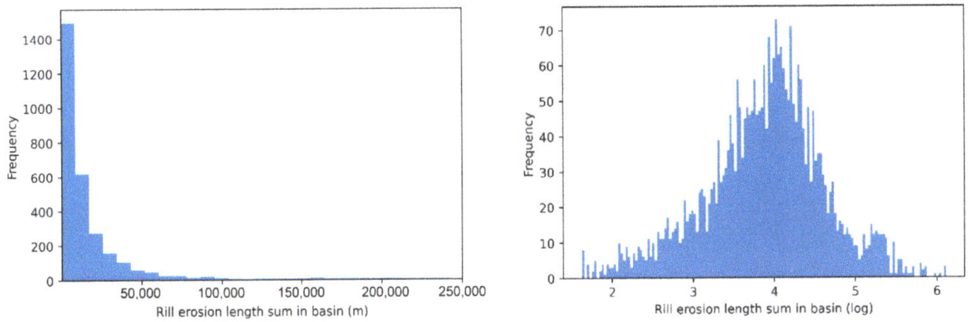

(A) (B)

Figure 8. Distribution of total rill erosion length (m) (**A**) and normalized distribution of total rill erosion length (**B**).

For most of the selected factors, no strong correlations were found with the total length of rill erosion. Only some factors revealed statistically significant ($p < 0.001$) weak relationships ($r > |0.3|$). For example, an inverse correlation describes the relationship between total rill erosion length and forest cover, hydrothermal coefficient, mean annual precipitation, mean May-August precipitation, and mean warm-season precipitation. The inverse relationship with forest cover is more than obvious and only states a well-known fact—the less forest cover of the territory, the greater the risk of soil and gully erosion. The inverse relationship with the other indicators is not so obvious. Selyaninov's hydrothermal coefficient of wetness (GTK) is a characteristic of the level of moisture availability of the territory. It is widely used in agronomy for general climate assessment and allocation of zones with different levels of moisture availability to determine the usefulness of the cultivation of various crops. The factor is closely related to the amount of precipitation in the territory. The inverse relationship here is explained not so much by the very nature of the dependence "the more precipitation, the less erosion", which cannot be, as by the influence of climatic conditions in general. This is partly confirmed by a direct relationship between the total length of the rill network and such factors, such as the sum of active temperatures and the average annual maximum temperature for the year. Apparently, with increasing temperature and climate change due to global warming, the character of precipitation changes—they may fall rarely, but with great intensity [70–74]. Unfortunately, it is impossible to check this hypothesis directly due to the lack of data on mean annual precipitation intensity changes. However, the same conclusions are drawn by other researchers who have worked in the study area [31,32,75,76]. Nevertheless, there is still no clear answer to this question.

Another question is how classical methods of relationship analysis can assess the effect of a single factor on the dependent variable. Linear dependencies are rare in nature, and polynomial dependencies of high degrees are even rarer. The analysis demonstrated poor applicability of classic correlation and regression analysis for revealing the relationships between natural–anthropogenic factors and the length of the erosion network. Some examples (Figures 9–11) demonstrate that the dependencies are close to the logistic or exponential type but more complex. Therefore, it was decided to use generalized additive models (GAM) for dependence analysis [77]. These models allow estimating the dependencies in the form of spline functions, which allows one to make better approximations and comprehensively study the dependence of a factor on a variable [24,78].

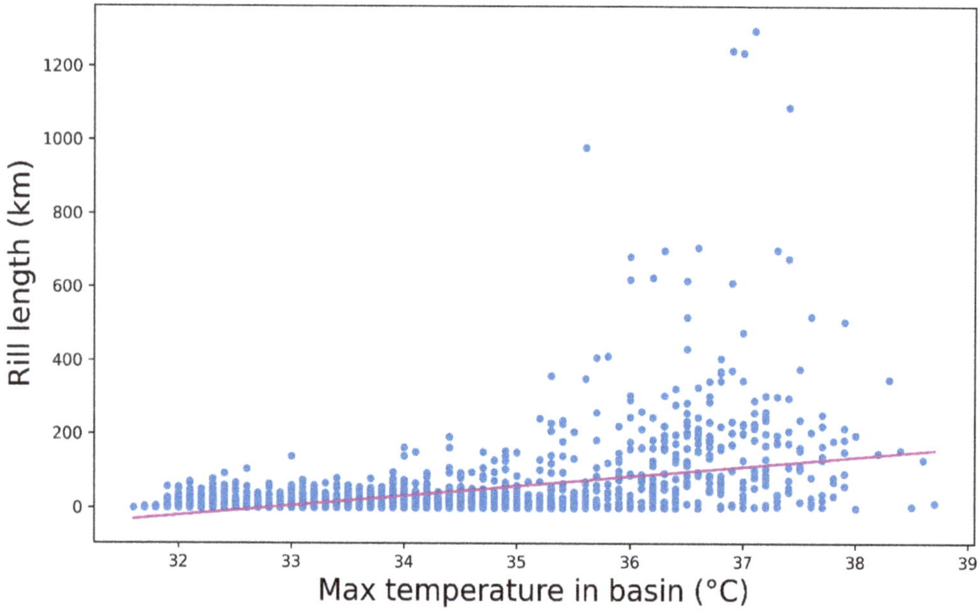

Figure 9. Linear relation of the total length of rill erosion to maximum air temperatures.

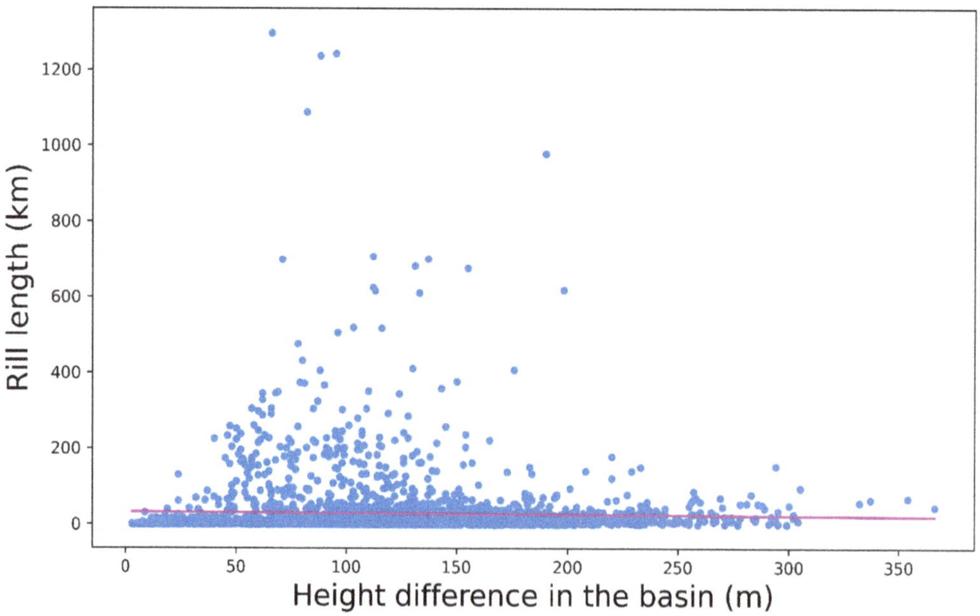

Figure 10. Linear relation of the total length of rill erosion to the elevation variation in the basin.

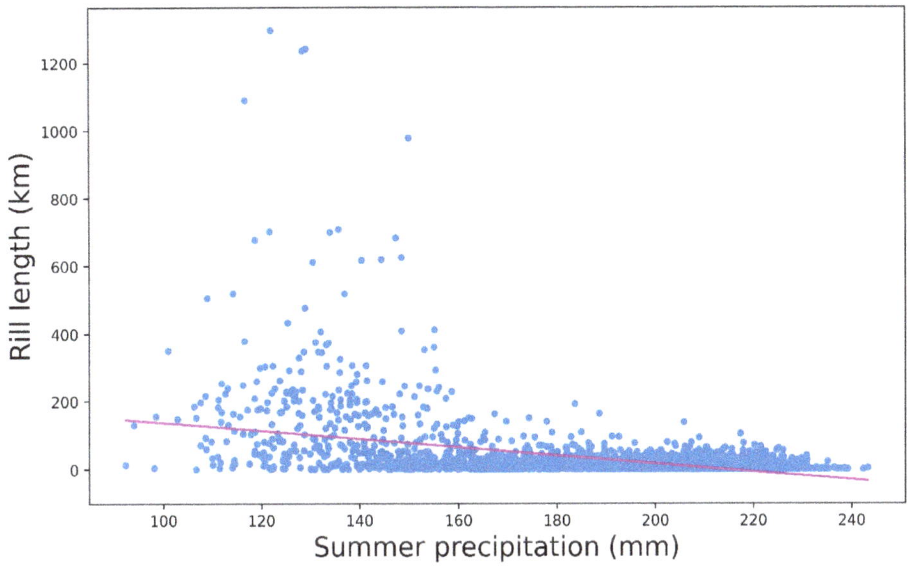

Figure 11. Linear relation of the total length of rill erosion to mean annual precipitation from May to August.

Modeling of generalized relationships was performed in a package for building generalized additive models in the Python language [79]. During modeling, statistically significant ($p < 0.001$), stable, and strong relationships were found between the density of rill erosion and such natural factors as mean basin elevation (Figure 12A), mean basin slope (Figure 12B), basin elevation range (Figure 13A), basin forest cover (Figure 13B), average air temperature in January (Figure 14A), average annual long-term maximum temperature (Figure 14B), average warm-season precipitation (Figure 15A), average basin hydrothermal coefficient (Figure 15B), longitude (Figure 16A), and latitude (Figure 16B) of the basin centroid.

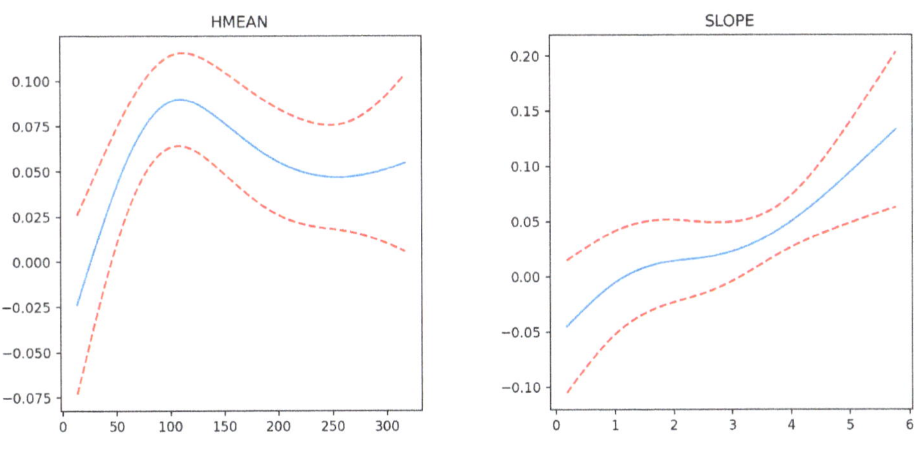

(**A**) (**B**)

Figure 12. Dependence of rill erosion density on topography factors—average height (m) (**A**) and average slope steepness (deg) (**B**) in the basin.

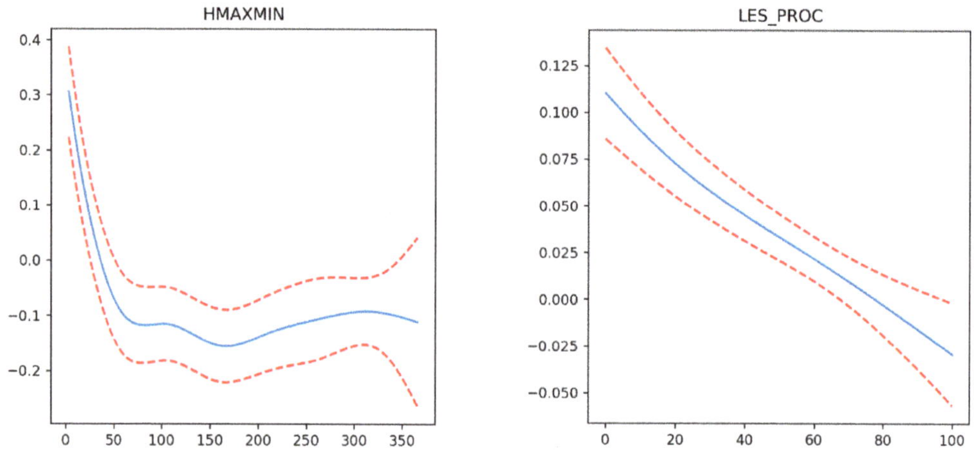

(A) **(B)**

Figure 13. Dependence of rill erosion density on elevation range (m) (**A**) and average forest cover (%) (**B**) in the basin.

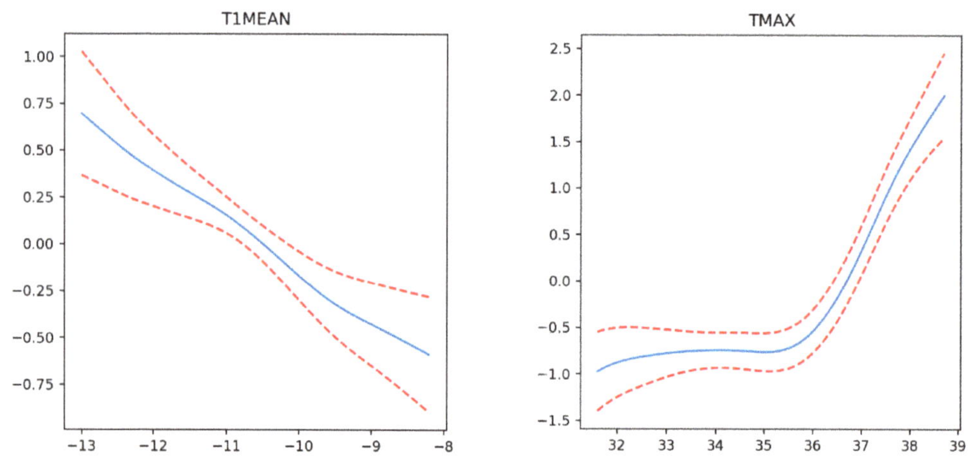

(A) **(B)**

Figure 14. Dependence of rill erosion density on temperature factors (°C)—mean annual temperature in January (**A**) and mean annual maximum temperature in the warm season (**B**).

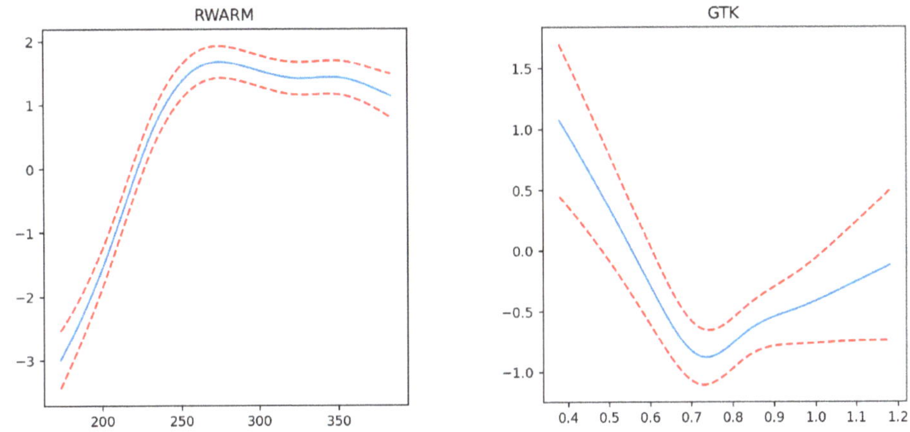

(**A**) (**B**)

Figure 15. Dependence of rill erosion density on mean annual precipitation in the warm period (mm) (**A**) and the value of the hydrothermal coefficient (GTK) (**B**) in the basin.

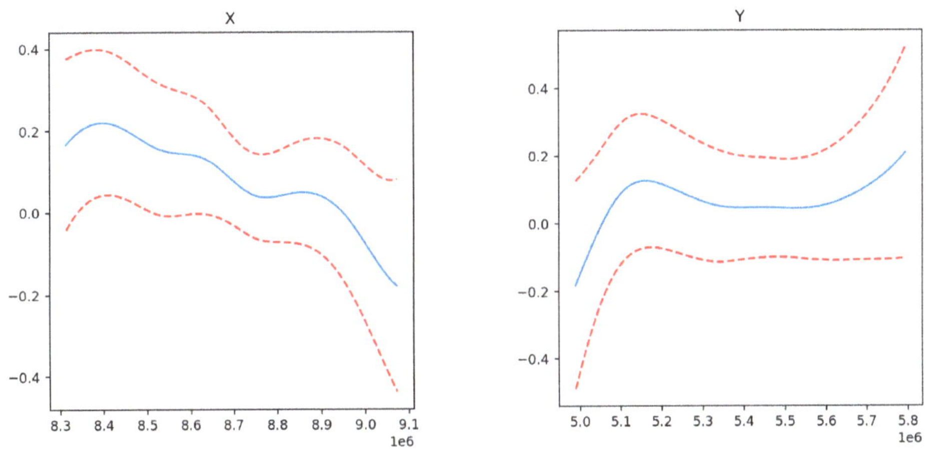

(**A**) (**B**)

Figure 16. Dependence of rill erosion density on geographical location: (**A**)—longitude (m), (**B**)—latitude (m).

Generalized models confirmed the known relationships between soil erosion and natural factors and revealed complex nonlinear relationships. It is observed that the density of the erosion network is stable with an increase in the average elevation of the basin, but up to 100 m, after which the growth of density is replaced by a decrease in the density and its stabilization (Figure 12A). In our opinion, this dependence should be considered in combination with the effect of average slope steepness on the density of the erosion network. In this case, the relationship is more superficial and closer to a direct linear relationship—the higher the steepness, the greater the density (Figure 12B). However, when analyzing steepness in basins with an average elevation of less than 100 m, a pattern becomes apparent—steepness does not exceed 3 degrees, and the slopes of these basins are well suited for farming, particularly as arable land. Indeed, in these areas, the percentage of the plowed area exceeds 50%.

The effect of relief morphometry is also confirmed by the dependence of the density of rill erosion on the height dispersion index in the basin—the difference between the maximum and minimum heights. In general, the smaller the height difference, the lower the density, but fundamentally, the most significant decrease in rill erosion intensity is observed in basins with a small height difference (up to ~70 m), after which the intensity of density reduction is not as pronounced and essentially remains constant (Figure 13A). The influence of anthropogenic development on the intensity of erosion processes is also confirmed by the inverse relationship between forest cover and the density of the rill network (Figure 13B).

Climatic factors have no less influence on the character of erosion processes in the study area, partly complementing the topographic and economic predisposition of the area to the development of rills. The relationship between the density of rill erosion and the average air temperature in January is interesting—the lower the mean annual temperature in the coldest month of the year, the more intense the process's pattern is (Figure 14A). It seems to be related to the soil freezing capacity, which ensures the intensity of the erosion processes. However, it is impossible to verify this reliably due to the lack of adequate models of soil freezing in the study area. Temperature influences the character of erosion processes in summer as well—there is a strong direct correlation between the intensity of the process and the maximum air temperature starting from 36 degrees Celsius (Figure 14B). Initially, the authors attributed this to the potentially high evaporation capacity at such temperatures and the possible high precipitation intensity. However, analysis of the dependence of erosion processes on the precipitation layer in the warm season and the zoning map of the study area by the maximum temperature rejected this hypothesis. The maximum air temperatures are associated with erosion processes indirectly through the prevailing soil types, as the conditions for the emergence of one or another type. This is confirmed because the areas with maximum temperatures over 36 degrees are located in the transition zone of semi-deserts and dry-steppes with prevailing chestnut and solonetz soil types. These types of soils are characterized by a large number of cracks reaching a depth of 5–6 cm [80]. Even with an aggregate not large annual precipitation layer, the passing rains create a continuous surface runoff on the dry soils, which erodes the soils along the available cracks, forming stable washouts.

In general, the relationship between the intensity of rill erosion and moisture is direct; there is a steady trend to an increase in the density of the erosion network with increasing precipitation in the warm period up to ~270 mm, after which the role of precipitation decreases while maintaining the rule—"much precipitation—more intense erosion" (Figure 15A). Stabilization of the trend, in this case, is explained by the fact that 270 mm is a kind of boundary between steppe and forest-steppe zones, in which there is a clear seasonality of precipitation. By the time of the most intensive precipitation (summer-autumn period), vegetation with good soil-protective properties develops on these territories [24], which becomes the dominant factor of soil erosion intensification. The influence of moisture factors most clearly shows the relationship between the density of rill erosion and the value of the hydrothermal coefficient of the territory (Figure 15B). The density of the erosion network decreases with increasing GTK up to the value of 0.7—a clear border between arid area and wetted area. The intensity of soil-protecting vegetation increases with the increase of the SCC, but the precipitation becomes sufficient for the vegetation to perform its soil-protecting functions less effectively when moving into the wetted zone.

The analysis of the influence of the geographical location of the basin on the intensity of erosion processes also demonstrated statistically significant relationships. The density of rill erosion decreases linearly as one move to the east of the study area (Figure 16A), which is explained by the fact that the western part of the study area has a higher right bank of the Volga River and, as a result, higher values of slope steepness, which, as shown earlier, directly affect the intensity of the process. The latitudinal pattern is not so linear and more complex (Figure 16B)—with moving from the south to the north up to the zone of typical steppes, there is a systematic increase in the density of rill erosion, after which there is some decrease in the intensity of soil erosion and its stabilization, after which, starting from the northern part of the forest-steppe zone there is another gradual increase in the intensity of erosion activity. The reasons for this, as previously shown, lie in the zonal features of the study area, changes in the degree of moisture, as well as the types of plant communities, including agricultural ones, growing in a particular area.

It is worth separately considering the influence of soil-geological conditions of erosion processes. A one-factor analysis of variance was conducted to analyze the relationship between the prevailing soil type and total rill erosion length (Figure 17).

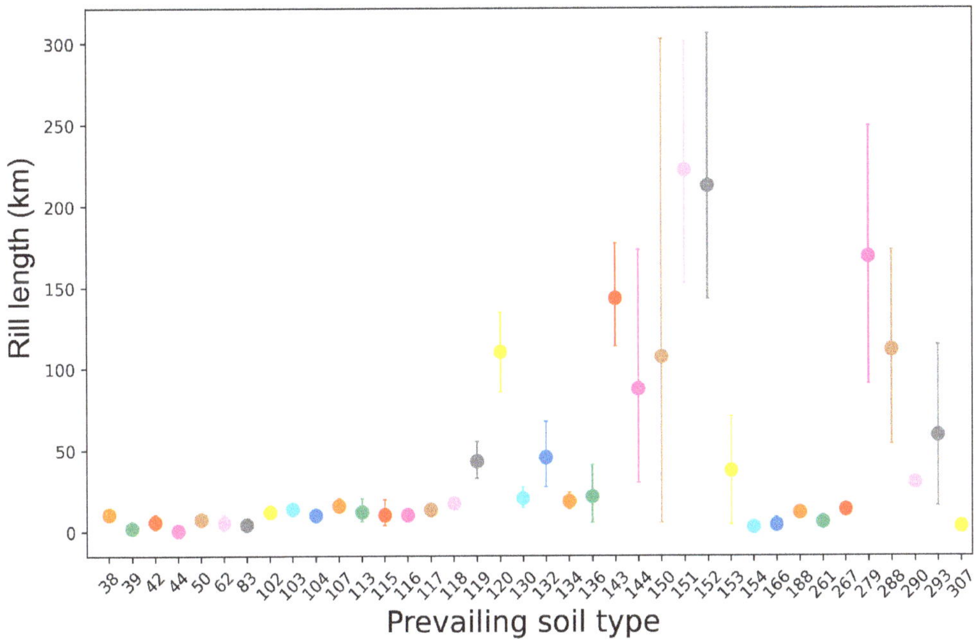

Figure 17. Dependence of the length of the erosion network on the prevailing soil type. Soil types are given in the Supplementary Materials.

The basic soils of taiga and coniferous-broadleaved forests, as well as hydromorphic soils, are least susceptible to erosion activity. The smallest total length of rills is observed on sod-podzol soils (shallow-podzol and deep-podzol). These taiga soils, as a rule, are covered by mixed forests with a dense canopy and good protection against the formation of surface runoff. Moving southward, there is a slight increase in the total length of the rill network on gray forest soils of various subtypes. These soils are actively involved in agricultural management and have suitable hydrometeorological conditions. Weak susceptibility to erosion of territories with hydromorphic soils is, in our opinion, evident and connected, by and large, with land use on such soils—these are meadows with maximum soil-protective properties. The next group of soils is more affected by erosion processes—steppe soils. These are all kinds of chernozems, starting from leached chernozems. Southern chernozems, which are peculiar "record-breakers" by erosion susceptibility among typical steppes soils, are distinguished. Southern chernozems are in the cluster with the highest density of erosion network (Figure 7). Southern chernozems are spread in the southern part of the steppe zone. They are formed in the conditions of semiarid climate under soddy-cereal medium steppes. Grass cover is sparse, clearly expressed summer semi-rest period for most dominant cereals. Soil-forming rocks are represented mainly by loess and loess-like loams, often containing easily soluble salts, as well as eluvial-deluvial deposits. Water conditions of soils are solid. Agricultural development of southern chernozems is high: in the European part of Russia, it exceeds 50%, while moving to the east, plowing decreases and the number of pastures increases. The main crops grown are cereals (wheat and corn) and legumes; considerable areas are occupied by industrial crops (sugar beet and tobacco), vegetable, and melon crops. Plowed soils are prone to water and wind erosion, structural degradation, and slitization under irrigation. The last group is soils with a maximum total length of erosion network—soils of dry steppes and semi-deserts. These are different subtypes of chestnut soils and solonetz soils, a description of which was given earlier.

The one-factor analysis of variance was also applied to analyze the relationship between erosion activity and the predominant type of soil-forming rocks (Figure 18). In this case, grouping types is problematic due to implicit clustering; however, certain patterns are still observed. The increase in total length of rill erosion directly depends on clay content in the soil-forming rock and—in its water-holding capacity, drainability. The minimum, near-zero erosion network length is observed on weakly erodible shales. Analyzing the rocks on which erosion is observed, the minimum length of the erosion network falls on the group of rocks conditionally united into the so-called "sands"—these are sandy rocks and sandstones. These rocks have the maximum infiltration capacity, not allowing the formation of washout surface runoff and coming to the soils with minimal humus horizons. Slightly better eroded areas are composed of clayey and loamy rocks, underlain by sandy and sandy loam rocks, as well as light loam rocks themselves. Medium loamy rocks and rocks of different textures with a predominance of loam and clay are better eroded. On average, easily erodible chalky rocks—limestone and other carbonate rocks—are even better erodible. They usually form gray forest soils and leached chernozems, heavily involved in agricultural production. The most significant total length of the erosion network is expected to be in water-resistant clay and heavy loam rocks, on which surface runoff is formed, eroding fertile soil layers. Separately, sandy loam rocks, for which we could not find a reliable relationship—these rocks, draining well the surface runoff, can be resistant to erosion processes and, folding typical chernozems, due to active economic activity falls in the area of maximum rill erosion.

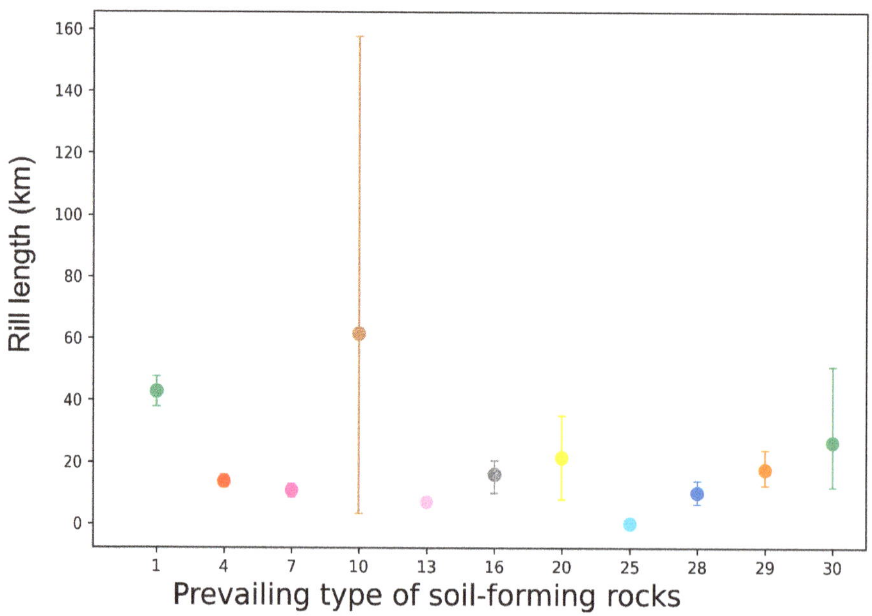

Figure 18. Dependence of the length of the erosion network on the prevailing type of soil-forming rocks. The rock types are given in the Supplementary Materials.

The analysis of variance was conducted separately by landscape zones, and the result can be observed in Figure 19. The results confirm the earlier conclusions—the zones of dry and typical steppes and semi-deserts are most susceptible to erosion processes.

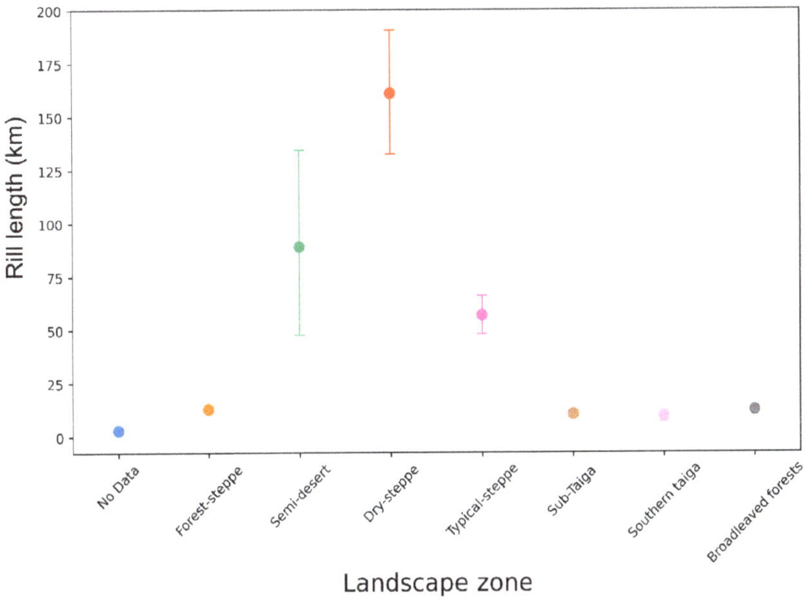

Figure 19. Relationship between the rill erosion network length and landscape zone.

The analysis of the relationship between erosion processes and anthropogenic load also confirms the previous conclusions (Figure 20). The score of the anthropogenic load is given based on the development of the territory, the presence of settlements, industry, and roads of various types, as well as population density.

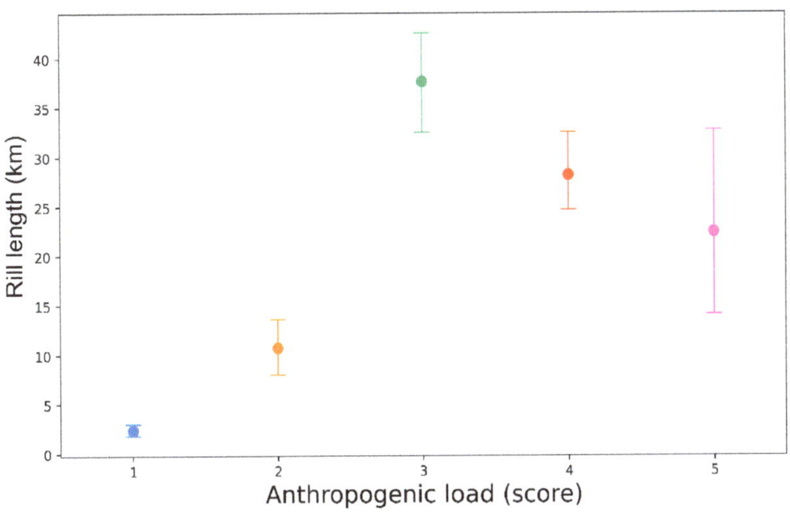

Figure 20. The dependence of the length of the rill erosion network on the anthropogenic load.

The intensity of erosion processes increases with increasing anthropogenic load, but not linearly. The maximum intensity of erosion is observed under moderate anthropogenic load. Decrease of the total length of erosion forms with further increase of anthropogenic load is related to decrease of the area not occupied by cities, manufactories, and other anthropogenic objects.

The cumulative effect of anthropogenic load, forest cover, prevailing soil type, and soil-forming rocks has the most significant contribution to the development of erosion processes in the study area (Figure 21).

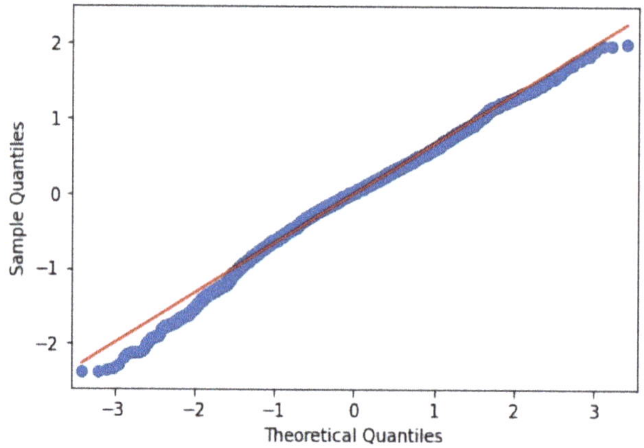

Figure 21. QQ–plot of MANOVA between rill network and forest cover, anthropogenic load, prevailing soil type, and soil-forming rocks.

5. Conclusions

Deep neural networks can effectively solve geomorphological problems, allowing large-scale mapping of erosion processes. For the first time in the world, a detailed assessment of the density of rill erosion and its relation to the main natural–anthropogenic factors is given. In this case, preparation of the training sample is the most time-consuming stage, taking 2/3 of the whole time of the study. The trained model allows highly efficient and at an acceptable level of recognition of the erosion network on high-resolution images, in our case using Sentinel-2 images. We investigated which Sentinel-2 bands allow the most efficient visual recognition at the preliminary stage. As a result of numerous experiments, the best "readability" of the image was demonstrated by a separate near-infrared band, winning in comparison with RGB-synthesis, synthesis of artificial colors in various combinations, and other channels, separately. Another sensitive point of the research is the choice of neural network architecture for semantic segmentation. Experiments were carried out with the most common architectures for such tasks, including the well-known U-Net architecture, as well as FPN, PSPNet, and LinkNet in combination with the most popular encoder groups, such as VGG, ResNet, SE-ResNet, ResNeXt, SE-ResNeXt, SENet154, DenseNet, Inception, MobileNet, and EfficientNet. The best recognition accuracy on the test set showed the FPN + EfficientNet combination; however, when applying the trained model, the recognition results were poor. The best results in terms of real-world applications were achieved using LinkNet + DenseNet (DenseNet201). In the future, it is planned to use the trained model to recognize the erosion network for 2019–2020 to assess the dynamics of erosion processes in time.

The analysis made it possible to assess the key factors influencing the intensity of rill erosion processes at a high level of confidence. In some cases, existing knowledge about the influence of certain factors on soil erosion was confirmed. Particular interest is the possibility of mapping the territories with the highest risk of erosion processes due to natural–anthropogenic conditions, which will allow producing precise and effective measures to minimize the risks. Under the conditions of intensive agricultural development, no reduction of anthropogenic load can be discussed; however, proper crop rotations, regular sowing of perennial grasses, and proper plowing will reduce the existing risks many times.

The use of generalized additive models (GAM) to analyze dependencies made it possible to describe complex relationships between the density of rill erosion and climate, geomorphological, and other geographic factors. Unfortunately, not all parameters could be used in the analysis; in the future, detailed modeling of the dependence of erosion processes on factors of soil freezing, and the intensity of precipitation, especially changes in intensity in time series, should be conducted.

Supplementary Materials: The following supporting information can be downloaded at: https://www.mdpi.com/article/10.3390/ijgi11030197/s1, Table S1: Prevailing soil type list; Table S2: Prevailing type of soil-forming rocks list.

Funding: This research was funded by grant number MK-2005.2021.1.5.

Institutional Review Board Statement: Not applicable.

Informed Consent Statement: Not applicable.

Data Availability Statement: The data presented in this study are available on request from the corresponding author and partially available on the http://bassepr.kpfu.ru/ accessed on 15 December 2021.

Acknowledgments: The author would like to thank Yandex Cloud (https://cloud.yandex.com/ accessed on 30 December 2021) for providing computing resources for the project within the Yandex Datasphere product. The author would like to thank the reviewers for their helpful comments and suggestions, which improved the manuscript.

Conflicts of Interest: The authors declare no conflict of interest.

References

1. Yang, Y.; Shi, Y.; Liang, X.; Huang, T.; Fu, S.; Liu, B. Evaluation of Structure from Motion (SfM) Photogrammetry on the Measurement of Rill and Interrill Erosion in a Typical Loess. *Geomorphology* **2021**, *385*, 107734. [CrossRef]
2. Di Stefano, C.; Ferro, V.; Pampalone, V.; Sanzone, F. Field Investigation of Rill and Ephemeral Gully Erosion in the Sparacia Experimental Area, South Italy. *Catena* **2013**, *101*, 226–234. [CrossRef]
3. Murgatroyd, A.L.; Ternan, J.L. The Impact of Afforestation on Stream Bank Erosion and Channel Form. *Earth Surf. Process. Landf.* **1983**, *8*, 357–369. [CrossRef]
4. Nearing, M.A.; Norton, L.D.; Bulgakov, D.A.; Larionov, G.A.; West, L.T.; Dontsova, K.M. Hydraulics and Erosion in Eroding Rills. *Water Resour. Res.* **1997**, *33*, 865–876. [CrossRef]
5. Yermolaev, O.P.; Gafurov, A.M.; Usmanov, B.M. Evaluation of Erosion Intensity and Dynamics Using Terrestrial Laser Scanning. *Eurasian Soil Sci.* **2018**, *51*, 814–826. [CrossRef]
6. Kociuba, W.; Janicki, G.; Rodzik, J.; Stępniewski, K. Comparison of Volumetric and Remote Sensing Methods (TLS) for Assessing the Development of a Permanent Forested Loess Gully. *Nat. Hazards* **2015**, *79*, 139–158. [CrossRef]
7. Vinci, A.; Brigante, R.; Todisco, F.; Mannocchi, F.; Radicioni, F. Measuring Rill Erosion by Laser Scanning. *Catena* **2015**, *124*, 97–108. [CrossRef]
8. Usmanov, B.; Yermolaev, O.; Gafurov, A. Estimates of Slope Erosion Intensity Utilizing Terrestrial Laser Scanning. *Proc. Int. Assoc. Hydrol. Sci.* **2015**, *367*, 59–65. [CrossRef]
9. Hu, T.; Sun, X.; Su, Y.; Guan, H.; Sun, Q.; Kelly, M.; Guo, Q. Development and Performance Evaluation of a Very Low-Cost UAV-Lidar System for Forestry Applications. *Remote Sens.* **2021**, *13*, 77. [CrossRef]
10. Nex, F.; Remondino, F. UAV for 3D Mapping Applications: A Review. *Appl. Geomat.* **2014**, *6*, 1–15. [CrossRef]
11. Gafurov, A.M. Possible Use of Unmanned Aerial Vehicle for Soil Erosion Assessment. *Uchenye Zap. Kazan. Univ. Ser. Estestv. Nauk.* **2017**, *159*, 654–667.
12. Gafurov, A. The Methodological Aspects of Constructing a High-Resolution DEM of Large Territories Using Low-Cost UAVs on the Example of the Sarycum Aeolian Complex, Dagestan, Russia. *Drones* **2021**, *5*, 7. [CrossRef]
13. Pandey, A.; Himanshu, S.K.; Mishra, S.K.; Singh, V.P. Physically Based Soil Erosion and Sediment Yield Models Revisited. *Catena* **2016**, *147*, 595–620. [CrossRef]
14. Wischmeier, W.H.; Smith, D.D. *Predicting Rainfall Erosion Losses-a Guide to Conservation Planning*; U.S. Department of Agriculture: Betsville, MD, USA, 1978.

15. Renard, K.G.; Foster, G.R.; Weesies, G.A.; Porter, J.P. RUSLE: Revised Universal Soil Loss Equation. *J. Soil Water Conserv.* **1991**, *46*, 30–33.
16. Sadeghi, S.H.R.; Gholami, L.; Darvishan, A.K.; Saeidi, P. A Review of the Application of the MUSLE Model Worldwide. *Hydrol. Sci. J.* **2014**, *59*, 365–375. [CrossRef]
17. Knisel, W.G. *CREAMS: A Field Scale Model for Chemicals, Runoff, and Erosion from Agricultural Management Systems*; Department of Agriculture, Science and Education Administration: Baltimore, MD, USA, 1980.
18. De Roo, A.P.J.; Wesseling, C.G.; Jetten, V.G.; Ritsema, C. LISEM: A Physically-Based Hydrological and Soil Erosion Model Incorporated in a GIS. In *Application of Geographic Information Systems in Hydrology and Water Resources Management*; Kovar, K., Nachtnebel, H.P., Eds.; IAHS: Wallingford, UK, 1996; Volume 235, pp. 395–403.
19. Flanagan, D.C.; Ascough, J.C.; Nearing, M.A.; Laflen, J.M. The Water Erosion Prediction Project (WEPP) Model. In *Landscape Erosion and Evolution Modeling*; Harmon, R.S., Doe, W.W., Eds.; Springer: Boston, MA, USA, 2001; pp. 145–199. ISBN 978-1-4615-0575-4.
20. Favis-Mortlock, D.T.; Boardman, J.; Parsons, A.J.; Lascelles, B. Emergence and Erosion: A Model for Rill Initiation and Development. *Hydrol. Process.* **2000**, *14*, 2173–2205. [CrossRef]
21. Misra, R.K.; Rose, C.W. Application and Sensitivity Analysis of Process-Based Erosion Model GUEST. *Eur. J. Soil Sci.* **1996**, *47*, 593–604. [CrossRef]
22. Mahmoodabadi, M.; Ghadiri, H.; Rose, C.; Yu, B.; Rafahi, H.; Rouhipour, H. Evaluation of GUEST and WEPP with a New Approach for the Determination of Sediment Transport Capacity. *J. Hydrol.* **2014**, *513*, 413–421. [CrossRef]
23. Morgan, R.P.C.; Quinton, J.N.; Smith, R.E.; Govers, G.; Poesen, J.W.A.; Auerswald, K.; Chisci, G.; Torri, D.; Styczen, M.E. The European Soil Erosion Model (EUROSEM): A Dynamic Approach for Predicting Sediment Transport from Fields and Small Catchments. *Earth Surf. Process. Landf.* **1998**, *23*, 527–544. [CrossRef]
24. Mukharamova, S.; Saveliev, A.; Ivanov, M.; Gafurov, A.; Yermolaev, O. Estimating the Soil Erosion Cover-Management Factor at the European Part of Russia. *ISPRS Int. J. Geo-Inf.* **2021**, *10*, 645. [CrossRef]
25. Maltsev, K.A.; Golosov, V.N.; Gafurov, A.M. Digital Elevation Models and Their Use for Assessing Soil Erosion Rates on Arable Lands. *Uchenye Zap. Kazan. Universiteta Seriya Estestv. Nauk.* **2018**, *160*, 514–530.
26. Pielke, R.A.; Cotton, W.R.; Walko, R.L.; Tremback, C.J.; Lyons, W.A.; Grasso, L.D.; Nicholls, M.E.; Moran, M.D.; Wesley, D.A.; Lee, T.J.; et al. A Comprehensive Meteorological Modeling System—RAMS. *Meteorol. Atmos. Phys.* **1992**, *49*, 69–91. [CrossRef]
27. Sheffield, J.; Goteti, G.; Wood, E.F. Development of a 50-Year High-Resolution Global Dataset of Meteorological Forcings for Land Surface Modeling. *J. Clim.* **2006**, *19*, 3088–3111. [CrossRef]
28. Shruthi, R.B.V.; Kerle, N.; Jetten, V. Object-Based Gully Feature Extraction Using High Spatial Resolution Imagery. *Geomorphology* **2011**, *134*, 260–268. [CrossRef]
29. Conoscenti, C.; Angileri, S.; Cappadonia, C.; Rotigliano, E.; Agnesi, V.; Märker, M. Gully Erosion Susceptibility Assessment by Means of GIS-Based Logistic Regression: A Case of Sicily (Italy). *Geomorphology* **2014**, *204*, 399–411. [CrossRef]
30. Conoscenti, C.; Agnesi, V.; Cama, M.; Caraballo-Arias, N.A.; Rotigliano, E. Assessment of Gully Erosion Susceptibility Using Multivariate Adaptive Regression Splines and Accounting for Terrain Connectivity. *Land Degrad. Dev.* **2018**, *29*, 724–736. [CrossRef]
31. Golosov, V.; Gusarov, A.; Sharifillin, A.; Ivanova, N.N.; Gafurov, A.; Yermolaev, O.; Rysin, I. Using Bomb-Derived and Chernobyl-Derived 137Cs for the Assessment of Soil Losses Trends in Different Landscape Zones of the European Russia. In Proceedings of the 14th International Symposium on the Interaction between Sediments and Water, Taormina, Italy, 17–22 June 2017; p. 10.
32. Gusarov, A.V.; Golosov, V.N.; Sharifullin, A.G.; Gafurov, A.M. Contemporary Trend in Erosion of Arable Southern Chernozems (Haplic Chernozems Pachic) in the West of Orenburg Oblast (Russia). *Eurasian Soil Sci.* **2018**, *51*, 561–575. [CrossRef]
33. Golosov, V.; Koiter, A.; Ivanov, M.; Maltsev, K.; Gusarov, A.; Sharifullin, A.; Radchenko, I. Assessment of Soil Erosion Rate Trends in Two Agricultural Regions of European Russia for the Last 60 Years. *J. Soils Sediments* **2018**, *18*, 3388–3403. [CrossRef]
34. Gafurov, A.M. Using Unmanned Aerial Vehicles for Evaluation of Soil Erosion. *Belgorod State Univ. Sci. Bull. Nat. Sci.* **2019**, *43*, 182–190. [CrossRef]
35. Yermolaev, O.P.; Medvedeva, R.A.; Platoncheva, E.V. Methodological Approaches to Monitoring Erosion of Agricultural Lands in the European Part of Russia by Using Satellite Imagery. *Uchenye Zap. Kazan. Univ. Ser. Estestv. Nauk.* **2017**, *159*, 668–680.
36. Yermolayev, O.; Platoncheva, E.; Essuman-Quainoo, B. Spatial-Temporal Dynamics of the Ephemeral Gully Belt on the Plowed Slopes of River Basins in Natural and Anthropogenic Landscapes of the East of the Russian Plain. *Geosciences* **2020**, *10*, 167. [CrossRef]
37. Seutloali, K.E.; Dube, T.; Mutanga, O. Assessing and Mapping the Severity of Soil Erosion Using the 30-m Landsat Multispectral Satellite Data in the Former South African Homelands of Transkei. *Phys. Chem. Earth Parts A/B/C* **2017**, *100*, 296–304. [CrossRef]
38. Saadat, H.; Adamowski, J.; Tayefi, V.; Namdar, M.; Sharifi, F.; Ale-Ebrahim, S. A New Approach for Regional Scale Interrill and Rill Erosion Intensity Mapping Using Brightness Index Assessments from Medium Resolution Satellite Images. *Catena* **2014**, *113*, 306–313. [CrossRef]
39. Desprats, J.F.; Raclot, D.; Rousseau, M.; Cerdan, O.; Garcin, M.; Le Bissonnais, Y.; Ben Slimane, A.; Fouche, J.; Monfort-Climent, D. Mapping Linear Erosion Features Using High and Very High Resolution Satellite Imagery. *Land Degrad. Dev.* **2013**, *24*, 22–32. [CrossRef]
40. Fiorucci, F.; Ardizzone, F.; Rossi, M.; Torri, D. The Use of Stereoscopic Satellite Images to Map Rills and Ephemeral Gullies. *Remote Sens.* **2015**, *7*, 14151–14178. [CrossRef]

41. Kashtanov, A.N.; Vernyuk, Y.I.; Savin, I.Y.; Shchepot'ev, V.V.; Dokukin, P.A.; Sharychev, D.V.; Li, K.A. Mapping of Rill Erosion of Arable Soils Based on Unmanned Aerial Vehicles Survey. *Eurasian Soil Sci.* **2018**, *51*, 479–484. [CrossRef]
42. Karydas, C.; Bouarour, O.; Zdruli, P. Mapping Spatio-Temporal Soil Erosion Patterns in the Candelaro River Basin, Italy, Using the G2 Model with Sentinel2 Imagery. *Geosciences* **2020**, *10*, 89. [CrossRef]
43. Walter, V. Object-Based Classification of Remote Sensing Data for Change Detection. *ISPRS J. Photogramm. Remote Sens.* **2004**, *58*, 225–238. [CrossRef]
44. Zhang, F.; Li, J.; Zhang, B.; Shen, Q.; Ye, H.; Wang, S.; Lu, Z. A Simple Automated Dynamic Threshold Extraction Method for the Classification of Large Water Bodies from Landsat-8 OLI Water Index Images. *Int. J. Remote Sens.* **2018**, *39*, 3429–3451. [CrossRef]
45. Ghosh, A.; Maiti, R. Soil Erosion Susceptibility Assessment Using Logistic Regression, Decision Tree and Random Forest: Study on the Mayurakshi River Basin of Eastern India. *Environ. Earth Sci.* **2021**, *80*, 328. [CrossRef]
46. Dinh, T.V.; Nguyen, H.; Tran, X.-L.; Hoang, N.-D. Predicting Rainfall-Induced Soil Erosion Based on a Hybridization of Adaptive Differential Evolution and Support Vector Machine Classification. *Math. Probl. Eng.* **2021**, *2021*, e6647829. [CrossRef]
47. Liu, Y.; Zhou, J.; Qi, W.; Li, X.; Gross, L.; Shao, Q.; Zhao, Z.; Ni, L.; Fan, X.; Li, Z. ARC-Net: An Efficient Network for Building Extraction from High-Resolution Aerial Images. *IEEE Access* **2020**, *8*, 154997–155010. [CrossRef]
48. Cai, J.; Chen, Y. MHA-Net: Multipath Hybrid Attention Network for Building Footprint Extraction from High-Resolution Remote Sensing Imagery. *IEEE J. Sel. Top. Appl. Earth Obs. Remote Sens.* **2021**, *14*, 5807–5817. [CrossRef]
49. Luo, L.; Li, P.; Yan, X. Deep Learning-Based Building Extraction from Remote Sensing Images: A Comprehensive Review. *Energies* **2021**, *14*, 7982. [CrossRef]
50. Blais, M.-A.; Akhloufi, M.A. *Deep Learning for Low Altitude Coastline Segmentation*; Online Only; SPIE: Orlando, FL, USA, 2021; Volume 11752.
51. Aryal, B.; Escarzaga, S.M.; Vargas Zesati, S.A.; Velez-Reyes, M.; Fuentes, O.; Tweedie, C. Semi-Automated Semantic Segmentation of Arctic Shorelines Using Very High-Resolution Airborne Imagery, Spectral Indices and Weakly Supervised Machine Learning Approaches. *Remote Sens.* **2021**, *13*, 4572. [CrossRef]
52. Garg, R.; Kumar, A.; Bansal, N.; Prateek, M.; Kumar, S. Semantic Segmentation of PolSAR Image Data Using Advanced Deep Learning Model. *Sci. Rep.* **2021**, *11*, 15365. [CrossRef]
53. Dong, Y.; Li, F.; Hong, W.; Zhou, X.; Ren, H. Land Cover Semantic Segmentation of Port Area with High Resolution SAR Images Based on SegNet. In *2021 SAR in Big Data Era (BIGSARDATA)*; Institute of Electrical and Electronics Engineers (IEEE): Piscataway, NJ, USA, 2021; pp. 1–4. [CrossRef]
54. Wei, H.; Xu, X.; Ou, N.; Zhang, X.; Dai, Y. Deanet: Dual Encoder with Attention Network for Semantic Segmentation of Remote Sensing Imagery. *Remote Sens.* **2021**, *13*, 3900. [CrossRef]
55. Illarionova, S.; Trekin, A.; Ignatiev, V.; Oseledets, I. Tree Species Mapping on Sentinel-2 Satellite Imagery with Weakly Supervised Classification and Object-Wise Sampling. *Forests* **2021**, *12*, 1413. [CrossRef]
56. Song, G.; Wu, S.; Lee, C.K.F.; Serbin, S.P.; Wolfe, B.T.; Ng, M.K.; Ely, K.S.; Bogonovich, M.; Wang, J.; Lin, Z.; et al. Monitoring Leaf Phenology in Moist Tropical Forests by Applying a Superpixel-Based Deep Learning Method to Time-Series Images of Tree Canopies. *ISPRS J. Photogramm. Remote Sens.* **2022**, *183*, 19–33. [CrossRef]
57. Gafurov, A.M.; Yermolayev, O.P. Automatic Gully Detection: Neural Networks and Computer Vision. *Remote Sens.* **2020**, *12*, 1743. [CrossRef]
58. Du, B.; Zhao, Z.; Hu, X.; Wu, G.; Han, L.; Sun, L.; Gao, Q. Landslide Susceptibility Prediction Based on Image Semantic Segmentation. *Comput. Geosci.* **2021**, *155*, 104860. [CrossRef]
59. Yermolaev, O.; Usmanov, B.; Gafurov, A.; Poesen, J.; Vedeneeva, E.; Lisetskii, F.; Nicu, I.C. Assessment of Shoreline Transformation Rates and Landslide Monitoring on the Bank of Kuibyshev Reservoir (Russia) Using Multi-Source Data. *Remote Sens.* **2021**, *13*, 4214. [CrossRef]
60. Yermolaev, O.P.; Mukharamova, S.S.; Maltsev, K.A.; Ivanov, M.A.; Ermolaeva, P.O.; Gayazov, A.I.; Mozzherin, V.V.; Kharchenko, S.V.; Marinina, O.A.; Lisetskii, F.N. Geographic Information System and Geoportal «River Basins of the European Russia». *IOP Conf. Ser. Earth Environ. Sci.* **2018**, *107*, 012108. [CrossRef]
61. Gafurov, A.M.; Rysin, I.I.; Golosov, V.N.; Grigoryev, I.I.; Sharifullin, A.G. Estimation of the recent rate of gully head retreat on the southern megaslope of the East European Plain using a set of instrumental methods. *Vestn. Mosk. Univ. Seriya 5 Geogr.* **2018**, *2018*, 61–71.
62. Gorelick, N.; Hancher, M.; Dixon, M.; Ilyushchenko, S.; Thau, D.; Moore, R. Google Earth Engine: Planetary-Scale Geospatial Analysis for Everyone. *Remote Sens. Environ.* **2017**, *202*, 18–27. [CrossRef]
63. Ke, G.; Meng, Q.; Finley, T.; Wang, T.; Chen, W.; Ma, W.; Ye, Q.; Liu, T.-Y. LightGBM: A Highly Efficient Gradient Boosting Decision Tree. In *Advances in Neural Information Processing Systems 30 (NIPS 2017)*; Curran Associates, Inc.: Long Beach, CA, USA, 2017; Volume 30.
64. Chaurasia, A.; Culurciello, E. LinkNet: Exploiting Encoder Representations for Efficient Semantic Segmentation. In Proceedings of the 2017 IEEE Visual Communications and Image Processing (VCIP), St. Petersburg, FL, USA, 10–13 December 2017; pp. 1–4.
65. Ronneberger, O.; Fischer, P.; Brox, T. U-Net: Convolutional Networks for Biomedical Image Segmentation. *arXiv* **2015**, arXiv:1505.04597.
66. Iandola, F.; Moskewicz, M.; Karayev, S.; Girshick, R.; Darrell, T.; Keutzer, K. DenseNet: Implementing Efficient ConvNet Descriptor Pyramids. *arXiv* **2014**, arXiv:1404.1869.

67. Jaccard, P. The Distribution of the Flora in the Alpine Zone. *New Phytol.* **1912**, *11*, 37–50. [CrossRef]
68. Miller, R.G., Jr. *Beyond ANOVA: Basics of Applied Statistics*; CRC Press: Boca Raton, FL, USA, 1997; ISBN 978-0-412-07011-2.
69. Huberty, C.J.; Olejnik, S. *Applied MANOVA and Discriminant Analysis*; John Wiley & Sons: Hoboken, NJ, USA, 2006; ISBN 978-0-471-78946-8.
70. Groisman, P.Y.A.; Karl, T.R.; Easterling, D.R.; Knight, R.W.; Jamason, P.F.; Hennessy, K.J.; Suppiah, R.; Page, C.M.; Wibig, J.; Fortuniak, K.; et al. Changes in the probability of heavy precipitation: Important indicators of climatic change. In *Weather and Climate Extremes*; Karl, T.R., Nicholls, N., Ghazi, A., Eds.; Springer: Dordrecht, The Netherlands, 1999; pp. 243–283. ISBN 978-90-481-5223-0.
71. Dore, M.H.I. Climate Change and Changes in Global Precipitation Patterns: What Do We Know? *Environ. Int.* **2005**, *31*, 1167–1181. [CrossRef]
72. Zolina, O.; Simmer, C.; Gulev, S.K.; Kollet, S. Changing Structure of European Precipitation: Longer Wet Periods Leading to More Abundant Rainfalls: Changing Structure of European Rainfall. *Geophys. Res. Lett.* **2010**, *37*, 1–5. [CrossRef]
73. Olchev, A.; Novenko, E.; Popov, V.; Pampura, T.; Meili, M. Evidence of Temperature and Precipitation Change over the Past 100 Years in a High-Resolution Pollen Record from the Boreal Forest of Central European Russia. *Holocene* **2017**, *27*, 740–751. [CrossRef]
74. Zolotokrylin, A.; Cherenkova, E. Seasonal Changes in Precipitation Extremes in Russia for the Last Several Decades and Their Impact on Vital Activities of the Human Population. *Geogr. Environ. Sustain.* **2017**, *10*, 69–82. [CrossRef]
75. Golosov, V.; Gusarov, A.; Litvin, L.; Yermolaev, O.; Chizhikova, N.; Safina, G.; Kiryukhina, Z. Evaluation of Soil Erosion Rates in the Southern Half of the Russian Plain: Methodology and Initial Results. *Proc. Int. Assoc. Hydrol. Sci.* **2017**, *375*, 23–27. [CrossRef]
76. Sharifullin, A.; Gusarov, A.; Gafurov, A.; Essuman-Quainoo, B. Preliminary Estimating the Contemporary Sedimentation Trend in Dry Valley Bottoms of First-Order Catchments of Different Landscape Zones of the Russian Plain Using the ^{137}CS as a Chronomarker. *IOP Conf. Ser. Earth Environ. Sci.* **2018**, *107*, 012022. [CrossRef]
77. Hastie, T.; Tibshirani, R. *Generalized Additive Models*; Chapman & Hall/CRC: Boca Raton, FL, USA, 1999; ISBN 978-0-412-34390-2.
78. Yermolaev, O.; Mukharamova, S.; Vedeneeva, E. River Runoff Modeling in the European Territory of Russia. *Catena* **2021**, *203*, 105327. [CrossRef]
79. Servén, D.; Brummitt, C.; Abedi, H. *Hlink Dswah/Pygam: V0.8.0*; Zenodo: Genève, Switzerland, 2018.
80. Babayan, L.A.; Protopopov, V.M. The Fertility of Light Chestnut Soils on Different Elements of Watershed Topography. *Eurasian Soil Sci.* **1997**, *30*, 1113–1116.

International Journal of *Geo-Information*

MDPI

Article

DEM- and GIS-Based Analysis of Soil Erosion Depth Using Machine Learning

Kieu Anh Nguyen and Walter Chen *

Department of Civil Engineering, National Taipei University of Technology, Taipei 10608, Taiwan; t106429401@ntut.edu.tw

* Correspondence: waltchen@ntut.edu.tw; Tel.: +886-2-27712171 (ext. 2628)

Abstract: Soil erosion is a form of land degradation. It is the process of moving surface soil with the action of external forces such as wind or water. Tillage also causes soil erosion. As outlined by the United Nations Sustainable Development Goal (UN SDG) #15, it is a global challenge to "combat desertification, and halt and reverse land degradation and halt biodiversity loss." In order to advance this goal, we studied and modeled the soil erosion depth of a typical watershed in Taiwan using 26 morphometric factors derived from a digital elevation model (DEM) and 10 environmental factors. Feature selection was performed using the Boruta algorithm to determine 15 factors with confirmed importance and one tentative factor. Then, machine learning models, including the random forest (RF) and gradient boosting machine (GBM), were used to create prediction models validated by erosion pin measurements. The results show that GBM, coupled with 15 important factors (confirmed), achieved the best result in the context of root mean square error (RMSE) and Nash–Sutcliffe efficiency (NSE). Finally, we present the maps of soil erosion depth using the two machine learning models. The maps are useful for conservation planning and mitigating future soil erosion.

Keywords: soil erosion; erosion pin; machine learning; morphometric factor; Shihmen Reservoir watershed

Citation: Nguyen, K.A.; Chen, W. DEM- and GIS-Based Analysis of Soil Erosion Depth Using Machine Learning. *ISPRS Int. J. Geo-Inf.* **2021**, *10*, 452. https://doi.org/10.3390/ijgi10070452

Academic Editor: Wolfgang Kainz

Received: 18 May 2021
Accepted: 29 June 2021
Published: 1 July 2021

1. Introduction

The United Nations General Assembly adopted 17 sustainable development goals (SDGs) in September 2015, which apply to all countries on the planet. Soil science is intertwined with a number of the SDGs. Among them, soils especially play an essential role in SDGs 2, 3, 6, 7, 12–15 [1].

Soil erosion is a form of land degradation and a severe threat to sustainable development. It is the process of moving surface soil with the action of external forces such as wind or water. Tillage also causes soil erosion. Among them, water erosion is the most tangible form of soil erosion in Taiwan. Soil erosion and sediment movement caused by rainfall and flooding, intense and persistent winds, agricultural activities, grazing, logging, mining, and construction result in significant damage to properties and potentially result in loss of lives, not to mention the livelihood support the land provides for communities. Therefore, it is a global challenge by 2030 to "combat desertification, and halt and reverse land degradation and halt biodiversity loss," as outlined by SDG 15.

Although the soil erosion process may seem to be slow at times, it dramatically impacts soil fertility, agriculture, and the ecosystem. Globally, it is estimated that the average soil erosion from agriculture is 75 billion tons/year ([2,3], as cited in [4]). Other scholars point out that about 85% of the 2 billion hectares of worldwide surface soil degradation stem from wind and water erosion ([5], based on [6,7]). The economic costs of erosion and sedimentation are substantial. For example, the cost of removing sediments alone could be somewhere between USD 7 and USD 68/yard^3 (or USD \$9.16–USD 88.94/m^3) in the US ([8], as cited in [9]). In Iran, the economic costs associated with soil erosion are thought to be around 10 trillion rials or USD 23,750,148 ([10], as cited in [11]). As a result, soil erosion modeling is critical to understanding soil erosion processes and preventing future soil loss.

Recently, the Global Applications of Soil Erosion Modeling Tracker (GASEMT) database was developed using peer-reviewed soil erosion modeling research literature published between 1994 and 2017 and is used to help the UN's global soil erosion assessment. The database contains the most up-to-date information on soil erosion modeling applications from around the world. In total, the GASEMT database contains 435 distinct soil erosion models and model variants. Despite the numerous models available for soil erosion modeling in the GASEMT database, statistics show that entries for watershed-scale applications are the most numerous (59%), and the (revised) universal soil loss equation ((R)USLE) family of models is the most commonly used soil erosion prediction models in the world, at about 41% [12]. Moreover, if USLE-based models such as the water and tillage erosion model/the sediment delivery model (WaTEM/SEDEM), erosion-productivity impact calculator (EPIC), and soil and water assessment tool (SWAT) are included in the same group, then this value could rise to 55%. Since the (R)USLE family is limited to sheet and rill erosion, the great majority of the model applications estimate only sheet and rill erosion amounts. Other types of erosion, notably stream bank erosion, gully erosion, and wind erosion, only account for 3.6% of the model applications combined. Finally, according to the bibliometric analysis based on the enhanced version of the GASEMT database, the (R)USLE model alone also has the largest number of total citations [13].

Understandably, the (R)USLE-family of models are also the most widely used soil erosion prediction models in Taiwan. For example, Chen et al. [14] applied the universal soil loss equation (USLE) model to reduce the order-of-magnitude discrepancy of soil loss estimates in the literature. Liu et al. [15] used two variants of the USLE model (grid cells and slope units) to calculate the soil loss due to sheet and rill erosion.

Beyond the traditional soil erosion models ((R)USLE, EPIC, SWAT, etc.), whether they are classified as empirical, conceptual, or process oriented, a growing alternative is to use machine learning (ML) or multicriteria decision making (MCDM) to improve the modeling ability of soil loss [11,16–19]. However, there are three significant limitations of these ML and ML-like studies. First, many of them only evaluate the presence or absence of soil erosion similar to landslide susceptibility modeling without considering the quantitative amount of soil loss [9,16–18]. Second, they tend to focus on or include gully erosion [16–19]. Third, and most distinctively, some of these studies use subjective evaluation, such as expert opinions, or results from other soil erosion models, as their validation [11,19], thus equivalently training models from subjective judgment and not from objective data such as field measurements.

Nguyen et al. [20] were the first to create machine learning models from field erosion pin measurements, a critical difference from other ML studies on soil erosion. The analysis was improved and expanded to different ML algorithms, including ensemble learning methods [21–23]. However, because some of the environmental factors used in the studies mentioned above were point data, the resulting models could not be directly applied to the entire study area (watershed) without interpolation.

The current study aims to improve the past studies by incorporating more independent variables (factors) derived from the watershed digital elevation model (DEM) and eliminate the dependence on the point data. The purpose is to create a comprehensive ML model that applies to the entire watershed.

2. Materials

Shihmen Reservoir watershed is located in northern Taiwan, which plays a crucial role in the metropolitan and irrigation areas of Taipei and Taoyuan [14]. It is also the third-largest reservoir in Taiwan. Typhoons bring the majority of the annual rainfall of 2350 mm to the Shihmen Reservoir watershed between May and October [24].

2.1. Environmental Factors and Erosion Pin Measurements

The 10 environmental factors (or parameters, or features, or variables, or attributes) examined in this study are main subwatershed, distance to river, distance to road, type of slope, slope direction, rainfall amount, lithology, epoch, elevation, and slope class. Environmental factors were obtained from various GIS sources such as land use/land cover maps, geological maps, river maps, and road system maps. These factors and four additional factors (% sand, % silt, % clay, and % organic) were previously analyzed in Nguyen et al. [22]. However, the four additional factors were removed from this study because they were point data and could not be directly mapped to the entire study area (watershed). We used morphometric factors to replace the point data.

The erosion pin data used in this study came from field surveys conducted over three years (September 2008 to October 2011). The erosion pins were mounted on 55 slopes in 17 of the 93 subwatersheds of the study area (Figure 1). Each slope had 10 erosion pins mounted, and the average value of the 10 pins represents the slope's erosion depth [25]. The measurements of erosion pins were taken with a caliper, as shown in Figure 2.

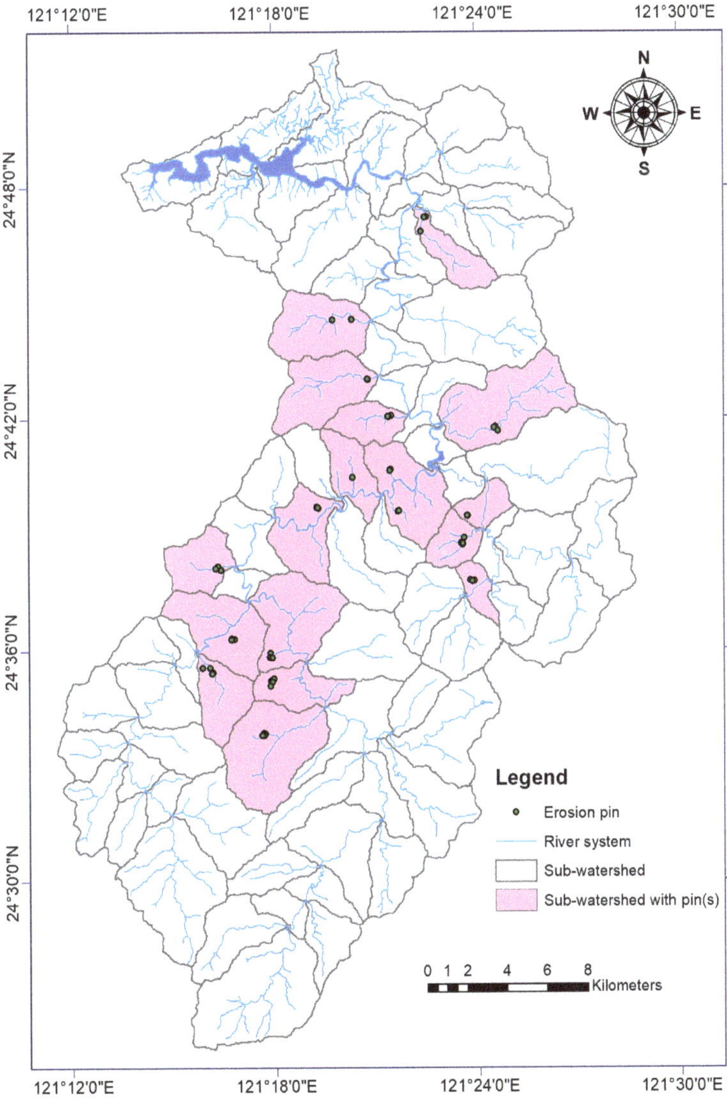

Figure 1. The study area of the Shihmen Reservoir watershed.

Figure 2. Illustration of field measurements in the Shihmen Reservoir watershed: (**a**) a natural slope, (**b**) an erosion pin and its label, and (**c**) measuring with a caliper ((**c**) from [22]).

2.2. Morphometric Factors

Morphometric analysis is the "quantitative description and analysis of landforms as practiced in geomorphology that may be applied to a particular kind of landform or to drainage basins and large regions generally" [26]. It is a technique for determining the scale and shape of watersheds, including two types of descriptive numbers: linear scale measurements and dimensionless numbers [27]. This approach can quantify the erosional growth of streams and their drainage watersheds, and compare geomorphic characteristics [28,29].

For this study, the Shihmen Reservoir watershed was divided into 93 subwatersheds to calculate the morphometric factors (or parameters, or features, or variables, or attributes) using the Central Geological Survey (CGS) DEM of Taiwan (10 m resolution) and ArcGIS 10.4.1. First, the DEM was filled in order to create flow paths and flow accumulations. Then, the stream networks were generated based on the flow accumulations of individual cells with a threshold value of 500. Finally, ArcGIS's Stream Link and Watershed functions were used to construct the subwatershed polygons. A total of 26 morphometric factors were calculated and described below (also see Table 1).

Subwatershed area (A) is the total area of a subwatershed. It ranged from 2.88 km^2 to 26.84 km^2 in this study. Research has indicated that total runoff or sediment yield is primarily determined by the subwatershed area [27].

Subwatershed perimeter (P) is the length of the boundary that surrounds a subwatershed. Its value varied between 10.70 and 37.29 km in the study area.

Stream order (U) indicates the complexity of a stream drainage system. The trunk river has the highest stream order and defines the order of a subwatershed [28]. An example of the stream order of a subwatershed is shown in Figure 3.

Number of streams (Nu) is the number of streams of a given stream order in a subwatershed. Figure 3 shows an example of the number of streams. The total number of streams (ΣN_u) is the summation of the number of streams of all orders.

Stream length (Lu) is the total channel length of a given stream order in this study for compatibility with the definition of the number of streams. It is not the cumulative channel length of a given order that includes all lesser orders, as sometimes defined [27]. The total stream length (ΣL_u) is the summation of the stream length of all orders.

Mean subwatershed slope (S) is the average slope of a subwatershed. It is calculated by the Slope function of ArcGIS and characterizes the steepness of a subwatershed.

Mean stream length (Lsm) is defined as the ratio between the stream length and the number of streams of a given stream order in a subwatershed in this study. We computed the average of the mean stream lengths as the characteristic mean stream length of the subwatershed.

Table 1. The morphometric factors used in this study.

	Morphometric Factor	Unit	Formula/Software
1	Subwatershed area (*A*)	km^2	ArcGIS (Calculate Geometry)
2	Subwatershed perimeter (*P*)	km	ArcGIS (Calculate Geometry)
3	Stream order (*U*)	-	ArcGIS (Calculate Geometry)
4	Total number of streams (ΣN_u)	-	ArcGIS (Stream Order)
5	Total stream length (ΣL_u)	km	ArcGIS (Calculate Geometry)
6	Mean subwatershed slope (*S*)	Degree	ArcGIS (Slope)
7	Mean stream length (*Lsm*)	km	$Lsm = avg(Lu/Nu)$
8	Subwatershed length (*Lb*)	km	ArcGIS
9	Stream frequency (*Fs*)	$1/km^2$	$Fs = \Sigma Nu/A$
10	Drainage density (*Dd*)	1/km	$Dd = \Sigma Lu/A$
11	Constant of channel maintenance (*C*)	km	$C = 1/Dd$
12	Length of overland flow (*Lo*)	km	$Lo = 1/(2Dd)$
13	Infiltration number (*If*)	$1/km^3$	$If = Fs \times Dd$
14	Subwatershed relief (*H*)	km	$H = h_{max} - h_{min}$
15	Relief ratio (*R*)	-	$R = H/Lb$
16	Melton index (*M*)	-	$M = H/\sqrt{A}$
17	Ruggedness number (*Rn*)	-	$Rn = Dd \times H$
18	Bifurcation ratio (*Rb*)	-	$Rb = avg(N_u/N_{u+1})$
19	Stream length ratio (*Rl*)	-	$Rl = avg((L_{u+1}/N_{u+1})/(L_u/N_u))$
20	Ratio Rho (ρ)	-	$\rho = Rl/Rb$
21	Elongation ratio (*Re*)	-	$Re = \left(2\sqrt{A/\pi}\right)/Lb$
22	Circularity ratio (*Rc*)	-	$Rc = 2\sqrt{\pi A}/P$
23	Form factor (*Ff*)	-	$Ff = A/Lb^2$
24	Shape factor (*Bs*)	-	$Bs = Lb^2/A$
25	Compactness coefficient (*Cc*)	-	$Cc = P2\sqrt{\pi A}$
26	Texture ratio (*T*)	1/km	$T = \Sigma Nu/P$

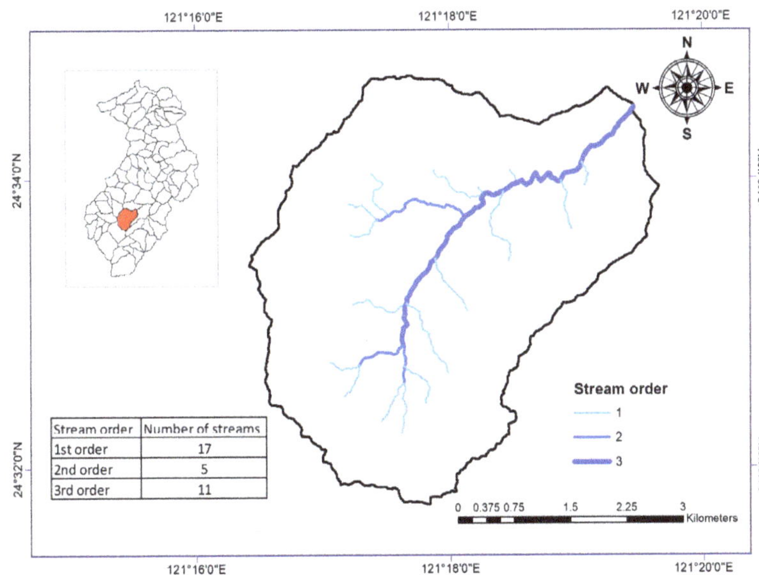

Figure 3. Stream order and the number of streams in a typical subwatershed.

Subwatershed length (Lb) in this study is defined as "the longest dimension of the basin parallel to the principal drainage line," as in the definition of relief ratio below [29]. The length is determined by ArcGIS 10.4.1.

Stream frequency (Fs) is the number of streams per unit area [28]. This value ranged from 0.47 to 2.46 streams/km^2 in this study.

Drainage density (Dd) is defined as the sum of the stream lengths divided by the subwatershed area. It is a crucial indicator of the linear scale of landform elements in a subwatershed [27].

Constant of channel maintenance (C) is defined as the inverse of drainage density. Along with drainage density, this value compares soil's erodibility or other factors influencing surface erosion [29]. Here, metric units were used, and the conversion factor of 5280 (from miles to feet) was ignored.

Length of overland flow (Lo) ranged from 0.32 km to 0.64 km in the study area. It is the length of runoff over the ground surface until it concentrates in definite stream channels and is half the reciprocal of drainage density [28].

Infiltration number (If) is the product of stream frequency and drainage density ([30], as cited in [31]). This value ranged from 0.44 to 2.98 in this study.

Subwatershed relief (H) is the difference in elevations between the lowest (h_{min}) and the highest (h_{max}) points in a subwatershed.

Relief ratio (R) is "the ratio between the total relief of a basin" and "the longest dimension of the basin parallel to the principal drainage line" [29]. For the study area, the relief ratio varied from 0.07 to 0.57.

Melton index (M), or the ruggedness of a subwatershed, is characterized by the dimensionless ratio between the subwatershed relief and the square root of the subwatershed area [32].

Ruggedness number (Rn) is known as the dimensionless product of drainage density and relief. As a result, high drainage density and low relief areas are just as rugged as low drainage density and high relief areas ([33], as cited in [34]).

Bifurcation ratio (Rb) is the average number of branchings or bifurcations of streams. It is defined as the number of streams of a given stream order to that of streams of the next higher order [28]. For a subwatershed, there are different bifurcation ratios for different stream orders. Following the example of Jothimani et al. [35], we computed the average of the bifurcation ratios as the characteristic bifurcation ratio of the subwatershed. For the 93 subwatersheds in the study area, the bifurcation ratio ranged from 0.50 to 8.00.

Stream length ratio (Rl) is defined by the average length of streams of a stream order to the next lower order [28]. Various stream length ratios exist for various stream orders. Therefore, we computed the average of the stream length ratios as the characteristic stream length ratio of the subwatershed, similar to Jothimani et al. [35]. For the 93 subwatersheds in the study area, the stream length ratio ranged from 0.46 to 5.86.

Ratio Rho (ρ) is the stream length ratio divided by the bifurcation ratio [28].

Elongation ratio (Re) is the ratio between the diameter of a circle with the same area as the subwatershed and the longest dimension of the subwatershed parallel to the main drainage line [29], as determined for the relief ratio.

Circularity ratio (Rc) is the circumference of a circle with the same area as the subwatershed divided by the subwatershed perimeter [29].

Form factor (Ff) is the ratio of the width to the length of a subwatershed and is defined as the subwatershed area divided by the square of the length of the subwatershed [36]. The subwatershed length is "measured from a point on the watershed-line opposite the head of the main stream" [36]. Here, we used subwatershed length (Lb) to be the length of the subwatershed.

Shape factor (Bs) is defined as the square of the length of a subwatershed divided by the area of the subwatershed, although other definitions have also been proposed ([37], as cited in [38]). The length of a subwatershed is defined as "the longest dimension from the mouth to the opposite side." Here, we used the subwatershed length (Lb) to represent the length of the subwatershed.

Compactness or compactness coefficient (Cc) is the ratio of the perimeter of the subwatershed to that of a circle with an equal area [36].

Texture ratio is the ratio of the number of crenulations on the contour with the maximum number of crenulations within the subwatershed to the length of the perimeter of the subwatershed [39]. Crenulations are chosen because they indicate streams too small to be shown on a topographic map [27]. The ratio is a measure of channel spacing closeness and thus is related to drainage density. For ease of computation, we used the total number of streams to replace the crenulations in this study. The texture ratio ranged from 0.16 to 1.27.

3. Methods

This study had five objectives: first, to identify and collect morphometric factors and environmental factors that affect soil erosion; second, to use feature selection to identify critical factors that can be used to model soil erosion depths; third, to apply machine learning algorithms to create models that can be used to predict soil erosion depth in the study area; fourth, to assess the validity of the models using statistical indices and threefold cross-validation; fifth and finally, to produce prediction maps of soil erosion depth for the study area.

3.1. Research Framework

Figure 4 depicts the five research steps of this analysis. First, we created an input dataset of 36 independent factors by combining 26 morphometric and 10 environmental factors. Second, we divided the dataset into three folds of roughly the same size based on the main subwatershed attribute to balance the class distribution from the five main subwatersheds [40]. We also used the erosion pin measurement as the target variable. Each time one of the three folds was held as the test data for testing the model, the remaining two folds were used as the training data. The whole process was repeated three times. Third, we applied the random forest (RF) and gradient boosting machine (GBM) to create erosion models based on the training data. Fourth, we assessed the models with the test data. In the process, we eliminated the unessential factors and kept the best models. Finally, we created the spatial distribution maps of soil erosion depth of the study area using the machine learning models.

3.2. Feature Selection

In order to identify the key factors that will generate the most credible soil erosion models, we used feature selection to rank the 36 morphometric and environmental factors in the study. Specifically, the Boruta algorithm was used to select the subsets of factors (predictors) for ML model building.

Boruta is a feature selection algorithm and feature ranking tool based on the RF algorithm and introduced by Kursa et al. [41]. It works by creating a randomized copy of the input dataset, merging it with the original dataset, and constructing the expanded system's classifier. Then, Boruta compares the importance of the factors in the original dataset to those of the randomized factors to identify the key factors. Only factors with greater importance than the randomized factors are considered essential. The advantage of Boruta is that it allows researchers to choose the most significant factors that influence the outcome. For this study, the Boruta package in the R software was used, and the maximum number of times the algorithm was run (maxRun) was set to the default value of 1000.

3.3. Machine Learning Models

In this analysis, two machine learning methods were used. They are the random forest and the gradient boosting machine.

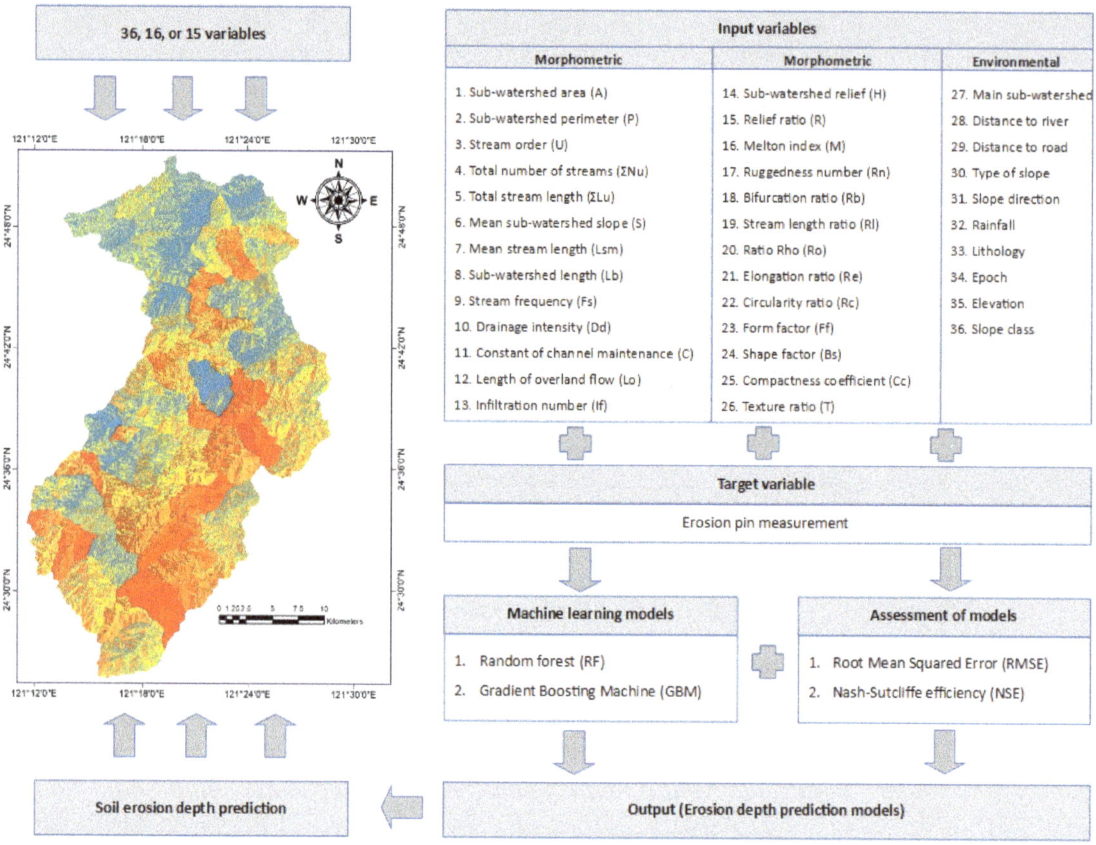

Figure 4. Research flowchart of this study.

Random forest (RF) was proposed by Breiman [42]. It is a supervised ML method that combines all tree-based results into the most appropriate model for the application. The RF algorithm runs many iterations and divides the training dataset (in terms of data and attributes) into many subsets at random to create many trees and produce better results than individual decision trees. The randomForest() package in the R software was used to implement random forest in this analysis, which uses the Gini index to separate data in order to minimize impurity at each node. Tsai et al. [23] provided a more detailed overview of the Gini index and random forest.

Friedman [43] proposed the gradient boosting machine as a simple and highly flexible machine learning tool. It is a widely used machine learning algorithm that has been shown to be effective in a variety of applications [44–46]. The basic idea behind GBM is to build a prediction model using a set of poor learning algorithms, most commonly decision trees. Unlike RF, which produces an ensemble of individual trees in parallel, GBM creates a sequenced tree ensemble. The knowledge gained from previously grown trees is used to grow new trees in a sequential manner. The GBM model was once used to model soil erosion [22]. It was implemented in this study using R software's "gbm" package.

3.4. Assessment of Models

In this study, the ML model performance was evaluated using two statistical indices. As shown in Equations (1) and (2), they are the root mean square error (RMSE) and the Nash–Sutcliffe efficiency (NSE).

$$RMSE = \sqrt{\frac{\sum(P-O)^2}{n}} \tag{1}$$

$$NSE = 1 - \frac{\sum(P-O)^2}{\sum(O-\overline{O})^2} \tag{2}$$

where P is the predicted value, O is the observed value, and $\overline{O}$ is the mean observed value.

RMSE was used to compare the difference between the expected values (model outputs) and the observed values (erosion pin measurements) in the two indices, while NSE was used to determine the effectiveness of the model against the average observed value [20–22].

4. Results

In this analysis, we used R version 4.0.5. In order to assess soil erosion in the Shihmen Reservoir watershed, this study employed two machine learning models, RF and GBM. To substitute four factors that were only point data, 26 morphometric factors were added to the original dataset of 14 environmental factors. In total, 36 variables were examined for their relationship with soil erosion depth (erosion pin measurement). The training data (used to create the ML models) made up two folds of the dataset, while the remaining fold was used to evaluate the models based on RMSE and NSE. Finally, through spatial mapping, machine learning models were used to predict the soil erosion depth for the entire Shihmen Reservoir watershed.

4.1. Feature Selection

Boruta was used as a feature selection tool to assess the relative importance of variables that influence soil erosion. Table 2 and Figure 5 depict the findings. It can be seen that Table 2 was divided into three categories based on decisions: rejected, tentative, and confirmed. They are also ranked by median importance. In total, 15 factors were identified as important, which includes texture ratio, subwatershed length, epoch, elongation ratio, lithology, subwatershed perimeter, form factor, relief ratio, total stream length, Melton index, the total number of streams, elevation, shape factor, subwatershed area, and type of slope. One factor was considered tentative, i.e., the main subwatershed. Moreover, 20 variables were ruled out, which consist of distance to river, mean stream length, ruggedness number, slope direction, ratio Rho, circularity ratio, distance to road, stream length ratio, stream frequency, rainfall, compactness coefficient, stream order, constant of channel maintenance, drainage density, length of overland flow, infiltration number, slope class, subwatershed slope, bifurcation ratio, and subwatershed relief. They should play no important role in the prediction of soil erosion. According to the Boruta analysis, the type of slope, subwatershed area, and shape factor are the three most significant variables among the factors that are shown to be important.

Boruta generates a corresponding "shadow" factor for each factor, whose values were obtained by shuffling the original factor's values across objects. The system then classifies these using all of the extended system's factors and calculates the importance of each factor [47]. Green is used to color the 15 factors listed as important in Figure 5. The 20 rejected factors are colored red, while the one tentative factor is colored yellow. To differentiate the variables, Figure 5 also shows the minimum, mean, and maximum of shadow factors. In general, factors ranked higher than the shadow maximum have been tested to be more significant than chance.

Table 2. Variable importance using Boruta feature selection.

	meanImp	medianImp	minImp	maxImp	Decision
Type of slope	8.162273	8.243055	3.952560	10.425924	Confirmed
Subwatershed area	5.019874	5.052494	2.343560	7.383705	Confirmed
Shape factor	4.976153	5.010055	2.159306	7.532354	Confirmed
Elevation	4.988711	5.000990	2.069170	7.789953	Confirmed
Total number of streams	4.907324	4.938526	2.276170	7.150461	Confirmed
Melton index	4.409082	4.469957	1.627233	7.206727	Confirmed
Total stream length	4.447973	4.438060	1.872146	6.785887	Confirmed
Relief ratio	4.348566	4.412615	0.893647	6.730493	Confirmed
Form factor	4.209047	4.230464	1.780136	6.776284	Confirmed
Subwatershed perimeter	4.064281	4.095632	0.925774	6.298489	Confirmed
Lithology	3.898596	3.915257	0.934244	6.630224	Confirmed
Elongation ratio	3.788302	3.794716	1.129358	5.898901	Confirmed
Epoch	3.517687	3.528155	0.622912	6.683233	Confirmed
Subwatershed length	3.421194	3.444065	0.453943	6.157326	Confirmed
Texture ratio	3.420789	3.421103	0.778195	5.719445	Confirmed
Subwatershed	2.952157	2.983119	−0.229199	5.487332	Tentative
Subwatershed relief	2.355625	2.315015	−0.156997	4.560039	Rejected
Bifurcation ratio	2.122521	2.217065	0.894857	3.185164	Rejected
Mean subwatershed slope	1.990054	2.069925	0.119557	3.807072	Rejected
Slope class	1.999188	2.035054	−0.586225	4.019329	Rejected
Infiltration number	1.761100	1.736976	−0.027713	3.775921	Rejected
Length of overland flow	1.757534	1.690870	−0.044952	3.617237	Rejected
Drainage density	1.674382	1.671754	0.212803	2.613970	Rejected
Constant of channel maintenance	1.412327	1.563865	−1.125096	2.720770	Rejected
Stream order	1.460165	1.537092	−0.199118	2.561430	Rejected
Compactness coefficient	1.492243	1.527261	−0.460479	2.733053	Rejected
Rainfall	1.601451	1.490486	0.124088	3.313065	Rejected
Stream frequency	1.359127	1.464677	−0.542830	2.702161	Rejected
Stream length ratio	1.090233	1.124972	−1.164121	2.851477	Rejected
Distance to road	0.998904	0.815334	−0.609436	2.881991	Rejected
Circularity ratio	0.975395	0.711048	−0.379865	2.311318	Rejected
Ratio Rho	0.518910	0.549485	−0.944962	2.122957	Rejected
Slope direction	0.692622	0.536967	−0.820490	2.884595	Rejected
Ruggedness number	0.183690	0.291990	−2.038411	1.565565	Rejected
Mean stream length	0.486127	0.288427	−1.123192	1.989272	Rejected
Distance to river	−0.937049	−0.925128	−2.557384	0.647234	Rejected

Among the green (important) factors, 4 are environmental factors, while the remaining 11 are morphometric factors. The percentage of the environmental factors in the confirmed group (4/15 = 27%) is slightly less than the overall percentage of the environmental factors in the dataset (10/36 = 28%). On the other hand, the environmental factors account for 100% of the tentative factor (1/1) and 25% (5/20) of the rejected factors. Furthermore, the environmental factors selected in the confirmed group are the type of slope, elevation, lithology, and epoch. Compared to the study by Nguyen et al. [22], which also reported the relative importance of environmental factors, we can see some similarities. The top four factors from Nguyen et al. [22] were slope direction, type of slope, % organic, and elevation. Two (type of slope and elevation) were also selected for this study, while one (slope direction) was not, and the other (% organic) was not included in this study because it is a point data. It is worth noting that Nguyen et al. [22] used 70% training and 30% test data, while this study used threefold cross-validation.

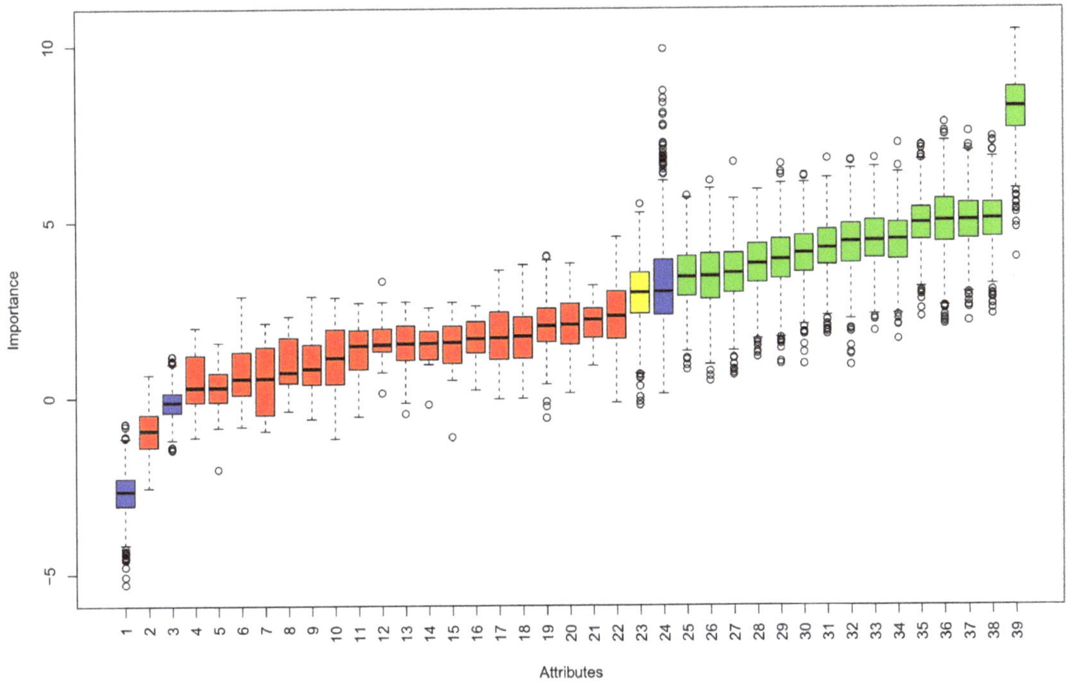

Figure 5. Feature selection with Boruta: 1: shadow min, 2: distance to the river, 3: shadow mean, 4: mean stream length, 5: ruggedness number, 6: slope direction, 7: ratio Rho, 8: circularity ratio, 9: distance to road, 10: stream length ratio, 11: stream frequency, 12: rainfall, 13: compactness coefficient, 14: stream order, 15: constant of channel maintenance, 16: drainage density, 17: length of overland flow, 18: infiltration number, 19: slope class, 20: subwatershed slope, 21: bifurcation ratio, 22: subwatershed relief, 23: main subwatershed, 24: shadow max, 25: texture ratio, 26: subwatershed length, 27: epoch, 28: elongation ratio, 29: lithology, 30: subwatershed perimeter, 31: form factor, 32: relief ratio, 33: total stream length, 34: Melton index, 35: total number of streams, 36: elevation, 37: shape factor, 38: subwatershed area, 39: type of slope.

4.2. Machine Learning

Based on the results of feature selection, we performed machine learning on three sets of factors separately: (1) all 36 factors, (2) 15 confirmed factors, and (3) 15 confirmed factors plus 1 tentative factor. Using threefold cross-validation in each set of factors, the dataset was divided into three, roughly equal folds. Then, two folds were used as the training data, and the other fold was used as the test data. The process was repeated three times so that every fold was used as the test data in the analysis. Both RF and GBM were used to analyze the same data. Finally, the results (RMSE and NSE) of three attempts were averaged. They are shown in Table 3 and Figure 6.

Table 3. Performance comparison of machine learning models using threefold cross-validation.

Model and Factors	No. of Factors	Average RMSE (mm/yr)		Average NSE	
		Training	Test	Training	Test
RF (all)	36	0.96	2.01	0.83	0.25
GBM (all)	36	0.88	1.84	0.84	0.39
RF (confirmed)	15	1.08	1.91	0.79	0.31
GBM (confirmed)	15	0.79	1.50	0.88	0.59
RF (nonrejected)	16	1.09	1.96	0.79	0.27
GBM (nonrejected)	16	0.82	1.52	0.87	0.57

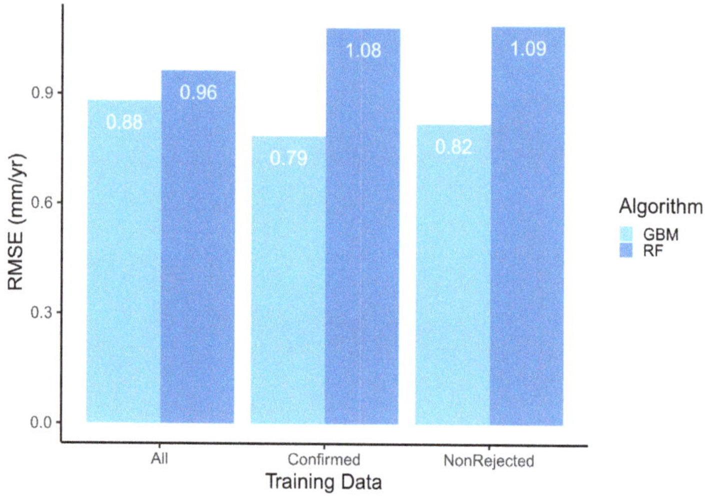

(**a**)

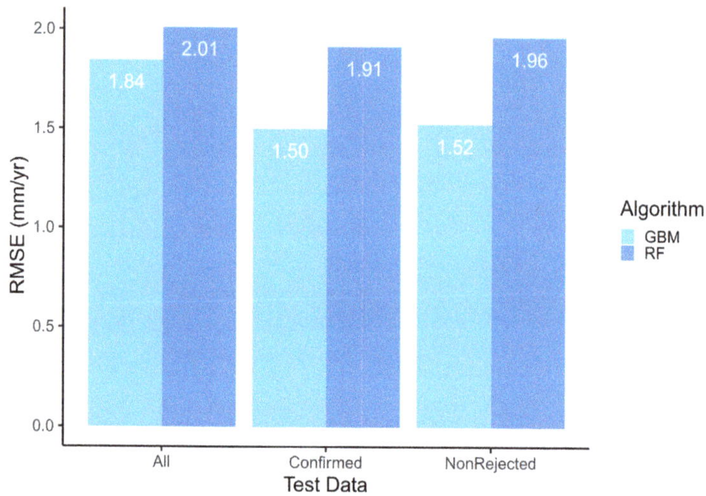

(**b**)

Figure 6. Comparison between parameter selections: (**a**) training data and (**b**) test data (lower is better).

The findings (Table 3) reveal that the ML models delivered good results. Both the average values of RMSE and NSE in Table 3 exhibit the same trend. The smaller the RMSE and the higher the NSE were, the better the model was. As shown in Figure 6, GBM consistently outperforms RF in both training data and test data. GBM also edges out RF in all three datasets that used different factors (all, confirmed, and nonrejected). For the training data, the best RF model result was obtained with the all-factor group, followed by the confirmed group and then the nonrejected group. However, for the test data, the confirmed group is the best, followed by the nonrejected group and then the all-factor group. This shows that the RF models were overfitted with more factors, and that feature selection indeed contributes to improving the ML models when facing unknown data.

On the other hand, the GBM model does not exhibit an overfit bias. For both the training and test data, the confirmed group is the best, followed by the nonrejected group and then the all-factor group.

Overall, the best test result obtained in this study is 1.50 mm/yr (GBM) and 1.91 mm/yr (RF). Both of them are from the confirmed group. Compared to the previous study [22], which used a 70/30 split and only 14 environmental factors, the results are mixed. In terms of RF, the Nguyen et al. [22] result was 1.75 mm/yr, which is better than the current study (1.91 mm/yr). However, in terms of GBM, the present study (1.50 mm/yr) is better than the previous study (1.72 mm/yr). If we only consider the best model, which is GBM in this case, this study is better than the previous study.

4.3. Model Prediction

Using the RF and GBM models, we predicted the soil erosion depth of the entire study area, as shown in Figure 7. The data of the whole Shihmen Reservoir watershed were investigated and then entered into the R software after the preparation of the machine learning models for predicting the soil erosion depth. The results were transferred to the ArcGIS software to create the soil erosion depth maps. Figure 7 showed the spatial distribution of soil erosion depth (in mm/yr) over the Shihmen Reservoir watershed produced by each model's three sets of factors: all, confirmed, and nonrejected. The red area represents a high erosion depth, whereas the blue area has a low erosion depth. Due to the morphometric factors used in the ML models, it is clear that the individual subwatershed has a significant impact on the soil erosion depth distribution.

Figure 7 shows that the all-factor group's maps (a and b) have more variance within individual subwatersheds than the confirmed group's (c and d) and the nonrejected group's maps (e and f). This is most likely due to the fact that there are more variables used in the mapping of all factors (36). The confirmed and nonrejected maps, on the other hand, appear to be more uniform in color throughout each subwatershed. They both have a similar appearance because they used a similar number of variables (15 and 16).

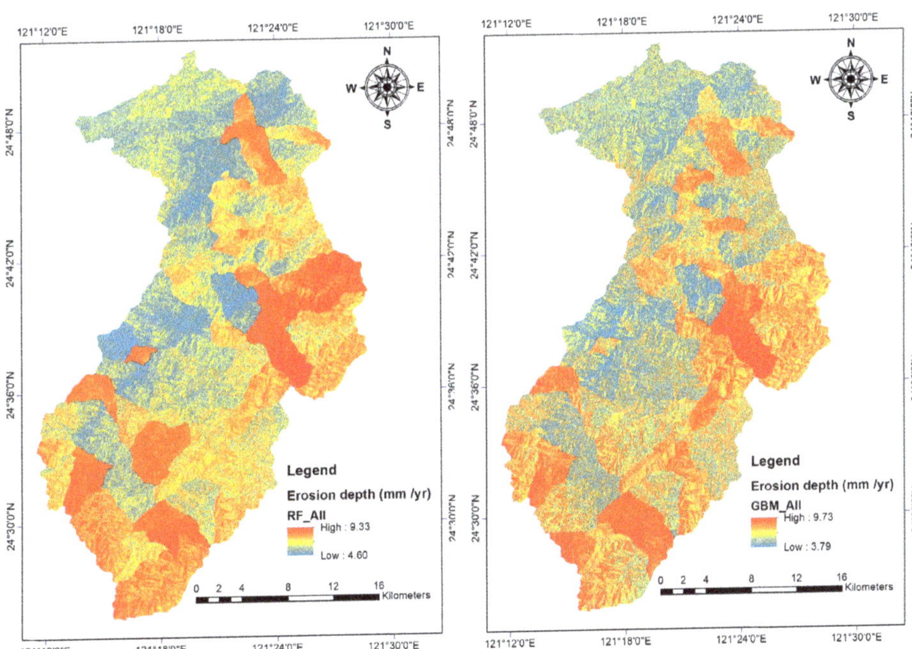

(a) (b)

Figure 7. *Cont.*

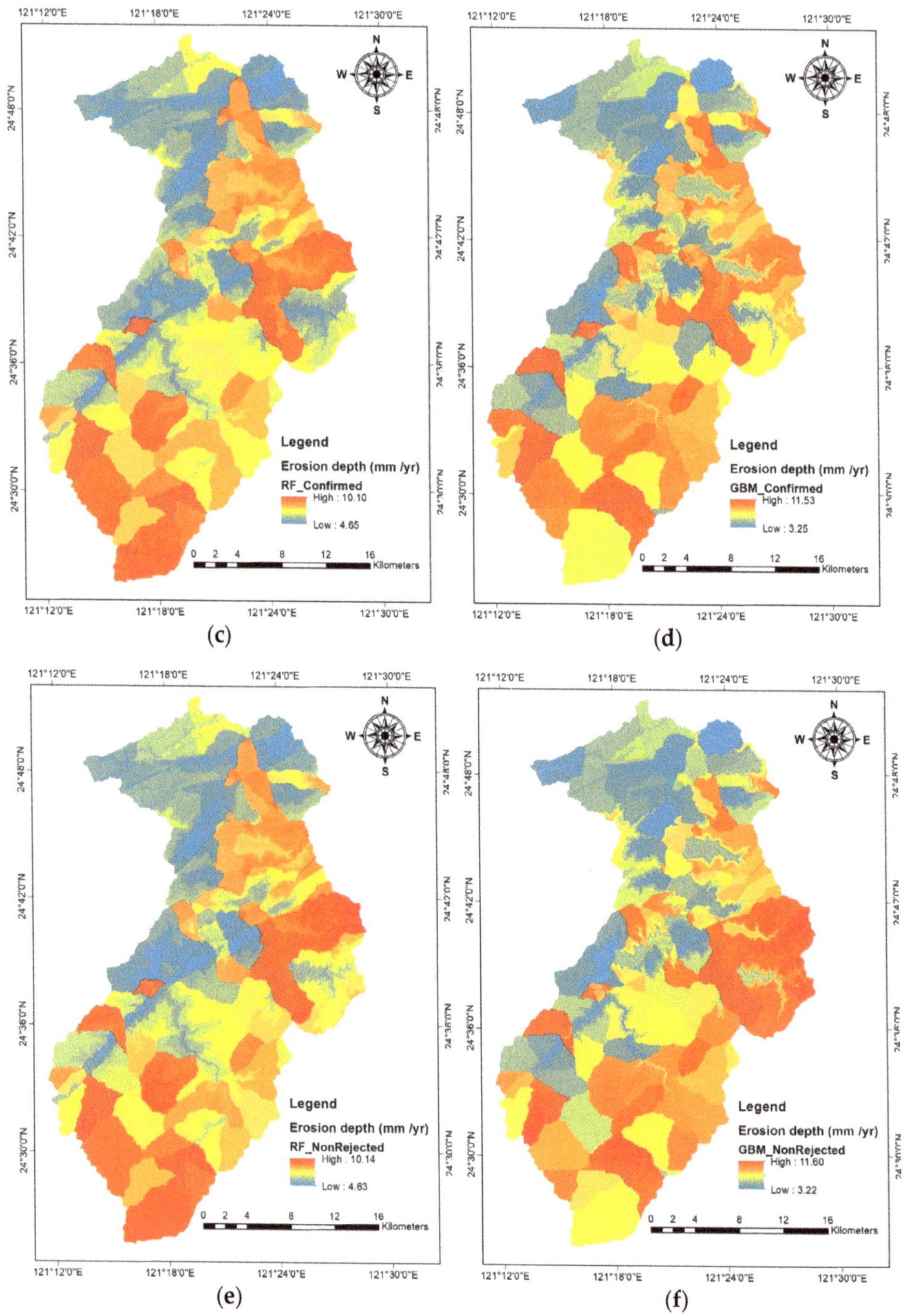

Figure 7. Prediction of soil erosion depth of the entire Shihmen Reservoir watershed using machine learning: (**a**) RF (all), (**b**) GBM (all), (**c**) RF (confirmed), (**d**) GBM (confirmed), (**e**) RF (nonrejected), and (**f**) GBM (nonrejected).

The minimum, mean, and maximum erosion depths expected for the entire Shihmen Reservoir watershed from different model/factor combinations are shown in Table 4. The table also includes field measurements of erosion pins for comparison. The table shows that the averages of various model/factor combinations are quite similar to the average of erosion pins. However, no model/factor combination accurately forecasts the extreme values of real-world measurements. The predictions are too high for the minimum value and too low for the maximum value.

Table 4. Comparing ML model results with erosion pin measurements.

Erosion Depth (mm/yr)	Min (mm/yr)	Mean (mm/yr)	Max (mm/yr)
RF (all)	4.60	6.77	9.33
GBM (all)	3.79	6.73	9.73
RF (confirmed)	4.65	6.68	10.10
GBM (confirmed)	3.25	6.68	11.53
RF (nonrejected)	4.63	6.67	10.14
GBM (nonrejected)	3.22	6.67	11.60
Erosion Pin measurements	2.17	6.50	13.03

5. Discussion

This study continues to model the soil erosion depth as measured by erosion pins in the Shihmen Reservoir watershed because of the watershed's significance and the degree to which it is affected by soil erosion [20–22]. Since the morphometric features of a watershed influence surface runoff and water erosion, they were included in this research to create a complete picture of the erosion activity in the study region and to improve the ML models. However, due to the overlapping (and sometimes conflicting) nature of some of the morphometric features, the overwhelming number of factors extracted from the morphometric analysis may be a deterrent to further analysis. As a result, feature selection was performed in this study before machine learning modeling. The widely used Boruta algorithm was used to separate the important from the nonimportant factors. In the end, 11 morphometric factors were identified as influential in estimating the soil erosion depth. They are texture ratio, subwatershed length, elongation ratio, subwatershed perimeter, form factor, relief ratio, total stream length, Melton index, total number of streams, shape factor, and subwatershed area. Overall, the morphometric factors were chosen in 42 percent (=11/26) of the cases. On the other hand, only four environmental factors (slope type, elevation, lithology, and epoch) were chosen as important. They account for 40% (=4/10) of the overall environmental factors.

Note that the point data (% sand, % silt, % clay, and % organic) in the original 14 environmental factors had to be removed because they were not available watershed-wide and cannot be used for model prediction of the entire Shihmen Reservoir watershed. Therefore, the lower selection rate of the environmental factors than the morphometric factors in this study could be attributed to the removal of these point data because some of them were shown to be important in the previous study [22].

Another aspect that distinguishes this study from the previous studies [20–22] is the use of threefold cross-validation instead of the 70/30 split with stratified random sampling. The threefold cross-validation divides the dataset into three roughly equal folds with a balanced class distribution. Therefore, each class (stratum) is adequately represented, as with the stratified random sampling. However, in the threefold cross-validation, two folds were used as the training data, and the third fold was used as the test data. The procedure was replicated three times so that the algorithm takes turns using two-thirds of the data as the training data, and each fold was used as the test data only once. The 70/30 split with stratified random sampling, on the other hand, did not rotate the training and test data. To find the average answer, the 70/30 split had to be repeated three times from the beginning using different random seeds.

Regardless of which set of factors was used (all, confirmed, or nonrejected), our analysis shows that GBM consistently outperforms RF in terms of RMSE and NSE. As compared to the previous study [22], the best RMSE value was noticeably reduced from

1.72 mm/yr to 1.50 mm/yr (GBM with confirmed factors). This demonstrates that, despite the elimination of potentially valuable point data, the inclusion of morphometric factors improves the soil erosion modeling.

Additionally, unlike the previous study that used point data [22], this study does not need to interpolate the modeling prediction for the entire research area. Instead, complete maps of the spatial distribution of soil erosion depth can be produced from the ML models directly. The resulting maps (Figure 7) show finer resolution of change with more features in color variation. There is densely packed information not present in the previous maps. It is a huge step forward for soil erosion control and prioritization.

6. Conclusions

To sum up, previous studies built machine learning models for the Shihmen Reservoir watershed using point data that were only available at individual slopes monitored with erosion pins. The current research improved upon past studies by incorporating new independent variables (morphometric factors) derived from the watershed digital elevation model and eliminating the dependence on the point data. A dataset of 36 predictive factors and one target factor was created. Feature selection was performed to remove redundant factors and to avoid the overfitting of models. In the end, 15 important factors were identified that include 4 environmental factors and 11 morphometric factors. Two ML algorithms, RF and GBM, were used in the analysis. Despite the removal of four environmental factors used in previous studies (point data that were not available watershed-wide), the new GBM model in this study shows an improvement in RMSE, which was reduced from 1.72 mm/yr to 1.50 mm/yr. Consequently, we were able to create the most accurate ML model to date of the distribution of soil erosion depth in the study area. This proves the value of adding morphometric factors to soil erosion analysis. Furthermore, the ML models were used to create prediction maps of soil erosion depth of the entire Shihmen Reservoir watershed, which were not possible, and only interpolation approximation was achieved previously (due to the point data issue). The new maps show great details of what needs attention for soil erosion control and prioritization. It is a valuable advancement of our understanding and future study of soil erosion modeling. Since the ML models are data-driven and rely on sufficient monitoring data, it is crucial to improve our data collection methods and use the latest technologies to record information. Solar-powered Internet of Things (IoT) devices that can monitor the change of slope surfaces are currently being experimented with in the Shihmen Reservoir watershed. The inexpensive and large amount of data generated by these devices will likely be the key driver for future research on this topic.

Author Contributions: Conceptualization, Walter Chen; data curation, Kieu Anh Nguyen and Walter Chen; formal analysis, Kieu Anh Nguyen and Walter Chen; funding acquisition, Walter Chen; investigation, Walter Chen; methodology, Walter Chen; project administration, Walter Chen; resources, Walter Chen; software, Kieu Anh Nguyen and Walter Chen; supervision, Walter Chen; validation, Walter Chen; visualization, Kieu Anh Nguyen and Walter Chen; writing—original draft preparation, Kieu Anh Nguyen and Walter Chen; writing—review and editing, Kieu Anh Nguyen and Walter Chen. All authors have read and agreed to the published version of the manuscript.

Funding: This study was partially supported by the Ministry of Science and Technology (Taiwan) Research Project (Grant Number MOST 109-2121-M-027-001).

Institutional Review Board Statement: Not applicable.

Informed Consent Statement: Not applicable.

Data Availability Statement: Data sharing is not applicable to this article.

Conflicts of Interest: The authors declare no conflict of interest.

References

1. Keesstra, S.D.; Bouma, J.; Wallinga, J.; Tittonell, P.; Smith, P.; Cerdà, A.; Montanarella, L.; Quinton, J.N.; Pachepsky, Y.; Van Der Putten, W.H.; et al. The significance of soils and soil science towards realization of the United Nations sustainable development goals. *Soil* **2016**, *2*, 111–128. [CrossRef]
2. Eswaran, H.; Lal, R.; Reich, P.F. Land Degradation: An overview. In *Response to Land Degradation, Proceedings of the 2nd International Conference on Land Degradation and Desertification, Khon Kaen, Thailand, 25–29 January 1999*; Bridges, E.M., Hannam, I.D., Oldeman, L.R., Pening de Vries, F.W.T., Scherr, S.J., Sompatpanit, S., Eds.; Oxford University Press: New Delhi, India, 2002. Available online: http://soils.usda.gov/use/worldsoils/papers/land-degradation-overview.html (accessed on 1 August 2013).
3. Myers, N. *Gaia: An Atlas of Planet Management*; Anchor/DoubleDay: Garden City, NY, USA, 1993.
4. Pimentel, D.; Burgess, M. Soil erosion threatens food production. *Agriculture* **2013**, *3*, 443–463. [CrossRef]
5. Weil, R.R.; Brady, N.C. *The Nature and Properties of Soils*, 15th ed.; Pearson: Harlow, UK, 2017; 1104p.
6. Oldeman, L.R. The global extent of soil degradation. In *Soil Resilience and Sustainable Land Use*; Greenland, D.J., Szabolcs, I., Eds.; CAB International: Wallingford, UK, 1994; pp. 99–118.
7. Daily, G. Restoring value to the world's degraded lands. *Science* **1997**, *269*, 350–354. [CrossRef]
8. U.S. Environmental Protection Agency. *Comparative Costs of Erosion and Sediment Control Construction Activities*; EPA-430/9-73-016; U.S. Government Printing Office: Washington, DC, USA, 1973.
9. Gray, D.H.; Sotir, R.B. *Biotechnical and Soil Bioengineering Slope Stabilization: A Practical Guide for Erosion Control*; John Wiley & Sons: New York, NY, USA, 1996; 378p.
10. National Geosciences Database. 2017. Available online: www.ngdir.ir (accessed on 12 February 2019).
11. Arabameri, A.; Tiefenbacher, J.P.; Blaschke, T.; Pradhan, B.; Bui, D.T. Morphometric analysis for soil erosion susceptibility mapping using novel gis-based ensemble model. *Remote Sens.* **2020**, *12*, 874. [CrossRef]
12. Borrelli, P.; Alewell, C.; Alvarez, P.; Anache, J.A.A.; Baartman, J.; Ballabio, C.; Bezak, N.; Biddoccu, M.; Cerdà, A.; Chalise, D.; et al. Soil erosion modelling: A global review and statistical analysis. *Sci. Total Environ.* **2021**, *780*, 146494. [CrossRef] [PubMed]
13. Bezak, N.; Mikoš, M.; Borrelli, P.; Alewell, C.; Alvarez, P.; Ayach Anache, J.A.; Baartman, J.; Ballabio, C.; Biddoccu, M.; Cerdà, A.; et al. Soil erosion modelling: A bibliometric analysis. *Environ. Res.* **2021**, *197*, 111087. [CrossRef]
14. Chen, W.; Li, D.-H.; Yang, K.-J.; Tsai, F.; Seeboonruang, U. Identifying and comparing relatively high soil erosion sites with four DEMs. *Ecol. Eng.* **2018**, *120*, 449–463. [CrossRef]
15. Liu, Y.-H.; Li, D.-H.; Chen, W.; Lin, B.-S.; Seeboonruang, U.; Tsai, F. Soil erosion modeling and comparison using slope units and grid cells in Shihmen reservoir watershed in Northern Taiwan. *Water* **2018**, *10*, 1387. [CrossRef]
16. Angileri, S.E.; Conoscenti, C.; Hochschild, V.; Märker, M.; Rotigliano, E.; Agnesi, V. Water erosion susceptibility mapping by applying Stochastic Gradient Treeboost to the Imera Meridionale River Basin (Sicily, Italy). *Geomorphology* **2016**, *262*, 61–76. [CrossRef]
17. Chakrabortty, R.; Pal, S.C.; Sahana, M.; Mondal, A.; Dou, J.; Pham, B.T.; Yunus, A.P. Soil erosion potential hotspot zone identification using machine learning and statistical approaches in eastern India. *Nat. Hazards* **2020**, *104*, 1259–1294. [CrossRef]
18. Pourghasemi, H.R.; Yousefi, S.; Kornejady, A.; Cerdà, A. Performance assessment of individual and ensemble data-mining techniques for gully erosion modeling. *Sci. Total Environ.* **2017**, *609*, 764–775. [CrossRef]
19. Svoray, T.; Michailov, E.; Cohen, A.; Rokah, L.; Sturm, A. Predicting gully initiation: Comparing data mining techniques, analytical hierarchy processes and the topographic threshold. *Earth Surf. Process. Landf.* **2012**, *37*, 607–619. [CrossRef]
20. Nguyen, K.A.; Chen, W.; Lin, B.-S.; Seeboonruang, U.; Thomas, K. Predicting sheet and rill erosion of Shihmen reservoir watershed in Taiwan using machine learning. *Sustainability* **2019**, *11*, 3615. [CrossRef]
21. Nguyen, K.A.; Chen, W.; Lin, B.-S.; Seeboonruang, U. Using machine learning-based algorithms to analyze erosion rates of a watershed in Northern Taiwan. *Sustainability* **2020**, *12*, 2022. [CrossRef]
22. Nguyen, K.A.; Chen, W.; Lin, B.-S.; Seeboonruang, U. Comparison of Ensemble Machine Learning Methods for Soil Erosion Pin Measurements. *ISPRS Int. J. Geo-Inf.* **2021**, *10*, 42. [CrossRef]
23. Tsai, F.; Lai, J.-S.; Nguyen, K.A.; Chen, W. Determining Cover Management Factor with Remote Sensing and Spatial Analysis for Improving Long-Term Soil Loss Estimation in Watersheds. *ISPRS Int. J. Geo-Inf.* **2021**, *10*, 19. [CrossRef]
24. Huang, C.L.; Hsu, N.S.; Wei, C.C. Coupled Heuristic Prediction of Long Lead-Time Accumulated Total Inflow of a Reservoir during Typhoons Using Deterministic Recurrent and Fuzzy Inference-Based Neural Network. *Water* **2015**, *7*, 6516–6550. [CrossRef]
25. Lin, B.S.; Thomas, K.; Chen, C.K.; Ho, H.C. Evaluation of soil erosion risk for watershed management in Shenmu watershed, central Taiwan using USLE model parameters. *Paddy Water Environ.* **2016**, *14*, 19–43. [CrossRef]
26. Encyclopaedia Britannica. Morphometric Analysis. 2009. Available online: https://www.britannica.com/science/morphometric-analysis (accessed on 4 May 2021).
27. Strahler, A.N. Quantitative analysis of watershed geomorphology. *Eos Trans. Am. Geophys. Union* **1957**, *38*, 913–920. [CrossRef]
28. Horton, R.E. Erosional development of streams and their drainage basins; hydrophysical approach to quantitative morphology. *Bull. Geol. Soc. Am.* **1945**, *56*, 275–370. [CrossRef]
29. Schumm, S.A. Evolution of drainage systems and slopes in badlands at Perth Amboy, New Jersey. *Bull. Geol. Soc. Am.* **1956**, *67*, 597–646. [CrossRef]
30. Faniran, A. The index of drainage intensity: A provisional new drainage factor. *Aust. J. Sci.* **1968**, *31*, 326–330.

31. Arango, M.I.; Aristizábal, E.; Gómez, F. Morphometrical analysis of torrential flows-prone catchments in tropical and mountainous terrain of the Colombian Andes by machine learning techniques. *Nat. Hazards* **2021**, *105*, 983–1012. [CrossRef]
32. Melton, M.A. The geomorphic and paleoclimatic significance of alluvial deposits in southern Arizona. *J. Geol.* **1965**, *73*, 1–38. [CrossRef]
33. Melton, M.A. *An Analysis of the Relation among Elements of Climate, Surface Properties and Geomorphology*; Tech. Rep. II; Office of Navy Research, Department of Geology, Columbia University: New York, NY, USA, 1957; 102p.
34. Patton, P.C.; Baker, V.R. Morphometry and floods in small drainage basins subject to diverse hydrogeomorphic controls. *Water Resour. Res.* **1976**, *12*, 941–952. [CrossRef]
35. Jothimani, M.; Abebe, A.; Dawit, Z. Mapping of soil erosion-prone sub-watersheds through drainage morphometric analysis and weighted sum approach: A case study of the Kulfo River basin, Rift valley, Arba Minch, Southern Ethiopia. *Model. Earth Syst. Environ.* **2020**, *6*, 2377–2389. [CrossRef]
36. Horton, R.E. Drainage-Basin Characteristics. Trans. *Am. Geophys. Union* **1932**, *13*, 350–361. [CrossRef]
37. *Corps of Engineers*; Civil Works Inv., Project CW 153; Department of the Army, Washington District, The Unit Hydrograph Compilations: Washington, DC, USA, 1949.
38. Morisawa, M. Measurement of Drainage-Basin Outline Form. *J. Geol.* **1958**, *66*, 587–591. [CrossRef]
39. Smith, K.G. Standards for grading texture of erosional topography. *Am. J. Sci.* **1950**, *248*, 655–668. [CrossRef]
40. Chen, W.; Chen, A. A statistical test of erosion pin measurements. In Proceedings of the 39th Asian Conference on Remote Sensing (ACRS 2018), Kuala Lumpur, Malaysia, 15–19 October 2018; Volume 4, pp. 2439–2443.
41. Kursa, M.B.; Jankowski, A.; Rudnicki, W.R. Boruta—A system for feature selection. *Fundam. Inform.* **2010**, *101*, 271–285. [CrossRef]
42. Breiman, L. Random forests. *Mach. Learn.* **2001**, *45*, 5–32. [CrossRef]
43. Friedman, J.H. Greedy function approximation: A gradient boosting machine. *Ann. Stat.* **2001**, *29*, 1189–1232. [CrossRef]
44. Chen, C.; Yang, D.; Gao, S.; Zhang, Y.; Chen, L.; Wang, B.; Mo, Z.; Yang, Y.; Hei, Z.; Zhou, S. Development and performance assessment of novel machine learning models to predict pneumonia after liver transplantation. *Respir. Res.* **2021**, *22*, 94. [CrossRef] [PubMed]
45. Kim, J.; Park, Y.; Park, S.; Jang, H.; Kim, H.J.; Na, D.L.; Lee, H.; Seo, S.W. Prediction of tau accumulation in prodromal Alzheimer's disease using an ensemble machine learning approach. *Sci. Rep.* **2021**, *11*, 5706. [CrossRef]
46. Mamun, O.; Wenzlick, M.; Hawk, J.; Devanathan, R. A machine learning aided interpretable model for rupture strength prediction in Fe-based martensitic and austenitic alloys. *Sci. Rep.* **2021**, *11*, 5466. [CrossRef]
47. Kursa, M.B.; Rudnicki, W.R. Feature selection with the boruta package. *J. Stat. Softw.* **2010**, *36*, 1–13. [CrossRef]

International Journal of *Geo-Information*

MDPI

Correction

Correction: Nguyen, K.A.; Chen, W. DEM- and GIS-Based Analysis of Soil Erosion Depth Using Machine Learning. *ISPRS Int. J. Geo-Inf.* 2021, *10*, 452

Kieu Anh Nguyen and Walter Chen *

Department of Civil Engineering, National Taipei University of Technology, Taipei 10608, Taiwan; t106429401@ntut.edu.tw

* Correspondence: waltchen@ntut.edu.tw; Tel.: +886-2-27712171 (ext. 2628)

Citation: Nguyen, K.A.; Chen, W. Correction: Nguyen, K.A.; Chen, W. DEM- and GIS-Based Analysis of Soil Erosion Depth Using Machine Learning. *ISPRS Int. J. Geo-Inf.* 2021, *10*, 452. *ISPRS Int. J. Geo-Inf.* **2021**, *10*, 724. https://doi.org/10.3390/ijgi10110724

Received: 27 August 2021
Accepted: 1 October 2021
Published: 27 October 2021

The authors of the published paper [1] would like to make the following corrections:

(1) The last four numbers in the second column (No. of Factors) of Table 3 should read 15, 15, 16, and 16 (instead of 16, 16, 19, and 19)

Original:

Table 3. Performance comparison of machine learning models using threefold cross-validation.

Model and Factors	No. of Factors	Average RMSE (mm/yr)		Average NSE	
		Training	Test	Training	Test
RF (all)	36	0.96	2.01	0.83	0.25
GBM (all)	36	0.88	1.84	0.84	0.39
RF (confirmed)	16	1.08	1.91	0.79	0.31
GBM (confirmed)	16	0.79	1.50	0.88	0.59
RF (nonrejected)	19	1.09	1.96	0.79	0.27
GBM (nonrejected)	19	0.82	1.52	0.87	0.57

Corrected:

Table 3. Performance comparison of machine learning models using threefold cross-validation.

Model and Factors	No. of Factors	Average RMSE (mm/yr)		Average NSE	
		Training	Test	Training	Test
RF (all)	36	0.96	2.01	0.83	0.25
GBM (all)	36	0.88	1.84	0.84	0.39
RF (confirmed)	15	1.08	1.91	0.79	0.31
GBM (confirmed)	15	0.79	1.50	0.88	0.59
RF (nonrejected)	16	1.09	1.96	0.79	0.27
GBM (nonrejected)	16	0.82	1.52	0.87	0.57

(2) The last parentheses on page 13 (4.3 Model Prediction) should contain the numbers 15 and 16 (instead of 16 and 19)

Original:
They both have a similar appearance because they used a similar number of variables (16 and 19).

Corrected:

They both have a similar appearance because they used a similar number of variables (15 and 16).

The authors apologize for any inconvenience caused and state that the scientific conclusions are unaffected. The original article has been updated.

Reference

1. Nguyen, K.A.; Chen, W. DEM- and GIS-Based Analysis of Soil Erosion Depth Using Machine Learning. *ISPRS Int. J. Geo-Inf.* **2021**, *10*, 452. [CrossRef]

MDPI
St. Alban-Anlage 66
4052 Basel
Switzerland
www.mdpi.com

ISPRS International Journal of Geo-Information Editorial Office
E-mail: ijgi@mdpi.com
www.mdpi.com/journal/ijgi

www.ingramcontent.com/pod-product-compliance
Lightning Source LLC
Chambersburg PA
CBHW061547220926
51850CB00005B/270
9783036585888